Physical Science:
What the Technology Professional Needs to Know

C. Lon Enloe • Elizabeth Garnett
Jonathan Miles • Stephen Swanson

WILEY

JOHN WILEY & SONS, INC.

New York / Chichester / Weinheim / Brisbane / Singapore / Toronto

Technical Illustration: Richard J. Washichek, Graphic Dimensions, Inc.
Research: Judy Sullivan

This book is printed on acid-free paper. ♾

This material is based on work supported by the National Science Foundation under Grant No. DUE97-51988. Any opinions, findings, and conclusions or recommendations expressed in this material are those of the author(s) and do not necessarily reflect those of the National Science Foundation.

Published by John Wiley & Sons, Inc. All rights reserved.
Published simultaneously in Canada.

This publication is designed to provide accurate and authoritative information in regard to the subject matter covered. It is sold with the understanding that the publisher is not engaged in rendering legal, accounting, or other professional services. If legal advice or other expert assistance is required, the services of a competent professional person should be sought.

Library of Congress Cataloging-in-Publication Data:

ISBN: 0-471-36018-x

10 9 8 7 6 5 4 3 2 1

Contents

Table of Contents

Preface

Life in the past appears to have been simple. If you were fortunate, you received a basic education. You then set out to become a professional or you accepted an apprenticeship as a cobbler, carpenter, or blacksmith to learn the trade. This system provided society with a few knowledgeable professionals and many skilled craftsmen.

In the years following, things became a bit more complicated. The span of knowledge and years of education required for the professional increased. Most of the goods once provided by tradesman were either no longer needed in the emerging throwaway society, or they were replaced with equipment and materials generally unknown less than a half-century earlier.

The twenty-first century holds great promise for students preparing to enter or re-enter the job market. The technician generally receives less training than the skilled tradesman of the past, and yet in the future they will be expected to understand the complexities of the conveniences upon which we have grown to rely. There is a pressing need for a new, highly trained technology professional who understands how to assemble, operate, and repair these conveniences. These professionals will replace the tradesmen and technicians of yesteryear, but they will also need to have a basic understanding of the scientific principles involved to keep pace with the changes.

Consider, for a moment, the automotive industry. Today, the industry is supported by technicians who perform services that are vastly different from those of the blacksmith or the shade tree mechanic of the past. Vehicles powered by alternate fuels, electricity, and fuel cells are just the beginnings of other changes on the horizon of that industry. Technology professionals preparing themselves to maintain these vehicles will need not only basic mechanical skills, but an understanding of the foundational science principles on which they operate.

This book has been written to provide that knowledge. It is an introductory text that will provided a broad-based understanding of the foundational chemistry and physics principles that are necessary for many different areas of technology.

To assist the student in mastery of the material, several learning devices have been provided. Each chapter starts with learning objectives to help the reader focus on the broad concepts to be mastered. Technical words that are critical to understanding have been bolded the first time they are used and included in the extensive glossary.

Each section in the chapter is followed by several questions to focus the reader on the important points in the section and give the student an opportunity to check their understanding. To increase reader interest, Technology Boxes that demonstrate the practical applications of basic science concepts, and Special Topics Boxes that provide additional information, have been included.

At the end of each chapter is a summary followed by a number of questions that are intended to provide the student with an opportunity to apply the knowledge gained. The text also includes an index to assist in the quick location of topics of special interest.

Technology is the practical application of scientific knowledge. It is our hope that anyone who studies this text will develop a deeper appreciation for the foundational role that science plays in current and future technologies.

Finally, a laboratory manual accompanies the text; *Physical Science: What the Technical Professional Needs to Know – A Laboratory Manual.* Included in the manual are additional practice and drill exercises that are applicable for both lecture and laboratory practice as well as a variety of laboratory experiments for those who wish to make the experience meet the general education, laboratory science requirements.

Acknowledgements

The development of this textbook and its accompanying laboratory manual has been no small undertaking. The project would have never been completed, were it not for a network of dedicated individuals. I would like to first acknowledge the contributions of each of the members of the National Academic Council (NAC). As PETE regional representatives, they assisted in the design and review of this textbook and its laboratory manual. They are:

- Ann Boyce, Bakersfield College

- Eldon Enger, Delta College

- Douglas A. Feil, Kirkwood Community College

- Christine Flowers, Shasta College

- Ross Marano, El Centro Community College

- Andrew J. Silva, South Dakota School of Mines

- Larry Stewart, Highland High School

- Brent Wurfel, Arkansas State University – Mountain Home

Next, I would like to acknowledge each of the contributing authors, who also assisted in the design and review process. They are:

- Dr. C. Lon Enloe, James Madison University

- Elizabeth Garnett, Kansas State University – Salina

- Dr. Jonathan Miles, James Madison University

- Dr. Stephen Swanson, Kansas State University – Salina

Once the manuscript was completed, it was through the efforts of INTELECOM staff, lead by Sally V. Beaty, INTELECOM President and project Principal Investigator, that provided the central coordination and support. In particular, I would like to thank the special and tireless efforts of Judy Sullivan, INTELECOM's Senior Editor. Her countless hours of effort, checking of every detail, as well as her polishing of the final manuscript to remove ambiguities and to improve student understanding have contributed tremendously to the production of this book.

For their ability to create a highly illustrated and attractive text and to meet stringent timelines, I extend many thanks to Dick and Mary Ann Washichek of Graphic Dimension, Inc.

For the initial efforts, recognition is due to the Partnership for Environmental Education (PETE) and Sally V. Beaty for orchestrating the writing and management of the grant. The final thank you is due to the National Science Foundation's Department of Undergraduate Education (DUE), which had the vision and provided the funding to make the completion of this project possible.

Howard H. Guyer, Lead Academic
Preserving the Legacy

What is this book about?

The Nature of Physical Science

It has been said that curiosity killed the cat. Curiosity about our world is natural and perhaps one of the more important aspects of life. Children are well known for asking "Why?" Perhaps childish curiosity is what drives scientists to seek answers. For some, the questions may not seem important enough to spend a lifetime searching for their answers, while others have dedicated their lives to finding just one of the answers.

Are there questions that you have considered – even for a moment? Have you ever wondered why a stick looks bent at the point where it enters the water? Do you wonder why some things burn, but others do not? Perhaps you know the answers to these questions, but for the people who asked them hundreds of years ago, satisfactory answers did not exist.

You may ask, "Are there questions that remain unsolved?" Definitely, yes! Can we predict when or where the next earthquake will occur? What are the chances that an asteroid will hit the Earth sometime during our lifetime? Will the average human life span eventually exceed 100 years? These are just a sample of the questions we ask today that still need answers.

Preparing yourself to answer such questions will require an understanding of the principles of science that form our knowledge base. In other words, you must learn what is known to be true in order to explore beyond these concepts. This is the essence of science: to build on current knowledge to answer questions no one has answered before.

1-1 Defining the Areas

Objectives

Upon completion of this section, you will be able to:

- Define physics, chemistry, engineering, and technology.
- Explain the relationship between science and technology.
- Discuss the role of the technology professional.
- Describe the scientific method.
- Explain the difference between observation and experimentation.
- Define pseudoscience.

This text presents the broad concepts of physical science; those typically included in the study of physics and chemistry. Technology Boxes are used to tell you how science concepts have been applied to produce a technological device that has become an every-day part of life. For the person whose interest has been piqued or who desires more information, occasional Special Topics Boxes have been included to provide additional depth on a topic. Before continuing, however, it is important to understand the focus of each of the broad areas of study and its relationship to technology.

Physics and Chemistry

What we call science today was once lumped under the title of **natural philosophy.** The physical sciences were initially separated for academic reasons. As more information became known about each subject, it finally became impossible to include it all in a single discipline. **Physics** became the area of physical science that deals with matter and energy and their interactions. Physics further seeks to predict how matter will behave when it interacts with other matter. **Chemistry** became the study of the composition of matter and the changes it can undergo. As your study in the physical sciences progresses, it will become increasingly apparent that although physics and chemistry have been academically separated, the concepts continue to overlap. It is, therefore, nearly impossible to study one without introducing concepts from the other.

Technology and Engineering

Engineering is the application of physical science principles and mathematics to the properties of matter and sources of energy to make useful things. Over the years engineers have provided us with many electronic devices, including motors, phonographs, radios, telephones, televisions, microwave ovens, facsimile machines, computers, cell phones, pagers, remote controls, CDs, and DVDs as well as a vast array of new fabrication materials. Collectively, such items are referred to as **technology.** Therefore, technology is the product of engineering in the form of the conveniences used in everyday life to facilitate work and play. As can be seen, science, mathematics, and engineering are the threads that weave the tapestry called technology. Each is dependent on the other and all are interconnected. A simple way of describing this connection is to say that science is the process of observing nature and developing an organized understanding of its forces and workings. It is through engineering and mathematics that these understandings are applied to produce technology.

It is instructive to review the preparation and education that scientists and engineers receive in order to understand the tasks that will be expected of them. Science education typically involves exposure to the three major scientific disciplines – chemistry, physics, and biology – and includes a grounding in mathematics and communication skills. A student then explores the specific science of interest, which may be one of the above or it may include geology, astronomy, meteorology, archeology, anthropology, etc. Typically, as one's education continues, there is a tendency to specialize in a specific science.

Engineering education usually begins with extensive mathematics preparation. Mathematics is used throughout the engineer's education and career as a basis for understanding and applying the principles that make devices work. For instance, it is vital that an engineer understand the nature of materials; how they can be shaped, how much stress they can withstand, the temperatures at which their properties change. This is all knowledge that has been collected through scientific investigation, and has been quantified and applied using mathematics.

With the emergence of today's technology, there is yet another group of highly skilled individuals, who work alongside the scientists, engineers, and researchers but do not have advanced degrees. These individuals, through very focused education and work experience, have developed specific skills in a narrowly defined part of a technology. They are the hands-on workers who set up or repair the equipment, take the measurements, operate the complex systems, analyze the data, etc., but they often work under the direct supervision of others. In some instances, they are the ones that take the x-ray, analyze the sample, or locate the glitch in the computer network, and they are called **technicians**. In other occupational areas, they perform similar responsibilities, but they are called **technologists**. To avoid confusion, in this text we will collectively refer to this group of one-year certificate, two-year associate degree, and four-year bachelors degree individuals as **technology professionals**.

A technology education usually prepares one to operate a specific device, perform a specific task, or collect a specific type of measurement. It is important to recognize that a person preparing to be a technology professional may feel science-based topics are very challenging. The training required varies widely, but generally involves taking some mathematics and science courses. If it can be said that science is theory and engineering is design, then the technology professional is the one trained in its application. Therefore, technology education is a marriage of knowledge and hands-on experience with the device or principle with which they will be working.

Experience has also shown that effective communication skills are nearly as important as understanding the theory and application. It is of little value to make accurate measurements, for example, if you cannot convey this information in an effective way to those who wish to use it. Have you ever attempted to assemble an item you purchased or install software on your computer while following the enclosed instructions? Unfortunately, there are many examples of this kind of information that apparently made sense to the writer, but cannot be understood by the person trying to use it

The Scientific Method

Have you ever observed something that sparked your interest and caused you to stop, even for just a moment, and wonder why? If you have, then you understand what **scientific inquiry** is about. Most of the time the observation is simply forgotten, but sometimes we remember to ask someone about it. Scientists frequently observe things no one can explain or ask questions that no one can answer. So, what do they do then?

In science, a **problem** is anything that causes you to stop and wonder why. For a scientist, the next step would be to gather all the known information about the problem by asking others, reading, or doing a Web search. If the information gathering

Interview Box 1–1 ■ Pete Zimmerman, Geotechnical Engineer

 Pete Zimmerman has a MS degree in civil engineering and has been working as a geotechnical engineer for Geoprobe® Systems, for almost three years. He likes his job, especially when it involves international travel. He states, "The challenge of working with people who speak different languages and have a totally different cultural background is always interesting."

Q. What is a typical day in your job?

A. I work in the research and development area. My main responsibilities are designing new tools, bringing new tools into production, managing our cone penetration testing product line, and providing technical support and training for our geotechnical tools. My days are quite varied, but can be a mix of outdoor and indoor work, with some time spent on the telephone with current or potential customers. When I am not traveling, I spend at least half of my time designing or testing new products.

Q. What type of training did you have to have to do your job?

A. Most of my hands-on training has been on the job, but many of my activities require knowledge that is based on topics I studied in school.

Q. Was basic science course work required in order to get the job?

A. A pre-requisite for my job is an engineering degree, but engineering is just applied science. Most of my engineering knowledge is ultimately based on my understanding of chemistry, physics, and geology. Basic science course work is quite important for all types of engineering.

Q. How often does the job require making and recording measurements and interpreting data?

A. I make and record measurements every day. I frequently use calipers for quality control and design of manufactured parts. I use strain gauge load cells and pressure transducers to calibrate our soil probes. In addition, I use a soil probe called a cone penetrometer to determine soil strength and soil type with depth. When I make any sort of measurement, I am typically the one responsible for its interpretation. I am also responsible for training our customers on how to interpret the data that they acquire with our tools.

Q. What advice would you give to someone considering a job of this type?

A. Technical expertise is important, but perhaps just as important are such skills as public speaking and problem solving. I work with electrical, mechanical, and chemical engineers, as well as people from our sales and marketing departments, welders, and machinists. It is important to be able to relate to people who do not have the same technical background as you do.

Interview Box 1-1 ■ Pete Zimmerman, Geotechnical Engineer (Continued)

Q. Would a course that covers basic science (physics and chemistry) concepts be helpful in doing your job?

A. Since an understanding of basic science concepts is important in understanding engineering concepts, this sort of course would be helpful early in an engineering curriculum.

Q. If you were to move to a different job within the organization, would retraining be required?

A. Many of the skills that I have learned on this job would be useful in other positions. I would say that my range of skills from my education and job are rather broad. There are some jobs within the company that I could do without retraining. However, since Geoprobe® Systems is also a manufacturing company, I would not qualify, without additional education and retraining, for many jobs.

Q. Does the company employ technical professionals and, if so, what are some of their duties?

A. Geoprobe® Systems employs a number of technical professionals. We have several mechanical engineering technical professionals, a couple of electrical engineering technicians, and a draftsman working in production and in research and development. I frequently rely on our electrical engineering technicians for their expertise in electrical repair and assembly and seek their advice on trouble-shooting electrical systems. I consult our mechanical engineering technical professionals about many aspects of tools design ranging from the properties of the materials used to the most efficient designs for machining and the proper heat treating techniques.

   Our draftsman uses computer aided drafting programs to convert my sketches into dimensioned drawings that are used in production. He also draws most of the pictures that are used in our marketing materials. Most of our engineering technical professionals hold a Bachelor of Science degree in engineering technology, or they have earned an Associate Degree in engineering technology and are studying for the Bachelors Degree. Our draftsman holds a vocational technology degree in computer aided drafting and he is studying for a degree in mechanical engineering technology.

Geoprobe® Systems is a registered trademark of Kejr, Inc., Salina, Kansas.

fails to provide an answer, then you make a reasonable guess. In science, this guess is called a **hypothesis**. The hypothesis needs to be stated in such a manner that you can devise an experiment to test it. For example, "taking 5,000 mg of vitamin C per day will prevent you from catching a cold" is a properly stated hypothesis. If the devised experiment shows the hypothesis to be incorrect, then it is scrapped in favor of a new one. As shown in Figure 1-1, this step may be repeated many times, until some success begins to be realized. If this seems hard to imagine, just think about all the research that has been done in the last 100 years on what causes cancer!

Once the experimental results seem to support a hypothesis, a theory may be formulated. A **theory** is a broader statement that, based on the previous investigations, explains everything you now know and that appears to be true about the problem. A series of experiments may be conducted to test the theory. If, for any reason, the theory fails, then it may have to be modified or scrapped. Based on these findings it is sometimes necessary to state a new hypothesis and repeat the previous steps. Years ago scientists proposed an explanation for why things burn based on an invisible substance, phlogiston. After teaching the phlogiston theory for over 100 years, oxygen gas was discovered and the phlogiston explanation for burning discarded.

If no exceptions are found after repeatedly testing a theory, then it may become a **law**. There are fewer laws in chemistry than there are in physics, but

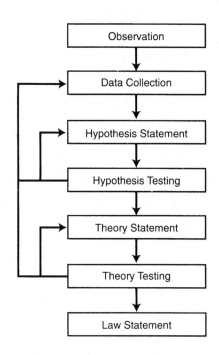

Figure 1–1: The scientific method.

one is the law of conservation of matter. It states that the sum of the weight of the reactants must be equal to the sum of the weight of the products. It seems so logical now, but it took chemists hundreds of years to find that during ordinary chemical reactions it is true.

Although not every theory or law currently held has been arrived at through strict adherence to the process described, the difference between the terms hypothesis, theory, and law is very important. In science, these terms tell you how well a problem is understood. One of the most important theories in chemistry is called the atomic theory. There are those who argue that enough is known for the atomic theory to become the atomic law!

The following is a hypothetical example that illustrates the development of a law of nature. Early in history, people noted the phases of the moon and the regularity with which it waxed and waned. After keeping track of these cycles, someone developed a method that could predict when the moon would next be full. When the person's hypothesis predicted the next full moon correctly, he boldly began to predict other phases of the moon far into the future. If he shared his hypothesis with others and it worked for them too, then it would have been elevated to a theory. If further testing found that the theory was always true and might even be applied to other cyclic celestial happenings, then it would have become a law of nature. This same process has the advantage that it can be applied to a wide range of problems, from the very simple to the very complex.

Reproducibility is the ability to repeatedly perform the same experiment and obtain the same results. If others cannot reproduce the experiment and obtain the same result, it likely means that there is a problem with the statement or the design of the experiment. Ideally, the scientists would check to see that the results are reproducible before announcing a hypothesis or a theory in a public forum. However, there are many examples of scientific breakthroughs being presented before they were repeated by others. Some examples of where this has happened are the so-called cures we see on the evening news. How many times have you seen a story involving a possible cure for arthritis or cancer, only to find that it isn't quite as effective as first reported?

Sometimes scientists become overzealous and leak information to the press, before it has undergone review in the scientific community. This behavior can be embarrassing for the individual and for the other members of the scientific community.

Observation vs. Experimentation

A distinction needs to be made here between observation and experimentation because they are two different things. When nature is observed, data may be taken, observations recorded, and patterns recognized. An experiment is a simplification of nature that is contrived to seek the answer to a specific question. The answer is generally yes or no or perhaps "I still can't tell." For example, you may observe that people must wait to take their seats until those ahead of them have stowed their luggage.

An airline might decide to perform an experiment to determine how to most efficiently load an airplane. On some flights, they may choose to load the airplane by letting everyone with a ticket get on in random order. On other flights, they orchestrate the loading by calling the passengers with the seats in the back to get on first and loading the plane from back to front. On still other flights, they choose to load the plane by allowing the passengers with window seats to get on first and load the plane from outside to inside, so to speak. If the airline recorded data on the time it took to load the planes in the various manners, they would have performed an experiment. Because they controlled how the plane was loaded, there was a simplification or regulation of the natural process, making this is an experiment.

In some experiments, a change in only one of the variables is made. In others, changes are made in several of the variables. However, the experimenter must be careful. If more than one variable is changed at a time, it may be impossible to determine which change caused the results. Very often, proper control of the variables is what causes the scientific process to move slowly. In fact, the experiment above would not be a well-controlled experiment, because the same people were not used each time the plane was loaded. Perhaps not even the same type of plane was used each time. Maybe one of the planes was loaded during a heat wave; thus the passengers simply moved more slowly due to the heat. Poorly designed experiments that do not account for uncontrolled variables are likely to produce misleading results. How does the scientist control the variables to ensure that only one variable is changing at a time? This is one of the many challenges that the experimental design must overcome.

In small particle physics and chemistry, it is conceivable the results of an experiment can be affected by changes in the magnetic field of the earth, the climate in the laboratory, or other seemingly inconsequential conditions. These variables may be difficult to recognize, let alone control.

Pseudoscience

In recent years, great emphasis has been placed on students developing critical thinking skills. During this same time, the Internet has become a vast distribution system for people to publish their "scientific" works. These works can range from the carefully detailed paper published in peer-reviewed electronic journals to articles put together by amateurs to describe their personal beliefs.

It is important to be able to distinguish between true science and pseudoscience. **Pseudoscience** is defined as a set of ideas put fourth as scientific, when, in fact, they are not. Historically, people have been led to believe that certain claims are scientific, when they have not been properly tested. A simple test – "Can someone else do the same experiment and get the same results, within a reasonable margin of error?" – can be used to determine whether something is true science or pseudoscience. The five percent rule serves as a useful rule-of-thumb. If the results are reproducible within five percent, they are often considered close enough. Obviously, there are experiments where even this small amount of error would be intolerable.

Science deals with facts and data that can be reproduced in support of a given hypothesis or theory. Pseudoscience deals with the unknowable, indisputable, and unprovable, and so cannot be called science. There are no hypotheses or theories that can be tested. History is not science. It is impossible to go back in time and see what really happened. We cannot always tell if past events were recorded accurately. Even the historical aspect of the Bible, while credible, cannot be proven. Because something is not science, however, does not mean that it is useless or misleading.

Astrology is pseudoscience. It gives the impression that it is scientific but it is not. If you were to compare several astrologers' predictions would they agree? Check one astrological calendar against another. Do they make the same predictions? If the statements made are not testable, then they are not hypotheses.

Scientists have a particularly hard time dealing with sloppy reasoning, fringe science, and pseudoscience. Pseudo-scientists typically claim to use the scientific method and to base their theories on experimental evidence. Often their understanding of what it takes to properly control the variables is inadequate. A theory that is in conflict with the facts is obviously not a good theory, but a theory that is consistent with only a few facts is not necessarily a good theory, either. Pseudo-scientists want to claim the authority of science, but are unwilling to abide by the rules that has earned science its reputation for finding the truth.

Throughout history, theories have been developed that use everything from the "magic of numbers" to the structure of a person's skull (phrenology) to explain observed phenomena. All are ways of attempting to explain the world and to predict what will happen in the future. All fail to fit into the category of science for one or more of the previously stated reasons. Newton's law of gravitation predicts that a pencil will fall to the ground when dropped. It happens every time the pencil is dropped, not just some of the time. Being right sometimes is not good enough. A scientific hypothesis or theory must prove itself every time. Reproducibility is key to science. Pseudoscience harms the credibility of science and, unfortunately, much of the public cannot discern one from the other.

Many statements have been made and publicized and then found to be in error, or even dangerous. Claims of diets and medical cures are some of the most disturbing. Recently a large number of people turned to a combination of drugs called fen-phen to lose weight. They later found that the safety of the combination of drugs had not been adequately tested and reported and some people developed devastating health problems. There is a difference between a hypothesis that was not properly tested and a scientific hypothesis that is presented and later proven wrong.

Check Your Understanding

1. What are the definitions of physics, chemistry, engineering, and technology?

2. Describe the relationships that exist among the terms defined in question 1.

3. What is the difference in understanding between a theory and a law in science?

4. Explain the difference between observation and experimentation.

5. Explain the difference between science and pseudoscience.

1–2 The Language of Science

Objectives

Upon completion of this section, you will be able to:

■ Discuss the origin and proper use of units.

■ Explain the origin of the common measuring systems.

■ Discuss the relationship between matter and energy.

Although it is not a formal definition, the term "matter" is used in science to include all physical things. This would include rocks, soil, bones, salt, water, alcohol, oxygen, carbon dioxide, and air, to name only a few. At room temperature, each of these exists in one of the three so-called states of matter. The three commonly recognized states of matter are solids, liquids, and gases. Sometimes other observations, measurements, or quantities that help describe the amount or condition of matter are needed. The quantities most often included are mass, volume, pressure, and temperature.

Much of the description of matter, therefore, involves measuring and recording quantities. Whether expressing its mass or volume, tracking a falling object, monitoring a toxic waste site, or using temperature to help diagnose disease, research and the role of the technical professional often involve collecting, recording, and understanding the importance of large amounts of data. If the measurement process is flawed or the data incorrectly recorded, the information is useless.

Something that affects the data collection process is the **uncertainty principle**. It acknowledges that the act of making an observation sometimes alters the system being measured. The impact of this principle was first suggested in very small-scale measurements of the direction and momentum of subatomic particles. The principle, however, also applies to other, larger experiments. For instance, when you put a thermometer into a liquid, the thermometer causes a change in the temperature of the liquid. Likewise, an environmental technology professional must be very careful when sampling river-bottom sediment so the material is not disturbed by the sampling process.

Another consideration involved when making and reporting measurements is that everyone must agree on the system of units used. Without this understanding, each measurement would be just one person's experience and could not be shared by others. Whether using the Metric system, English system, or the International system (SI), the units must be known and accepted.

Numbers and Units

All measurements are composed of two parts, a number and a unit. The number can be any odd or even digit or fraction thereof, but the unit is determined by what is being measured. For example, 3 apples and 3 oranges both have the same number, but they differ in the kind of fruit. As you have undoubtedly been told, you cannot add apples and oranges. In a similar way 3 inches and 3 feet have the same number, but the units differ and, again, you cannot add them together unless you first convert one of the measurements into its equivalent in the other unit.

In everyday practice, we often rely on the units to tell us what is being measured. If someone says they keep their car tires at 32 psi (pounds per square inch), we understand that they are reporting air pressure, not weight. If, however, someone reports they were going 50 when they passed the radar unit, we must *assume* they mean 50 miles per hour, since that is the unit used most often for speed. However, if the same statement were made in a country where the Metric system is used, then our conclusion would likely be different. The point here is that if the units are omitted from a number, the other person can only guess. This becomes particularly risky when the measurement involved is less familiar or when more than one measuring system may be involved.

Over the years, collections of units for expressing lengths, weights, volumes, pressure, etc. have been grouped together forming measuring systems. Of these, you may be most familiar with the English system. You have heard of miles, yards, feet, and inches. Any of these may be used to express distances, such as your height. You have also heard of gallons, quarts, pints, cups, fluid ounces, tablespoons, and teaspoons. These are units used to express liquid volumes. Potatoes are not purchased by the gallon or pint, but by the less known dry volume called pecks.

The point here is that within any measuring system, only particular units are appropriate for expressing lengths, volumes, weights, and time. With those basic units, it is then possible to derive combinations of these units to express other kinds of measurements. For instance, the density of water is 8.31 lb/gal. Here a combination of weight and volume units is being used to indicate the weight of 1 gallon of water. The same is true in the previous example of speed. When you report 50 miles/hour, you are telling someone the distance you would travel in one hour. Avoid mixing units from two different measuring systems. Expressing the density of water as 3.78 kg/gal, although mathematically correct, would not be the preferred method. Since the mass is in the SI system, the volume should be in liters.

Development of Systems of Units

The United States is one of possibly only two remaining countries in the world that routinely use two different systems of measurements. Commerce within our country continues to be done using primarily the English system, where tons, pounds, gallons, quarts, miles, feet, and inches are common units. Within the sector of international commerce and throughout the scientific community the metric system or the more recent System International (SI), which has its roots in the metric system, is used. This poses an interesting question. Why would a country use two different measuring systems? The answer is tangled in history.

Since the dawn of civilization, groups of peoples adopted systems of weights and measures based on need and the availability of some type of measuring standard. Often these standards involved human dimensions. Native people used the palm of their hand to measure the height of their ponies. An ancient unit of mass is the carat. It is based on the mass of a seed and is still used as the unit of mass for precious stones, such as diamonds. Its weight is approximately 200 milligrams.

When different groups of people started trading, there was a need to develop conversion factors or tables to equate a weight in one system into that of another. Over time, the measuring systems of the countries that were more active in commerce became favored and the others mostly faded away. One of the countries that had a powerful naval fleet and introduced commerce to many different nations was England. Over the centuries, the English people had developed their own measuring system. It, too, had many of its units based on human dimensions. The foot, for instance, was based on the length of the king's foot. The yard was the distance from the middle finger of the king's outstretched arm to his nose, etc. One of the problems with this system was that each time a new king was crowned, it was necessary to replace the existing rulers with ones corresponding to the length of his foot and arm. A second problem with the English system was that no attempt had been made to systemize the units for length, weight, and volume.

Interview Box 1–2 ■ Stacey Grosh, Fuel Consultant and Energy Analyst

Since completing a Bachelor's Degree two years ago at James Madison University in the Integrated Science and Technology (ISAT*) program, Stacy Grosh has been employed as an Analyst for Pace Global Energy Services, an energy consulting and management company based out of Fairfax, VA. She has found work in consulting both exciting and challenging. Projects change quickly and are diverse in nature. Her favorite project to date was working as part of an independent fuel consultant team for a co-generation financing project. During construction of the facility, Stacy participated in a plant tour, explaining the facilities fuel arrangements to multiple lenders while seeing first hand how the parties in a power project development project work together.

Q. What is a typical day in your job?

A. There is no such thing as a typical day in consulting. I work on two to three projects at a time, most of which are completed within a couple of months. As an analyst, I do extensive research, data analysis, and report writing.

Q. What type of training did you have? For example, are basic science principals and data interpretation involved in your job?

A. By completing the ISAT program, I gained a background in energy, economics, governmental policy, and analytical analysis. I was able to begin my job with a technical understanding of energy, fuel units and conversion, and power generation fundamentals. On the job, I interpret mostly market and plant operational data on a daily basis. I use multiple databases and data sources and make extensive use of the spreadsheets and models I develop.

Q. What advice would you give to someone considering a job of this type?

A. I would recommend, as encouraged by the ISAT program, "...you should know a little bit about a lot of things so you're not afraid of anything." Basic science concepts help develop the analytical mind necessary for doing this type of work.

Q. If you were to move to a different job within the organization, would retraining be required?

A. I do not think outside formal training would be necessary. The skills I am currently acquiring are very broad. Although there would be a transition period, I think I could come up to speed relatively quickly in another position.

*The Integrated Science and Technology (ISAT) program at JMU is an innovative effort to create a new kind of education for students interested in the details of science and technology and their impact on society. Its curriculum integrates the study of science, mathematics, technology, society, and business to develop a graduate with unique professional qualifications. Computer use is a central feature of the curriculum, along with collaborative (team) learning, and hands-on experience.

It is believed that King Henry I (1068–1135 AD) was the last king to decree that the proportions of his body be used as the standard for length measurements. By the late thirteenth century, the English had stabilized the yard, foot, and inch by legally enforcing the use of marks that had been carefully placed on a single bar of metal. This legal act was the first time in history that there were exactly 3 feet in a yard and 12 inches in a foot.

In the latter half of the eighteenth century, the National Assembly of France embarked on a systematic remedy to the growing confusion among the various measuring systems. After scientific consideration a new unit of length, based on one ten-millionth of the length of the arc from the equator to the North Pole was recommended as the new, unchanging, unit of length. In May 1793, this unit was given the name meter, from the Greek word *metron,* meaning a measure. From the same root came the name metric system, which was adopted in France in 1795.

Not all went smoothly, however. In 1812, Napoleon, Emperor of France, reinstated the old units of measure. In 1840, the government again mandated the use of the metric system in all of France and it has remained so ever since. Its use also slowly expanded, initially including Germany, Italy, Greece, the Netherlands, and Spain. By 1850, growing international trade and commerce expanded the use of the metric system as a common language of measurements, and by 1880, all major European countries and most of South America had adopted it. In 1960, a restricted version of the metric system, called the name International System of Units (SI), was adopted.

The things that have made the metric system so successful are the following. First, the scientists developing the system attempted to remedy many of the problems that existed in the other measuring systems. For example, at the heart of this system is a set of base units; the meter for length, the gram for weight or mass, and the liter for fluid volume. Each of these base units is, in turn, linked to the other, using the properties of water at 4°C.

Second, rather than using smaller and larger units, each carrying different values, they used a system of decimal prefixes. For example, the prefix kilo means 1,000 times the base unit, whether it is 1 kilogram, 1 kilometer, or 1 kilowatt. This eliminated the need for rote memory and cumbersome usage of such measures as 5,280 feet in a mile. Think how difficult it is to change a length from miles to inches in the English system. In the metric system, however, to change a distance from kilometers to centimeters would be no more difficult than changing $25.00 to 2,500¢ – a mere two decimal place move that can be done without pencil or calculator.

During the early history of the United States, immigrants, many of whom had received their science training in Europe, came to America and continued their work using the familiar metric system. Consequently, science in this country has always used the metric system. Although the metric system missed being nationalized by one vote in Congress, it was legalized, but not made mandatory, for use in the United States in 1866. By this time, however, the English system was fully entrenched in American commerce.

Today, nearly 150 years later, the United States remains the only major industrialized nation uncommitted to the mandatory use of the metric system. England, Canada, and Australia have now adopted and use only the metric system, while we continue to use two systems. Certain areas of engineering and technology use the English system or even a mix of the two. Why is this important to you? As a person preparing to become a technology professional, you face an immediate hurdle that individuals from most other countries would not. You must become familiar with, and be prepared to work within, two different measuring systems. It is not difficult, but it will require your careful study. Two chapters of this book, therefore, have been devoted to helping you understand these systems and gain this ability.

Matter and Energy

Energy is another property that all matter possesses, but how do we know it is there? What does it mean to say something has a lot of energy? For one thing, matter with a lot of energy tends to move around more vigorously, but often that movement is at an atomic level and cannot be observed directly. For another thing, energy comes in many forms, making it difficult to identify. One thing is certain, however. Anything that has energy has the ability to perform work.

Kinetic energy is the energy of motion. The amount of kinetic energy depends on two things: mass and velocity. It can be demonstrated, for instance, that a fast moving small object can have the same amount of energy as a slower moving larger object. The same kind of thinking goes into a wide range of other energy concepts. The amount of mass is important when quantifying the amount of energy. For instance how much heat energy an object has depends on its mass and its temperature. So, once again a small object at a high temperature can have as much energy as a larger object at a lower temperature.

The amount of gravitational potential energy also depends on the mass of matter. When an object is lifted from the surface of the Earth to a higher point, the work done is stored in the form of potential energy. The water behind a dam, for example, possesses a great deal of potential energy due to its mass and position. It can be used to do mechanical work on a turbine to produce electric energy as it passes through on its way to the stream below.

If matter isn't moving, it can still contain a form of energy. It isn't obvious, for example, that a can of gasoline contains a great deal of energy. Yet, anyone who has driven a car has some idea of how much chemical energy the combustion of the gasoline can deliver. In a similar way, the amount of nuclear energy locked within a radioactive element isn't apparent until it is released.

One of our greatest discoveries was that energy can be converted from one form to another. The industrial age was not launched until ways had been developed to capture kinetic, potential, chemical, and electrical energy and put it to use. Inventors in the nineteenth century built on the concept of energy conversion and through clever methods found ways to transform energy into more useful forms. Thomas Edison, Alexander Graham Bell, and others are remembered today for the technologies they developed making direct use of electrical energy.

Heat, once thought to be a fluid, is now understood as the transfer of energy. It provides many energy conversion examples. The steam engine made use of heated water to drive pistons and turn wheels, so that some of the heat was eventually converted into kinetic energy. Steam from both nuclear and chemically fueled power plants is currently being used around the world to convert thermal energy into electric energy.

Now that you are thinking about all of the different forms of energy around you, it is interesting to notice that many energy changes we observe are gradual, rather than cataclysmic. The universe is gradually cooling. We also know that the change is so gradual that we won't have to be concerned with it in our, or our children's, children's lifetime. Through a process called photosynthesis, light energy from the Sun produces chemical energy in grain to feed thousands.

The term conservation (preservation) has again become popular when considering the environmental impact of humans on the Earth. You may know that the conservation of mass and energy ($E = mc^2$) is one of the basic laws of nature. The amount of matter and energy never really change. It is sometimes hard to imagine, but when water evaporates, that mass of water becomes a part of the air, so the quantity is preserved. When you burn gasoline, you not only release its chemical energy as heat and light, but you also release matter in the form of carbon dioxide, carbon monoxide, carbon, and water. Every bit of the original matter is conserved.

The total amount of energy released by burning gasoline is also conserved, or more correctly stated, it is balanced. The amount of energy you have at the beginning of an event is the same amount of energy you have at the end. Energy may change forms or be transferred from one place to another. The energy may do work, but the energy that goes into an event must also come out of the event in the same amount. The sunlight striking the Earth can be absorbed by plants or by solar arrays, or by the ocean; or the light can be reflected. Whatever processes occur, all of the energy striking the Earth can be accounted for as either stored energy, energy that is doing work, or energy that has been reflected. Because of the importance of these relationships, the remaining chapters have been dedicated to further exploration of the relationships between matter and heat, chemical, electrical, electromagnetic, sound, and nuclear energies.

Check Your Understanding

1. What is matter?

2. What does the uncertainty principle warn us about?

3. What are the two parts of all measurements?

4. What is the name of the basic unit of length in the metric system?

5. What are two advantages of the metric system over the English system?

6. What is another name for the energy of motion?

7. To what does the equation $E = mc^2$ refer?

1-3 The Working Technology Professional

Objectives

Upon completion of this section, you will be able to:

- Discuss the nature of the work performed by technology professionals.

- Discuss the educational requirements for becoming a technology professional.

- Discuss the future of the job market for science-based technology professionals.

One has only to scan a newspaper to find that the job market today contains hundreds of job titles that did not exist a few years ago. At the end of the twentieth century, the United States was enjoying the lowest unemployment rate in 30 years. The range of job titles and qualifications has continued to soar. Many individuals working today are performing jobs in which they have little or no formal training. Individuals preparing to enter the job market today are likely to encounter future opportunities that are currently unknown. What kind of educational and skills background should one have to enter this market? What are the underlying foundational skills that everyone should have?

In the remainder of this chapter, some of the findings reported in the Department of Labor's, *Occupational Outlook Handbook*, and interviews with several working individuals are presented. As you will see, each person has a different level of education and work experience. Most are in jobs that require specific training and if they were to leave their current area of employment, would need to have additional training before they could enter a different specialty, even within the same field.

Nature of the Work

Most technology professionals specialize and, therefore, have occupational titles that tend to follow the same structure as professionals working in the field. Technology professionals in the areas of science use the principles and theories of science and mathematics to solve problems. For example, individuals working in an area that requires them to perform routine laboratory and field tests to monitor environmental contaminants and sources of pollution would likely have a job title of environmental technology professional or environmental technician. A nuclear technology professional might monitor radiation, operate remote control equipment, and assist nuclear engineers and physicists in maintenance or research activities.

Regardless of the specific job title, technology professionals will find that their jobs tend to be practically oriented or hands-on positions. They may be expected to set up, operate, and maintain instrumentation; observe, calculate, and record results; and sometimes develop conclusions. They may be required to keep detailed logs of their work-related activities. They may also be involved in ensuring quality control of the work and products.

In recent years, electronic instrumentation has become more commonplace. While working under the direction of a supervisor, a technology professional may be asked to perform routine tasks, develop procedures, collect and interpret data, and devise solutions to problems. In some areas, robotics is now used to perform routine tasks that were once performed by technicians. Consequently, it is valuable for many technology professionals to have extensive knowledge of computer, computer-interfaced equipment, robotics, and other high-technology industrial applications.

Interview Box 1–3 ■ Marvin Foster, Automotive Technician

 Marvin Foster has worked as a mechanic for the past 30 years. Following high school, he completed a $2\frac{1}{2}$ year Diesel Mechanics Program at Merced College. Since serving as an aircraft mechanic in the Air Force, he has worked as a civilian diesel technician, owned his own shop, and is currently employed as a light automotive technician. Marvin finds his current job to be very interesting because he seldom has to do the same exact thing.

Q. What type of training did you have to have to do your job?

A. I guess I've been a mechanic all my life. After I completed high school, I went through a Diesel Mechanic Program to become a diesel technician.

Q. Are basic science principles involved in doing your job?

A. Yes. For example, the way the tires on the front of your car lean makes a difference on the way your car wants to go down the road. Also we use different types of pry bars and levers to take things apart and tighten them. We use torque a lot. We have a wrench that automatically measures it and the book tells what the torque should be.

Q. How often does this job require you to make and record measurements? Do you have to interpret data?

A. When you first put a car on the front-end alignment machine, you get a printout of the car's camber, caster, and toe. For example if the camber is too high you need to know which way to change it to bring it back to zero. You have to calculate how many shims to take out or put into make the adjustments necessary so the vehicle will go straight down the road when your steering wheel is straight.

Q. Would a course that covers basic science be helpful in doing your job?

A. Yes. I had some science when I was in high school. It helps to understand why using a cheater bar increases the amount of torque applied.

Q. If you were to move to a different job would, retraining be required?

A. Yes. To specialize in any one thing, like wheel alignment, you need to go to alignment school.

Most technology professionals will likely find employment opportunities on regular shifts and indoors. However, others may prefer positions that allow them to work outdoors and/or to have varying shifts. Some technology professionals will choose to work with hazardous equipment or toxic chemicals. Those working in the nuclear industry, for example, may be exposed to radiation, while biological technicians sometimes work with disease-causing organisms. Although these conditions may seem alarming at first, remember that with proper training and safety procedures, these working conditions will likely pose little long-term risk.

Training and Other Qualifications

Many different paths can lead to a position as a technician or a technology professional. In the past, many individuals with a minimal amount of formal education started as trainees in routine positions. Under the direct supervision of a manager or more experienced worker, they gained hands-on experience, demonstrated an expertise and interest in the area, and gradually accepted more responsibility as they became more specialized in the specific area. This "grow your own" plan has been more successful in the past than it will likely be in the future. Grow your own plans work best when employee-employer relationships are long-term; the goods or service produced retain their marketability; and the training required to do the job is relatively low. In the job market today, however, the trends are toward a higher employee turnover rate, goods and services having shorter market-lives, and technical professional positions requiring greater amounts of education and training. These trends are particularly true in all high technology sectors of the job market.

Most employers prefer to hire applicants who have at least 2 years of specialized training or an associate degree. For other positions the employer may require an individual with a bachelor's degree to fill a technology professional position. Many community and technical colleges offer associate degrees in a specific technology. Some are designed to transfer easily to a nearby 4-year program. Community and technical college programs typically offer less theory, but require more hands-on training than 4-year programs. They also have the advantage of offering 1-year certificate as well as 2-year associate degree programs. This allows the students the opportunity to enter the job market in less time, a factor that may be of critical importance, especially if they must be self-supporting. Later, they can re-enter the program and complete the second year of training. Many 2-year programs offer internships in cooperation with local employers, while attending classes. Participation in such programs can significantly enhance both the employee's marketability and the employer's chances of finding workers with the desired skills.

Individuals seeking higher levels of responsibility, and perhaps management opportunities, should complete 4-year college or university degrees. Articulation agreements have been developed and are in place for many of the 2-year associate degree programs. In these arrangements, the 2-year program meets the required general education requirements and the first two years of education required for the bachelor's degree program. The 4-year programs are typically more theoretical, require higher levels of both science and mathematics training, and may provide less real-world training. Of the hands-on training received, most is typically in the form of laboratory-based courses, or from summer job placement in research or an area business.

The individual intending to complete a 4-year degree program should start in high school by building a broad foundation containing as many science, mathematics, and computer related courses as possible. As they near high school graduation, they should work with a counselor to carefully select their next step. There are many factors to be considered including high school academic record, course work completed, strengths, financial backing available, and personal strengths and interest. If you elect to attend a community or technical college prior to entering a 4-year college, make sure you understand the transfer requirements for the 4-year program. Visit the area 4-year college first, and discuss your transfer options. College counselors can give you valuable assistance on what courses transfer easily and also make recommendations on others that may need to be taken as a part of the program. Don't enter any program without doing some research and talking to professionals whose job it is to help you navigate your way through the academic maze. Academic history is strewn with examples of students who could have saved several semesters of work, had they asked the appropriate questions before planning their course schedule.

Interview Box 1–4 ■ Galen Manners, Manager of Cellular Operations/Kansas

Galen Manners has a Bachelor's Degree in Electrical Engineering Technology and has worked for ALLTEL for two years. He likes his job, which involves deploying, testing, and optimizing a statewide digital network. "We added digital TDMA capability to all of our cell sites. Testing and optimizing the new system has been most interesting." (TDMA stands for time division multiple access and is the digital transmission technology that allows a number of users to access a single radio frequency (RF) channel by allocating unique time slots to each user within each channel.)

Q. What is a typical day in your job?

A. I manage 18 field technical professionals and provide a second level of technical support for them. I am also responsible for budgeting and ordering all equipment for the new cell sites and expanding existing sites.

Q. What type of training do you need to do your job?

A. You need a strong technical background. It is necessary to understand and interpret mechanical and electrical data for all types of measurements and units. In addition to the science courses required for my engineering degree, knowledge of both basic and advanced electronics is also necessary.

Q. How often does the job require making and recording measurements and interpreting data?

A. When I was in the field, it was quite regular. Now that I am primarily a manager, I only get involved when helping a technical professional troubleshoot a specific problem.

Q. Would a course that covers basic science (physics and chemistry) concepts be helpful in doing your job?

A. For a person wanting to enter this field, Yes. They should also attempt to obtain a broad range of knowledge and experience.

Q. If you were to move to a different job within the organization, would retraining be required?

A. My experience and knowledge is very broad, but I may be somewhat limited to the electronic and communication fields. I do believe that I could step into several other jobs and be productive in a very short time, since my existing job requires me to wear multiple hats. I find this frustrating at times, but it does provide me with valuable experience and helps keep the job interesting.

Interview Box 1–5 ■ Larry J. Wacker, Environmental Technician

Larry J. Wacker has worked for nearly five years as an environmental technician at Rockwell/Collins in Cedar Rapids, Iowa. Upon completing the Water and Wastewater Technology program at Kirkwood Community College, he earned his Grade II Water/Wastewater Certification. In addition, he completed the Hazardous Materials Technology program at Scott Community College.

Three years after joining Rockwell, a project was started to remodel the wastewater pretreatment facility. "I was able to apply knowledge I had gained at Kirkwood, from previous employment, and from talking to other professionals. As a result, we have a good facility that meets the needs of the company and the municipal treatment plant."

Q. What is a typical day in your job and what advice would you give to some considering this type of work?

A. We deal with a lot of heavy metals, cyanide, acids, and bases in electroplating wastes. We do metal recovery, so there is a lot of sampling, analysis, and treatment involved in my workday. Keeping accurate records is key to making sure that we comply with the regulations. I spend a lot of time at the computer following the processes and tracking the treatment. I believe that working in industry is much different than working in the municipal sector. To be satisfied, you have to be able to adapt and enjoy the industrial aspects of wastewater treatment.

Q. Was basic science course work required in order to get the job and would a course that covers basic chemistry and physics be helpful?

A. I was required to take a basic chemistry course to get the job, and I use it every day. Industrial wastewater treatment is a mixture of chemical and physical procedures. So chemistry is extremely important, but some understanding of microbiology is also needed. The analytical procedures and terminology learned as a part of science are important.

Q. If you were to change jobs, do you think retraining would be required? In other words, are your skills broad or very narrow?

A. I believe my skills are very broad, but whenever you make a move there is a learning curve that has to be met. My field of expertise is wastewater treatment and for me to move to another position would require retraining. I would bring a lot of knowledge and skills with me, but I would still need specific training to do a new job.

Q. Any further comments?

A. For me, the environmental technician career has been very gratifying and a continual educational experience. I really enjoy the hands-on part of my job. I should have taken more computer training when in the program; it would have been a big help in my current job. I have also gained a lot of expertise by networking with my fellow operators.

Regardless of your 2-year or 4-year option, it is important to remember that communication skills are ranked as most important by nearly every employer. Technical professionals are often required to report their findings both orally and in writing. Improve your keyboarding ability. Computer skills, including basic word processing and spreadsheets, are nearly as important as good oral and writing skills. Attempt to gain experience or take course work designed to improve your teamwork skills. Most technical professionals find that they are a part of a larger working group.

Earnings

The *Occupational Outlook Handbook*, published by the Bureau of Labor Statistics, U.S. Department of Labor is one of the most comprehensive, up-to-date, and reliable sources for labor information. It profiles information on over 250 occupations and includes information about the nature of the work, typical working conditions, requirements for entry, and the opportunities for advancement. For example, the most recently produced *Handbook*, covering the period of 1998–2008, predicts that there will be an overall 14% increase in jobs during that time period. Of the 30 fastest growing occupations, most will require an associate or bachelor degree and will be in either the computer or health occupations.

Technical professionals in science related areas held about 227,000 jobs in 1998. Nearly one-third of these were in manufacturing, about 12% in education services, and 15% in research and testing services. In 1998, the Federal government employed about 14,000 science-based technical professionals, mostly in the Departments of Defense, Agriculture, and Interior. The projection for science-based technical professionals through 2008 is expected to grow more slowly than for other occupations. Although continued growth of scientific and medical research, as well as the development and production of technical products, are expected to stimulate the demand.

Starting salaries for technical professionals vary widely depending on the nature of the position and the part of the country in which one works. The reported average starting salaries paid by the Federal Government in 1999 for all technician jobs was $30,300. The breakdown by area is $41,000 for mathematics, $38,200 for physical science, $48,800 for geodetic, $36,000 for hydrologic, and $45,200 for meteorologic technical professionals.

For more information about careers as a technology professional, meet with your job counselor or do a web search at http://stats.bls.gov/blshome.htm.

Check Your Understanding

1. Why is it likely that most employers will attempt to hire technology professionals who have specialized training in the future, rather than "grow" their own?

2. Why is it important to discuss your future educational plans with a job or school counselor?

3. What are three skill areas that are important to all technology professionals?

4. What is one good source of occupational information and who publishes it?

Summary

Humans have been curious about their surroundings and have observed, experimented, and recorded their findings since the beginning of time. Those individuals who had similar interests in the physical world, engaged themselves in the study of natural philosophy. When the body of known information became overwhelmingly large, it was subdivided into physics and chemistry, although the two areas remain intertwined. The area of physics includes our understandings of the interactions between matter and energy while chemistry is the study of the composition of matter and the changes it can undergo.

Engineering is the merger of science and mathematics in the application of knowledge to design such daily conveniences as refrigerators, cars, cellular telephones, televisions, computers, CDs, and the Internet. Technology is the product of science, mathematics, and engineering.

As in the past, it is recognized that the systematic pursuit of scientific knowledge involves the recognition of a problem, the collection of data through observation and experimentation, and the formulation and testing of various hypotheses and theories. If repeated experimentation demonstrates that there are no exceptions and the theory explains all that is known about the problem, it becomes a law of nature. These laws are the cornerstones on which all other scientific investigations are based. Peer review and reproducibility are the two other cornerstones of all scientific endeavors. As in the past, pseudoscience still flourishes. One of the best tests to discriminate between true science and pseudoscience is to ask, "Can someone else do the same experiment and obtain the same results?" If the answer is no, then it is not good science.

As science has developed, so has the need for an understandable vocabulary and uniform measuring system. Matter is defined as anything that has mass and occupies space. When we wish to specify more about the extent of matter, we need a measuring system that can be used to express such things its mass, length, and volume. Although the English system is the most commonly used measuring system in the United States for commerce, the metric system is used for scientific investigation and reporting.

Energy and its interaction with matter is another area that affects all areas of science. Kinetic energy is the energy of motion. However, matter at rest may also possess energy due to either its position. Many substances contain stored chemical energy that can do useful work. One of science's greatest discoveries was how to harness the energy released as matter is converted from one form to another. Throughout all energy transformations, it is important to remember that mass and energy are always conserved. This understanding remains one of our most important and useful laws of nature.

Today's working technology professionals report that their basic educational skills were very important in obtaining and performing on the job. Their professional training, however, is based on or enhanced by an understanding of scientific principles. They also believe that anyone preparing for a career as a technology professional should attempt to gain as much basic sciences knowledge as possible, which will serve as the foundation for a better understanding of their selected technology. In preparing for any job, be sure to emphasize development of oral, written, team building, and computer skills.

Make use of your school or occupational counselor early in your training. Consult the *Occupational Outlook Handbook*, published by the Bureau of Labor Statistics, U.S. Department of Labor to obtain the most comprehensive, up-to-date labor information. A small amount of time spent planning at the start of your career can pay large dividends later.

Chapter Review

1. Write a paragraph explaining the difference between science and technology.

2. Select and research a technology area of interest. Write a short paper describing the technology and the educational background required to work as a technology professional in that area.

3. Write a short paper discussing scientific inquiry. Explain why this process results in explanations of natural phenomena that can be trusted.

4. Write a short explanation of the difference between observation and experimentation.

5. Select and make a copy of a magazine or Internet article that you consider pseudoscience. Write an explanation for why you believe it to be pseudoscience and what kind of experiment you would devise to test its claim(s).

6. What are two items that must be included in any measurement and what is the purpose of each item?

7. Write a paragraph or two explaining what feature of the metric system makes it easier to use than the English system to use.

8. The rod, gill, stone, fortnight, and furlong are all units of measurement in the English system. Using the appropriate conversion factors, express your age, height, weight, or some other feature using these units.

9. Using the Internet or library, research and write a short history of the metric system. What were the reasons leading to the development of a new system of weights and measure?

10. Select a possible career target and use the *Occupational Outlook Handbook* to write a description of the job, its future projections for growth, and salary.

11. Visit with the school or a career counselor and develop a training program for a particular career choice.

12. Arrange an interview with a local company. Determine the qualifications for a position that interests you. Write a brief description of what you have learned.

How many measurements do you make in a typical day?

Making Measurements

We depend on many measurements every day. Clocks indicate when it is time to get up, go to school, and go to bed. Bathroom scales and rulers tell us our weight and height. Cars are equipped with many measuring devices and warning lights that indicate if the oil pressure, temperature, and alternator voltage are within specifications. The speedometer and maybe even a radar gun tell us if we are driving within the speed limit. Gauges measure the level of oil and fuel in the car. The car's oil, however, is sold in pre-measured amounts. Other products, like fruits and vegetables, are sold by the pound; by the yard, like fabrics; or by the dozen, like eggs. Electricity is sold by the kilowatt-hour and natural gas by the cubic foot. In fact, it was commerce that, many centuries ago, created the need for a measuring system.

2-1 Systems of Units

Objectives

Upon completion of this section, you will be able to:

■ Identify the units that are appropriate for expressing a quantity.

■ Recognize the importance of reliable standards for units of measurement.

■ Identify the two most commonly used measuring systems.

When a group of people have an agreed upon method for expressing measured quantities, they have a **measuring system**. Two of the more common systems are the **English** and **metric systems**. Measuring systems have specific units that are used for expressing such quantities as length, mass, and volume, to name only a few. But what is a unit? A **unit** is an agreed upon quantity that is used as a standard against which all other quantities of the same kind are measured. Each time a measurement is taken, two things must be reported: the number and its units. As noted in Chapter 1, when someone reports the length of a rope as 10 feet, the 10 tells us how many and feet tells us the unit of comparison. So, 10 meters and 10 feet have the same number, but different units of comparison. When two measurements have the same units, they can be compared. For example, if one rope is 10 feet and another is 6 feet long, we can determine which is longer and by how many feet. In order to compare a rope 10 feet long with a rope 10 meters long, both measurements would have to be expressed in the same unit of length.

We rely on units to tell us what is being measured. For that reason, only certain units may be used for a particular type of measurement. If someone reports that a rope is 17 "pounds long," you would immediately know something is wrong. The appropriate units for expressing length can be inches, feet, miles, millimeters, meters, etc. If the previous report stated that the rope was 17 inches long or even 17 miles long, then you would understand something about the length of the rope. In every measuring system, specific units have been defined for expressing specific kinds of quantities.

Although modern society is very dependent on making and reporting accurate measurements, the process has an ancient heritage. In the United States, for example, road distances are measured in miles, a unit that has its origins in the Latin word *mille*, meaning one thousand. The Roman equivalent of our mile was the *mille passus* or the distance traveled by a Roman soldier after marching one thousand paces. A milestone was a unit; literally, a stone placed by the side of the road at intervals of one thousand paces.

Some Ancient Units of Measurement

In the ancient world, the human body was often used as a point of comparison. The units were convenient, but they had a drawback. Take the example of the *mille passus*. People have different strides, so one thousand paces by two different people will not be the same distance. Consequently, these units are not very reproducible, even when made by the same person. The Biblical cubit, which was defined as the distance from the point of a person's bent elbow to the tip of the middle finger, is another example (try question 1 in the Chapter Review). While Egyptian craftsmen always had a ruler with them, the unit varied from person to person. In short, using the human body as a standard isn't a standard at all. A **standard** is something against which a measurement can be compared in order to determine its accuracy.

A Roman Pace   One Cubit

Figure 2–1: Many of the oldest measurement units were based on people, such as the *mille passus* (the Roman mile, equivalent to 1,000 paces) and the ancient Egyptian unit of the cubit.

Nature provides many standards of measurement. The year and day are natural time divisions, which provided the bases for our present-day calendar: hours, minutes, and seconds. As technology advanced, more accurate measurements revealed slight irregularities in the accepted length of the year, and the second was adopted as the new natural standard. Today, the second is defined in terms of the radiation emitted by a specific isotope of the element cesium under controlled conditions. To use this as a time standard, however, takes highly sophisticated equipment (see the Technology Box 2-2: The Atomic Clock). Such standards have the advantage of being reproducible – unlike the length of the human arm. Every atom of this isotope is identical in its fundamental properties, so every atomic clock in the world "ticks" at exactly the same rate.

Modern Systems of Units

Previously it was noted that measuring systems are a **system of units**. If speed is reported in miles per hour, the units come from the English system. If, however, you choose to report speed in meters per second, then the units are from the metric system. Although most of the world uses the metric system, American industry and some branches of engineering and technology continue to use the English system.

There are two separate self-consistent sets of units within the metric system. A **self-consistent set of units** being defined as those that belong to a set and, if used for the measurements and equations, will always produce an answer with the appropriate units – guaranteed. One of these self-consistent set of units uses centimeters for length, grams for mass, and seconds for time – often called the **CGS system**. The other set uses the meter for length, the kilogram for mass, and the second for time. At one time, this set of units was referred to as the **MKS system**. Now it is known as the **SI system** (from *Système International*, which is French for International System).

At first glance there appears to be little difference, other than multiples of ten, between the two systems. It is important to remember, however, that if you want the answer to have the appropriate units, you cannot mix units from two different self-consistent sets of units.

For a measurement to be in the SI system the units must be one of the seven **SI fundamental units** or one of the many units derived directly from the fundamental SI units. Each of the SI units will be introduced in the following section, followed by some of the more commonly used derived units.

Check Your Understanding

1. In a measuring system, what is a unit?

2. Define a standard as used within a set of units.

3. Name two systems of measurement in use today.

4. What value comes from using all input units in the same system of units?

Fundamental Units of Measurement

Objectives

Upon completion of this section, you will be able to:

■ Identify the fundamental quantities that are measured by all unit systems.

■ Recognize the fundamental units of the SI, CGS, and English systems.

There are many properties that could be used to describe an object; however, there are only seven **fundamental quantities** that are required to measure any quantity in the world. As shown in Table 2-1, each of these quantities has a unit associated with it. There are units that express length, mass, time, electrical charge, the motion of electrical charges, temperature, number of atoms, and luminous intensity. Any other property, such as speed, acceleration, or density, can be derived by using a combination of these fundamental units.

Quantities	SI system	CGS system	English system
Length	Meter (m)	Centimeter (cm)	Foot (ft)
Mass	Kilogram (kg)	Gram (g)	Slug (slug)
Time	Second (s)	Second (s)	Second (s)
Electric current	Ampere (amp)	Ampere (amp)	Ampere (amp)
Temperature	Kelvin (K)	Kelvin (K)	Rankine (R)
Luminous intensity	Candela (cd)	Candela (cd)	–
Number of atoms	Mole (mol)	Mole (mol)	Pound-mole (lb-mole)

Table 2–1: Fundamental units for quantities in self-consistent sets of units.

Length

We live in a world surrounded by three-dimensional objects. To describe even a simple box requires information about its length, width, and height. Some objects, such as the one shown in Figure 2-2, have complicated shapes that require many measurements to describe it completely. Each of these measurements, however, have one thing in common – they measure a **length** that tells us how long the object is in that

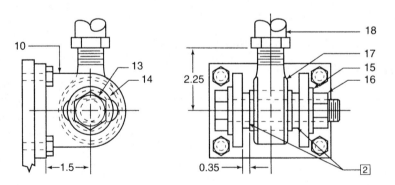

Figure 2–2: Although this mechanical drawing contains many dimensions, they all measure the same fundamental quantity, length. Mechanical parts in the United States are still made in English units, so the dimensions on this drawing are in inches.

direction. Length is the fundamental quantity for measuring all physical space, including area and volume.

The SI fundamental unit for length is the **meter**, m. It is about 3.3 feet, or slightly more than half the height of an average adult. The English system uses feet and the CGS system uses the centimeter for the unit of length. It is common to find combination yard/meter sticks with inches and feet on one side and metric measurements on the other.

Mass

Another fundamental quantity that defines matter is **mass.** The mass of an object is a measure of how much the object resists being accelerated by an applied force. The more mass something has, the harder it is to get it moving, and once moving, the harder it is to deflect from its path.

One mass attracts another mass through the phenomenon known as gravity. Because we spend most of our lives on the surface of the Earth, we tend to think of mass with weight as being the same, but they are not. Mass is more fundamental. For example, an object still has mass even if it is in a weightless environment, like earth orbit. NASA astronauts had to deal with the mass of a large satellite they retrieved by hand on one space shuttle mission (see Figure 2-3). Because of its large mass, it was dangerous to handle and took them several hours to maneuver into the payload bay. If the astronauts had not handled it slowly and carefully, it could have slammed into the shuttle and done enormous damage.

Figure 2–3: Mass is a more fundamental quantity than weight. Here, NASA astronauts carefully handle a satellite that, although weightless, has considerable mass. Photo courtesy NASA.

The fundamental SI unit for the quantity of mass is the **kilogram**, kg, literally meaning 1,000 grams. It is the only fundamental SI unit that includes a prefix. It has an equivalent weight of approximately 2.2 pounds in the English system.

The unit for mass in the CGS system is the **gram**. In the English system it is the derived unit, **slug**, which is defined as the mass of a body whose acceleration is 1 ft/s^2, when the resultant force on the body is 1 pound. This unit, which is equivalent to 32.2 pounds or 14.606 kg, has been largely abandoned in favor of SI units.

Time

Time is fundamental to every part of life, yet it is difficult to define. Perhaps one definition of time would be, "...the stuff we never seem to have enough of." There does seems to be a direction to the flow of time that we cannot change. Much science fiction has been written about time travel. Yet, as far as we currently know there is only one direction we can go in time – from past, through present, to the future.

The **second**, s, is the fundamental unit of time in all measuring systems. A second is about the length of time it takes the average person to say "one-one-thousand." Olympic racing events are typically timed to the nearest 0.01 of a second. Such precision was impossible in the days of human timekeepers and stopwatches. Today sophisticated electronic timekeeping equipment is used to eliminate the variability of human reaction time.

Technology Box 2–1 ■ The Web of Systems and Units

As we begin defining types of units and the quantities they measure, one of the more difficult concepts is the difference between weight and mass. We use the term weight on a regular basis, but is it different from the term mass? One thing that adds to the confusion is that in the English system the unit pound is often used for both weight and mass. The English-based American engineering system, however, does make a distinction between pounds of mass (**pounds-mass**, lb_m) and pounds of force (**pounds-force**, lb_f).

So, when you step on the bathroom scale, are those numbers pounds of mass or pounds of force? If you remember that your body has mass and the Earth's gravitational force is acting on that mass, then the scale must be measuring pounds of force, so we say that the scale measures our weight.

Fortunately, here on Earth a one pound force equals a one pound mass, so the numbers are the same. If you took your bathroom scale to the planet Mars, which has a gravitational field only 0.39 times as strong as the earth's, however, your weight would be less. Although your body still has the same mass, m, the reading on the scale would be less because your mass is being pulled downward by a lesser gravitational force, g, ($f = m g$). For example, a 150 pound-mass person would weigh about 58.2 pounds-force on Mars.

When using pounds-force and pounds-mass, the difference is actually obvious. A tire gauge reads the air pressure as pounds per square inch, psi. Surely, these are not pounds of mass. In American engineering units, pressure is measured as pounds-force per square inch.

One of the advantages of the SI system is that it reports mass and force in different units: the kilogram is used for expressing mass and the newton for expressing force. Your mass (kilograms) is the same no matter what planet you are on, but your weight (newtons) will vary with the gravitational force acting on your mass.

Electric Charge

Most of the modern conveniences we enjoy, such as computers, stereos, streetlights, telephones, and microwave ovens use electricity. Electricity, however, is not a scientific term. Scientists are interested in more specific quantities like voltage and current. However, all electrical phenomena depend on a single property of matter: **charge.**

All of the objects around us have charge as well as mass. This is not obvious because they typically contain equal amounts of positive and negative charge, so they are neutral overall. Like charges repel, while unlike charges attract each other. An imbalance between the amount of positive and negative charge is likely to get our attention. For example, walking across a carpet on a dry day causes a charge imbalance to build up on the body, so that when something else is touched – like a door knob or another person – a shock results as the charge difference equalizes. The SI unit of charge is the **coulomb**, C.

Current is the phenomenon of charges in motion, such as those that flow through the wiring of our homes. The SI fundamental unit of electrical current is the **ampere**, which is defined as the quantity of charge that passes through a length of wire per unit time. Consequently, while charge remains the fundamental electrical *property* of matter, it yields to current as the fundamental *unit* in the SI system.

Technology Box 2-2 ■ The Atomic Clock

A Regulator clock (left) and an atomic clock (right) perform the same function – keeping time by counting repetitions. In the case of the Regulator clock, the event is the swinging of a pendulum. In the case of the atomic clock, the event is the atomic transition of a cesium atom from one quantum mechanical state to another. Atomic clock photo courtesy National Institute of Standards and Technology (NIST).

At first glance, the Regulator clock on the left-hand side and the atomic clock on the right-hand side seem to have nothing in common. In fact, they perform essentially the same function. The process of keeping time is the process of counting some event that repeats at a reliable time interval. In the case of the Regulator clock, that event is the swinging of a pendulum from side to side. Physics tells us that if gravity is constant, the time it takes for the pendulum to repeat its motion depends only on its length. Usually the pendulum in such a clock is sized so that one back-and-forth swing takes one second. This motion moves a series of gears that move the minute hand 1/60 of the way around the circle. Through additional sets of gears, the hour hand moves at 1/60 the speed of the minute hand. The position of the hands on the clock face is the readout of this timekeeping instrument. The pendulum loses accuracy, however, because the length of the pendulum changes ever so slightly as the heat and humidity in the room changes.

The atomic clock solves this accuracy problem by using a "pendulum" that does not change with heat and humidity – atoms of cesium. In the atomic clock, an oven evaporates cesium atoms into a vacuum region, where a laser "pumps" them with energy. The atoms give up this energy a certain amount of time later, so that the "tick-tock" of the Regulator clock is replaced with the oscillation of a microwave cavity and the movement of the hands on the clock face is replaced with an electronic readout of these oscillations. The principle is the same.

Atomic clocks are too expensive and too delicate for everyday timekeeping, but anyone in the world can access these clocks at the U.S. Naval Observatory (USNO) via the World Wide Web. There's nothing like having the final word when someone asks, "What time is it?"

The SI unit for quantity of charge is the coulomb, C. It is a big unit, defined as the quantity of electricity transferred by a current of one ampere in one second, so most of the time we deal with only a fraction of coulomb. For example, shuffling across the carpet moves about 1×10^{-6} coulombs of charge, yet this is sufficient to generate a noticeable shock. The smallest amount of charge that has been isolated is called the **fundamental charge**, e. It is the amount of charge held by an electron.

Temperature

Temperature is at once a simple and a complicated concept. Some of the first descriptive words each of us probably learned as children were, "hot" and "cold." We quickly developed an intuitive grasp of the difference between the two. Later in this text, you will learn that temperature is related to the energy stored in a material in the form of the random motion of its atoms or molecules.

The fundamental SI unit for the quantity of temperature is the **Kelvin**, K, (not °K, although it is sometimes incorrectly used). The Kelvin scale is not one of the temperature scales that we use on a daily basis. Its zero is **absolute zero**, meaning the lower limit of how cold matter can be, or the total absence of heat. Compared to the temperature scales we use for reporting weather and body temperature, readings on the Kelvin scale are much higher, with 273 K being the freezing point and 373 K the boiling point of water. A temperature of 300 K corresponds to the usual definition of room temperature.

Rather than use these larger numbers, we continue to use the **Fahrenheit** and **Celsius** (Centigrade) scales, °F and °C, respectively, in our daily activities. It is customary to describe these temperatures in terms of water. Water freezes at 32 degrees and boils at 212 degrees on the Fahrenheit scale. The Celsius scale breaks this same range into 100 parts, so that water freezes at 0°C and boils at 100°C.

Another absolute temperature scale that is used in some engineering applications is the **Rankine** scale. This scale has degree divisions equal in size to the Fahrenheit scale and 0 R is equal to 0 K. Like the Kelvin scale, the Rankine scale does not use the degree symbol. Table 2-2 compares some common measurements on these four scales, along with the formulas that can be used to convert temperatures between them. Note that there are no negative temperatures on either the Kelvin or Rankine temperature scales.

Although temperatures approaching absolute zero exist in the dark of outer space, they must be artificially created on Earth. Today, it is possible to buy off-the-shelf **cryogenic** coolers – sophisticated refrigerators – that can cool to as low as 10-15 K (see Technology Box 2-3). At these temperatures, nitrogen and oxygen freeze into solids, so these coolers can be used to remove most of the components of the air from an enclosed area. The resulting high vacuum has become essential for the production of such items as contamination-free microchips.

Luminous Intensity

Any source of light sends out a stream of energy in small packets called **photons**. Our eyes can detect part of this energy as visible light and we can feel another part of this energy on the surface of our skin as heat. We will discuss this process in more detail in subsequent chapters. For now, it is important to realize that there is a unit to measure this energy flow. The amount of energy per unit of time that a source emits into space is called the **luminous intensity**. The fundamental SI unit for luminous intensity is the **candela**, cd. The candela may be the least-used of all the SI fundamental units, but we include it here for the sake of completeness.

Temperature	Kelvin	Celsius	Rankine	Fahrenheit
Absolute zero	0 K	−273.15°C	0 R	−459.67°F
Freezing point of oxygen, O_2	54.8 K	−218.4°C	98.5 R	−361.1°F
Freezing point of nitrogen, N_2	63.3 K	−209.86°C	113.9 R	−345.75°F
Melting point of water	273.15 K	0°C	491.67 R	32°F
Boiling point of water	373.15 K	100°C	671.67 R	212°F
Melting point of lead	600.65 K	327.5°C	1,081.17 R	621.5°F
Melting point of Iron	1,808 K	1,535°C	3,255 R	2,795°F
Conversion, Celsius to Fahrenheit	°F = 9/5°C + 32			
Conversion, Fahrenheit to Celsius	°C = 5/9(°F − 32)			
Conversion, Celsius to Kelvin	K = °C + 273.15			
Conversion, Fahrenheit to Rankine	R = °F + 459.67			

Table 2–2: Common temperature measurements and conversions to different scales.

Number of Atoms

At first glance, the final fundamental unit in the SI and CGS systems appears to be just a number. The **mole**, mol, is the unit for quantity, whether atoms, molecules, ions or other kinds of particles. Since these particles are extremely small, it would be rare to work with them one at a time. Both systems define the mole as the number of carbon atoms in exactly 12 grams or 0.012 kilograms of carbon-12 (carbon-12 is the most common atomic form, consisting of atoms having 6 protons and 6 neutrons). The actual number of particles in a mole is called **Avogadro's number**, N_A, after the Italian chemist and physicist Amedeo Avogadro (1776–1856). Avogadro's number is approximately 6.02×10^{23} particles per mole.

From the turn of the twentieth century, there was general disagreement between physicists and chemists on the definition of the **atomic mass unit,** u, which had been done in a manner that hydrogen, the lightest known element, had a mass of 1 u. In 1960, an agreement was reached for the atomic unit to be defined as one-twelfth the mass of a carbon-12 atom. Experiments measured this value to be approximately 1.66 $\times 10^{-24}$ g. In actuality, this number is simply 1.00 g divided by Avogadro's number.

Note that since Avogadro's number has no units, when the mole is used the elementary particles – such as atoms, molecules, or ions – must be specified. Since we cannot actually count a mole of atoms, molecules, or ions, this connection between their mass and the number is important.

The **formula mass** of any substance is the sum of the masses of the atoms or ions of each substance present in a formula. The mass of each element or ion is found on the periodic table inside the front cover of the text. One formula mass, in grams, of any substance is the amount that contains 6.02×10^{23} or 1 mole of that substance's particles (atoms, molecules or ions). For example, from the periodic table we find that the formula mass of the element copper is 63.546, so 63.546 grams of copper would contain one mole or 6.02×10^{23} copper atoms. For a molecular substance such as water, H_2O, its formula mass is 18. Therefore, 18 g of water contains one mole or 6.02×10^{23} water molecules.

In the English system, the pound-mole, abbreviated lb-mole, is used by American engineers. Since one pound equals 454 grams, a pound-mole contains 454 times more atoms, molecules or ions than a gram-mole.

Technology Box 2–3 ■ Cryogenic Pumps: What does temperature have to do with a vacuum?

One particularly useful kind of vacuum pump is the cryogenic pump. It is used in many applications, especially in areas where very high vacuums are required. A vacuum is any region with an absence of particles, including those that make up the air.

As you move from the Earth there is a gradual reduction in the density of particles. In space the density of particles is about one trillionth, 1×10^{-12}, that of the air at the surface of the Earth. Orbiting weather satellites operate farther from the earth than the space shuttle; therefore the particle density in that region is less. In deep space the density of particles approaches zero; therefore, most of the universe is nearly a perfect vacuum.

To create a volume devoid of particles on Earth requires specialized equipment. The chamber must be made with thick walls so the surrounding air pressure will not crush them. A means of removing the particles from within the chamber is also necessary. There are many types of vacuum pumps, but most rely on some type of mechanical device to remove the particles. Cryopumps have the advantage of having no moving parts. They take advantage of the fact that every pure substance has a freezing point.

Commercial cryogenic pumping units routinely operate at 10–15 kelvin. At these low temperatures, oxygen and nitrogen freeze solid.

We are accustomed to seeing water in its solid, liquid, and gaseous form, but we rarely think of elements such as oxygen and nitrogen freezing into a solid. In a cryogenic pump, a refrigeration system cools a piece of metal to between 10 and 15 Kelvin. Oxygen and nitrogen molecules in the air inside the chamber condense on the metal and freeze into a block of solid oxygen and nitrogen ice. This is the same process by which water vapor freezes on a cold windshield overnight forming frost, except that the temperatures are much colder. Eventually, most of the molecules in the chamber have been frozen onto the metal surface, resulting in a nearly perfect vacuum.

Creating a vacuum using a cryogenic pump is the preferred method in the semiconductor industry. It is a clean process, leaving none of the contaminants common to other types of vacuum pumps. Contaminants could destroy the microscopic features of a modern microprocessor chip or other integrated circuit.

Technology Box 2–4 ■ Scientific Notation: How do you write a number in scientific notation?

When working with very large or very small numbers, it is often convenient to express them in **scientific notation**. Simply put, scientific notation involves breaking the number into two parts. The first part is a number that is determined by moving the decimal point to the left or right until the number falls between one and ten. For example, to get 6,000,000 to a number between one and ten would require moving the decimal six places to the left (6.000 000). The second part of the scientific notation is the number 10 raised to a power. The power of the ten is determined by how many places you move the decimal to the left or right. If you move the decimal to the left, the power is positive. If you moved the decimal to the right, the power is negative. In the previous example we had to move the decimal six places to the left, so the power on the ten would be +6 or 10^6. Putting this together then, makes the scientific notation for the number $6,000,000 = 6 \times 10^6$. Had the number been smaller than one, say 0.00000056, then the decimal would have to be moved seven places to the right. The number 0.00000056, therefore equals 5.6×10^{-7}.

Here are two more examples:

$$6,290 = 6.29 \times 10^3$$

$$0.00795 = 7.95 \times 10^{-3}$$

To change a scientific notation back into the number it represents, is just the reverse process. The power on the ten tells you the number of places to move the decimal left or right. For example Avogadro's number, 6.02×10^{23} would require you to move the decimal 23 places to the right. Avogadro's number, therefore, would be 602,000,000,000,000,000,000,000. It is probably obvious now, why we elected to express it and the fundamental charge, e, $(1.6 \times 10^{-19}$ C) in scientific notation.

Check Your Understanding

1. What two physical descriptions of space are based on length?

2. What are the seven fundamental quantities that are measured in the physical world and what do each describe?

3. What is the difference between the Kelvin and Celsius temperature scales?

4. What is the value of Avogadro's number?

5. Why are large and small numbers often expressed in scientific notation?

2-3 Application of the Systems of Units

Objectives

Upon completion of this section, you will be able to:

- Use letter prefixes to convert between SI units and other metric units.
- Multiply and divide fundamental units to derive new units for measuring other quantities.
- Convert measurements from one system of units to another.

Even when measuring quantities such as length and time that have fundamental units, you will frequently encounter units that are metric units, but not SI fundamental units. The nanometer, for example, is equal to 1×10^{-9} m, and is more appropriate for measuring distances on the atomic scale, such as the wavelength of light, than its counterpart, the meter, in the SI system.

Multiples of SI Fundamental Units

There was a time in this country when wrenches and bolts were produced only in English units (1/2 inch, 9/16 inch, 5/8 inch, etc.) Today, imported products are so common that most people own both English and metric tools. It is important to make a clear distinction, however, between metric units and SI units. Centimeters and kilometers are metric units of length, but they are not the fundamental SI unit of length. If you determine the length of anything in a multiple of the meter, like centimeters or kilometers, then you must remember to convert it into meters.

There are many times when use of the SI fundamental units result in either very large or small numbers. To avoid the inconvenience of writing them in scientific notation or counting and writing many zeros, shorthand symbols have been agreed upon that stand for multiplying or dividing by powers of 10. You have already encountered one of these, the "k" in kg, which is for the prefix kilo-.

Prefix	Multiple or Fraction	Scientific Notation	Pronunciation
G	1,000,000,000	10^9	giga
M	1,000,000	10^6	mega
k	1,000	10^3	kilo
h	100	10^2	hecto
da	10	10^1	deka
*	1.	10^0	*
d	0.1	10^{-1}	deci
c	0.01	10^{-2}	centi
m	0.001	10^{-3}	milli
μ	0.000001	10^{-6}	micro
n	0.000000001	10^{-9}	nano
p	0.000000000001	10^{-12}	pico

Table 2–3: Prefixes used with SI fundamental units.

In the metric system, these prefixes are used as a matter of convenience. The kilometer (1 km = 0.621 miles) is useful for longer measurements, like the distance between cities, and the centimeter (1 cm = 0.3937 inches) is useful for measuring distances on a piece of paper. Computer chip manufacturers now worry about how to make parts of their integrated circuits with sizes less than 1 micrometer, μm, which is too small to see without the aid of a microscope. The *mi*-cro-meter (accent on the first syllable) is often called the **micron**, probably to avoid confusion with the mi-*cro*-meter (accent on the second syllable), which is a device for measuring lengths very precisely.

The prefix pico- representing one-trillionth or 10^{-12}, has become very important in electronics, where the timing of electronic signals is often measured within picoseconds (ps), and the electrical property of capacitance is frequently measured in picofarads (pf), commonly called **puffs** in the technical slang.

The kilo- prefix means times 1,000. It can, therefore, be interpreted in two different ways:

$$1 \text{ kg} = 1(1,000) \text{ g} = 1,000 \text{ g, or}$$

$$1 \text{ kg} \times \frac{1,000 \text{ g}}{1 \text{ kg}} = 1,000 \text{ g}$$

The second way of writing it seems unnecessary, but it is very important to realize that this is the underlying logic. We need to think of the prefix as a placeholder for a multiplying factor. The list of the multipliers from one billion to one trillionth, their placeholders, and how they are pronounced, is given in Table 2-3. Unless you are familiar with the Latin language, from which these are derived, there will seem to be little rhyme or reason to how the prefixes are assigned. The only way to remember them is to simply memorize them. Start by memorizing the bold ones in the shaded rows first because these are the ones most commonly used.

The best way to handle prefixes in a calculation is to treat them as placeholders for numbers. For example, to determine how many meters there are in 10 kilometers, simply replace the k with 1,000 and multiply.

$$10 \text{ km} = 10 \ (1,000) \text{ m} = 10,000 \text{ m} = 1.0 \times 10^4 \text{ m}$$

A second method for replacing the k with 1,000 is to introduce a fraction, in which the numerator and denominator terms are equal. For example,

$$10 \text{ km} \times \frac{1,000 \text{ m}}{1 \text{ km}} = 10,000 \text{ m} = 1.0 \times 10^4 \text{ m}$$

In this method of converting km to m, the fact that 1 km = 1,000 m was used to create a fraction. The choice as to which measurement to put in the numerator and denominator was determined by the need to be able to divide out the km unit.

This second method is also useful for determining the number of centimeters in 10 kilometers.

$$10 \text{ km} \times \frac{1,000 \text{ m}}{1 \text{ km}} \times \frac{100 \text{ cm}}{1 \text{ m}} = 1,000,000 \text{ cm} = 1.0 \times 10^6 \text{ cm}$$

The use of fractions, where the numerator and denominator terms are equal, as conversion factors may seem cumbersome to use at first, but as you become more accustomed to the method, it is guaranteed to produce the correct results.

Derived Units

At this point, a fair question might be, "What good is a system that can measure only seven things?" The answer would be, "Just because there are only seven fundamental units does not mean that only seven different things can be measured." On the contrary, the fundamental units are merely building blocks from which **derived units** can be constructed to measure other quantities. For example, the derived SI unit of area is the square meter, m^2, which is built on the fundamental SI unit of length, the meter. To determine the surface area of a table, therefore, you would take its length, say 1.5 m, and multiply by its width, say 0.8 m. The result would be

$$\text{area} = \text{length} \times \text{width}$$

$$\text{area} = 1.5 \text{ m} \times 0.8 \text{ m} = 1.2 \text{ m}^2$$

Similarly, the derived SI unit for volume is the cubic meter, m^3. It, too, is built on the fundamental SI unit of length, the meter. To determine the volume of a warehouse 40 m wide by 60 m long by 4 m high would be

$$\text{volume} = \text{length} \times \text{width} \times \text{height}$$

$$\text{volume} = 40 \text{ m} \times 60 \text{m} \times 4\text{m} = 9{,}600 \text{ m}^3$$

Another example is the derived SI unit for speed. Just as mile per hour is a length divided by a time, the SI unit of speed is meter per second (m/s). If an object is moving 1 km every minute, then what would be its speed, expressed in SI units? Since the measurements are not in SI units, they must first be converted to get the answer in the appropriate units.

$$\text{speed} = \frac{\text{length}}{\text{time}}$$

$$\text{speed} = \frac{1 \text{ km}}{\text{min}} \times \frac{1{,}000 \text{ m}}{1 \text{ km}} \times \frac{1 \text{ min}}{60 \text{ s}} = 16.7 \text{ m/s}$$

As the above two examples demonstrate, derived units are built by simply multiplying and dividing fundamental SI units. Some derived SI units have even been given their own names. For example, the SI unit of force is the fundamental unit of mass (kg) times the unit of acceleration (m/s^2), which is named the **newton** in honor of Sir Isaac Newton. The newton also has its own symbol, N, so that:

$$N = \frac{\text{kg} \cdot \text{m}}{s^2}$$

The derived SI unit for energy is another unit with its own name, the **joule**, J, named for James Prescott Joule, where:

$$J = \frac{\text{kg} \cdot \text{m}^2}{s^2}$$

Many derived units are used in the world and in this text. A list of them appears in Appendix C. The best way to learn them is by using them in context. There are three rules, however, that you should always keep in mind:

1. In expressions with **exponents**, units get squared, cubed, etc. right along with their values.

   Example: The formula for the area A of a square with sides of length L is $A = L^2$. If the sides are 3 m in length, then

   $$A = (3 \text{ m})^2 = (3)^2 (\text{m})^2 = 9 \text{ m}^2$$

   Notice that the meter unit gets squared right along with the 3.

2. The second rule is similar to the first. In expressions with exponents, multipliers (centi-, milli-, etc.) get squared, cubed, etc. right along with their units.

Example: If a square has sides that are 3 cm in length, then

$$A = (3 \text{ cm})^2$$
$$A = (3)^2 (c)^2 (m)^2$$
$$A = (9)(0.01)^2 (m)^2 = 0.0009 \text{ m}^2$$

Remember that the prefixes c-, m-, k-, and so forth are merely placeholders for the factors (0.01), (0.001), (1,000), etc. The use of prefixes makes the metric system convenient, but it can also be a minefield. One way to avoid difficulty when doing calculations is to replace the prefixes with their appropriate multiplying factors, which gets all of the measurements into true SI units. For example, if a model train goes 20 centimeters in 10 seconds, what is its speed in SI units? The SI unit of speed is meters per second, so one way to get this calculation correct is as follows:

$$\text{speed} = \frac{\text{length}}{\text{time}} = \frac{20 \text{ cm}}{10 \text{ s}}$$

$$\frac{20 \ (0.01) \ m}{10 \text{ s}} = 0.02 \ \frac{m}{s}$$

An alternate way of converting the unit centimeter to the desired SI unit of length, meter, is:

$$\text{speed} = \frac{\text{length}}{\text{time}} = \frac{20 \text{ cm}}{10 \text{ s}}$$

$$\frac{20 \ \cancel{cm}}{10 \text{ s}} \times \frac{1 \ m}{100 \ \cancel{cm}}$$

$$\frac{20 \ m}{1,000 \text{ s}} = 0.02 \ \frac{m}{s}$$

When using these methods, it is important to remember that the SI unit of mass is the kilogram, kg, where the prefix k- must be kept to maintain the true SI unit.

A similar application using English units would be to measure a beach ball in inches, then calculate its volume in cubic feet. As an example, let us say that the ball has a diameter of 32 inches and the formula for finding the volume is $V = 4/3 \ \pi r^3$ or $V = 4/3 \ \pi \ (d/2)^3$. The value of pi is approximately 3.14, the volume of the ball, therefore, is:

$$V = \frac{4}{3}\pi \times \left(\frac{32 \ \cancel{in}}{2} \times \frac{1 \ ft}{12 \ \cancel{in}}\right)^3 = 9.93 \ ft^3$$

3. The third rule for derived units states that, although quantities of different units can be multiplied and divided freely, they may not be added or subtracted.

Example: The equation $5 \text{ m} + 12\frac{m}{s} = \underline{\hspace{1cm}}$ is meaningless.

Technology Box 2–5 ▪ Unit Analysis: How can you solve a problem if you have forgotten the formula?

Here is a scenario – you are facing a test question such as, "How long does it take a bullet traveling 100 m/s to cover a distance of 25 m? You know that the formula you need has distance, x, time, t, and velocity, v, but you can't remember the exact formula. Are you stuck? Not at all! You simply need to use the fact that both sides of an equation must have the same units. This is the apples and oranges argument all over again. You could try three possibilities and compare the units of the right- and left-hand sides:

$t = xv$	$s = m \times \dfrac{m}{s} = \dfrac{m^2}{s}$	wrong!
$t = v/x$	$s = \dfrac{\cancel{m}/s}{\cancel{m}} = \dfrac{1}{s}$	wrong!
$t = x/v$	$s = \dfrac{\cancel{m}}{\cancel{m}/s} = s$	right!

Unit analysis is an excellent way to check more complicated calculations, to make sure that you have not inverted or omitted a factor in your calculations.

If you find yourself adding or subtracting two quantities that have different units, you have made a mistake in your calculations. You must go back and find where your error has occurred. As a matter of fact, the technique called **unit analysis** (see Technology Box 2-5) can point you in the direction of your mistake. Simply look for the point in your calculation where the units are wrong – where your mass comes out in kilograms per second, for example.

Multiples of Derived SI Units

You will encounter units that are multiples of derived SI units as well as those that are multiples of fundamental SI units. The SI unit of energy, the joule (J), is fine for some measurements, but for others, the millijoule (mJ) may be more appropriate. On the other end of the energy scale, the megajoule (MJ) may be the unit of choice. For each of these units, the usual caution applies: be sure to realize that the prefix is a stand-in for a power of 10.

As previously mentioned, there are some metric units that have their own symbols. One of these is the cubic centimeter. It is perfectly appropriate to use the symbol cm^3 for this unit. But in medical applications, cc will be used for this unit of volume. Either is appropriate. Since the cubic centimeter is not the SI unit of volume, let's do one last conversion to put this unit in perspective:

$$1\ cm^3 = (1)^3\ (c)^3\ (m)^3 = 1\ (0.01)^3\ (m)^3 = 1\ (0.000001)(m)^3 = 1 \times 10^{-6}\ m^3$$

With a million-to-one difference between a cm^3 and its true SI counterpart, m^3, it is clear to see why this might be a unit of choice for reporting some medicine dosages.

Converting Between Systems of Units

In the beginning of this chapter, we noted that commerce was the initial driver behind mankind's development of systems of measurement. In one of the *Star Trek* movies, the crew of the *Enterprise* time-warps into twentieth century San Francisco. Needing money, Capt. Kirk pawns a pair of antique glasses that Dr. McCoy had given (and will give) him in the future. The pawnshop owner offers the captain $100. Capt. Kirk stares at the owner for a moment, dumbfounded, then asks, "Is that a lot?" The owner's offer was meaningless, since Capt. Kirk did not understand the unit. It is necessary to be able to answer questions such as, "How big?" and "How fast?" before moving on to a greater understanding of the physical world.

A **conversion factor** equates a measurement in one unit or system of units to its equivalent in another. For example, 1 inch = 0.0254 m or 1 pound = 0.454 kg. The last column in Table 2-4 gives conversion factors between each of the English measurements listed and their equivalent SI units.

Remember that calculators work only with numbers and it is your responsibility to keep track of the units. The easiest way to get the correct units for your answers is to work exclusively within a unit system. The penalty for mixing systems of units was dramatically demonstrated in the events that led to the loss of the $125 million Mars Climate Orbiter probe in 1999. One group was using English units, while everyone else was using SI units. For the navigational instructions to work, the English units had to be converted to SI units – a step that was overlooked, resulting in the loss of the orbiter as it came in too low, too fast in the thin Martian atmosphere and broke up. Had both groups been working with the same self-consistent system of units, this loss could have been avoided.

English Unit	Quantity Measured	SI Unit	Conversion
Inch	Length	Meter	1 inch = 0.02540 meter
Foot	Length	Meter	1 foot = 0.3048 meter
Yard	Length	Meter	1 yard = 0.9144 meter
Mile	Length	Meter	1 mile ≅ 1609 meters
Second	Time	Second	no conversion necessary
Minute	Time	Second	1 minute = 60 seconds
Hour	Time	Second	1 hour = 3600 seconds
Day	Time	Second	1 day = 86,400 seconds
Ounce	Weight/Mass*	Kilogram	1 ounce ≅ 0.02835 kilogram
Pound	Weight/Mass*	Kilogram	1 pound ≅ 0.4536 kilogram
Ton	Weight/Mass*	Kilogram	1 ton ≅ 907.2 kilograms
Square inch	Area	Square meter	1 square inch = 0.0006452 square meter
Square foot	Area	Square meter	1 square foot = 0.09290 square meter
Square yard	Area	Square meter	1 square yard = 0.8361 square meter
Mile per hour	Speed	Meter per second	1 mile per hour = 0.4470 meters per second

*The conversion of English units of weight to metric units of mass is often convenient, but they measure different physical properties of matter.

Table 2–4: Everyday English units and their equivalent in the SI system.

Let's try several examples to demonstrate the differences. Suppose, for example, you measured a standard sheet of typing paper and found it 8.50 inches by 11.0 inches. To calculate the area of the paper you use the formula for finding the area of a rectangle, which is:

$$\text{area} = \text{length} \times \text{width}$$

$$\text{area} = 11.0 \text{ in} \times 8.50 \text{ in} = 93.5 \text{ in}^2$$

If you again measured the sheet, but this time using the metric side of the ruler, you would find that the page measures 27.9 cm by 21.6 cm. When these numbers are applied to the area formula, it results in:

$$\text{area} = 27.9 \text{ cm} \times 21.6 \text{ cm} = 603 \text{ cm}^2$$

Although the answers 93.5 in^2 and 603 cm^2 are each correct for expressing the area in English and metric units, respectively, Table 2-4 shows that they are not the correct units for expressing area in SI units. Area in the SI system must be expressed in square meters, m^2. If the answer is to have the proper SI units, then all the inputs must also be in SI units. This will eliminate the later need to use a conversion factor.

Since there are 100 cm in a meter, the dimensions of the paper can easily be changed to 0.279 meters by 0.216 meters. When these numbers are applied to the formula, it results in:

$$\text{area} = 0.279 \text{ m} \times 0.216 \text{ m} = 0.0603 \text{ m}^2$$

The answer and units are now correct for reporting the area of the sheet of paper in SI units.

Another way to determine the area of a sheet of paper in SI units is to use the conversion factor given in Table 2-4 to convert square inches to square meters. The calculation would be as follows:

$$93.5 \text{ inches}^2 \times \frac{0.000645 \text{ m}^2}{1 \text{ inch}^2} = 0.0603 \text{ m}^2$$

Conversion factors can also be used to change English units of mass to its equivalent SI unit, kilogram. For example, what would be the mass, in SI units, of a baseball bat that has a weight of 40 ounces? From Table 2-4, we see that 1 ounce = 0.02835 kilogram, so its mass in SI units is:

$$40.0 \text{ ounces} \times \frac{0.02835 \text{ kg}}{1 \text{ ounces}} = 1.13 \text{ kg}$$

If a car is traveling at 55 miles per hour, what would be its equivalent speed in SI units of meters per second? Referring to Table 2-4 we find 1 mile per hour = 0.447 meters per second. Therefore,

$$55.0 \text{ mph} \times \frac{0.447 \text{ mps}}{1 \text{ mph}} = 24.6 \text{ mps}$$

In each of the above examples, an English measurement is converted into its equivalent SI units. Conversion factors, however, work equally well when converting SI measurements into metric or English equivalents. The only thing that is required for making these calculations is the appropriate conversion factor.

Check Your Understanding

1. What are the most commonly used prefixes in the SI system?

2. How do derived units differ from fundamental units?

3. Name two derived units that have their own names.

4. Name three rules for using derived units.

5. What is the value of unit analysis?

6. What is the conversion factor and when is it used?

Summary

There are many systems of units for measuring physical quantities in nature, some of them quite ancient, but the universal system of measurement for science and technology is the SI system. The most fundamental quantities (and their associated units) in the SI system are length (the meter), mass (the kilogram), time (the second), charge (the coulomb), temperature (the Kelvin), luminous intensity (the candela) and number of atoms (the mole). SI units for measuring other quantities can be found by multiplying and dividing combinations of these fundamental units, such as the unit of the meter per second (m/s) for measuring velocity. Some of these derived units have their own names. For example, the SI unit for energy is the joule, which equals $1 \text{ kg m}^2/\text{s}^2$. For convenience, it is common to refer to multiples of these units using a prefix notation such as 1 centimeter = 0.01 meter. Remember that multiples of SI units are metric units, but not all metric units are SI units. Through the use of the appropriate conversion factor it is possible to convert measurements in other systems to SI units.

Chapter Review

1. As a classroom activity, clear a portion of your desk and measure one cubit from the edge using your arm as the measuring tool. Then, use a meter stick to measure the distance from the edge of the desk to the mark you made. Ask several of your friends or classmates to repeat this exercise, and record their results. What is the average distance? How far from the average are the shortest and longest measurements? If you have a computer or calculator, use the built-in function to calculate the standard deviation of the measurements. How many of the measurements fall within one standard deviation on either side of the average?

2. The text implies that the best standards are based on fundamental quantities in nature. The international standard for the kilogram is a cylinder of metal kept under double vacuum jars in Paris. Write an explanation for why this standard does/ does not qualify as being the base for such a fundamental quantity? What could go wrong with this standard?

3. How many of the fundamental SI units in Table 2-1 could you use to describe yourself? Explain.

4. Explain, in your own words, how astronauts can be weightless but still have mass.

5. The formula mass of water (H_2O) is 18. A mole of water molecules is equivalent to how many grams of water? If a drop of water has a mass of 0.1 grams, how many moles of water molecules are in the drop? How many water molecules would that be?

6. A one-degree change on the Celsius scale is equivalent to what change in temperature on the Fahrenheit scale? On the Kelvin scale?

7. An iron bar at room temperature has a temperature of 300 K. What is its temperature in °C, °F, and R?

8. Convert the following to seconds, the fundamental SI time unit:
   a. a minute
   b. an hour
   c. a week
   d. a year (assume 1 year = 365.25 days)
   e. a century

9. Express your answers to Question 8 in scientific notation.

10. What units do the symbols Mm and Gg stand for?

11. Einstein's famous equation, $E = mc^2$, relates energy (E) to mass (m) and the speed of light (c). Of these three quantities, E, m, and c, which can be measured in fundamental SI units and which require derived SI units?

12. How many...
    a. millimeters are there in a kilometer?
    b. centigrams are there in a kilogram?
    c. micrometers are there in a meter?
    d. microseconds are there in a millisecond?

13. A unit that you would not encounter very often, but that is perfectly valid nonetheless, is the megasecond. How many megaseconds are there in a) a year, b) a decade, and c) a century.

14. One inch is 2.54 centimeters, exactly. Express the height of a 5 foot, 7 inch person in a) meters, b) centimeters, and c) millimeters.

15. Dollar figures associated with the United States federal budget are often expressed in megabucks. To get a feeling for how enormous these numbers are, convert the following costs to megabucks:
    a. a $0.10 piece of gum
    b. a $2.00 loaf of bread
    c. a $15.00 compact disk
    d. a $30,000 automobile

16. Table 2-4 gives conversion factors between English and SI units. Give the conversion factors between two different English units of a) length, and b) time (you choose the units).

17. The Massachusetts Avenue Bridge connects Boston, where many of the fraternity houses at the Massachusetts Institute of Technology are located, to the city of Cambridge, home of the MIT campus. One year during the 1970s, fraternity pledges were given the task of measuring the length of the bridge in a unique unit of their own devising. They chose the "smoot," which was the height of one of their pledge brothers. They measured the bridge and found its length to be "364.4 smoots plus one ear," which they painted at each end of the bridge, along with marks every 10 smoots along the length of the bridge. Subsequent pledge classes were given the task of repainting these marks. The decoration became so

traditional that when the bridge was completely resurfaced during the 1990s, the city engineers allowed the smoot markings to be repainted on the bridge.

Your task is to devise a similar unit based on yourself (your height, the width of your palm or little finger, etc.). Measure the length of five common objects, one of which must be less than a tenth of your unit and one of which must be more than 20 of your units long. Give your unit a name (the "smithpalm," for example) and give the conversion factor between your unit and the nearest English unit. State the length of the objects you have measured in terms of both English units and your new unit.

18. Determine the conversion between the unit of length you determined in Question 4 and the English unit of the foot. Using this factor, determine the following quantities using your unit of length:
    a. the height of a standard ceiling, 8 feet
    b. the length of a football field, 100 yards
    c. the length of a mile, 5,280 feet

19. One inch is defined as exactly 0.0254 m. Convert each of the quantities in Question 5 into meters.

20. One of the joke units in physics is the "furlong per fortnight." The furlong (still used in horse racing) is an old English unit equal to 1/8 of a mile, while a fortnight is two weeks. The speed of light in English units is 186,000 miles/second. What is the speed of light in furlongs per fortnight?

21. You are a medical technician. The dose for a new experimental medicine is not to exceed 1 tablet for every 30 kg of body mass. What is the maximum number of tablets that you can safely administer to a 185 lb patient?

22. Which is thicker, a 1/4-inch-diameter bolt in an American automobile chassis or an 8-mm diameter bolt in a Japanese automobile? Show your calculations.

23. Your company has a choice of two suppliers for sheet metal. An American supplier quotes you a price of $1.25 per square yard. A European supplier quotes you a price of 1.40 Euros per square meter. The current exchange rate is 1.05 US dollars for one Euro. Which supplier offers you the better deal? (Assume that these prices include shipping and handling.)

24. Convert the following to units that are fundamental SI units or combinations of fundamental SI units:
    a. 24 cm^2 (square centimeters)
    b. 0.1 mm^3 (cubic millimeters)
    c. 10 g/cm^2 (grams per square centimeters)

25. What is the speed of an automobile going 60 miles per hour in SI units (see Table 2-4)?

What tools do we use when making measurements?

Measuring with Instruments and Reporting Data

There is an old physics joke that goes like this: a professor handed each of four students a barometer (a device for measuring air pressure) and asked them to use it to determine the height of a nearby apartment building. The professor assumed that the students would measure the air pressure at the bottom and top of the building and use the difference to calculate the height of the building. That's exactly what the first student did, handing her barometer back to the professor. The second student used a different method – he measured the height of the barometer, the length of the shadow of both the barometer and the building, and from that information calculated the height of the building. The professor was quite annoyed when the third student returned with the shattered pieces of her barometer. She had taken the barometer to the top of the building, dropped it and timed its fall. From the time, she calculated the height of the building. Finally, the fourth student returned empty-handed. "What did you do with your barometer?" the professor asked. "Oh," replied the student, "I just went to the building supervisor and told him, 'I'll give you this neat barometer if you tell me the height of the building.'"

This story has a serious point – there are few things we measure directly. Most of what we know about the world we know from the effects that one thing has on another. If we can quantify these effects, we can build an instrument to measure them. Some instruments are simple, and others are more complex. Some are inexpensive, while others cost enormous amounts of money. Most importantly, some are better suited for a given task than others. The keys to using an instrument to make a measurement in physics or chemistry are 1) selecting the instrument that gives the most suitable result at the least cost, and 2) using the instrument correctly.

A Brief History of Instrumentation

The earliest instruments – balances and measuring sticks – performed the most basic type of measurement: direct comparison. Sometimes this is still the most effective method. For example, rulers and tape measures are used to determine length. Pan balances are used by placing the item to be measured on one pan and a known mass on the other. When the two pans are balanced, the masses are equal. This direct comparison method was the common way of determining mass for centuries.

General stores at the turn of the century had another type of instrument that was used to determine mass or weight. The mechanical scale had linkages that connected a needle, a pan, and a spring through a series of gears. When a mass was placed on the pan, it stretched the spring until the force of the spring matched the force of gravity on the mass. The stretching of the spring also turned a gear that was connected to a needle. The output was the position of the needle on the outer edge of a dial that indicated the mass or weight of the object. The physical position of the needle was proportional to the mass in the pan – the greater the mass, the greater the needle deflection.

It is common for measuring instruments to appear to be measuring one thing when in reality they are measuring something proportional to it.

Figure 3–1: The pan balance determines mass in a simple way, by directly comparing a known and unknown mass. Photo copyright © Tim Davis, Photo Researchers.

This characteristic is common to most measuring instruments. For example, the thermometer claims to measure temperature. Looking at a thermometer, however, it is clear to see that what is actually being measured is the length of a column of liquid in a thin capillary tube connected to a larger bulb of liquid. To use another example, a car's speedometer doesn't measure speed at all – it measures the rotation of a car's tires. This is why the speedometer registers a high speed when the wheels of a car are spinning but the car isn't actually moving.

Figure 3–2: Simple modern instruments such as mechanical scales measure one quantity by measuring another, proportional quantity. Photo copyright © Paul Biddle, Photo Researchers.

Estimation

Objectives

Upon completion of this section, you will be able to:

- Explain the usefulness of an estimate.

- Demonstrate that your own senses can be used as instruments for making estimated measurements.

An **estimate** is an approximation based on measurements that are known to have some uncertainties, but are based on the best information available. For example, can you estimate how many tons of aluminum soft drink cans are thrown away each year in a given city? There are several inputs that are needed to make this estimate. What is the population of the city? How many soft drinks does the average person drink each day? How much does a single can weigh? Each of these may also require an estimate. But be careful of estimates that will skew the data. For instance, as a student you probably drink more soft drinks than the average person, so your consumption is probably not a good average to use.

While grossly exaggerated estimates (high or low) must be avoided, it is also true that if an estimate is a little high on one input, chances are it will be a little low on another. The effect of these errors will often average out. An estimate for the repair of a car's fender, for example, should be very accurate because it is based on an extensive list of prices for parts and labor.

One tool of estimation is **sampling**. For example, take a count of friends who smoke versus those who do not, and from the numbers deduce the number of students who smoke on your campus. This is obviously an estimate – to get the exact number, would require polling every student. On the other hand, if the group of friends polled is typical, the estimate is likely to be reasonable, without the time and expense of conducting a complete survey. Sampling is a very common technique used in quality control work. A few products are pulled from a large number that were manufactured at the same time. The assumption is that there are the same relative number of defects in the sample as in the whole lot.

Another estimation tool is to infer one quantity from the measurement of another. As an exercise, the physicist Enrico Fermi would have his students estimate the number of pianos in Chicago from the number of piano tuners listed in the Chicago phone book. This estimate relies on several assumptions, but does result in a reasonable number. One tool of social analysis is the technique of garbology – inferring the behavior of people in the past from an analysis of their garbage.

It is possible to refine estimates based on sampling with various statistical techniques that are beyond the scope of this book. On election night, for example, you may hear news agencies report, that "With 1% of the polling places reporting, we project the winner…" as if the remaining 99% of the votes are irrelevant. As annoying as this practice may be, it can be surprisingly accurate, if the proper statistical analyses are applied.

Directly estimated measurements can also be made. The word "instrument" probably conjures up thoughts of an electronic device. In fact, the five senses can also be thought of as types of instruments. They provide important information about the world. The eye, for example, is able to distinguish between fine variations in the wavelength of light, which the brain perceives as color. The color of an indicator is one way to determine pH. The sense of smell can be used to distinguish between substances that have odors. Be careful about using this sense to identify chemicals, however, since some are very poisonous.

The senses can also provide information to make more quantitative estimates. The description of a criminal by height and weight is common, but the person giving the description did not use a ruler or a scale. When used by a trained observer, however, such estimates may be surprisingly close to the true value. Often, a good estimate can indicate whether it is worthwhile to make a more detailed measurement or abandon the idea. For example, it is not necessary to know the exact mileage between New York and Los Angeles to realize that driving from city to city in a single day is unrealistic.

Check Your Understanding

1. What is an estimate?

2. What is the value of sampling?

3. Describe a situation in which a reliable estimate would be as valuable in terms of decision making as an accurate measurement.

3–2   Accuracy and Precision

Objectives

Upon completion of this section, you will be able to:

■ Explain the difference between accuracy and precision.

■ Demonstrate the difference between random and systematic errors.

When applying for a driver's license, the applicant's height and weight are requested. How close must the information be? Probably most people give their height to the nearest inch and their weight to the nearest 10 pounds. In other words, most people probably answer as accurately as they know and as precisely as possible. The related concepts of accuracy and precision address the key question of all physical measurements: "How close is close enough?"

Accuracy vs. Precision

The words "accuracy" and "precision" may sound interchangeable, but as used in physics and chemistry, they are not. Determining the appropriate level of accuracy and precision are the key issues that must be addressed in order to choose the right instrument for a job.

Accuracy means how close a measurement is to the true value, and is an instrument's most important requirement. For example, a gasoline gauge that indicates more available gasoline than is really there will eventually leave you stranded by the side of the road.

Precision is the degree of exactness of a measurement. It is simply the smallest difference that can be detected between two measurements. The relationship between precision and accuracy can be summed up as follows: 1) precision without accuracy is of little value, and 2) accuracy is limited by precision. Therefore, the best instruments are those that are both precise and accurate.

It sounds as if a measurement is either accurate or not. There is, however, an absolute standard of accuracy. Whether we can measure it or not, we assume that under constant conditions a bar of metal has one true length, width, height, volume, and mass. If we say that a measurement is precise, what we really mean is that it is sufficiently exact for a given purpose. Measuring the length of the fence posts surrounding your house to the nearest centimeter is probably fine. This degree of precision, however, would be unacceptable if you were cutting the height of the door to fit your house. A door that is 1 cm too long would not fit into its frame and one that was 1 cm too short would leave a huge gap for the wind to blow through.

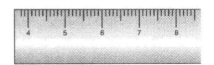   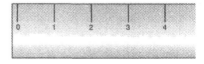

Figure 3–3: A ruler with its end cut off (left) may be precise, but it is not accurate. A ruler that is only marked to the nearest centimeter (right) may be accurate, but it is not precise.

The **tolerance** of a measurement is a statement of its precision by use of the plus or minus sign, ±. For instance, a measurement written as 45.0 ± 0.5 cm indicates a possible margin of error of 0.5 cm. In other words, the true value could be as high as 45.5 cm or as low as 44.5 cm.

There's a saying that you get what you pay for. This is especially true when it comes to the purchase of measuring instruments. A machinist's scale is more precise and expensive than a molded plastic ruler (see Figure 3-4). The amount you are willing to pay for an instrument should be based on the precision and accuracy that is appropriate for its intended use.

Random vs. Systematic Errors

One way of determining the precision and accuracy of an instrument is by repeatedly measuring an object and comparing the measurements to each other and to the known value. Comparing the measurements to each other indicates the precision of the instrument; comparing the measurements to the true value indicates the accuracy.

Let's look again at the rulers in Figure 3-3. If a bar of metal is known to be 10 cm long, but measurement with a ruler indicates it is 14 cm, it would quickly be discovered that 4 cm were missing from the ruler. Suppose the lengths of several other objects are measured using this ruler. After the error is discovered, each of the other measurements could easily be corrected, because the error in each is the same. It would merely be necessary to subtract 4 cm from each of the previous measurements. This is an example of a **systematic error**. Once discovered, the offset can be corrected in a systematic way.

Special Topics Box 3–1 ■ Determining the Accuracy and Precision of Measurements

The accuracy of a measurement is the difference between the average measured value your instrument gives and the true value. In mathematical terms, for any quantity x, the accuracy can be defined as

$$\Delta x_{accuracy} = \left| x_{true} - \frac{\sum x_{measured}}{N} \right|$$

where N is the number of measurements, and the summation symbol, Σ, means simply to add up all of the measured values. We take the absolute value of this number, because accuracy and precision are usually given as positive numbers. Similarly, a useful definition for precision is the standard deviation:

$$\Delta x_{precision} = \left(\sqrt{\frac{\sum (x_{measured} - x_{average})^2}{N-1}} \right)$$

There are two important things to notice about these equations. The first equation assumes that the true value is known. To determine an instrument's accuracy, you need to have a standard against which to compare its measurements. If the true value is not known, the best that you can do is use your average value and assume that its error is the same as the precision of the instrument. If you have only one data point, you can estimate the precision error from the manufacturer's specifications for the instrument. This assumes that the precision of the measurement is same as the precision with which you can read the instrument. Remember that it is always best to take multiple measurements.

Special Topics Box 3–2 ■ Throwing Away Data

Throwing away data sounds like cheating. In fact, many university honor codes forbid tampering with data and students are often reluctant to erase any data points that they take in lab. Removing a measurement from your data set is not something that you should do lightly, but it can be done from time to time, for the right reasons. Suppose that you did the same mass measurement ten times, and wrote down the following:

17.14 g	17.72 g
17.89 g	17.22 g
17.56 g	17.40 g
12.56 g	17.83 g
17.72 g	17.38 g

If these are all for the same object, then there is a problem with the 12.56 g measurement. When it is included in the average a value of 16.99 ± 1.56 g results, however, without it included the value becomes 17.52 ± 0.28 g. The problematic point, 12.56 g, is known as an **outlier**. It skews the results dramatically.

There are many reasons that the outlier could have become part of the data set. It could be a case of poor handwriting – mistaking a 7 for a 2, or someone bumping the balance at the wrong time. One way to recognize an outlier is that it is more than three standard deviations away from the mean (when both the mean and the standard deviation are calculated without the questionable point). For a normal set of data, such points are very rare, occurring less than 1% of the time.

Don't throw out a data point without first carefully considering whether there might be a good reason for it. Some of the most important discoveries in science have been unanticipated. On the other hand, don't let one bad data point spoil your results.

The other type of measuring error is a **random error**. In contrast to the systematic error, the random error is just what its name implies – a random variation around the true value. This is often referred to as the *scatter* in the data, because if the various measurements are plotted on a graph, they appear to be scattered around a point. Random errors are due largely to errors of usage or the limited precision of the measuring instrument.

It is possible to correct for random errors by taking multiple measurements of the same thing. If the errors are truly random, they will average out and the average or mean value will be closer to the true value than any single measurement. If only a single measurement is made, the random error will not be detected.

Check Your Understanding

1. Define precision and give an example of how it applies to a measuring instrument.

2. Define accuracy and give an example of how it applies to a measurement.

3. Give an example of an offset error and explain how it may be corrected.

4. Explain the difference between systematic and random errors.

5. Is throwing away data ever warranted and, if so, under what circumstances?

3-3 Significant Figures and the Digital Age

Objectives

Upon completion of this section, you will be able to:

■ Identify the number of significant figures in a measurement.

■ Choose the correct number of significant figures to use, based on the uncertainty in a measurement.

In the real world, every measurement has some uncertainty associated with it. How can we tell just how precise it is? The significant figures included in a measurement can tell you. If a person reports their age to be 25, they could be 25 years and one day or they could be 25 years, 11 months, and 29 days. Either way, 25 is accurate to within one year. If, on the other hand, the information comes from a youngster who proudly reports to be five-and-a-half years of age, it is understood that this reported age is to an accuracy of a half year. When a number is reported, the assumption can be made that all of the numbers are accurate except the last one. **Significant figures**, then, are all of the numbers in which a high degree of certainty is expected, and one that is an estimate.

Here's another example. Suppose you are asked to buy a pound of hamburger, but return empty-handed because you found packages marked 1.05 lbs, 0.98 lbs, and 1.12 lbs, but none that was exactly one pound. In reality, you would have bought one that was close to one pound and not worried about it. In everyday conversation, the meaning of a pound of hamburger implies some give-and-take. The way a grocery store marks packages of hamburger implies a certain precision in their measuring system. They are telling you that there is a difference between 1.12 lb and 1.13 lb packages. The more significant figures used when expressing a measurement, the more precision it implies.

Every measurement has an implied uncertainty, and significant figures indicate the precision of the measurement. If an object were measured using each of the instruments in Figure 3-4, each reading would be recorded using the smallest increment available on the particular device. In other words, each measurement would be accurate to the degree of precision allowed by the measuring device.

Finally, be careful when using a calculator to performing calculations with measurements that are expressed to the correct number of significant figures. Try an experiment. Using a calculator, enter (1 / 3.14 =) and see what you get. Write down the number for later reference. Now enter (1.000 / 3.14 =). If your calculator is typical you will get the same answer as you got before. A calculator does not differentiate between the number of significant figures in 1 and 1.000. When you write down 1.000, however, you imply a better precision, regardless of how your calculator interprets the number.

Figure 3–4: A cheap plastic ruler (left), a high-quality machinist's scale (center), and a micrometer (right) can measure the length of the same object to different levels of precision.

Counting Significant Figures

Sometimes it is easy to recognize what digits in a number are significant. The general rule is that a digit is significant, if it helps tell the difference between one value and another. Zeros, on the other hand, may cause confusion. The inclusion of a zero in a measurement may serve one of three different functions. It may be either a number, a leading zero or a trailing zero. **Leading zeros** are defined as any zero to the left of the first non-zero digit. **Trailing zeros** are the zeros on the right-hand-side of the number. To count the number of significant figures in a number, use the following guidelines:

— Count all of the non-zero digits in a number. For example, 123.4; 8562; and 1.476 each has four significant figures.

— Don't count leading zeros. The number 0.231 has one leading zero, while the number 0.00000846 has six. Each of these measurements has only three significant figures.

— Do count zeros in the middle of other, non-zero digits. The number 102.403 has six significant figures. All of the digits, including the zeros are meaningful.

— Do count trailing zeros if they are on the right-hand side of the decimal. For example, the numbers 890.00; 45.000; and 34.780 all have five significant figures. If there are trailing zeros without a decimal, let the context be your guide. Usually it is safe to assume that trailing zeros are not significant unless you have some reason to believe otherwise.

For example, if you encounter the number 750 by itself, you would assume that the 7 and 5 are significant, but not the zero. If you see this number is part of a list, however, such as 423; 846; 231; and 750, assume that all of the numbers in the list have three significant figures. To avoid ambiguity, express the number in scientific notational form, e.g., 7.50×10^2 (3 significant figures) vs. 7.5×10^2 (2 significant figures).

Reading Digital Instruments

The routines that digital instruments use to display their results follow the same rules that we have already learned for rounding numbers. If the number is less than halfway to the next higher number, round it down. If the number is halfway or more to the next higher number, round it up. So all of these numbers 3.14245; 3.14499; and 3.14001 would be rounded to 3.14, but the numbers 3.14875; 3.14501; and 3.14999 would all be rounded to 3.15.

The way to express the uncertainty in the above number is by writing 3.14 ± 0.005. Unless otherwise noted, assume that the uncertainty in a reported measurement is plus or minus one-half of the value of the position held by the last significant digit. In the above example, the last position that contained a significant figure was two places to the right of the decimal (the hundredths position). Taking one-half of the hundredths position produces 0.005, which is the expected uncertainty and the reason why the answer was reported as 3.14 ± 0.005. If for any reason the uncertainty is larger, it should be specified, such as 3.14 ± 0.07.

Calculation vs. Reporting

When performing calculations, a good rule of thumb is to report the same number of significant figures in the answer as there is in the data with the smallest number of significant figures. For example, if multiplying a number with eight significant digits

Special Topics Box 3–3 ■ Scientific Notation

When you type $(1 / 3.14 =)$ into your calculator, the answer may be expressed as 3.185 E – 01. If that is the case, your calculator is set to display in scientific notation. In scientific notation, answers are displayed as a number times a power of 10. The E part is for the term exponent and indicates the power of 10. Positive powers of 10 are easy, $10^1 = 10$; $10^2 = 100$; $10^6 = 1,000,000$; etc.

Remember that anything to a negative power is one divided by that quantity to the positive power, so $10^{-3} = 1/10^3 = 1/1000 = 0.001$. The calculator display 3.185 E –01 stands for $3.185 \times 10^{-1} = 3.185 \times 0.1 = 0.3185$. There are two advantages to scientific notation. First, it lets us express very large and small numbers compactly. For example, Avogadro's number, N_A, or the mole is 602,000,000,000,000,000,000,000 or 6.02×10^{23} in scientific notation. The other advantage to having a number expressed in scientific notation is that it makes it easy to count significant figures. All of the digits in the number multiplied by the exponent are significant! Therefore, in Avogadro's number as written above, there are only three significant figures.

When you write a number in scientific notation, the number multiplied by the exponent should be a number that is between 1 and 10. For example, 0.3185 should be written as 3.185×10^{-1} rather than as 31.85×10^{-2}. There are two exceptions to this practice. In the first instance, all numbers are written so they have the same powers of ten. This makes it obvious how the numbers in the list are different and it makes addition easy. It would not be possible to add the figures on the left. In the second instance, use of the closely-related engineering notation requires that all exponents be in multiples of three. Using this system, 46.7×10^3 would be preferred over 4.67×10^4.

4.67×10^4	4.67×10^4
7.40×10^3	0.74×10^4
2.95×10^4	2.95×10^4

(2.3456243) by a number that has two significant digits (1.8), the answer should be reported with only two significant digits. This simple rule can be easily justified by realizing that 1.8 is assumed to mean 1.8 ± 0.05. The number 1.8, therefore, could be as large as 1.85 and as small as 1.75. Multiplying 2.3456243 by each of these numbers produces the following results:

$$1.85 \times 2.34562435 = 4.33940505 \approx 4.3$$

$$1.75 \times 2.34562425 = 4.10484244 \approx 4.1$$

The average of these answers is 4.2 and includes only two significant figures.

Although multiplication was used in this example, the rule holds true for division as well. The precision of a number is reduced when it is multiplied by a number with a lower precision. Note that in Figure 3-5, the number of significant figures in the stated amount is the same, regardless of the system used.

Be careful, however, of numbers that are meant to be exact. For example, suppose the known diameter of an object is 5.40 meters. It is perfectly appropriate to state that the radius is 2.70 meters, even though the formula $r = d / 2$ was needed to get this answer. The 2 in the denominator of this equation is not a measured quantity – it is an exact number, representing the ratio of the diameter to the radius.

Typically, three significant figures are used when reporting answers to calculations, such as on homework. Although you can often assume that the numbers given in the problem are exact, this implies a precision of between 0.1% and 1%, which in practice, would take a well-calibrated set of instruments to distinguish.

Figure 3–5: Each of these labels specifies the amount of product in the package. Both 12.5 oz and 354 g have three significant figures, but 12 oz and 340 oz have only two significant figures.

Check Your Understanding

1. What are three different uses for the digit zero?

2. What is the difference between a leading and trailing zero?

3. Explain the rules for rounding a number.

4. What are two advantages to expressing a number in scientific notation?

5. How can the degree of uncertainty be included as a part of the answer?

6. What rule guides the number of significant figures that should be included in an answer that is derived by multiplication and division?

Presenting Data Graphically

Objectives

Upon completion of this section, you will be able to:

- Explain how graphical displays are used to present a large amount of data in a compact and intuitive way.

- Give an example of and explain how to read an analog scale.

- Identify a logarithmic scale and explain its advantages and disadvantages.

- Explain the difference between interpolation and extrapolation of data.

Graphical displays are the method of choice for displaying large amounts of data concisely. We don't often consider the amount of information necessary to compile something as simple as a road map. You probably see a display of more complex data every day on television or in the newspaper – a weather map. The colored bands of temperature are compiled using data from hundreds of weather stations across the country. It would be impractical and ineffective to just print these numbers on a map. A trained meteorologist might be able to see a pattern, but even then the process would be time-consuming. Color-coding bands of temperature allows both the expert and novice to see the thermal patterns across the entire country at a glance. Figure 3-6 also shows how graphical displays of complicated things like airflow patterns over a newly designed aircraft are becoming common.

Not every graphical display of data is complex. For example, there was a time when it was common for cars to have gauges that monitored the engine temperature, oil pressure, and so forth. Although it is relatively easy to convert a pressure reading from pounds per square inch (psi) to kiloPascals (kPa), what is really important is whether 30 psi of oil pressure is good or bad. Today you do not need to know, because your car has an indicator light that goes on if the oil pressure falls below an acceptable range.

These indicator lights provide information in the most basic of forms, called binary. As the name implies there are only two values: yes/no, acceptable/unacceptable, on/off. As shown in Figure 3-7, the control room of a sophisticated operation may contain hundreds of binary indicator lights that allow operators to quickly make

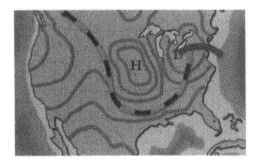

Figure 3–6: Common weather maps (left) combine the output of hundreds of measurements in an easy-to-grasp graphical format. Even the extremely sophisticated output of high-end computers (right) comprising hundreds of thousands of data points is becoming common place. Photo (right) courtesy NASA / Ames Research Center.

Figure 3–7: Binary (yes/no) indicators are the simplest form of data display, but they are also a powerful way to give people lots of information about sophisticated systems. Photo courtesy Bureau of Reclamation, Boulder City, Nevada.

decisions by using the best tool possible to quickly understand what is happening in the system.

Analog Scales

Between the extremes of a color map and a binary graphical display is the analog scale. An **analog scale** is one in which we compare the position of a mark or a needle to a fixed scale to make the measurement.

Reading an analog scale is among the earliest experiences most people have with instruments. In elementary school, for example, use of a type of analog scale – called a ruler – is taught to measure distance. Other common analog measuring devices include thermometers and speedometers. Interestingly, when digital readouts became inexpensive enough to put into consumer products, digital speedometers became a popular option on expensive automobiles. This fad quickly faded and most automobiles today use the traditional needle to indicate speed. Although analog indicators are sometimes thought of as an old-style technology, they do have one inherent advantage: it is easy to get a sense of whether a reading is large or small, near one end of the scale or the other, before you actually read the numerical value. This makes them a particularly good choice for such items as a speedometer, enabling the driver to keep his or her eyes on the road.

Linear Scales

The most common type of analog scale – represented by the ruler – is the **linear scale**. Every centimeter along the scale counts the same and there is a one-to-one correspondence between the length of the scale and the length of the object. In the case of the speedometer, every degree of needle deflection corresponds to a given change in speed. The relationship between the quantity being measured (speed), y, and the position of the indicator (needle), x, is the equation of a straight line ($y = mx + b$). This means that the increase in speed and the movement of the indicator needle are directly proportional.

A common metric ruler (shown in Figure 3-8) is divided into a number of evenly marked centimeter divisions. The scale is then divided into smaller millimeter subdivisions. There are ten of these smaller subdivisions to every centimeter. Psychologists tell us that we can glance at a small number of objects and tell how many there are almost instantly, whereas we have to count objects in large numbers. The dividing line between a small number and a large number is between five and ten for most people. To aid in reading the smaller subdivision, the metric ruler highlights every fifth one. When we place the ruler alongside an object, we can quickly and easily tell the length of the object by comparing it to the scale on the ruler.

Figure 3–8: A common ruler is a good example of a well-designed analog scale, in which the divisions and subdivisions are clearly marked.

Scales typically read from the lowest value on the left to the highest value on the right. The ruler is a common example of a well designed instrument, and illustrates two basic concepts: the limitations of a scale and what each of its divisions mean. People often fail to notice when a scale doesn't start at zero. For example, the speedometer shown in Figure 3-9 measures only down to 10 km/h, which limits the usefulness of this scale for slow speeds. Careful examination of a machinist's scale, as shown in Figure 3-9 illustrates the many levels of subdivisions, down to small fractions of an inch.

Figure 3–9: A speedometer (left) is a common example of a scale that doesn't start at zero. This is important when reading the low end of the scale. A machinist's scale (right) is a very accurate ruler that includes subdivisions down to 1/32 inch for precision.

The way a scale is marked is not a trivial matter since it ultimately determines the precision to which it can be read. It is assumed that the scale can be clearly read either at a particular mark or half way in between two marks. Therefore, the precision of a scale is one-half of the smallest division on the scale.

Given this information, it can be determined that the metric ruler in Figure 3-8 is precise to ± 0.5 mm. But this says nothing about the accuracy of the ruler, which is impossible to determine without comparing it to a known standard.

Logarithmic Scales

So far, we have dealt only with linear scales, which are the most common type of analog scale. Some instruments, however, generate readings that are not in linear proportion to their inputs. The most common scale of this type is the logarithmic, or log scale. This type of scale, as shown in Figure 3-10, is based on the common, or base-10, logarithm. The log of a number is the power of 10 that would give that number. Therefore, log 10 = 1, log 100 = 2, log 1000 = 3, and so forth. By using the log values, we can create a scale that is evenly divided in powers of 10. If we plot the position of the logarithm of a number, but mark the scale according to the value of the number rather than the $\log_{10}$, we get the scale shown in Figure 3-10. When reading a log scale, it is very important to keep in mind that the subdivisions of the scale change dramatically from one end to the other, so it is especially important to understand what the subdivisions mean. It is also important to understand that log scales can never reach a true value of zero, they can only get closer and closer (0.1, 0.01, 0.001, etc.) to zero by powers of 10. This is both their greatest advantage and disadvantage. Log scales are very useful for displaying measurements that vary over a wide range of values.

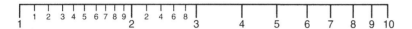

Figure 3–10: The logarithmic scale is the most common type of non-linear scale. Note that the marks for 1, 2, 3, etc. are unevenly spaced.

X-Y Graphs

The most important aspect of the X-Y graph is that it displays, in a visual way, the relationship between two quantities. Entering data into most commercial spreadsheet programs, then clicking the graph icon, will generate a menu of options – bar graphs, line graphs, pie charts, and others. For the purposes of physics and chemistry, however, the most commonly used option is the X-Y graph, or **scatter plot**. No attempt will be made here to explain a computer spreadsheet program. However, the principles behind making a good graph are the same whether you use a computer or a pencil and straightedge.

The relationship of two quantities represented on a graph may be purely **temporal** – that is, relating to the sequence of time – such as the measurement of the depth of a lake taken on the first day of June every year. Or the relationship may be **causal** – meaning arising from a cause – such as the changing pressure in a cylinder of gas as the temperature of the cylinder is increased. The basic X-Y graph can handle both types of relationships. Figure 3-11 is an X-Y graph illustrating a temporal relationship that has generated significant controversy. It is a graph of the average temperature at the Earth's surface as a function of time.

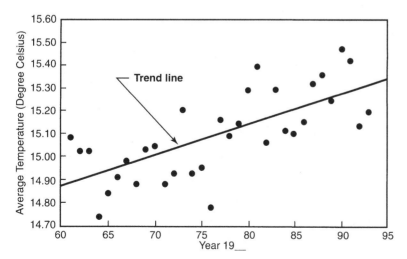

Figure 3–11: A graph of average yearly temperature at the surface of the Earth shows the phenomenon of global warming. Whether the data are accurate or not is still open for debate.

The terminology "temperature as a function of time," comes from the precise language of math. The term **function** means a specific mathematical operation that, given an input, returns one and only one output. For example, the mathematical expression $y = x^2$ is a function. If the input, x, is 3, then the output, y, is 9, guaranteed. The input is known as the independent variable, because we can use any input we wish and, by convention, its value on a graph placed on the x, or horizontal, axis. To make the position along the axis meaningful, it must include a scale.

The output of this function is known as the dependent variable, because its value depends on the value of the input. By convention, its value is represented on a graph on the vertical, or y axis. Again, vertical position is meaningless without a scale. When both the input and the output of the function are known, a data point can be located at the intersection of the horizontal and vertical positions.

Referring again to Figure 3-11, in the context of global warming the average yearly temperature at the surface of the Earth depends on the time frame being referenced. When a year is specified, the question can be answered. Time is the independent variable in this context, and temperature is the dependent variable. Therefore, time goes on the horizontal axis and temperature goes on the vertical axis. Each pair of measurements, time and temperature, is represented by a single point on the graph. Are the data in Figure 3-11 predicting global warming accurate? The debate rages on, but accurate or not, the graph is a good example of the effectiveness of the X-Y graph in presenting such data.

When data points are far apart, the connect-the-dot approach is usually not effective. In Figure 3-11, the important point about the graph is that the trend is upward, not that in some years the average temperature seems to go up rapidly and in others it appears to decline. The line through the data points is a best-fit line, determined by a linear regression (see Special Topics Box 3-4). It not only helps the viewer to follow the trend in the data, but it is a mathematically rigorous interpretation of the trend.

Figure 3-12 shows data from a different source – the cooling of a cup of coffee – with temperature again shown as a function of time. In this graph, there are so many data points that they appear as a smooth line. The pattern of temperature decrease is clear (quickly at first, more slowly later), and actually follows a well-defined mathematical relationship: the exponential decay function. When data points are numerous, as in this example, it is usually best to simply plot them as if they were a line.

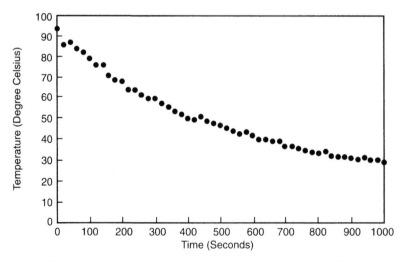

Figure 3–12: When many data points are taken in succession, the results can look like a smooth line.

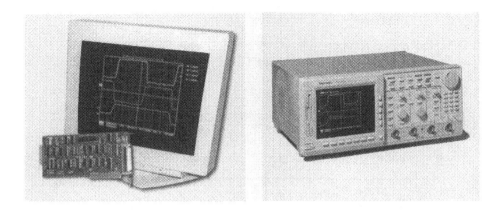

Figure 3–13: A digital data acquisition card (left) that fits in a personal computer performs the equivalent as a stand-alone instrument (right). Photo (right) copyright © Tektronix, Inc. Reprinted with permission. All rights reserved.

Data of the quality used to produce Figure 3-12 cannot be generated without special equipment. The digital oscilloscope used to collect these data consisted of a card in a personal computer, and software to generate the digital display on the computer's front panel. Oscilloscopes are also available as stand-alone instruments, both types of instrumentation are shown in Figure 3-13. Oscilloscopes differ from most instruments in that they display the results of two measurements simultaneously. High-end oscilloscopes can display the effect of a single independent variable on multiple dependent variables.

Interpolation and Extrapolation

From the graph shown in Figure 3-11, it is possible to predict the average temperature for the year 2020 – a data point that has not yet happened. The danger in predicting the future or the past, however, is that it assumes they follow the known data. This is not always a good assumption.

The most effective way of using data to make predictions is to use a best-fit line. In fact, some spreadsheets use these lines, known as **trend lines,** because they show the trend that the data are taking. If the mathematical formula for a trendline is available, it is the most reliable way to predict data because it effectively accounts for all of the data at once. For example, suppose that only the following two global temperature data points were known:

1972	14.93°C
1991	15.42°C

Since two points define a straight line (see Figure 3-14), the formula $y = mx + b$ applies. Temperatures in the intervening years or in the future can be determined through the two procedures known as linear interpolation and extrapolation. **Interpolation** finds points between known values, while **extrapolation** allows prediction beyond known data. The procedure for linear interpolation and extrapolation is essentially the same. Refer to Figure 3-14 for a graphic illustration of the steps that follow.

1. Find the slope of the line connecting the two points using the equation that follows, where – by convention – x_1, y_1 (1972, 14.93°C) is the left-hand and x_2, y_2 (1991, 15.42°C) is the right-hand known points:

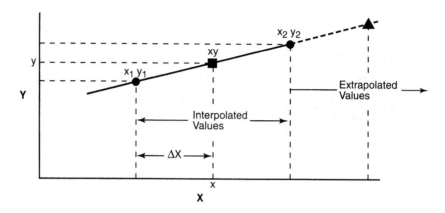

Figure 3–14: Linear interpolation and extrapolation assume that the trend shown by the data we have continue outside the range of the data we have, as well as between the known data points.

$$m = \frac{\Delta y}{\Delta x} = \frac{y_2 - y_1}{x_2 - x_1}$$

2. Use point (x_1, y_1) as your starting point, then move to the right along the independent variable axis until you reach the x-value for the point you wish to estimate the value of the dependent variable. Determine the difference, Δx, between this x-value and x_1, so that

$$\Delta x = x - x_1$$

3. Find the value of the dependent variable, y, at the new location by using the following relationship:

$$y = y_1 + m\,\Delta x$$

For the above example of Earth's temperature in 1972 and 1991, the rate of global warming implied by the two data points can be found by using the equation for slope given in step 1 above:

$$m = \frac{(15.42 - 14.93)°C}{(1991 - 1972)\ \text{years}} = \frac{0.0258°C}{\text{year}}$$

At this rate, Earth's average temperature in 2020 could be found by applying the relationship given in step 3 above, where y = temperature and Δx is the difference in years:

$$T = 14.93°C + (0.0258°C/\text{year})\,[(2020 - 1972)\ \text{years}]$$
$$T = 14.93°C + (0.0258°C/\text{year})(48\ \text{years})$$
$$T = 14.93°C + 1.24°C = 16.17°C$$

Remember that the farther from the known data range the prediction gets, the more suspect the extrapolation becomes. For example, extrapolation of these same two data points back to 1776 gives a temperature of 9.87°C. Since we know there was no ice age during the Revolutionary War, we know this extrapolation to be incorrect. Attempting to extrapolate 200 years based on 30 years of data is very unlikely to give reliable results. In general, it is best to avoid extrapolating more than 10–20% outside the data range. There are more sophisticated extrapolation schemes available, using more than two data points, that take into account the curvature in the data. But even these are unreliable far from known data. Interpolation is more reliable, because it is bounded by known data. Even known data, however, does not guarantee that no unusual event occurred.

Special Topics Box 3–4 ■ Linear Regression

The definition of a **best-fit line** seems obvious: the line that best represents the trend in the data. What, then, is the definition of best? In this case, the best line is the one that gets closest, on average, to each of the data points. It is defined mathematically, by 1) defining a function that measures how far away the line is from all of the data points, and 2) choosing the line that minimizes this function. This function is called "Chi-squared," (symbolically, χ^2, where χ is the Greek letter "chi"), and is given by

$$\chi^2 = \sum \left[\frac{y_{fit}(x_{measured}) - y_{measured}(x_{measured})}{\sigma_{measured}} \right]^2$$

where the summation means that this function is evaluated for every data point and added up. Because the uncertainty, σ, for each data point appears in the denominator, less reliable data points have less influence on Chi-squared. Often, the best we can do is associate the same uncertainty with each measurement, and in such cases all data points are given equal weight. The function y_{fit} is not limited to straight lines, but if the straight line is the best choice, the function y_{fit} would be given by

$$y_{fit}(x) = mx + b$$

where m and b are called parameters, in this case the slope (m) and the y-intercept (b). If you know the functional form (in other words the equation and all of its parameters) then you have all of the information you need to plot it. A straight line has two parameters, while more complicated functions may have many parameters, but the process is the same for all functions: adjust the parameters to minimize Chi-squared.

A straight line has the advantage that you can solve for the parameters a and b that minimize Chi-squared directly by matrix methods. These techniques are straightforward enough to be included in the programming of spreadsheets and high-end calculators, and so are usually available at the push of a button or the click of a mouse. Consult the instructions for your calculator or spreadsheet for details on how to input the data.

For other (nonlinear) functions, it is necessary to try various combinations of parameters to get the best fit. This could be a process of trial and error, but there are mathematical techniques to automate this as well. These algorithms are involved and usually available only on specialized scientific software packages. Nevertheless, it is possible to do a reasonable job of fitting a line to a complicated function using a spreadsheet by looking at the data and fitted function side by side on a graph, while varying the parameters and keeping an eye on a direct calculation of Chi-squared.

Check Your Understanding

1. What are two reasons for using graphical representations to display data?

2. What type of indicator lights are used in cars and in what types of applications are they valuable?

3. Why is a ruler considered a linear scale?

4. What are logarithms and when are they used for displaying data points?

5. What types of data can be displayed using X-Y graphs?

6. What is the value of interpolation and extrapolation of data?

Summary

It is possible to estimate the value of a quantity that describes a system, such as mass or length, but the use of an instrument to make a physical measurement produces more reliable results. The goal is to obtain measurements that are both accurate (that faithfully represent the true value of a quantity) and precise (that have little uncertainty in their value). Accuracy and precision are limited by systematic and random errors.

For the measurement taken by an instrument to be useful, the instrument must display the information in a way that can be easily interpreted. If the instrument uses an analog scale, the value of the measurement is determined by the position of an indicator (such as a needle or pointer) against a printed scale. Many modern instruments have digital readouts, which display measurements as an electronically generated number, such as 12.34. It is important to include all significant figures, but no more than that, when reporting measurements and calculations.

A graphical display of data makes it more compact and understandable. Very complex data may even be displayed in color-coded or three-dimensional graphics. The most useful format for displaying the relationship between one quantity and another, however, is the X-Y graph, in which a data point is located by its values on two different axes.

Chapter Review

1. (Suggested for use in the classroom.) Using the aluminum can example from Section 3-1, enlist your instructor's help to collect estimates from each student and the important inputs on which they were based. Write estimates on the board, and make a list of the inputs. Compare inputs and estimates for each to available data. For example, the city's population can be found in an atlas or on the city's Web page and someone can easily weigh an empty aluminum soda can. Take the average of these numbers and, using a spreadsheet program or a programmable calculator, calculate the standard deviations for each. How reliable do the estimated answers appear to be?

2. Fill in the following blanks for some product or food as directed by your instructor. If the number of _____ consumed in the United States in one year were placed end to end, they would stretch across the country _____ times.

   Make a list of the inputs used to estimate your answer and identify each source. You may use information from almanacs, newspapers and magazines, the World Wide Web, or your own experience. Which input do you believe to be the most suspect? Based on the quality of your inputs, what would you say is the margin of error in your estimate?

3. In the movie *Down and Out in Beverly Hills,* one of the characters has become rich (and settled in one of America's most expensive neighborhoods) because he has cornered the market on wire clothes hangers.

   Is this a reasonable premise? Assume that he can sell clothes hangers for a penny each and makes a 3% profit on each. What is his annual income, if he has a 30% market share across the nation? As in question 2, describe the inputs used to arrive at your estimate.

4. If you held the patent on a widget that went into every automobile sold in the United States and if you made a penny in royalties on each one, how much would your yearly income be? As in question 2, describe the inputs used to arrive at your estimate.

5. This is another good activity for the classroom with the help of your instructor. Make a ruler that is marked every two centimeters, either by scraping off the marks of a commercial ruler or by marking a strip of cardboard. Pass an object of known length around the room, along with the ruler. Each student measures the length of the object and records the data. After everyone has had a chance to measure the object, compare answers.

   Determine the average reading and, if a spreadsheet program or a programmable calculator is available, calculate the standard deviation of the measurements. Compare the average answer with the known length of the object. If the standard deviation was calculated, compare it with the difference between the average answer and the known answer.

6. The edge of a protractor is a type of linear scale, although it is physically curved. What are the units of the divisions and subdivisions of the scale on the protractor? What are the values indicated by points A, B, and C?

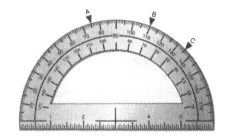

7. Look around your kitchen and bathroom. Read the fine print on packages of food, medicine. What precision is implied by the numbers you find? Look for packages that give weights and measures in English and SI units. Do the significant figures match?

8. As shown in the figure below, the temperature measured by a thermometer is actually determined by measuring the length of a liquid column. How do we know that the scale is a linear scale? For example, if *y* is the temperature in degrees Fahrenheit and *x* is the length of the column in centimeters, what is the linear equation that relates *y* and *x*?

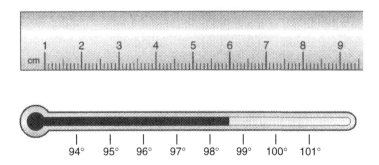

9. The figure contains some unusual scales. What reading is indicated on each scale?

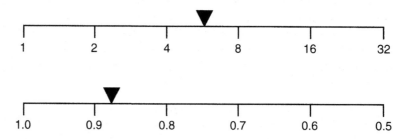

10. For the logarithmic scale shown in Figure 3-10, with what precision could you read the scale at the lowest end? With what precision could you read the scale at the highest end? What would the percentage of error at each end of the scale be?

11. Plot the following data points on an X-Y graph. Include a best-fit line.

x	y
0.0	0.9
1.0	3.2
2.0	4.8
3.0	7.1
4.0	8.9
5.0	11.2

12. Using the following two measurements of global temperature,

1970	14.96°C
1988	15.32°C

determine what temperatures these measurements imply or predict for the years 1776, 1959, 1983, and 2020.

13. Explain the following statement in terms of accuracy and precision: "A working clock is never right, but a clock that is stopped is right twice a day."

14. Before pre-packaging was used for meats, it was common for a customer to get meats cut to order. The butcher would then weigh the meat to determine the price. The icon of dishonest business was the butcher with his thumb on the scale. Is this an example of systematic error, random error, or both? Explain your answer.

15. The speedometer of a car reads 55 mph when the actual speed of the car is 58 mph. What would the true speed of the car be at 25 mph, 35 mph and 45mph, if:

    a. the error is due solely to an offset in the speedometer?

    b. the error is due to the speedometer being off by a given factor or multiplier?

16. If the 25 mph reading of the speedometer in Question 15 corresponds to a true speed of 27 mph, what true speed do the 35 mph and 45 mph readings correspond to? Assume that the response of the speedometer is linear.

17. Before delivering a digital balance, the quality control department places a known 2 kg mass on the balance and records the results. If the reading is within ± 40 g of the true reading, the balance passes. At the end of the day, the quality control

clerk has recorded the following values. What should the manufacturer advertise as the precision and accuracy of the balance?

2.007 kg	1.990 kg	2.032 kg	2.003 kg	1.975 kg
1.988 kg	2.025 kg	2.006 kg	1.979 kg	2.011 kg

18. Ten students pass a metal bar around the laboratory and measure its length. Examine the lowest and highest measurements. Should you consider either of these to be outliers? Justify your answer based on the statistics of the measurements.

10.9 cm	9.9 cm	10.1 cm	9.7 cm	10.4 cm
9.8 cm	10.1 cm	10.2 cm	10.2 cm	9.6 cm

19. The number of significant figures in the English and SI measures on this product label do not match. How would you recommend that the manufacturer change this label to be consistent? Explain your answer.

20. Determine the number of significant figures in each of the following:
    a. 12.39
    b. 40580
    c. 0.000358
    d. 7020000
    e. 0.0035840
    f. 23.00

21. Express each of the numbers in Question 20 in proper scientific notation. Make sure you include the appropriate number of significant figures in your answer.

22. If a car travels 15.34 meters in 0.52 seconds, what would you quote for the speed of the car in meters/second? Knowing what you can assume about the uncertainty in each of these numbers, show that your quoted value for the speed has the correct number of significant figures.

23. The conversion factor from inches to centimeters is 1 inch = 2.54 centimeters exactly. Convert each of the following measurements in inches to centimeters, including the correct number of significant figures in your answer.
    a. 10 inches
    b. 1.4590 inches
    c. 10. inches

24. Write an explanation for why the edge of a protractor (Question 6) is a type of linear scale, although it is physically curved?

25. An analog watch with a sweep second hand can be considered a combination of three linear scales. Explain what each of the scales are and how they work.

Just what is all this stuff that surrounds us?

The Nature of Matter

The scientific method was introduced in Chapter 1. It involves identifying a problem, gathering information, and stating and testing of hypotheses. If they prove to be a satisfactory explanation, they become a theory, which is tested further. One of the most important theories to the study of chemistry was developed using this method. It is known as the atomic theory. The original observations and hypothesis were made by the ancient Greek philosophers, who were trying to understand the nature of matter. Nearly 2,000 years later, John Dalton, an English chemist, expanded the hypothesis and proposed the atomic theory.

In the theory, Dalton stated that each element is composed of tiny indestructible particles called atoms. He also stated that all atoms of an element are chemically alike and that the atoms of different elements are different. He further stated that during chemical reactions elements combine only in ratios of small whole numbers, e.g., 1:1, 2:1, 1:2, etc., to form chemical compounds. Finally, he stated that some elements such as carbon and oxygen can combine in two or more ratios, forming a different compound with each combination; CO and CO_2, for example. Although some of these statements were later found to be incorrect, the atomic theory served as the guiding light for probing the nature of matter throughout the nineteenth century.

4-1   Description of Matter

Objectives

Upon completion of this section, you will be able to:

■   Define matter.

■   Discuss the differences between pure substances and mixtures.

■   Identify and explain the three common states of matter.

■   Discuss the conditions that result in transitions between the states of matter.

■   Name one unusual state of matter.

Description of Matter

The ancient Greek philosophers are remembered for being good observers who tried to explain their observations. One of their conclusions was that everything is composed of some combination of four basic elements: earth, air, water, and fire. It turns out that this explanation, although much over-simplified, resembles many of our modern ideas.

An example of one of their observations was that it takes earth and water to grow grass. They reasoned, then, that if a cow eats that grass and breathes air, then the cow must be composed of earth, water, and air. They further observed that when the cow dies, its body decomposes back into air, water, and earth.

Today, we still believe that solids (like earth), gases (like air), and liquids (like water) are the basic components of all physical things. We call these components **matter** and define matter as anything that occupies space and has mass. In our discussion, we will subdivide matter by the various ways it occurs in nature and examine the properties of each group. Whether solid, liquid, or gas, all matter can be placed, as shown in Figure 4-1, into one of two large groups: pure substances and mixtures.

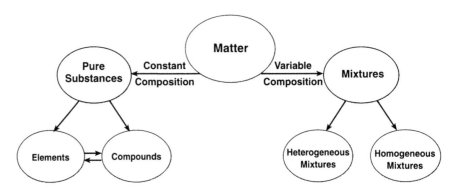

Figure 4–1: The subdivisions of matter.

Pure Substances

Matter that has a constant composition is a **pure substance**. This group, in turn, can be subdivided into two smaller groups: elements and compounds. By definition, an **element** cannot be chemically broken down into a simpler substance. There are 92 naturally occurring, and a growing number of man-made elements. Each has been given an internationally agreed upon name and symbol. Many names are familiar, such as carbon, C; iron, Fe; oxygen, O; and aluminum, Al. The smallest part of an element that retains all of its properties is an **atom**.

Compounds are pure substances that are composed of two or more elements in a specific ratio. This category of pure substance is comprised of over 20 million known chemicals. Living organisms, including our own bodies, make many thousands of chemical compounds daily for very specific purposes. In the United States, over 70,000 chemical compounds are made and used in industry every day – and the list is still growing. A few examples of compounds include water, H_2O; glucose, $C_6H_{12}O_6$; ammonia, NH_3; and carbon dioxide, CO_2. By definition, a **molecule** is the smallest amount of a compound that retains all characteristics of that substance. A **chemical formula**, such as $C_6H_{12}O_6$, shows the number of each kind of atom contained in a molecule.

Special Topics Box 4–1 ■ Is there a difference between natural and synthetic vitamin C?

Although the symptoms were recorded as early as 1500 BC, it was Aristotle (ca. 450 BC) who first described a malady characterized by the lack of energy, gum inflammation, bleeding problems, and tooth decay. During the 1700s, many British sailors died from a condition called scurvy. It was not until James Lind (1747) discovered that the juice of citrus fruits could prevent and cure the deadly disease that British ships started carrying limes and her crews became known as "limeys."

The specific compound in citrus that prevents scurvy was isolated in 1932 by Charles G. King. It is known as ascorbic acid or vitamin C (C for citrus). Today, many other natural sources are known, including rose hips, acerola cherries, papayas, cantaloupes, and strawberries, to name but a few. Most animals can also synthesize vitamin C, but primates, including man, and guinea pigs cannot and, therefore, must rely on dietary sources.

Vitamin C $C_6H_8O_6$

During the 1970s, Nobel Prize winning chemist, Dr. Linus C. Pauling published a book entitled *Vitamin C and the Common Cold*. In this book, he claimed that 1,000 to 5,000 milligrams of vitamin C taken per day would prevent or cure the common cold. The recommended daily allowance (RDA) for vitamin C is 60 – 70 mg per day. Fortunately, because it is a water-soluble vitamin, any unused vitamin C is excreted from the body in the urine, so that the most common side effect of these megadoses is a urinary tract irritation due to acidic urine.

It was difficult to meet the market demand for such large amounts of vitamin C during this craze. No natural sources contain megadose amounts, so companies started making it and supplying it as a dietary supplement. The manufacture of vitamin C from glucose by the bodies of most animals enables them to compensate for a dietary lack of the vitamin. Vitamin C is, after all, a chemical compound with a definite composition ($C_6H_8O_6$), so manufacturers turned to an inexpensive source of glucose (corn syrup) and started making it. The resulting "synthetic" vitamin C (ascorbic acid) contains the exact same chemical compound found in natural sources, is easy to concentrate in tablets, and less expensive for the consumer. There is no chemical difference between natural and synthetic vitamin C, although the best way to get your RDA remains eating a well-balanced diet.

Mixtures

Mixtures are the second major division of matter and differ from pure substances by having variable compositions. For example, if you examine a fruit salad or piece of concrete, different components are clearly visible. These are examples of **heterogeneous** mixtures, which are not the same throughout. There are times, however, when mixtures are less obvious. If several students were asked to each make a glass of salt water, the amount of salt and water used would likely vary. Yet, each student would have made salt water and they all would look alike. Each glass of salt water is an example of a **homogeneous** mixture, which means that it is just as salty at the bottom of the glass as it is at the top. Appearance alone, then, cannot be used to distinguish between a pure substance and a mixture.

Even though the salt has been dissolved by the water, it has retained its salty taste. It is true of all mixtures, that the individual substances retain their original characteristics. If you were to make a mixture of sand and salt, for example, water will still dissolve the salt, but not the sand. The fact that individual components retain their original properties when in a mixture can sometimes be used to separate them.

Properties that are helpful when describing matter include their physical and chemical properties. **Physical properties** are those characteristics that help us describe the substance, such as color, odor, hardness, density, electrical conductivity, specific heat, and melting and boiling points. They are generally the same types of characteristics we would use to describe an animal: floppy ears, bright eyes, short fur, and long tail. It is not necessary to change the nature of a substance in order to determine its physical properties.

The **chemical properties** of a substance, on the other hand, describe how it reacts with other substances. Does it burn? Does it react violently with water? Is it inert? Chemical properties can be compared to an animal's disposition. Is it playful, lazy, or always getting into fights? In the process of determining its chemical properties, the substance is chemically changed. Burning a piece of paper tests its flammability, but it also changes the paper into chemically different substances.

Two other terms that are helpful when describing matter are chemical changes and physical changes. **Chemical changes** result in the production of a new substance with its own set of chemical and physical properties. The identification of a rock by observing whether gas bubbles form when hydrochloric acid is dripped on it, is a test of its chemical properties. The rusting of a piece of iron is a chemical process that results in the formation of a reddish-brown powder that no longer looks or acts like iron. **Physical changes** do not produce a new substance, but may change the physical form of the substance. You still have wood, for example, even after it has been converted to a pile of sawdust; and converting ice into steam does not alter the fact that chemically it is H_2O.

States of Matter

There are three **states of matter** in which all pure substances, both elements and compounds, can exist: solid, liquid, and gas. It may, however, take conditions other than room temperature and normal air pressure to effect change from one state to another. The transitions water makes as it changes from ice to water to steam are familiar ones. The conditions required to transform matter from one form to another (a phase change) are characteristic of the substance and can be used to help identify it.

The temperature at which a material makes the transition between solid and liquid is called its **melting point** or **freezing point**, depending on the direction. The temperature responsible for the transition between the liquid and gaseous states is called either the **boiling point** or **condensation point**. Figure 4-2 shows the temperatures required for the phase changes of water

Under certain circumstances, some substances (carbon dioxide, iodine, p-dichlorobenzene) can change directly from a solid to a gas, without passing through the liquid state. This transition is called **sublimation**. Perhaps you have seen this process, in the form of dry ice that appears to be boiling and creating a cloud. The cloud you see is not colorless CO_2, but water vapor condensing when it comes in contact with the very cold carbon dioxide gas that is escaping from the solid.

Changes in state also result in a change in density. **Density** is the ratio of mass to volume. The mathematical formula for calculating density, ρ, is:

$$\rho = \frac{m}{V}$$

where mass, m, is divided by volume, V.

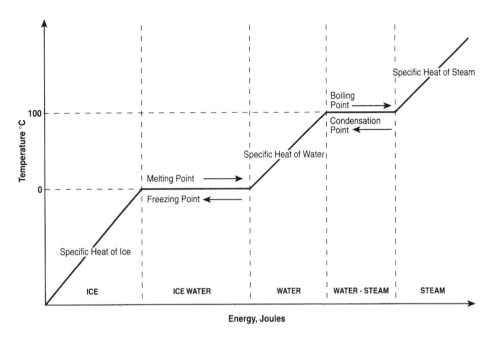

Figure 4–2: Phase diagram for water. The temperatures shown on the vertical axis are correct at standard pressure (1 atmosphere).

As pure substances change from solid to liquid to gas, they typically become less dense. That is, they increase in volume, while their mass stays the same. The notable exception is water. Although ice is solid water, its density is less than liquid water – which is why it floats. So, lakes freeze from the top rather than the bottom and there is still liquid water below where fish can live. If water behaved like most substances, only a few feet on the surface of the Earth's oceans and lakes would thaw during the summer months and the rest would remain frozen. Life forms, as we know them, might never have evolved.

Solids

Compared to liquids and gases, solids come in a fantastic diversity of forms. Houses, roads, tables, and steel are solids; but so are cotton, polystyrene, and feathers. Solids can feel either hard or soft, rough or smooth, but there are certain properties they all have in common. **Solids** are noncompressible substances that have a definite shape and volume. That is to say, solids are rigid and will only change their shape when an external force is applied. The fact that they do not change their volume tells us that they are non-compressible. Solids like cotton and polystyrene also contain a lot of air. They feel light and airy, but if you removed the air, the result would be a small hard block.

Solids show resistance to compression because the particles they are made of are packed tightly together. Attractive forces between these particles hold them in a rigid arrangement. In some solids, the arrangement is a recognizable, repeating pattern, or **crystal**. The characteristic shape of the crystal is called its **crystal structure** and is sometimes used to identify the substance. Other solids are non-crystalline and composed of seemingly random arrangements of particles.

Carbon, for example, can commonly be found in two different crystalline and one non-crystalline forms. (See Figure 4-3.) When carbon atoms form tetrahedral crystals, it is called diamond. But when carbon atoms form crystals that are flat hexagonal sheets, it is graphite. These hexagonal sheets can slide on each other, making graphite a good dry lubricant. More recently, more complicated carbon structures containing

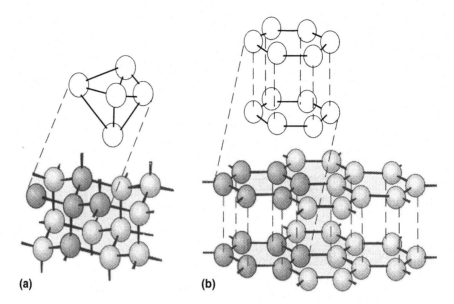

Figure 4–3: The (a) tetrahedral structure of diamond and (b) hexagonal structure of graphite.

60 atoms arranged as if on the surface of a sphere, have been discovered. Such crystals resemble the geodesic domes designed by visionary architect Buckminster Fuller, and have been given the name Buckminster fullerences, or "Bucky balls" for short. Carbon atoms can also exist in a non-crystalline or random arrangement, known as its **amorphous** form. Soot and charcoal are examples of amorphous carbon.

Numerous other terms are used to describe the properties of solids such as hard, soft, strong, elastic, or breakable. Scientists employ more specific terms. If the shape of a solid material can be changed without breaking, it is said to be **ductile**. This is the characteristic that allows most metals to be drawn into wire. A measure of the amount of effort it takes to change the shape of a solid is called its **tensile strength**.

Most metals are also **malleable**; that is, they can be shaped by hammering, pressing, or bending. A blacksmith takes advantage of the malleability of iron when he hammers it into a horseshoe. The metal parts on your car, stove, and refrigerator were likely shaped by either a bending or a pressing process.

Liquids

A **liquid** is a non-compressible substance that occupies a definite volume, but assumes the shape of its container. Unlike solids, the particles of a liquid, although in mutual contact, aren't in any definite arrangement. They just slide past each other.

Despite their obvious differences, liquids and gases are both referred to as **fluids** because both flow and can be poured. Influencing the ability of a fluid to be poured is its **viscosity**, which is a measure of its resistance to flow. Cold molasses is an example of a liquid with a high viscosity.

Energy, usually in the form of heat, is required to convert a substance from a solid to a liquid. The energy that is added tends to loosen the particles in the crystalline structure. Although they are still in contact, they are no longer part of a crystal. The melting point is a reflection of the amount of energy required to disrupt the crystal arrangement, turning the crystalline substance into a liquid. Table 4-1 lists the melting points of various fluids.

Technology Box 4–1 ■ What are liquid crystals?

Liquid crystals are a phase of matter whose order is intermediate between that of a liquid and a crystal. They are typically composed of rod-shaped molecules, whose arrangement changes with temperature. These chemicals will flow like liquids, yet their molecules fall into organized arrangements, tending to orient parallel to one another.

Different types of liquid crystals can be formed based on whether the molecules are parallel, with like ends pointing the same way; or antiparallel, with like ends pointing in alternating directions; or stacked in parallel layers. The correlation between organization and temperature is the basis for making the numbers on liquid crystal thermometers appear. The most common use of liquid crystals recently, though, is in liquid crystal displays (LCDs) for computers and other kinds of electrical equipment.

An LCD monitor was used in writing this chapter. Here, small electrical currents cause some small parts of the computer screen to be ordered in one direction and others in a different direction. This ordering results in written characters that can be read from the screen.

Material	Melting Point	
	°C	°F
Corn Oil	–20	–4
Lard	30.5	86.9
Sugar (sucrose)	185	365
Table Salt (NaCl)	801	1,474
Paraffin wax	49–63	120–145
Sulfur	112.8	235
Mercury	–38.9	–37.8
Lead	327.5	621.5
Gold	1,064.4	1,948
Iron	1,535	2,795

Table 4–1: Melting points of some common materials.

Gases

A **gas** is compressible, but it lacks both a definite shape and volume. Unlike a liquid, a gas tends to fill its container. In contrast to liquids and solids, where the particles are in contact with each other, in gaseous form they only occasionally come into contact. The gas particles in a container fly about, bouncing off the walls and colliding with each other. This continual motion is how gases manage to scatter throughout the entire volume of a container. The process of one gas spreading through another gas is called **diffusion**, and is how the aroma of coffee or perfume disperses throughout a room.

As heat is added to a liquid, a point is reached where the faster moving particles start rapidly escaping from the surface of the liquid, thereby becoming a gas. This is what happens at the boiling point. The boiling point is a characteristic of a substance and depends on the surrounding gas pressure acting against the surface of the liquid. Pressure is defined as a measure of force exerted over an area. That is, to calculate pressure, P, you must divide force, F, by the area, A, that it is acting on; or, stated mathematically:

$$P = \frac{F}{A}$$

The gas pressure we measure is the result of many particles hitting against the surface of a container or instrument. The pressure depends on the speed of the particles colliding against it, the mass of the individual particles, and how many such collisions occur during a given time interval. Temperature plays a significant part in this, because temperature is related to the average speed of the particles. As the temperature increases, so too does the average speed of the particles. Our eardrums are sensitive to very small changes in the number of particles hitting them on opposing sides. This is why our ears can hurt or pop when we are on an airplane and the amount of pressure in the cabin becomes different from what is in our inner ears. We feel the pressure on our eardrums and swallow. The process of swallowing briefly opens the eustachian tube connecting the throat to the inner ear and allows the number of particles hitting the eardrum from the two directions to equalize. This equalization is frequently accompanied by a popping sound as the eardrum returns to its normal shape.

If you place a container of water on a table, it will evaporate slowly. That is, all the water molecules will eventually escape and enter the gaseous phase. This transition from liquid to gas happens at all temperatures. At room temperatures, where the transition is gradual, we refer to it as **evaporation**. As the liquid is heated, the transition happens faster. The temperature at which bubbles start to form in the liquid is called its boiling point. Boiling is defined by bubble formation, or the point at which the vapor pressure equals the atmospheric pressure. Liquid nitrogen, for example, boils at −195.8°C under normal atmospheric pressures.

Both evaporation and boiling happen faster when the surrounding gas pressure is reduced. These surrounding gas particles frequently crash into particles at the surface of the liquid, thereby blocking their escape. At the top of some of the world's highest mountains, you can actually drink boiling water. Closer to sea level, we have to let our coffee or tea cool before drinking it. The reason for the difference is that there is less air pressure at higher elevations, so water boils at a lower temperature. For example, the air pressure in New Orleans averages 1 atmosphere or 14.7 pounds per square inch (psi); in Denver it averages a little more than 0.8 atmosphere (11.8 psi); and at 10,000 ft it is 0.7 atmosphere (10.3 psi). Consequently, water boils at 100°C (212°F) at sea level; at about 95°C (203°F) in Denver; and at 90°C (194°F) at an altitude of 10,000 ft.

Plasmas

Occasionally referred to as the fourth state of matter, **plasma** is a relatively new discovery. A plasma is an ionized gas consisting of a neutrally charged cloud of positively and negatively charged particles, which can be contained (kept in circular motion) by a strong magnetic field. Such strong fields are found in stars and can also be created in a laboratory. There is evidence that an Earthly example of plasma can be seen in ball lightning, which is rather like an aftershock of very intense lightning.

Plasmas exhibit unique properties when compared to solids, liquids, and gases. Because of the high temperatures typically required to produce them, they may be difficult to contain and control. The importance of understanding and working with plasmas is increasing, however, since plasma technology is used to make semiconductors, and may play a part in future advanced rocket flight.

For most conditions, the ordinary three states of matter – solid, liquid, and gas – are all that need to be considered.

Technology Box 4-2 ■ Glass: Why is it called a supercooled liquid?

We think of glass as being a modern building material, yet the art of glass making is ancient. The chemistry of glass is still being studied today, yet some of its properties remain a mystery.

Simple glass can be made by just melting pure sand. Small amounts of other ingredients are often added to change its color, refractive index, thermal coefficient of thermal expansion, or strength. Pure sand is composed of small, hard, transparent white silica crystals. It doesn't resemble glass and you can't see through a pile of it. Glass, on the other hand, is transparent. It feels and acts like a rigid crystalline solid. There are other rigid crystalline substances – like diamonds and other gemstones – that are hard and transparent, but they would make very small expensive windows!

Scientists are still debating the internal structure of glass. Today, it is generally referred to as being a supercooled liquid. Although the sand is melted allowing its smallest components to move around during the glassmaking process, when it cooled, these components fail to form a crystalline solid. The small components simply freeze in the disorganized pattern that is more typical for liquids. Therefore, while glass internally resembles a cold liquid, outwardly it manifests the physical characteristics of a typical crystalline solid.

Check Your Understanding

1. Explain the difference between chemical and physical properties of matter.

2. Describe each of the three states of matter.

3. Refer to Figure 4-2, the phase diagram for water, and explain what is occurring at each of the phase transition points.

4. How do plasmas differ from ordinary states of matter?

5. Distinguish among compounds, elements, and mixtures.

Atomic Theory

Objectives

Upon completion of this section, you will be able to:

■ Describe the structure and parts of atoms.

■ Discuss the importance of the Bohr model.

■ Identify and describe the characteristics of elements found on the periodic table.

Developing a Model

Did you ever shake a present trying to figure out what it was? If you have, then you were trying to develop a mental picture of something that you couldn't see. **Models** are useful for understanding phenomena that we observe only indirectly. Through the combination of experimentation, imagination, and mathematics, scientists have attempted to develop models representing the behavior of matter.

These models provide a view of reality, but are imperfect. They are like maps of a city, but are not the city itself. Like a good map, a good model allows us to understand the reality that we are dealing with, and the quality of a model is a measure of how closely it represents the real situation. Working with models can be frustrating. However, an even bigger problem can arise if we start believing that the models are exactly true.

Democritus

The idea of the atom is quite old. The Greek philosopher Democritus, born about 460 BC, proposed a theory of matter that is surprisingly close to our modern concepts. He reasoned that if you repeatedly cut a piece of some substance in half, eventually you would have a piece so small that you could not cut it in half again. He called this last smallest piece *atomos* – meaning "not cuttable." From this reasoning we inherited the word "atom."

None of the writings of Democritus have survived. We know of his ideas only because they were quoted by others – especially Aristotle a century later. Aristotle, who summarized Greek scientific thought in a great encyclopedia of learning, dismissed the atomic idea as worthless. During the centuries that Aristotle's ideas remained dominant in western civilization, the concept of atoms was largely ignored.

Modern Atomic Theory

By the late 1700s, 33 elements were known and various attempts had been made to organize them based on their chemical properties. But it wasn't until the early 1800s that an English chemist, John Dalton, again proposed atoms as the fundamental unit of all elements. He arrived at this on the basis of his discovery of the **law of multiple proportions**. He observed, for example, that when carbon and oxygen are reacted, they can unite in either a 1:1 ratio forming carbon monoxide, CO, or in a 1:2 ratio forming carbon dioxide, CO_2. He wondered why one of these ratios used exactly twice as much oxygen as the other.

You can follow Dalton's logic by thinking in terms of adding sugar to tea. Suppose you were able to measure the amount of sugar people use in their tea, and found it always to be in multiples of 10 grams per cup. You might conclude that sugar cubes weighing 10 grams each were being used and that different people were using different numbers of sugar cubes. On the other hand, if you found the weights were not simple multiples – you might conclude it was being spooned from a bowl. Dalton realized that the combining of carbon and oxygen is like the use of sugar cubes in that elements combine in simple multiple proportions. In other words, elements come in small pieces. This reminded him of the concept of atoms proposed by Democritus centuries before.

Today we define an atom as the smallest part of an element that retains all the properties of that element, meaning an element can be subdivided only down to atoms. It does not mean, however, that an atom can't be broken into its smaller parts. A little more than a century after Dalton announced the atomic theory, physicists learned that atoms could be split.

Dalton's suggestion that all elements are composed of atoms ignited a search to test the theory. Although atoms were not immediately identified, the search led to many other interesting discoveries; some based on findings that predated Dalton's work. For example, in 1666, Newton discovered that when a narrow beam of sunlight passed through a prism, it spreads out into a rainbow of colors ranging from red at one end to violet at the other. And in 1854, Robert Bunsen and Gustav Kirchoff found that when the light from an energized element is passed through a prism, a series of brightly colored lines resulted. The instrument they developed for studying these lines is called a **spectroscope** (see Figure 4-4).

After carefully examining and recording the positions of the bright-line spectra of the known elements, scientists found that each was uniquely different. Since this discovery, the **bright-line spectra** and the corresponding **absorption spectra** have been used to identify both known and unknown elements. See Technology Box 4-3: Elemental Fingerprints, for an example.

A number of attempts were made to develop a system for organizing the elements. In the early 1800s, German scientists noticed that when they organized elements by their chemical properties, some of the elements appeared to fall into groups

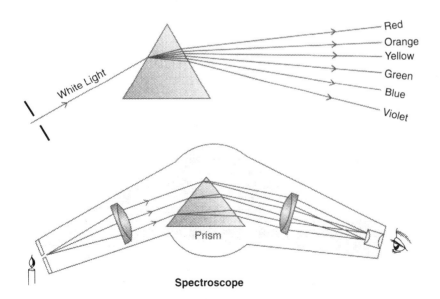

Figure 4–4: Light passing through a prism and a spectroscope. The shorter wavelengths of light (toward the violet end of the spectrum) interact more with the medium they pass through and therefore are bent more than the long wavelengths.

Physical Science: What the Technology Professional Needs to Know

of three, or **triads**, such as fluorine, bromine, and iodine. However, this plan did not work for all elements. In 1847, a system was developed based on the mass of each element. This worked fairly well, but didn't account for their chemical properties.

Technology Box 4–3 ■ Elemental Fingerprints: How do we know the composition of distant stars?

The combined use of a telescope and spectroscope during a solar eclipse in 1868, resulted in a bright-line spectrum. Since no element on Earth was known to have the same spectra, the assumption that it was present only in the sun was made, and it was given the name helium, from the Greek word *helios*, meaning sun. It was not until 1895, that the terrestrial (Earthly) helium was found as a by-product of a nuclear reaction.

As telescopes and spectroscopes become more sophisticated, we are able to learn increasingly more about distant stars. Though the light coming from them appears white (or off-white) to us, when it is passed through a spectroscope, the specific elements that are present can be identified by the unique lines in their spectra.

Row	Group						
	I	**II**	**III**	**IV**	**V**	**VI**	**VII**
1	H (1)						
2	Li (7)	Be (9.4)	B (11)	C (12)	N (14)	O (16)	F (19)
3	Na (23)	Mg (24)	Al (27.3)	Si (28)	P (31)	S (32)	Cl (35.5)
4	K (39)	Ca (40)	?[1] (44)	Ti (48)	V (51)	Cr (52)	Mn (55)
5	Cu (63)	Zn (65)	?[2] (68)	?[3] (72)	As (75)	Se (78)	Br (80)
6	Rb (85)	Sr (87)	Yt (88)	Zr (90)	Nb (94)	Mo (96)	? (100)
7	Ag (108)	Cd (112)	In (115)	Sn (118)	Sb (122)	Te (125)	I (127)
8	Cs (133)	Ba (137)	Di (138?)	Ce (140?)	?	?	?
9	?	?	?	?	?	?	?
10	?	?	Er (178)	La (180?)	Ta (182)	W (184)	?
11	Au (199)	Hg (200)	Tl (204)	Pb (207)	Bi (208)		

The number in parentheses is the atomic weight assigned by Mendeleev.
The elements predicted by Mendeleev:
[1] named Eka-Boron by Mendeleev, named Scandium when discovered;
[2] named Eka-Aluminum by Mendeleev, named Gallium when discovered;
[3] named Eka-Silicon by Mendeleev, named Germanium when discovered.

Table 4–2: Mendeleev's 1871 periodic table of the elements.

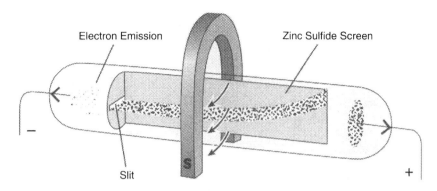

Figure 4–5: Effect of a magnetic field on a cathode ray.

In 1869, the Russian scientist Dmitri I. Mendeleev proposed an organization of the 63 known elements, which not only accounted for their chemical properties, but accurately predicted new elements and their chemical properties. Mendeleev's original plan is shown in Table 4-2 and is known as the **periodic table of the elements.** Over the years, the table has followed the basic logic of the original. However, in the late 1890s, another column had to be added to the right-hand side to accommodate the discovery of the inert gases. We will have more to say about the organization of the periodic table later in this Chapter.

A few years later, in 1879, Sir William Crookes was carrying out an experiment in which he put a high electrical charge on two electrodes that had been sealed into the opposite ends of a glass tube. He found that, as the tube was evacuated (air pumped out), an eerie bluish light appeared to be originating from the **cathode** or negative electrode. Fascinated by these **cathode rays**, he found that a magnet placed near the tube would deflect the rays into a curved path. (See Figure 4-5.) This implied that the rays contained charged particles – because a charged particle moving across the magnetic field lines is deflected into just such a curved path.

Later investigations with cathode rays led J. J. Thomson, in 1897, to conclude that cathode rays are beams of particles from within the atom. He gave these particles the name **electrons** and, based on their behavior, assigned them a negative electric charge. But, since atoms are electrically neutral, Thomson hypothesized that an atom must consist of a sphere of positive charge that balanced the negative charge of the electrons, with the electrons embedded in this positive sphere like plums in a pudding. (See Figure 4-6.)

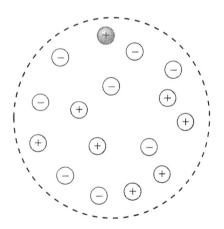

Figure 4–6: The plum pudding model of the atom.

The plum pudding model failed to explain a series of experiments later conducted by Ernest Rutherford in 1911. In his experiments, as shown in Figure 4-7, a very thin sheet of gold foil was bombarded with **alpha particles** coming from a radioactive source. Alpha particles were known to be much more massive than electrons, and to bear a positive electrical charge. Rutherford's intention was to use the alphas to further probe the structure of atoms – rather like a high-tech game of "Battleship."

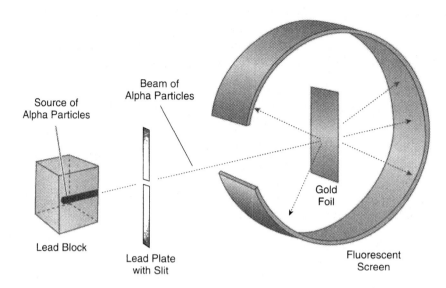

Figure 4–7: Gold foil bombarded by a stream of alpha particles.

Rutherford and his assistant, Ernest Marsden, found that most of the positively charged alpha particles went straight through the foil with only minor deflections, and that some penetrated the foil but emerged with sharply bent paths. They were totally amazed to find some of the alpha particles bouncing back toward the source. After months of experimentation and analysis, Rutherford and Marsden concluded that each atom must have a very small, dense, positively-charged **nucleus** that repelled the positively-charged alpha particles. This would explain the sharp deflection of the occasional alpha particle that neared the more massive gold nucleus. Although they are oppositely charged, the lighter electrons do not change the path of the heavier alpha particles.

Therefore, Rutherford abandoned the plum pudding model in favor of a new one. In his new model of the atom, there was a large open region occupied by orbiting negatively-charged electrons and a much smaller region, the nucleus, filled with positively-charged **protons**. Since atoms were known to be electrically neutral, the charges on the electrons and the protons had to balance. But the deflections of the alpha particles showed that most of the mass was in the protons – that is, in the nucleus of the atom.

Analysis of his experimental results allowed Rutherford to estimate the diameter of atoms to be about 1×10^{-8} cm, compared to a diameter of 1×10^{-13} cm for the nucleus. This means that if the nucleus were vastly enlarged to measure two inches in diameter, the outer electrons would be orbiting more than three miles away.

In 1913, a young Oxford scientist, Henry Moseley, conducted a series of experiments in which he observed the X ray spectra of the elements. X rays, as we will see later, are a very energetic, penetrating "color" of light. He found that the frequencies of these X ray emissions varied in a systematic way from element to element. He was able to assign an **atomic number**, A, to each of the elements, such that, for each element, the square root of the frequency of the most prominent X ray line was directly proportional to A. This proved to be an important contribution because, in addition to a name, each element could now be identified by its atomic number, which corresponded to the number of protons in its nucleus.

Secondly, it revealed that the elements had actually been placed on the periodic table in order of increasing atomic numbers, which justified the placement of several elements. Finally, since the atomic number equals the number of protons in the nucleus, it was suddenly very easy to see where elements were missing.

The mass of both the electron and proton were calculated from the earlier experiments of J. J. Thomson and the later, 1913, oil-drop experiments of R. A. Millikan. In Millikan's experiment, droplets of oil were allowed to pick up very small amounts of electric charge and were then allowed to fall between the two electrically charged horizontal plates. The charges on the plates slowed the fall of the drops, depending on how much charge was on them. As with Dalton's results on simple multiple proportions, it was found that the oil drops always carried a charge that was a simple multiple of a basic charge. Millikan concluded that this basic charge was the charge of a single electron.

From Thomson's work on the path of cathode rays as they crossed an electrical field, the electron's charge to mass ratio, e/m, had been estimated. Combining this value with the results of the oil-drop experiment made it possible to estimate the electron's mass as 9.109×10^{-28} grams. Parallel studies showed the mass of the proton to be significantly heavier at 1.673×10^{-24} grams. This means that it takes 1,837 electrons to equal the mass of a single proton. This finding further substantiated Rutherford's conclusion that most of the mass of an atom is concentrated in the very small nucleus.

Rutherford's experiments resulted in a working model in which atoms are similar in appearance to the sun and the planets in our solar system, with the nucleus in the center and the electrons orbiting around it. To this day, Rutherford's **planetary model** of the atom remains useful, explaining and clarifying many phenomena in chemistry and physics even though we now know that this is not what the atom looks like.

Since the mass of atoms and other subatomic particles is very small, the **atomic mass unit**, u, was developed as a matter of convenience. On this scale, 1.66×10^{-24} grams, was selected and made equal to 1.00 u. Due to the great difference in mass between a proton and electron, the contribution of the electron to the total mass of an atom is typically considered to be negligible.

There was a serious flaw in the planetary model of the atom. It had been clearly demonstrated in the mid 1800s, by the electromagnetic experiments of Michael Faraday and the electromagnetic theory of James Clerk Maxwell, that a charged particle circling in an electric field will lose energy. According to that theory, electrons should lose energy and spiral into the nucleus, giving off a continuous spectrum of radiation like a rainbow. In other words, matter should be unstable and should have all collapsed long ago. In fact, being that unstable, atoms should never have been able to exist in the first place.

In 1913, Niels Bohr, a Danish physicist, offered a unique solution to this problem. He hypothesized that only certain "orbits" are allowed for electrons, and that electrons, rather than following the classical behavior predicted by Maxwell's theory, could jump between these allowed orbits. According to this model, the various orbits around the atomic nucleus correspond to a series of concentric **energy levels** or **shells**. Each of these energy levels was given a letter, starting with K.

Bohr postulated that as long as an electron stays within an energy level, it neither gains nor loses energy. However, if the atom becomes energized through heating, absorbing light, or electrical stimulation, then the electrons can jump to one of the higher energy levels farther from the nucleus. Later, the electron will drop back to its lowest level, or **ground state**, due to the constant pull of the nucleus.

When electrons make transitions between lower and higher energy levels, they must either absorb or give back specific amounts of energy. The energy given up by the electron returning to a lower energy level, as shown in Figure 4-8, may be emitted as energy in the form of visible light. The longer the electron jump, the higher the frequency of light emitted. Hence, the allowable energy level jumps are what determine the unique spectra of that element. In this way, Bohr's model explained the source of Bunsen's and Kirchhoff's bright-line spectra of the elements.

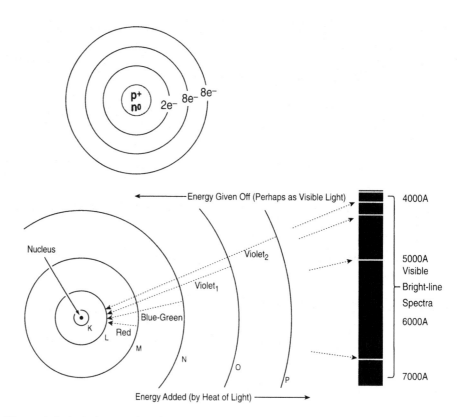

Figure 4–8: In Bohr's model of the atom, only certain electron orbits are allowed (top). Electrons emit or absorb energy when they jump between orbits (bottom).

Still more energetic jumps result in emission or absorption of ultraviolet radiation and X rays. These highly energetic X ray emissions are those that Moseley used in his work on atomic numbers.

It wasn't long before scientists developed other models to improve on the Bohr model. In one of these models, electrons can be either particles orbiting the nucleus or they can be wave patterns rippling around the nucleus. Quantum theory allows both of these interpretations to be true, and it provides a broad framework within which Maxwell's electromagnetic theory and Bohr's energy jumps – which seem totally incompatible – both hold true. However, the Bohr model does an adequate job of explaining much of chemistry and physics. More importantly, it is much easier to understand and use than quantum theory and other complex models. For an explanation of the quantum mechanical model, see Appendix A.

Before we leave the development of atomic models, there is still one major piece of the atom missing – the **neutron**. In 1919, Francis William Aston developed the first mass spectrograph. It was designed to separate elements by their masses, on an atom by atom basis. But when Aston tested the device on a pure sample of neon, it kept separating the neon atoms into two different groups. Finally, he was convinced there must be atoms of the same element that differed in their masses. For this to be true, however, there would have to be a third subatomic particle, having no charge but with the same mass as the proton. It was this hypothetical particle that was given the name "neutron," a name reflecting that it was electrically neutral.

The existence of the neutron was confirmed in 1932 by James Chadwick. It is accepted that the neutron contributes to the total mass of the atom; however, what other role it plays in nuclear stability is still being explored (see Chapter 11).

Isotopes

Although the number of protons in the nucleus determines the identity of the atom, the number of neutrons is quite variable. Near the top of the periodic table, among the lighter elements, the number of neutrons is roughly equal to the number of protons, but atoms of the same element exist that have different numbers of neutrons. Having a different number of neutrons does not affect the chemical properties or electrical balance of an atom. It affects only the mass of the atom.

For example, chlorine has 17 protons, and most chlorine atoms have a mass of 35 u. That means that these chlorine atoms must have 18 neutrons (35 − 17 = 18). However, careful measurements indicate that one quarter of all chlorine atoms have a mass of 37 u. These chlorine atoms must have 20 neutrons (37 − 17 = 20). Atoms of the same element that have different masses are called **isotopes**. The element chlorine has two naturally occurring isotopes.

The fact that most elements have two or more isotopes immediately creates another problem. If atoms of the same element can have different masses, then which one should be put on the periodic table? The decision was made to use the **weighted average** of an element's naturally occurring isotopes. What this means can be demonstrated using the information given above for the isotopes of chlorine. Measurement indicates that 75 percent of all naturally occurring chlorine atoms are of the chlorine-35 isotope and the remaining 25 percent are the chlorine-37 isotope. The general formula for calculating the weighted average is:

(Atomic Mass × %) + (Atomic Mass × %) + (Atomic Mass × %) + ... = Average

For the two chlorine isotopes, we would have:

(35 u × 75%) + (37 u × 25%) = 26.25 u + 9.25 u = 35.5 u

If you examine the periodic table, you will find the slightly more accurate 35.453 u. It is important to note that there cannot be such a thing as a single atom of chlorine with a fractional mass. Nearly all of the atomic masses listed on a periodic table are weighted averages. (Exceptions to this convention have the mass of the most stable isotope enclosed in parentheses.)

Artificial Isotopes and Elements

Since the beginning of modern atomic physics, scientists have experimented with making new isotopes for many elements. They have found that for each element there appears to be a limit to the number of neutrons that can be added to the nucleus before it becomes unstable and breaks apart. The process by which atoms spontaneously decompose or disintegrate is called **radioactive decay**. Those elements that decompose in this way are called either **nuclides** or **radioisotopes**.

The tendency toward radioactive decay seems to be related to both the size of the nucleus and its ratio of neutrons to protons. All elements with atomic numbers above $A = 83$ (Bi) have unstable nuclei and are therefore radioactive. In addition, various isotopes of many smaller elements and all artificial elements are radioactive.

Figure 4-9 shows the ratio of neutrons to protons for both stable and radioactive isotopes. A line in the figure marks the 1:1 proton and neutron ratio. It is apparent that as the nucleus of the atom gets larger, it can deviate farther from the 1:1 ratio and remain stable. As an example, in the inset for carbon ($A = 6$), we find that carbon has three naturally occurring isotopes. Of these three isotopes, two are stable (carbon-12 [1:1 ratio] and carbon-13 [1:1.17]), while the third, carbon-14 (1:1.33), is unstable.

In 1933, Irene Curie and her husband Frederic Joliot made the important discovery that radioactive elements can be artificially prepared. They bombarded boron

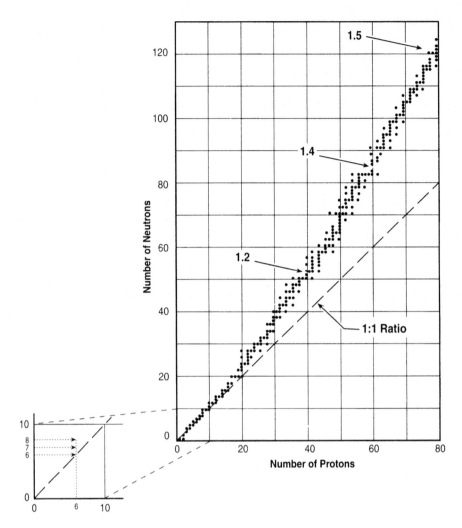

Figure 4–9: Ratio of neutrons to protons for the elements.

atoms with alpha particles. These were the same positively charged particles used by Rutherford and Marsden in their experiments that established the nuclear atom. By the time Curie and Joliot carried out their experimental work, an alpha particle was known to consist of two protons and two neutrons. In other words, it was identical to a helium-4 nucleus.

The collisions between the alpha particles and the boron atoms had changed some of the boron atoms into nitrogen atoms, creating a radioactive form of nitrogen. For this contribution to nuclear research, the team of Joliot and Curie was awarded the 1935 Nobel Prize in chemistry. This was the first radioactive isotope to be artificially created.

Some radioisotopes have proven their value through their use as important diagnostic tools in the field of medicine. For example, when iodine-127 is bombarded with neutrons, a small number of iodine-131 atoms result. Iodine-131 is a radioisotope whose chemical properties remain the same as stable iodine-127. When we consume foods or water containing traces of iodine, the thyroid gland systematically converts it into a needed substance called thyroxin. Therefore, if iodine-131 is introduced diagnostically into a patient's body, a radioactive form of thyroxin results. Because of its detectable radioactivity, the amount and location of the thyroxin within the body can be traced. This has given medical researchers a powerful tool for studying a patient's thyroid metabolism.

Technology Box 4–4 ■ Radioisotope Dating: How do we date artifacts?

When archeologists find remnants from an ancient civilization, one of the first questions asked is "How old are they?" In the 1940s, Dr. Willard F. Libby developed a method known as radiocarbon dating, for which he received the 1960 Nobel Prize in chemistry.

The reasoning behind radiocarbon dating is that all living organisms continue to assimilate carbon into their structures until they die. Of the carbon incorporated, a very small amount (about one out of every one million carbon atoms) is radioactive carbon-14. Starting at the death of the organism, the carbon-14 present starts its first half-life of radioactive decay. If, for example, there were 6.02×10^{23} of carbon-14 atoms present, during the first half-life, one-half of those atoms, or 3.01×10^{23}, will disintegrate. It would require about 80 half-lives for all of them to disintegrate. Since the half-life for carbon-14 is 5,700 years, some of it will remain for up to 456,000 years. By measuring the amount of carbon-14 remaining in the artifact, scientists can therefore determine its approximate age.

Half-life

What exactly do we mean by nuclear stability? We know that some isotopes are stable, while others disintegrate. A small amount of any element contains vast numbers of atoms. If it is a radioactive element, some of its atoms are disintegrating at any given moment. We don't know exactly which atom will disintegrate next, but we can determine how many atoms will disintegrate in a given length of time.

The term **half-life** was defined by Ernest Rutherford to help describe the instability of a group of atoms. He said that the half-life of an element is the time it takes for one-half of its atoms to disintegrate.

At the end of the first half-life, you will have only one-half of the original atoms remaining. During the second half-life, one-half of the remaining atoms will disintegrate and so on. (See Figure 4-10.) Theoretically, this would mean that some of the original atoms in a given sample will always be present, because only half of the remaining atoms disintegrate during a half-life. However, this reasoning breaks down when you get down to a very small numbers of atoms remaining.

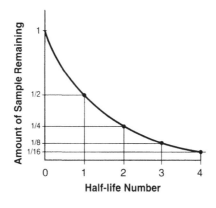

Figure 4–10: Radioactive decay and half-life.

The half-lives of the various isotopes vary dramatically. For example, the half-life of potassium-40 is 1.25 billion years, carbon-14 about 5,700 years, and phosphorous-30 about 2.5 minutes.

Check Your Understanding

1. What is the difference between a pure substance and a mixture? Matter is composed of elements and elements are composed of atoms. Define the terms matter, element, and atom.

2. Describe the Bohr model of an atom. Identify each of its components and their locations.

3. Why has the Bohr model of an atom remained important to scientists?

4. What is the origin of a bright-line spectrum?

5. What is the structural difference between two isotopes of the same element?

6. What are artificial isotopes and elements and how are they produced?

4–3   Periodic Chemical Relationships

Objectives

Upon completion of this section, you will be able to:

■ Describe the major subdivisions of elements on the periodic table.

■ Use the periodic table to predict the chemical behavior of elements.

■ Explain the importance of the octet rule.

■ Write formulas for common ionic compounds.

■ Describe covalent bonds and write formulas for covalent compounds.

Periodic Table Subdivisions

Chemists have found many elements unknown to the ancients by chemically decomposing matter into its simpler parts. As the list grew, scientists wrestled with determining what relationships existed between them. By 1869, the list of elements had reached 63 and Mendeleev proceeded to arrange them into his table (see Table 4-2) based primarily on increasing atomic mass. In doing so, however, he took into consideration the physical and chemical properties of the surrounding elements. If an element did not appear to fit based solely on its atomic mass, he simply moved it and suggested that perhaps its atomic mass was in error. This stirred up some criticism, but it was soon determined that, indeed, the atomic masses of seventeen of the elements were wrong. Slowly the criticism faded as his table continued to correctly predict the properties for each new element added.

From gaps in the table, Mendeleev boldly predicted the existence of the elements scandium, gallium, germanium, technetium, rhenium and protactinium, including their atomic weights, melting points, reactions with acids and bases, as well as several other properties based on their position in the table. The critics were finally silenced when the first of these predicted elements was discovered. The periodic table introduced by Mendeleev has proven to be one of the most important tools used by chemists. On the inside of the front cover of this text is a copy of this valuable tool. Tables 4-3 and 4-4 also show this periodic arrangement. The only difference between these tables is that they highlight different relationships between the elements.

Each element in a modern periodic table is identified by an internationally agreed upon symbol. Most often the first letter comes from its English, Latin, or agreed upon name. In cases where that letter has already been used, a second letter in the name is added. The first letter of the symbol is always capitalized. If there is a second letter, it is always lower case. So we have: H, hydrogen; He, helium; Hf, hafnium; etc. Above the symbol is its atomic number, indicating the number of protons in its nucleus. Below the symbol is the average atomic mass of the element.

The arrangement of elements in the periodic table by increasing atomic number also tends to put them in order of increasing atomic mass. There are three exceptions: argon and potassium (Ar and K), cobalt and nickel (Co and Ni), and tellurium and iodine (Te and I). Although these exceptions were originally puzzling, the explanation is now known to be based on the ratio of their naturally occurring isotopes.

As shown in Table 4-3, there are 92 elements that are naturally occurring; that is, elements found in the matter of the universe. At room temperature and pressure, only 2 of the elements in the periodic table are liquids, 11 are gases, and the remainder solids.

Most artificial elements have been made only in small quantities and have very short half-lives. So far, over 20 artificial elements have been made and room has been left for more to be added.

Physical Science: What the Technology Professional Needs to Know

Periodic Table of the Elements

IA (1)	IIA (2)	IIIB (3)	IVB (4)	VB (5)	VIB (6)	VIIB (7)	(8)	VIIIB (9)	(10)	IB (11)	IIB (12)	IIIA (13)	IVA (14)	VA (15)	VIA (16)	VIIA (17)	VIIIA (18)
1 H 1.008																	2 He 4.003
3 Li 6.941	4 Be 9.012											5 B 10.811	6 C 12.011	7 N 14.007	8 O 15.999	9 F 18.998	10 Ne 20.180
11 Na 22.990	12 Mg 24.305											13 Al 26.982	14 Si 28.086	15 P 30.974	16 S 32.066	17 Cl 35.453	18 Ar 39.948
19 K 39.098	20 Ca 40.08	21 Sc 44.956	22 Ti 47.88	23 V 50.942	24 Cr 51.996	25 Mn 54.9380	26 Fe 55.847	27 Co 58.933	28 Ni 58.69	29 Cu 63.546	30 Zn 65.39	31 Ga 69.723	32 Ge 72.61	33 As 74.922	34 Se 78.96	35 Br 79.904	36 Kr 83.80
37 Rb 85.468	38 Sr 87.62	39 Y 88.906	40 Zr 91.224	41 Nb 92.906	42 Mo 95.94	43 Tc 98.9062	44 Ru 101.07	45 Rh 102.906	46 Pd 106.42	47 Ag 107.868	48 Cd 112.41	49 In 114.82	50 Sn 118.71	51 Sb 121.75	52 Te 127.60	53 I 126.904	54 Xe 131.3
55 Cs 132.905	56 Ba 137.33	57 La 138.906	72 Hf 178.49	73 Ta 180.948	74 W 183.85	75 Re 186.21	76 Os 190.23	77 Ir 192.22	78 Pt 195.08	79 Au 196.966	80 Hg 200.59	81 Tl 204.38	82 Pb 207.2	83 Bi 208.98	84 Po (209)	85 At (210)	86 Rn (222)
87 Fr (223)	88 Ra 226.025	89 Ac 227.03	104 Rf (261)	105 Db (262)	106 Sg (263)	107 Bh (262)	108 Hs (265)	109 Mt (266)	110 * (269)	111 * (272)	112 * (277)		114 * (285)		116 * (289)		118 * (293)

58 Ce 140.11	59 Pr 140.91	60 Nd 144.24	61 Pm (145)	62 Sm 150.36	63 Eu 151.96	64 Gd 157.25	65 Tb 158.92	66 Dy 162.50	67 Ho 164.93	68 Er 167.26	69 Tm 168.93	70 Yb 173.04	71 Lu 174.97
90 Th 232.04	91 Pa 231.04	92 U 238.03	93 Np 237.05	94 Pu (244)	95 Am (243)	96 Cm (247)	97 Bk (247)	98 Cf (251)	99 Es (252)	100 Fm (257)	101 Md (258)	102 No (259)	103 Lr (260)

*Names not officially assigned.　　□ Naturally occurring elements　　□ Artificial elements

Table 4–3: Periodic table showing naturally occurring and artificial elements.

Periodic Table of the Elements

IA (1) Alkali Metal	IIA (2) Alkaline Earth Metals	IIIB (3)	IVB (4)	VB (5)	VIB (6)	VIIB (7)	VIIIB (8)	IXB (9)	XB (10)	XIB (11)	XIIB (12)	IIIA (13) Boron Family	IVA (14) Carbon Family	VA (15) Nitrogen Family	VIA (16) Oxygen Family	VIIA (17) Halogens	VIIIA (18) Noble Gases
1 H 1.008																	2 He 4.003
3 Li 6.941	4 Be 9.012					Transitional Metals						5 B 10.811	6 C 12.011	7 N 14.007	8 O 15.999	9 F 18.998	10 Ne 20.180
11 Na 22.990	12 Mg 24.305											13 Al 26.982	14 Si 28.086	15 P 30.974	16 S 32.066	17 Cl 35.453	18 Ar 39.948
19 K 39.098	20 Ca 40.08	21 Sc 44.956	22 Ti 47.88	23 V 50.942	24 Cr 51.996	25 Mn 54.9380	26 Fe 55.847	27 Co 58.933	28 Ni 58.69	29 Cu 63.546	30 Zn 65.39	31 Ga 69.723	32 Ge 72.61	33 As 74.922	34 Se 78.96	35 Br 79.904	36 Kr 83.80
37 Rb 85.468	38 Sr 87.62	39 Y 88.906	40 Zr 91.224	41 Nb 92.906	42 Mo 95.94	43 Tc 98.9062	44 Ru 101.07	45 Rh 102.906	46 Pd 106.42	47 Ag 107.868	48 Cd 112.41	49 In 114.82	50 Sn 118.71	51 Sb 121.75	52 Te 127.60	53 I 126.904	54 Xe 131.3
55 Cs 132.905	56 Ba 137.33	57 La 138.906	72 Hf 178.49	73 Ta 180.948	74 W 183.85	75 Re 186.21	76 Os 190.23	77 Ir 192.22	78 Pt 195.08	79 Au 196.966	80 Hg 200.59	81 Tl 204.38	82 Pb 207.2	83 Bi 208.98	84 Po (209)	85 At (210)	86 Rn (222)
87 Fr (223)	88 Ra 226.025	89 Ac 227.03	104 Rf (261)	105 Db (262)	106 Sg (263)	107 Bh (262)	108 Hs (265)	109 Mt (266)	110 2 (269)	111 2 (272)	112 2 (277)		114 2 (285)		116 2 (289)		118 2 (293)

Lanthanide Series	58 Ce 140.11	59 Pr 140.91	60 Nd 144.24	61 Pm (145)	62 Sm 150.36	63 Eu 151.96	64 Gd 157.25	65 Tb 158.92	66 Dy 162.50	67 Ho 164.93	68 Er 167.26	69 Tm 168.93	70 Yb 173.04	71 Lu 174.97
Actinide Series	90 Th 232.04	91 Pa 231.04	92 U 238.03	93 Np 237.05	94 Pu (244)	95 Am (243)	96 Cm (247)	97 Bk (247)	98 Cf (251)	99 Es (252)	100 Fm (257)	101 Md (258)	102 No (259)	103 Lr (260)

[2]Names not officially assigned.　　□ Metals　　▨ Metalloids　　□ Nonmetals

Table 4–4: Periodic table showing metals, nonmetals, and metalloids.

Periodic Properties

Before discussing the detailed relationships that can be determined about an element from its position on the periodic table, some vocabulary and broad generalizations about the periodic table are needed. First, the elements in vertical columns, called **groups** or **families**, have similar chemical properties. As shown in Table 4-4, some of these groups have also been given names.

The elements in each horizontal row, also known as a **period** or **series**, have sequential atomic numbers and, generally, increasing atomic masses. By moving from left to right and down the chart, this arrangement makes the properties of the elements periodically repeat. This simple relationship is known as the **periodic law** and is credited to the work of Mendeleev. The power of the periodic table is that not only can it be used to determine similarities among elements, but it can even be used to predict the properties of as yet undiscovered elements.

Group Properties

Although the periodic table was developed more than 40 years before Bohr proposed his model, there is a strong correlation between the two. The reason is that Mendeleev used the properties of an element to help position it on the periodic table. When the Bohr model was later applied to the periodic table, it became apparent that each element in a group had the same number of electrons in its outermost energy level.

Today, we recognize that elements exhibiting similar chemical properties are **isoelectronic**. That is, they have the same number of electrons in their outermost energy level, although the distance to the outer level is different for each period. For example, each element in the left-most column of the periodic table, from H through Fr, has only one electron in its outer shell. Each element in the right-most column, with the exception of He, has eight electrons in its outer shell.

If you closely examine the periodic table shown inside the front cover, you will find a dark zigzag line that moves down and across the right side of the periodic table. This line separates the elements into two large groups: metals and nonmetals. The metals are found to the left of the zigzag line and along the bottom of the table. Elements to the right of the zigzag line are nonmetals.

Metals

The properties of metals are familiar to all of us. When they are freshly cut or polished they have a shiny appearance called a **metallic luster**. Only a few metals, like copper and gold, have colors other than silvery-white. Metals also tend to be malleable and ductile, making them easy to shape or draw into wires.

The use of metals has been extended by mixing metals to produce **alloys**. Often these homogeneous mixtures are less expensive than pure metals and exhibit unique properties. Some familiar alloys are solder (tin and lead), brass (copper and zinc), bronze (copper and tin), pewter (tin, copper, and lead), and stainless steel (iron, chromium, and carbon). In the case of gold, the price is based on its purity. Twenty-four karat, 24 k, gold is pure and expensive, but it is very soft and doesn't last long under constant wear. When gold is mixed with less expensive nickel or more expensive palladium (41% Au and 59% Ni or 90% Au and 10% Pd) the alloys are silver in color, more durable, and sold as "white gold."

Technology Box 4–5 ■ Transmutation of the Elements

One of the dreams of alchemy (ca. 400–1500 AD) was to find a process that would change a base metal, like lead, into gold. The process is called **transmutation**. Today we know the process for making new elements is straightforward. You simply need to find a way to change the number of protons in the nucleus of an atom and presto you have a different element. One way to accomplish this is to bombard a sample of the element with enough neutrons to cause it to undergo radioactive breakdown. As the nucleus splits into smaller pieces, you hope that one of the fragments has the number of protons desired. For example, to convert a lead atom ($A = 82$) into an atom of gold ($A = 79$), three protons must be removed. (In reality, it would cost far more to convert lead into gold than gold is worth.)

Another way to create new elements involves accelerating a small atom or fragment of matter to a very fast speed and then slamming it into another atom. If you are lucky, the two nuclei fuse into an atom with the desired number of protons. The details are not so simple, but with modern equipment, it is not difficult to create a series of atoms or fragments that are moving at great speeds. The challenge remains in getting them to stick together to form a new element. Even when formed, these new elements tend to quickly disintegrate, often in fractions of a second.

The first artificial or man-made element to be produced was technetium, Tc. It was made by Carlo Perrier and Emilio Segre in 1937, by bombarding the element molybdenum with a beam of subatomic particles. Mendeleev had predicted the existence of the element and a space was available in the periodic table; however, no natural technetium could be found. In subsequent years, elements with atomic numbers from 93 to 112 have been made. Research at the time of this writing indicates the existence of a few atoms of the elements 114, 116, and 118. Atomic theory suggests that starting with element 114 there should be an "island of greater stability." Preliminary experimental results indicate that it is true. The majority of synthetic elements to this point, however, disintegrate almost as soon as they are made.

Nonmetals

About the only thing nonmetals have in common is that they are not metal. Nonmetals can be in the solid, liquid, or gaseous state under normal room conditions. When they are solids they do not have a metallic luster, are neither ductile nor malleable, and they are not good conductors of heat or electricity.

As can be seen by examining the Table 4-4, four of the six columns that start with nonmetals cross the zigzag line making a transition to metals as you move toward the bottom of the column. This implies that the physical and chemical properties of the group gradually change as atomic number increases and the outer electrons move farther from the nucleus.

The nonmetals in group 7 are called the **halogens,** meaning "salt producers." These elements – fluorine, chlorine, bromine, iodine, and astatine – include the most reactive nonmetal elements in the table. Fluorine and chlorine are yellowish-green gases under normal conditions. Bromine is an orange liquid and iodine is a purplish-gray solid. Astatine is a grayish-black radioactive solid that has been used in medical research on the thyroid gland. The first four members of this family form compounds with nearly every other metal and nonmetal on the periodic table. The most common of these is common table salt, sodium chloride.

The halogens have also found wide use in the manufacture of synthetic organic chemicals known as **halogenated hydrocarbons** in the twentieth century. The success of innovations such as refrigeration and air conditioning is largely due to the introduction of a group of chemicals called **chlorofluorocarbons**. Many of these chemicals are now the focus of public concern due to their ozone-depleting capabilities. In addition, some of the halogenated hydrocarbons have also been shown to cause cancer, birth defects, and other negative health effects. Because of these negative impacts, several (CFCs, DDT, and PCBs) have been banned from use in North America and many other parts of the world.

The elements in the far right column play a very significant role in understanding the chemistry of all other elements. This family was not included in the original periodic table, because the elements had not yet been discovered. The reason for the delay is the fact that these colorless gases do not react with any other element under normal conditions. These elements were not found in the Earth's crust, in any ores, in water, in plants or animals, or in any of the other places where scientists were looking for new elements. They were accidentally discovered by William Ramsay after careful measurements of the amount of oxygen and nitrogen in air showed that one percent of air is a non-reactive substance. This family is known as the **inert** or **noble gases**. The reason for their lack of chemical reactivity will be a topic for later discussion.

Metalloids

As noted above, four of the families starting with nonmetals make a transition to metals near the bottom of the column. In each of these columns, one element on the zig-zag line exhibits some properties of both nonmetals and metals. Collectively, these elements are referred to as **metalloids** and include boron, B; silicon, Si; arsenic, As; tellurium, Te; and astatine, At. Arsenic, as an example, is a steel gray, brittle nonmetal, but it does have the metallic luster that is typical of metals.

Chemical Properties

The periodic law states that the properties of elements are periodic functions of their atomic numbers. The Bohr model provided an explanation for the repetition by suggesting that members of a group have isoelectronic outer energy levels. Let's examine the first few elements, starting with hydrogen, to determine how this repetition occurs. When we look at the periodic table, we find that hydrogen is element number one and has an average atomic mass of about 1 u. As shown in Figure 4-11, the Bohr model for hydrogen would have a nucleus containing a single proton surrounded by one electron in the lowest or K-energy level. Hydrogen is a chemically reactive gas and burns explosively when mixed with oxygen.

Hydrogen Atom   Helium Atom

Figure 4-11: Bohr orbits for hydrogen and helium.

The second and last element in the first period is helium. Helium has an atomic number of $A = 2$ and an atomic mass of about 4 u. This means that the helium nucleus contains two protons and $(4 - 2 = 2)$ two neutrons. Helium also has two electrons in the first or K-energy level. Chemically, helium is the first of the inert gas family. The two electrons in the K-energy level result in a very stable arrangement.

Moving to the next period, lithium, Li, is the first element directly below hydrogen in the alkali metal family. Lithium has an atomic number of $A = 3$ and an atomic mass of about 7 u The diagram of a lithium atom would contain a nucleus composed of three protons and $(7 - 3 = 4)$ four neutrons. There must also be three electrons to balance the electrical charge of the three protons. The next question is how are the three electrons arranged around the lithium nucleus?

To answer this question, scientists examined the properties of lithium. It is a silvery-white metal and, among other things, reacts quickly with oxygen forming lithium oxide, Li_2O, with a ratio similar to hydrogen in water, H_2O.

Based on its chemical behavior, it was decided that lithium's first two electrons must be in the K-energy level with the third electron one level higher in the L-level. This would make lithium's outer energy level contain one electron, and, therefore, isoelectronic with hydrogen. It would also explain the similar chemical reactivities of lithium and hydrogen.

Beryllium, $A = 4$, has an atomic mass of about 9 u. The diagram of a beryllium atom contains a nucleus composed of four protons and $(9 - 4 = 5)$ five neutrons. There must be four electrons to balance the electrical charge. If the first two electrons are in the K-energy level, then the two remaining electrons can be put into the L-energy level. This time, however, having two electrons in the outer energy level does not result in an inert gas. Although beryllium is much less reactive than lithium, it will react slowly with oxygen and form compounds with several other of the more reactive nonmetals.

For the element boron, $A = 5$ the atomic mass is about 11 u. The nucleus of a boron atom contains five protons and $(11 - 5 = 6)$ six neutrons. Two of the five electrons are in the K-energy level and the remaining three electrons are in the higher L-energy level. Apparently, the L-energy level has room for more than just two electrons. Boron is a chemically reactive element, forming compounds with many of the same elements as beryllium.

Table 4-5 shows that as you continue to move to the right across the second period, each new element adds one more electron to the growing L-energy level. As this process continues with carbon, nitrogen, oxygen, and fluorine, there is an increase in the chemical reactivity with each element. Oxygen, for example, is an extremely reactive element and combines with both metals and nonmetals forming compounds known as oxides. Fluorine, the first of the halogens, is the most reactive of all nonmetals. Suddenly, when the eighth electron is added to the L-energy level, the second inert gas results. Neon cannot be coaxed into reacting with any other element.

Based on these observations, chemists concluded that it must take eight electrons for the larger L-energy level to reach stability. This pattern of electron filling and repeating chemical properties occurs again in the third period of elements, only now it is the M-energy level that is filling from one to eight electrons.

Not long after the Bohr model provided an explanation for the periodicity of elements, chemists noted that each of the inert gases, except for helium, had eight electrons in its outer energy level. This resulted in the formulation of the **octet rule**, which suggested that the inert gases do not react with other elements because their outer energy shells are filled. The implication is that all other elements are chemically reactive in order to achieve a complete outer shell. This means that they must either gain or lose electrons from their outer, incompletely filled shell called the **valence shell**.

For elements above $A = 20$, calcium, the order of filling the energy levels becomes somewhat more complicated – although the basic principle remains the same. The elements in this area of the periodic table are known as the **transition elements.** Iron, cobalt, and nickel are examples of some of the elements in this area that have **valence electrons** appearing in the two outer energy shells. To avoid becoming further mired in these complexities, we will stop the discussion using this model with the element calcium.

Symbol	Atomic Number	Atomic Mass	No. of Protons, p^+	No. of Neutrons, n^0	Number and Location of Electrons, e^-			
					K	L	M	N
H	1	1	1	0	1	0	0	0
He	2	4	2	2	2	0	0	0
Li	3	7	3	4	2	1	0	0
Be	4	9	4	5	2	2	0	0
B	5	11	5	6	2	3	0	0
C	6	12	6	6	2	4	0	0
N	7	14	7	7	2	5	0	0
O	8	16	8	8	2	6	0	0
F	9	19	9	10	2	7	0	0
Ne	10	20	10	10	2	8	0	0
Na	11	23	11	12	2	8	1	0
Mg	12	24	12	12	2	8	2	0
Al	13	27	13	14	2	8	3	0
Si	14	28	14	14	2	8	4	0
P	15	31	15	16	2	8	5	0
S	16	32	16	16	2	8	6	0
Cl	17	35	17	18	2	8	7	0
Ar	18	40	18	22	2	8	8	0
K	19	39	19	20	2	8	8	1
Ca	20	40	20	20	2	8	8	2

Table 4–5: Distribution of electrons for the first 20 elements.

Atoms, Ions, and Ionic Compounds

Elements are made up of atoms. All atoms are neutral and must, therefore, have an equal number of protons (+) and electrons (–). If an atom gains or loses electrons from its valence shell, there is a charge imbalance and the resulting particle carries an electrical charge. These electrically charged particles are called **ions**. More specifically, if positively charged they are called **cations**, and if negatively charged, **anions**.

Consider the element sodium, which is a very reactive alkali metal. Examination of its electron configuration reveals that it has only one electron in its outer energy level. Chlorine is a very reactive nonmetal. It has seven electrons in its outer energy level. When sodium metal and chlorine gas are brought together, a violent reaction results with a release of energy and the production of a white crystalline solid. What happened?

Apparently, it takes less energy for each sodium atom to lose one electron and become isoelectronic with neon than to gain seven electrons. Conversely, it would appear that it takes less energy for each chlorine atom to gain one electron and become isoelectronic with argon than to lose seven electrons.

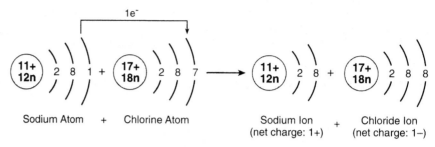

Figure 4–12: The formation of NaCl from sodium and chlorine atoms.

By examining Figure 4-12, you can see that the transfer of one electron from the sodium atom to the chlorine atom satisfies the needs of both. The loss of the outer electron from the sodium atom leaves it with only ten electrons, which is the same distribution as that of a neon atom. There is one major difference, however. The remaining particle still has 11 protons, but only 10 surrounding electrons. Some electrical bookkeeping quickly tells us that the resulting particle has an excess charge of plus one. The resulting particle is called, therefore, a sodium ion and its symbol now includes its charge, Na^+.

When the chlorine atom accepts the electron lost by the sodium atom, the chlorine atom becomes isoelectronic with argon. Again, however, an electrically unbalanced particle results. The number of electrons in this particle is one greater than its number of protons. The resulting particle is known as a chloride ion and its symbol now includes its charge, Cl^-. The rearrangement of the electrons for these two elements has resulted in a new substance, called sodium chloride, NaCl, or table salt. The release of chemical energy, in the form of heat and light, is evidence that this arrangement of electrons requires less energy and is, therefore, more stable.

In a similar way, each of the alkali metals could be shown to form the same type of compounds with members of the halogen family, e.g., LiF, KBr, RbI, and CsCl. Each of these metals has lost one electron from its valence shell during chemical combination. When an atom gives away electrons, the process is called **oxidation**. In each instance, the resulting ion formed (Li^+, K^+, Rb^+, and Cs^+) would have a charge or **oxidation number** of plus one. Oxidation numbers are used to indicate the number of electrons that an atom provides (positive) or needs to acquire (negative) during a chemical change.

The chlorine atom and the other halogens each gain one electron to become isoelectronic with the noble gas at the end of their periods. When an atom gains electrons, the process is called **reduction**. Their symbols and negative oxidation numbers are, therefore, F^-, Br^-, and I^-.

By knowing the oxidation numbers of two elements, it is possible to quickly determine the formulas for the resulting compounds. Suppose the elements potassium, K, and bromine, Br, are reacted. Knowing that potassium ions are K^+ and that bromine becomes bromide ions, Br^-, you can predict that the ratio of these two ions would be 1:1, resulting in the compound, KBr.

What happens when the metal magnesium, Mg, from the alkaline earth group reacts with chlorine? Examination of the elements in the alkaline earth family reveals that they all have two electrons in their valence shells. Since it is easier to lose two electrons than gain six, each of these elements are said to have oxidation numbers of 2+. (Note that the 2 indicates the number of valence electrons and the + indicates that the electrons are given away.) The chlorine atom still needs only one electron to become isoelectronic with the nearby noble gas. Therefore, each magnesium atom can supply enough electrons to convert two chlorine atoms to the more stable arrangement. The ratio of elements found in a combination of magnesium and chlorine is always two chloride ions to each magnesium ion, or $MgCl_2$.

There are two important things to note about the formula for magnesium chloride, which hold true for all compounds formed from ions. First, the symbol for the positive ion is always written first. Second, if the ratio of ions in a formula is other than 1:1, then a subscript will follow the more plentiful ion in the proper ratio.

Careful consideration of the position of the elements on the periodic table (see Table 4-4) reveals that those elements on the far left of the periodic table always form ions with a + or 2+ oxidation number. In fact, any time that an atom has less than four electrons in its outer valence shell, it will tend to lose those electrons during chemical reactions, forming positively charged ions. The elements previously identified as metals all tend to form positively charged ions.

All elements that have more than four electrons in their valence shell tend to gain electrons forming ions with 3–, 2– or – oxidation numbers. From this, it can be seen that the position of two elements on the periodic table can be used to determine the formulas for their resulting compound.

For example, if the element aluminum, Al, reacts with oxygen, O, then from their position in the periodic table it can be determined that aluminum has three electrons in its valence shell and that oxygen has six. If aluminum wants to give away three electrons, but oxygen wants only two, what ratio of the two must be used? Perhaps, the easiest way to approach the problem is to find that 6 is the smallest number divisible by both 3 and 2. Therefore, if two aluminum atoms join three oxygen atoms, a total of six electrons will be exchanged resulting in the compound, Al_2O_3.

A second way to approach this problem is simply to crisscross the oxidation numbers, without including the sign. In this example, $Al^{3+}\!\!-\!\!O^{2-}$ by moving the 3 down and behind the O and the 2 down and behind the Al, you also arrive at Al_2O_3 as the smallest whole number ratio.

In Table 4-6, the oxidation numbers have been listed for some of the more common elements. Try your hand at putting an element with a positive oxidation number with one that has a negative oxidation number. See if you can find the smallest numbers that will balance the electrical charges. For example, Mg^{2+} and O^{2-} would require a 1:1 ratio to balance the 2+ with the 2–, therefore, the formula for magnesium oxide is MgO. Each formula written using this simple method also has a chemical name. As can be seen in the above example, the names are little more than a combination of the positive and negative ion names. For a more complete explanation of how to name inorganic compounds, see Appendix B.

+								0
H	2+		3+	4±	3–	2–	–	He
Li	Be		B		N	O	F	Ne
Na	Mg		Al		P	S	Cl	Ar
K	Ca	Transition Elements				Se	Br	Kr
Rb	Sr						I	Xe

Table 4–6: Oxidation numbers listed by column.

Bonding

After each ion achieves its inert gas configuration, how are crystals formed? The obvious answer is that they are now charged particles and opposites attract. This type of attraction is called an **ionic bond**, which is one of the major types of bonding.

Technology Box 4–6 ■ Measuring Pollution by Conductivity

When ionic compounds dissolve in water the ions separate from each other and become surrounded by several water molecules. If two wires from an electrical source, like a battery, are placed in this solution, the positive ions are attracted to the electron-rich negative wire and the negative ions to the electron-deficient positive wire. Since the ions can move through the solution, they complete the circuit allowing electricity to flow. Solutions that contain ions and conduct electricity are known as electrolytes.

It is easy to make and calibrate a simple conductivity device that will allow you to relate the number of ions in the solution to the amount of electricity passing through the solution. 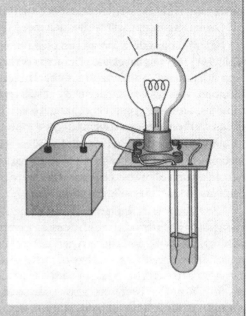 When doing environmental studies, measuring the conductivity of a body of water can be a way to determine its degree of contamination. Although the most pure natural water is rainwater, it can dissolve various air pollutants such as carbon dioxide, CO_2; sulfur oxides, SO_x; and nitrogen oxides, NO_x, forming acid rain. Each of these will dissolve in water forming ions.

In the crystalline state, the attraction isn't limited to a single oppositely charged ion. Instead, large populations of the positive and negative ions seek an arrangement in which each ion can have several oppositely charged neighbors. That arrangement is what defines the crystal structure for that substance.

Compounds held together by ionic bonds are called **ionic compounds** and they tend to have similar properties. For example, most have hard crystals with high melting and boiling points and they are not flammable. When melted they conduct electricity. They are not soluble in organic solvents, like gasoline or acetone (fingernail polish remover), but they tend to dissolve in water. Common table salt, NaCl, is the most familiar ionic compound.

Molecules and Covalent Compounds

So far, we have been rather selective of the elements used to explain ionic bonding. In each example, one element came from the left side of the periodic table and the other from the right – one metal and one nonmetal. Previously, we mentioned that if two metals are combined the result is a mixture called an alloy. But what about a compound, such as CO_2, that is composed of two nonmetals?

The success of the octet rule soon led to the realization that atoms may be able to attain a noble gas configuration through a sharing process, rather than losing or gaining electrons. Whenever atoms share one or more pairs of electrons, it is called a **covalent bond**. The resulting smallest group of atoms to form a stable arrangement is called a **molecule**.

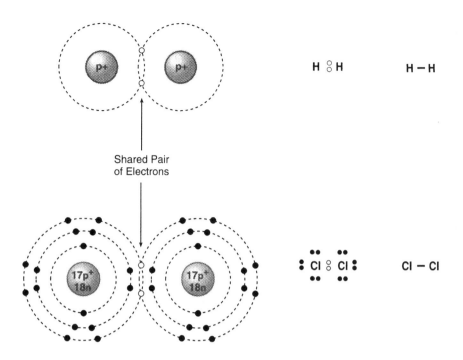

Figure 4–13: Electron dot formulas for hydrogen and chlorine gas.

Seven elements (H_2, N_2, O_2, F_2, Cl_2, Br_2, and I_2), for example, exist at room temperature and pressure as **diatomic** molecules. By themselves, atoms of these elements do not have noble gas configurations; however, when two atoms share electrons they can form molecules that do. In Figure 4-13, **electron-dot formulas** have been drawn for molecules of H_2 and Cl_2 to show how they achieve noble gas arrangements by sharing electrons.

Whenever two atoms share only a single pair of electrons to achieve a noble gas arrangement, the bond between those atoms is known as a single covalent bond. A single bond, as shown in Figure 4-13, can be represented by either two dots, :, or by a single straight line, –. The two atoms participating in the bond can be either two atoms of the same element or atoms of two different elements.

Each element in the periodic table has been assigned a value called an **electronegativity number**. This number is a measure of the element's attraction for electrons. The more electronegative the atom is, the more tightly it will hold to its bonded electrons.

There is a diagonal relationship between an element's position on the periodic table and its electronegativity number. It is apparent in Table 4-7 that those elements found in the upper right corner of the periodic table (fluorine, oxygen, chlorine, etc.) tend to have the highest values and those in the lower left corner the lowest. By comparing the electronegativity numbers of the two elements bonding, it is possible to predict whether they will form a **nonpolar covalent**, **polar covalent**, or ionic bond.

If the difference in the electronegativities is less than 0.5, then the electron pair will be shared nearly equally between the two nuclei and the bond will be nonpolar. Nonpolar molecules are found in gasoline, vegetable oil, and other substances that typically do not dissolve in water. These molecules are electrically symmetrical. This means that because the atoms are sharing electrons equally, the center of the positive charges will be found in the center of the negative charge.

Technology Box 4–7 ■ Polar and Nonpolar Molecules: How do soaps and detergents work?

Throughout history, cultures have had varying opinions and priorities regarding cleanliness. Many ancient cultures – Egyptians, Greeks, Romans, and Hebrews – had a very high regard for cleanliness. The real problem with staying clean is not the dirt. Things like soil, sand, and dust actually come off very easily – you can just brush them off. The problems are oil and grease. Oil and grease act like water insoluble glues that bind the dirt to your skin and clothing. That makes washing them off difficult.

There is evidence that the ancient Persians knew that they could mix animal fat with sodium or potassium salts to cleanse wounds. Today, we recognize this combination as being a form of soap. Animal fats are compounds with two major components: glycerol and fatty acids. Fatty acids are long strands of nonpolar carbon atoms that have a polar end. By boiling fats with ashes (containing sodium or potassium salts) the fatty acid end is converted into a hydrophylic (water loving) ionic salt, as shown in the figure below. The opposite end of this long molecule, however, remains nonpolar or hydrophobic (water fearing) and favors associating with the oils and greases. Molecules that have dual polar and nonpolar ends are called surfactants. As shown, the nonpolar end of the surfactant picks up the oil and grease and the ionic end follows the water molecules down the drain.

Since World War II, detergents – a new kind of surfactant – have replaced most household soap uses. Detergents are also composed of a long nonpolar portion and an ionic portion, so they perform in much the same way as soap. Detergents, however, are made from petroleum and their pH can be adjusted for various applications.

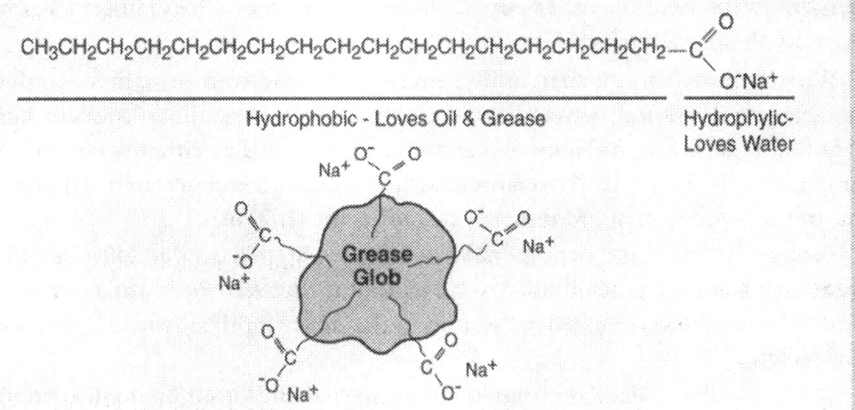

If, on the other hand, the difference between the electronegativities of the two atoms is between 0.5 and 1.7, then the electron pair will be unequally shared, forming a polar covalent bond. The best known polar covalent compound is water, but other compounds such as ammonia, alcohols, and sugars are also polar. In these molecules, the shared electrons are more closely bound to the more electronegative atom, so the center of the positive charges is no longer in the center of the negative charge. This results in molecules that have **dipoles** or partially charged separations.

1 H 2.1																	2 He —
3 Li 1.0	4 Be 1.5											5 B 2.0	6 C 2.5	7 N 3.0	8 O 3.5	9 F 4.0	10 Ne —
11 Na 0.9	12 Mg 1.2											13 Al 1.5	14 Si 1.8	15 P 2.1	16 S 2.5	17 Cl 3.0	18 Ar —
19 K 0.8	20 Ca 1.0	21 Sc 1.3	22 Ti 1.5	23 V 1.6	24 Cr 1.6	25 Mn 1.5	26 Fe 1.8	27 Co 1.8	28 Ni 1.8	29 Cu 1.9	30 Zn 1.6	31 Ga 1.6	32 Ge 1.8	33 As 2.0	34 Se 2.4	35 Br 2.8	36 Kr —
37 Rb 0.8	38 Sr 1.0	39 Y 1.2	40 Zr 1.4	41 Nb 1.6	42 Mo 1.8	43 Tc 1.9	44 Ru 2.2	45 Rh 2.2	46 Pd 2.2	47 Ag 1.9	48 Cd 1.7	49 In 1.7	50 Sn 1.8	51 Sb 1.9	52 Te 2.1	53 I 2.5	54 Xe —
55 Cs 0.7	56 Ba 0.9	57 La 1.2	72 Hf 1.3	73 Ta 1.5	74 W 1.7	75 Re 1.9	76 Os 2.2	77 Ir 2.2	78 Pt 2.2	79 Au 2.4	80 Hg 1.9	81 Ti 1.8	82 Pb 1.8	83 Bi 1.9	84 Po 2.0	85 At 2.2	86 Rn —
87 Fr 0.7	88 Ra 0.9	89 Ac 1.1	90 Th 1.3	91 Pa 1.5	92 U 1.7												

Key:
- 9 → Atomic Number
- F → Symbol
- 4.0 → Electronegativity Value

The difference in electronegativity between two atoms determines the character of the bond formed between them.
1. If electronegativity difference is larger than 1.7, the bond is considered to be ionic.
2. If electronegativity difference is less than 0.5, the bond is considered to be pure covalent.
3. If electronegativity difference is more than 0.5 but less than 1.7, the bond is considered to be polar covalent.

Table 4–7: Pauling's table of electronegativities.

Finally, if the electronegativity difference is greater than 1.7, then electrons will be transferred completely between the two atoms, resulting in an ionic bond.

In general, whenever an F–H, O–H, Cl–H or N–H bond is involved, a dipole results and a polar covalent bond is formed. In this bond, the electrons are shifted toward the nucleus with the greater electronegativity making that end of the dipole partially negative, symbolized by $\delta-$. The less electronegative atom, therefore, becomes partially positive, symbolized as $\delta+$.

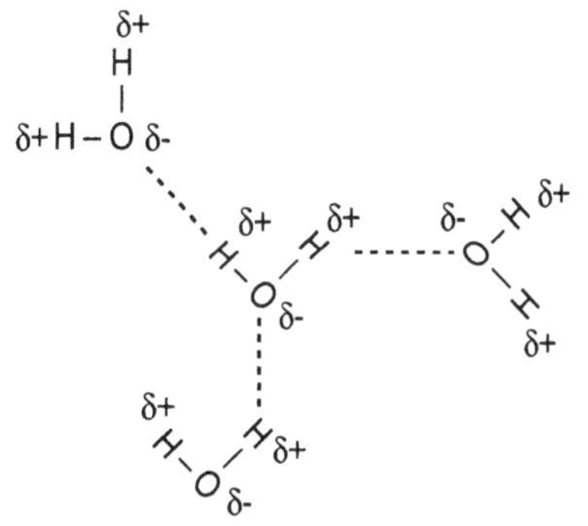

Figure 4–14: Polar water molecules and hydrogen bonding.

The fact that water contains two such H–O bonds, and therefore, has two partially positive sites and one partially negative site accounts for some of its important physical properties. Remembering that opposite charges attract, this means that neighboring water molecules tend to cling to one another by **hydrogen bonding**, as shown in Figure 4-14. The negative (oxygen) end of one water molecule tends to be attracted to one of the positive (hydrogen) ends of a neighboring water molecule.

Technology Box 4–8 ■ Physical Properties: Why do boiling points vary?

A boiling point is the temperature at which a pure substance makes a transition from liquid to gas. After years of investigation, it is believed that two primary factors determine these points. First, large molecules tend to require a higher temperature to boil than smaller molecules. The second contributing factor is the type of bonding involved.

The difference that molecular size makes can be illustrated by using the members of the organic family known is alkanes. They all have the same type of bonding; nonpolar covalent. The smallest member of the family is methane, CH_4. Each of the following members conforms to the general formula C_nH_{2n+2}. So ethane is $C_2H_{2(2)+2}$ or C_2H_6; propane $C_3H_{2(3)+2}$ or C_3H_8; etc. The name, formula, formula mass, and boiling point for the first five members are listed in the following table for standard pressure (1 atmosphere).

Formula	Formula Mass, u	Boiling Point, °C
Methane, CH_4	16	−164.00
Ethane, C_2H_6	30	−88.63
Propane, C_3H_8	44	−42.07
Butane, C_4H_{10}	58	−0.50
Pentane, C_5H_{12}	72	36.07
Water, H_2O	18	100.00

It should be noted that the boiling points of this family increase as the molecules get larger. The first four members are gases at room temperature and pressure. Methane is a natural gas that is used for heating homes. Propane and butane can be liquefied by putting them under higher pressures. Butane is the colorless liquid that you can see in some cigarette lighters. Pentane is the first member of this family that is a liquid at room temperature. It is used in gasoline during the winter months to promote faster starting of your car.

As you examine the table of boiling points, be aware that small nonpolar molecules have little attraction for each other. This is the reason they exhibit low boiling points. The data for water has been added to the table as a comparison of bonding differences. Both water and methane are very small molecules and both have single covalent bonds. But, unlike methane, the bonds in water molecules are polar covalent. Therefore, water is a polar molecule, which attracts its neighboring molecules through weak but effective hydrogen bonding. The boiling points of polar covalent substances are higher because of hydrogen bonding. If it weren't for hydrogen bonds, there would be no liquid water on this planet.

Hydrogen bonds are symbolized as dotted lines ••••, and are not strong bonds. In fact, a hydrogen bond has only about one-twentieth the strength of either an ionic or a covalent bond. That is not to say, however, that they are unimportant. Due to hydrogen bonding, the melting and boiling points of water are significantly elevated from what would be expected based solely on its molecular mass.

There are times when sharing of a single pair of electrons between two nuclei fails to result in the desired octet of electrons. In those situations, the atoms will tend to share two pairs of electrons. In a compound such as carbon dioxide, CO_2, the only satisfactory arrangement of the valence electrons is to have each of the oxygen atoms share two of its electrons with carbon. The resulting molecule can be shown as O::C::O, or, using the straight line notation for an electron pair, O=C=O. Whenever two pairs of electrons are required to satisfy the octet rule, it is called a double covalent bond. There are many examples of double covalent bonding in organic chemistry, which is the topic of a later chapter.

Finally, in a few situations, where a single or double covalent bond fails to provide the necessary octet of electrons, a bond will form using three pairs of shared electrons, called a triple covalent bond. Perhaps two of the most familiar examples of this type of bonding are found in the element nitrogen, N:::N or N≡N, and the organic compound known as acetylene gas, H:C:::C:H.

Compounds composed of single, double, and triple covalent bonds tend to form soft, waxy, greasy solids that have low melting and boiling points. They do not dissolve readily in water, but most are soluble in nonpolar solvents. Most are flammable when in contact with the air. Two familiar examples – lard and motor oil – are mixtures of these compounds.

In summary, there are two ways in which elements achieve a noble gas arrangement: gaining and losing electrons or sharing electrons. If they gain or lose electrons, it results in the formation of ions and an ionic compound. If they share one, two, or three pairs of electrons, they form a molecule and a covalent compound. If the electronegativities of the elements sharing the electrons are nearly equal, then the resulting molecules will be nonpolar. If the electronegativities are quite different, the molecules will be polar and attracted to one another by a weak force called hydrogen bonding, which greatly affects the physical properties of a compound.

Polyatomic Ions

A final consideration. There are clusters of atoms, called **radicals** or **polyatomic ions**, that are held together by covalent bonds, but have either a greater or lesser number of electrons than the total number of protons. One of the simplest examples is the hydroxide ion, OH^-.

If you refer to Table 4-6, you will find radicals listed along with the other common ions. It should be noted that most have negative oxidation numbers. The important exception is the ammonium ion, NH_4^+. The NH_4^+ and OH^- ions are two ionic components of ammonium hydroxide, NH_4OH, present in such everyday compounds as window cleaner.

If a polyatomic ion is present as a single ion within a compound, then no parentheses are placed around the ion. An example is the sulfate ion, SO_4^{2-}, which combines with two sodium ions to form, Na_2SO_4. If, however, the polyatomic ion is needed more than once, then the entire group must be enclosed in parentheses and the number of ions are shown as a subscript. An example of this is the combination of three sulfate ions SO_4^{2-} with two aluminum ions, Al^{3+} to form, $Al_2(SO_4)_3$.

POSITIVE Charge of the Ion			
(+)	(2+)	(3+)	(4+)
Hydrogen H^+	Calcium Ca^{2+}	Iron(III) Fe^{3+} (Ferric)	Tin(IV) Sn^{4+} (Stannic)
Lithium Li^+	Magnesium Mg^{2+}	Aluminium Al^{3+}	Lead(IV) Pb^{4+} (Plumbic)
Sodium Na^+	Barium Ba^{2+}		
Potassium K^+	Radium Ra^{2+}		
Silver Ag^+	Zinc Zn^{2+}		
Copper(I) Cu^+ (Cuprous)	Tin(II) Sn^{2+} (Stannous)		
Ammonium NH_4^+	Iron(II) Fe^{2+} (Ferrous)		
	Lead(II) Pb^{2+} (Plumbous)		
	Copper(II) Cu^{2+} (Cupric)		

NEGATIVE Charge of the Ion			
(–)	(–)	(2–)	(3–)
Flouride F^-	Acetate $C_2H_3O_2^-$	Oxide O^{2-}	Nitride N^{3-}
Chloride Cl^-	Chlorate ClO_3^-	Sulfide S^{2-}	Phosphide P^{3-}
Bromide Br^-	Permanganate MnO_4^-	Sulfite SO_3^{2-}	Phosphate PO_4^{3-}
Iodide I^-	Hypochlorite ClO^-	Sulfate SO_4^{2-}	Arsenate AsO_4^{3-}
Hydroxide OH^-		Carbonate CO_3^{2-}	
Nitrite NO_2^-		Dichromate $Cr_2O_7^{2-}$	
Nitrate NO_3^-			
Hydrogen Carbonate (Bicarbonate) HCO_3^-			

Table 4–8: Common ions.

Check Your Understanding

1. What are the two major subdivisions of the periodic table? How do the elements in these two divisions differ?

2. Explain how the periodic table can be used to predict the chemical behavior of an element.

3. What is meant by the term isoelectronic and what does it have to do with the properties of elements in families?

4. What is the oxidation number and what information does it tell us?

5. Explain the different types of bonding and their effects on the physical characteristics of a substance.

Summary

Matter is anything that occupies space and has mass. If it has constant composition, it is either an element or a compound. Matter that is variable in composition is either a heterogeneous or a homogenous mixture. All pure substances have both physical properties and chemical properties. Matter also exists in one of three physical states: solids, liquids, and gases. Plasmas are a recently recognized fourth state of matter, consisting of an electrically neutral mixture of ionized gases at a very high temperature.

The Greeks recognized four basic elements, earth, air, water, and fire. They also provided us with the word "atom", meaning "uncuttable." Modern atomic theory was stated in the early 1800s by John Dalton. In the attempt to prove the existence of atoms, many new discoveries were made. In 1869, Mendeleev found a way to organize the known sixty-three elements into a chart that allowed for the prediction of missing elements and their properties. This chart is the periodic table of the elements.

A series of discoveries led to the eventual theory that all matter is composed of atoms, which in turn, are composed of three subatomic particles. In addition to the negatively charged electron, there are the positively charged proton and the neutral (uncharged) neutron, that together comprise the small, dense nucleus. Bohr developed a model in which electrons can exist only in allowed energy shells. These orbiting electrons are capable of accepting energy and moving to energy levels farther from the nucleus or returning the energy as they move back to the ground state. If the returned energy is in the form of visible light, it results in the element's unique bright-line spectrum.

The periodic table continues to be a very valuable learning tool. It is divided into vertical columns called groups or families. The outer energy levels of the elements in a family are isoelectronic with each other and the elements tend to have similar properties. The horizontal rows, called periods or series, have elements with increasing atomic numbers, and generally, increasing atomic masses. A zigzag line separates the elements into two major groups: metals and nonmetals. At room temperature, nonmetals range from gases to solids. They are neither ductile nor malleable and they are not good conductors of heat or electricity. As solids, they tend to be brittle.

The octet rule has provided chemists with a way to understand why chemicals combine in definite ratios. In general, if an element has less than four electrons in its outer energy level, it tends to lose them forming ions with +, 2+ or 3+ charges. If an element has more than four electrons, it tends to gain electrons to become isoelectronic with the noble gas at the end of its row. Nonmetals, therefore, tend to form 3–, 2–, and – charged ions. By knowing the number of electrons an elements needs to lose or gain, it is possible to correctly predict the formulas for their combinations.

Ionic bonding is the attraction between oppositely charged ions with large differences in electronegativities. When combining elements have similar electronegativities, they tend to share one, two, or three electron pairs. These shared pairs are called a single, a double, or a triple covalent bond. Atoms that are covalently bonded are called molecules. When forming covalent bonds, the two participating nuclei do not always share the electrons equally, resulting in a polar covalent molecule. In those molecules, the electron pair is pulled away from the less electronegative nucleus, resulting in a partially positive and a partially negative site on the molecule. The weak attraction between the partially charged ends of nearby molecules is called hydrogen bonding. Hydrogen bonding has a dramatic effect on water solubility as well as its melting and boiling points.

Chapter Review

1. Select an object and write an explanation for why it is matter. Include in your description, whether you believe it to be a pure substance or a mixture. Explain your answer.

2. Describe the three states of matter and discuss what they have in common and how they differ.

3. Write an explanation for the terms physical properties, physical change, chemical properties and chemical change. Use examples in your explanation.

4. Write a description of the phase diagram for water (Figure 4-2). Explain what is occurring along each line segment.

5. Using an outside reference or the Internet, write a description of what must be done to convert the hexagonal form of carbon crystals into the tetrahedral form. If possible, also comment on the economics of making such a conversion.

6. Both liquids and gases are called fluids. Explain what properties they have in common.

7. Cathode rays are used to create the picture on the TV screen. What would be the effect of bringing a strong magnet near your TV? Why would it not be recommended?

8. If you have only an element's atomic number, A, what other information do you know?

9. The naturally occurring isotopes of bromine are: 50.54% bromine-79 and 49.46% bromine-81, using the formula:

(Atomic Mass × %) + (Atomic Mass × %) + (Atomic Mass × %) + ... = Average

What should be listed on the periodic table as its weighted average? Show your work.

10. The naturally occurring isotopes of iron are: 5.82% iron-54, 92.66% iron-56, 2.19% iron-57 and 0.33% iron-58. Using the above formula, calculate the weighted average for iron.

11. The half-life of plutonium-239 is 24,360 years. If you started with an extremely small sample of 1,000,000 atoms of plutonium, how many would be remaining after the first half-life? At the end of the second half-life? If you had 6.02×10^{23} atoms of plutonium, how long would it take for all of the plutonium atoms to be gone? Explain your answer.

12. Write a short paragraph explaining the periodic law and the placement of elements on the periodic table.

13. Write a description that could be used to decide if a newly discovered element is a metal or nonmetal.

14. Use information from the period table to draw Bohr model diagrams for the first 18 elements. Be sure to include the number of protons and neutrons in the nucleus and the number of electrons in each of its energy levels.

15. If the atomic mass of an element is 238 u, and it has an atomic number of 92, how many neutrons would be in the nucleus?

16. Write a description of the difference between an atom and a positive ion and also between an atom and a negative ion.

17. Write a correct formula for each of the following:
    a. Mg^{2+} and S^{2-} = _____
    b. Al^{3+} and Br^- = _____
    c. Li^+ and O^{2-} = _____
    d. Na^+ and SO_4^{2-} = _____
    e. Mg^{2+} and PO_4^{3-} = _____

18. Explain why some elements form ionic, while others form covalent bonds. Be specific and use examples.

19. Explain why some covalently bonded compounds dissolve in water and others not.

20. What is the octet rule and why is it important?

21. Draw electron dot representations for each of the following molecules:
    a. CH_4
    b. HCl
    c. H_2O
    d. Br_2
    e. C_2H_2

22. Using the information in Table 4-7, determine the type of bonding (nonpolar, polar, or ionic) that would be expected for each of the following combinations:
    a. H and N
    b. Na and F
    c. C and H
    d. H and O
    e. Ca and Cl

23. Select a polyatomic ion from Table 4-8. Explain why it is considered to be a polyatomic ion.

24. If you have access to the Internet, attempt to consult the site:
    http://www.pbs.org/wnet/hawking/strange/html/quarks.html
    Write a paragraph or two describing the construction of a carbon atom, starting with quarks and electrons.

Just how fast are we moving through space?

Matter in Motion and Newton's Laws

Beginning with your earliest exposure to sciences, you have been taught that the Earth is turning on its axis while simultaneously orbiting the Sun. It took centuries of study and hundreds of people to arrive at this basic conclusion. Scholars charted, hypothesized, tested, recorded, and finally, in 1513, Nicolaus Copernicus finished a paper proposing that the Sun – not the Earth – is in the center of our universe and that the Earth is in orbit around it. The theory was not published until months before his death in 1543, and was considered one of the most politically volatile theories ever proposed. In addition to the political opposition, some religious leaders believed that his hypothesis was in direct opposition to the teachings of the scriptures. Later supporters, including Johannes Kepler (1571 – 1630) and Galileo Galilei (1564 – 1642), found themselves at odds with both the religious and political powers of the day; a fact that put their jobs and personal freedom in jeopardy.

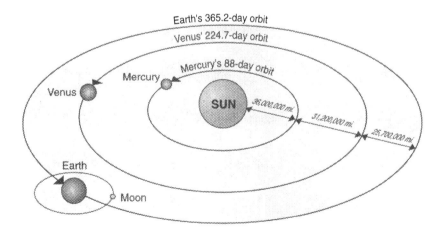

Figure 5–1: The average distance of the path of the Earth around the Sun is approximated using the equation for circumference, $C = 2\pi r$, where r is the distance from the Earth to the Sun.

The first successful demonstration that the Earth is turning on its own axis came in 1851 when the French physicist Jean Bernard Leon Foucault discovered that the plane of a swinging pendulum changes according to the rotation of the Earth.

Today almost everyone accepts the Sun-centered solar system as fact. The next question, then, is how fast the Earth is moving around the Sun. Before speed can be calculated, two pieces of information are necessary: distance and time. The average speed of the Earth can be approximated by dividing the distance traveled by the elapsed time, according to the following equation.

$$\text{average speed} = \frac{\text{distance}}{\text{time}}$$

Figure 5-1 shows that the path of the Earth around the Sun is not a straight line, but an ellipse. The average distance between the Sun and the Earth is known as one **astronomical unit**, AU, which is equal to 93,000,000 miles or 149,000,000 kilometers. The formula $C = 2\pi r$ can then be used to calculate the circumference of the orbit, giving the distance traveled.

The time is one year or 365.25 days, which is the time it takes for the Earth to make one complete orbit. So, to calculate the speed the Earth is traveling, substitute in the equation

$$\text{average speed} = \frac{2\pi\ 93{,}000{,}000\ \text{miles}}{1\ \text{year}} \times \frac{1\ \text{year}}{365.25\ \text{days}} \times \frac{1\ \text{day}}{24\ \text{hours}} \cong 66{,}700\ \text{mph}$$

This problem illustrates only the calculation of average speed. In the following sections, other motions of matter will be introduced, along with Newton's laws.

5-1   Measuring Motion

Objectives

Upon completion this section, you will be able to:

■ Explain the difference between distance and displacement.

■ Explain the difference between speed and velocity.

■ Calculate displacement, velocity and acceleration.

■ Explain straight line acceleration.

Distance, Speed, and Vector Quantities

In physics, distance and displacement are not the same thing. Distance indicates only how far an object has moved. **Displacement**, $\vec{s}$, indicates both the distance and the direction in which it has moved. The fact that it includes direction makes displacement a **vector quantity**, which is denoted by the arrow, $\rightarrow$, above the variable. The direction of a vector quantity is often represented as an angle, measured in degrees. Direction may also be given in either clockwise or counter-clockwise degrees from a principal (north, south, east, or west) reference direction. Often, the reference direction selected is north or east. Pilots, for example, measure direction in degrees, clockwise from north.

Distance and Displacement

To quickly determine the position of a traveler requires knowing not only the distance traveled, but also the direction. For example, if Sally leaves Kansas City and drives a) 50 miles east, and then b) 50 miles north, how far and in what direction is she from her starting point? The answer can be determined and expressed several different ways:

— The data can be plotted, to scale, on a graph, as illustrated in Figure 5-2. The point that represents 50 miles east and 50 miles north from the starting point is marked. Then the distance from the origin to that point is measured and calculated from the scale of the graph. This method shows that Sally is about 70 miles, 45° clockwise from north.

— The Pythagorean theorem can be used, where a and b are the 50-mile legs of a right triangle and c is the hypotenuse:

$$a^2 + b^2 = c^2$$

$$50^2 + 50^2 = c^2$$

$$2500 + 2500 = c^2$$

$$\sqrt{5000} = c$$

$$70.7 \text{ miles} = c$$

Sally's position from her starting point is represented by the angle θ. The value of this angle can be determined by the trigonometric function sin θ = side opposite/hypotenuse or sin θ = 50 miles/70.7 miles = 45°. Therefore, although Sally has traveled a distance of 100 miles (50 miles east and 509 miles north) her final location (displacement) is 70.7 miles, 45° clockwise from north of where she started.

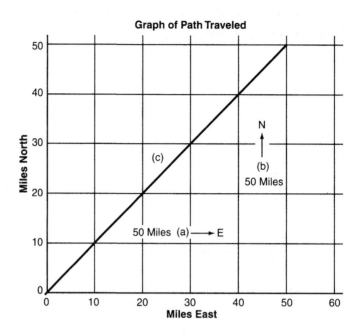

Figure 5–2: Graphing reveals that Sally is 70.7 miles and 45° north of east from her starting point.

— In the first two methods, the direction of the vector quantity has been expressed as an angle measured in degrees. As shown in Figure 5-3, angles may also be expressed in **radians**, where 360° = 2π radians, 180° = π radians, 90° = $\frac{1}{2}$π radians. Sally's position, therefore, could be expressed as 70.7 miles, 0.25π radians (0.785 radians) counter-clockwise from east.

Regardless of the method chosen to report displacement, it needs to be clearly stated so the information can be understood by others.

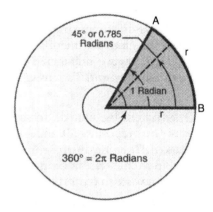

Figure 5–3: A unit of angular measurement that results in an arc (AB) equal in length to the radius, r, of the circle.

Speed and Velocity

Speed and velocity are related in much the same way as distance and displacement. Speed is how fast you are moving, regardless of direction. Velocity includes the direction of the motion. A car going north at 50 mph has the same speed as one going south at 50 mph, but they have different velocities. The speedometer measures a car's speed, but to measure its velocity you need another instrument – a compass.

Like displacement, **velocity**, $\vec{v}$, is a vector quantity. Rather than the distance and direction measured by displacement, velocity is a measure of speed and direction. Consider the difference between driving north at 50 mph for two hours and driving north at 50 mph for one hour then east at 50 mph for one hour.

In situations where direction is not important the term speed or the symbol, v, is used without an arrow, indicating that it is not a vector quantity. The numerical value and units used, however, are the same for both. Today most physicists prefer using meters per second to measure speed and velocity. It may be expressed, however, in miles per hour, kilometers per hour, feet per second, meters per second, inches per second, centimeters per second or any other combination of units expressing distance per unit of time.

Suppose Sally is now driving from Chicago to Kansas City. If it takes her 10 hours to travel the 650 miles, her average speed would be calculated as follows:

$$\text{average speed} = \frac{\text{distance}}{\text{time}}$$

$$v = \frac{d}{t}$$

$$v = \frac{650 \text{ miles}}{10 \text{ hours}} = \frac{65 \text{ miles}}{\text{hour}}$$

Using a map and protractor to determine the direction of travel, her average velocity, $\vec{v}$, therefore, would be reported as 65 miles per hour S 60°W. By comparing the two answers, it can be seen that the only difference is that velocity also includes the direction traveled. It is very common, however, to refer to velocity without specifying the direction. It is not strictly correct to do so, but it is commonly done when a direction is either unimportant or understood without being specified.

The equation relating speed, distance, and time can be used to find any one of the three quantities, if the other two are known. For instance, the speed of light, 3.0×10^8 meters/second, can be used to calculate the length of time required for light to travel from the Sun to the Earth, 1.49×10^{11} m. We merely solve the equation for time.

$$\text{time} = \frac{\text{distance}}{\text{speed}}$$

$$t = \frac{1.49 \times 10^{11} \text{ m}}{3.0 \times 10^8 \text{ m/s}}$$

$$t \cong 500 \text{ s}$$

Astronomy, the space program, and science fiction writers frequently use the term **light-years**. It is actually a measure of distance and is based on how far light travels in a year. To calculate the distance in meters, make the following substitutions:

$$\text{distance} = \text{speed} \times \text{time}$$

$$d = 3.0 \times 10^8 \text{ m/s} \times 365.25 \text{ days}$$

$$d = 3.0 \times 10^8 \frac{\text{m}}{\text{s}} \times 365.25 \text{ days} \times \frac{24 \text{ hours}}{1 \text{ day}} \times \frac{3600 \text{ s}}{1 \text{ hr}}$$

$$d = 9.47 \times 10^{15} \text{ m}$$

Note that, in the first step, the time units do not match. You cannot multiply meters per second (m/s) by days and get a meaningful answer. This unit mismatch is corrected in the second step by converting days to hours and hours to seconds.

Instantaneous Velocity vs. Average Velocity

In each of the previous examples, we have assumed an average velocity. **Average velocity**, v_{ave}, is defined as the change in distance divided by the corresponding change in time. When we speak of **instantaneous velocity**, we divide the change in distance by a very tiny, almost zero, change in time. For example, Sally's average speed from Chicago to Kansas City was 65 mph. However, she may have spent some of the trip driving 80 mph and some of it at 50 mph. While passing someone, she may have briefly hit a speed of 85 mph. The point here is that there may be quite a difference between average and instantaneous velocities.

Acceleration

We have all experienced acceleration. It may have been when riding in a car, an elevator, a roller coaster, or a jet plane. By definition, **acceleration**, a, is a change in velocity over time. Remembering that velocity is a vector, this means that acceleration can be a change in speed, direction, or both. To determine if an object is experiencing acceleration, we have to consider two possibilities. First, did the numerical value of its velocity change over time? Second, did the direction of the velocity change over time? The acceleration due to a change in speed is calculated by using the formula

$$\text{acceleration} = \frac{\text{change in } v}{\text{change in } t}$$

$$a = \frac{\Delta v}{\Delta t}$$

The triangular symbol, Δ, used in the above equation is the Greek letter delta. To mathematicians and scientists, it is shorthand for "change in." Delta v, then, simply means change in velocity. For example, what would be the average acceleration of a ball that has a speed of 3 m/sec at $t_1 = 10$ and 6 m/s at $t_2 = 20$ sec?

$$a = \frac{v_2 - v_1}{t_2 - t_1} = \frac{6 \text{ m/s} - 3 \text{ m/s}}{20 \text{ sec} - 10 \text{ sec}}$$

$$= \frac{3 \text{ m/s}}{10 \text{ sec}} = 0.3 \text{ m/s}^2$$

In the following example, we will calculate the amount of acceleration and distance covered by a smoothly accelerating wagon at several different time intervals (see Figure 5-4). By pushing with his foot, Johnny is able to accelerate his wagon at the rate of 0.2 m/s every second, or $a = 0.2$ m/s^2. If the wagon begins from a dead stop, how fast will it be going after 1, 2, and 3 seconds? How far will Johnny's wagon have

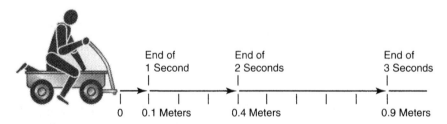

Figure 5–4: Distance traveled by a wagon accelerating at the constant rate of 0.2 m/s^2.

traveled at the end of each second? We will start by calculating the velocity at the end of each time interval using the equatoin $v = at$.

After one second	After two seconds	After three seconds
$v = at$	$v = at$	$v = at$
$v_1 = 0.2 \text{ m/s}^2 \times 1 \text{ s}$	$v_2 = 0.2 \text{ m/s}^2 \times 2 \text{ s}$	$v_3 = 0.2 \text{ m/s}^2 \times 3 \text{ s}$
$v_1 = 0.2 \text{ m/s}$	$v_2 = 0.4 \text{ m/s}$	$v_3 = 0.6 \text{ m/s}$

Note the use of subscripts. When dealing with many measurements of one quantity, it is helpful to differentiate one measurement from another by using subscripts.

Next, we need to calculate how far the wagon has traveled at the end of each second. If the wagon is accelerating uniformly, then its increase in speed occurs continuously and does not occur abruptly in one-second intervals. Put another way, the speed at 1.1 seconds will be slightly greater than it was at 1.0 second, etc. The average velocity for the wagon between each time interval can then be calculated using the following formula where v_1 is the beginning velocity and v_2 is the ending velocity.

$$v_{ave} = \frac{v_1 + v_2}{2}$$

When the acceleration is constant, the distance traveled can be calculated by multiplying the average velocity, v_{ave}, by the time traveled, Δt, at that velocity, $d = v_{ave} \times \Delta t$. After one second, for example, the velocity of the wagon was 0.2 meters per second. Its beginning velocity was zero.

$$v_{ave} = \frac{v_1 + v_2}{2}$$

$$v_{ave} = \frac{0.0 \text{ m/s} + 0.2 \text{ m/s}}{2} = 0.1 \text{ m/s}$$

The average velocity is 0.1 m/s; therefore, the distance traveled would be:

$$d_1 = 0.1 \text{ m/s} \times 1 \text{ s}$$

$$d_1 = 0.1 \text{ m}$$

During the next time interval, the wagon reached a velocity of 0.4 m/s.

$$v_{ave} = \frac{v_1 + v_2}{2}$$

$$v_{ave} = \frac{0.2 \text{ m/s} + 0.4 \text{ m/s}}{2} = 0.3 \text{ m/s}$$

The average velocity is 0.3 m/s; therefore, the distance traveled would be:

$$d_2 = 0.3 \text{ m/s} \times 1 \text{ s}$$

$$d_2 = 0.3 \text{ m}$$

During the last time interval, the wagon reached a velocity of 0.6 m/s.

$$v_{ave} = \frac{v_1 + v_2}{2}$$

$$v_{ave} = \frac{0.4 \text{ m/s} + 0.6 \text{ m/s}}{2} = 0.5 \text{ m/s}$$

The average velocity is 0.5 m/s; therefore, the distance traveled would be:

$$d_3 = 0.5 \text{ m/s} \times 1 \text{ s}$$

$$d_3 = 0.5 \text{ m}$$

When the distances are totaled ($d_1 + d_2 + d_3 = 0.9$ m), we find that the wagon traveled 0.9 meters in the three seconds. It is obvious that this method for calculating the distance traveled by a uniformly accelerating object is cumbersome. Fortunately, there is another way. When an object starts from rest, the average velocity is simply half of the final velocity, v_f. To find the total distance traveled, d, multiply half the final velocity by the time traveled.

$$d = \frac{v_f\, t}{2}$$

When this formula is applied to the previous example, it gives:

$$d = \frac{0.6 \text{ m/s} \times 3 \text{ s}}{2} = 0.9 \text{ m}$$

If, for any reason, we want to go from acceleration and time directly to the calculation of distance, we can. Remember, velocity is acceleration times the time.

$$v_f = at$$

Since we know that the final velocity, v_f, is the same as the acceleration times the time, we can substitute acceleration times time in place of final velocity.

$$\text{distance traveled} = \tfrac{1}{2}(\text{acceleration} \times \text{time}) \times \text{time}$$

$$d = \tfrac{1}{2}at^2$$

Now, let's repeat the calculations for the wagon, using only the information provided in the problem statement.

At the end of second one: $\quad d_1 = \tfrac{1}{2}at^2 = \tfrac{1}{2}(0.2 \text{m/s}^2)(1\text{s})^2 = 0.1\text{m}$

At the end of second two: $\quad d_2 = \tfrac{1}{2}at^2 = \tfrac{1}{2}(0.2 \text{m/s}^2)(2\text{s})^2 = 0.4\text{m}$

At the end of second three: $\quad d_3 = \tfrac{1}{2}at^2 = \tfrac{1}{2}(0.2 \text{m/s}^2)(3\text{s})^2 = 0.9\text{m}$

Continue making the above calculations for the wagon at 4 seconds, 5 seconds, 6 seconds, and 7 seconds. Use this information to complete Table 5-1.

Time	Average Velocity per 1 second interval	Total Distance
1 second	0.1 m/s	0.1 m
2 seconds	0.3 m/s	0.4 m
3 seconds	0.5 m/s	0.9 m
4 seconds		
5 seconds		
6 seconds		
7 seconds		

Table 5–1: Velocity and distance traveled by the constantly accelerating wagon.

Examine the relationship between the times and average velocities and the times and total distances traveled. If you were to plot the average velocity vs. time, what would be the shape of the curve? If you were to plot the total distance vs. time, what shape would the curve be? Why do the curves have different shapes?

When objects fall toward the Earth, they accelerate at a known constant rate. The same calculation may also be used, therefore, to determine a falling object's velocity, distance, or time. Knowledge of only the acceleration of gravity and one variable are needed to calculate the other two. At amusement parks, some rides involve free-fall. After completing the following section, you will be able to calculate how far you have fallen or the velocity you are traveling, from just the time of the fall.

As a final note, mathematics is often the language used in physics. Sometimes the calculations result in numbers that are very big, and at other times, very small. Scientists calculate acceleration the same way for objects they can see as for subatomic particles they cannot see. When making these calculations, one of the most common mistakes is neglecting to make the units of measurement compatible. Sometimes conversion factors must be used to make the units match. If you need to review unit conversion, refer to Chapter 2.

Check Your Understanding

1. What is the difference between distance and displacement?

2. What is the difference between speed and velocity?

3. If you got on the interstate at mile marker 235 and got off at mile marker 283, how far did you travel? From this information, can you determine your displacement? Why?

4. Louise flies from Kansas City, MO to Fargo, ND in 2 hours. What additional information do you need to calculate her speed? Her velocity?

5. A little red sports car will go from zero to a hundred mph in 15 seconds. What is its acceleration in miles per second?

5-2 Acceleration of Gravity

Objectives

Upon completion of this section, you will be able to:

■ Explain the effects of gravity.

■ Identify two parameters of gravity that are related and explain how each affects its magnitude.

■ Describe the motion of a freely falling body.

In space there is no dust, air, or sticky stuff to slow things down. Therefore, once an object is in motion it will continue traveling in a straight line forever, unless it is being acted upon by a force. The fact that the Earth is moving in an orbit around the Sun is due to the Sun's gravitational force.

In 1687, Isaac Newton published *The Mathematical Principles of Natural Philosophy*, or *The Principia* as it is commonly known, in which he identified gravity as a special kind of force. While scientists have still not determined its origin, **gravity** is the attraction or pull that two masses exert on each other. Although both objects are affected equally by the gravitational force, it is always the smaller object that appears to accelerate or fall toward the more massive object.

Based on its mass and diameter, the gravity for each planet in our solar system has been calculated. At sea level on the surface of the Earth, the **gravitational acceleration**, g, is 9.81 m/s^2. This means that an object falling toward Earth will gain speed by 9.81 m/s each second it falls. In English units, it is 32.2 ft/s or 22.0 mph for each second it falls. These values can be used with reasonable accuracy for freely falling objects under normal atmospheric conditions and no more than a few thousand feet above the Earth's surface.

Consider a heavy object with no air resistance, falling freely for 10 seconds. Table 5-2 shows how fast the object will be going at the end of each second. Since the acceleration of gravity is a constant (a = 9.81 m/s^2), the equation $v = a\,t$ can be used.

Time of Fall, in seconds	Velocity, m/s	Velocity, ft/s	Velocity, mph
1	9.8	32.2	22.0
2	19.6	64.4	44.0
3	29.4	96.6	66.0
4	39.2	128.8	88.0
5	49.1	161.0	110.0
6	58.9	193.2	132.0
7	68.7	225.4	154.0
8	78.5	257.6	176.0
9	88.3	289.8	198.0
10	98.1	322.2	220.0

Table 5–2: Velocity of falling object per second.

As shown in Figure 5-5, when two objects with different masses are placed in a vacuum, they fall at the same rate.

On Earth, however, objects do experience air resistance and can fall only so fast. This speed is called their **terminal velocity** and is defined as the speed that a free falling object cannot exceed because the frictional force (drag) of the air equals the gravitational force at terminal velocity. For example, a feather and an apple have different terminal velocities. A feather appears to float to the ground because it is lightweight with a large surface area and reaches its terminal velocity after falling a short distance. An apple, on the other hand, will have to fall several meters to reach its terminal velocity.

Since the acceleration of gravity is a constant, it is possible to calculate the distance something has fallen and its ve-

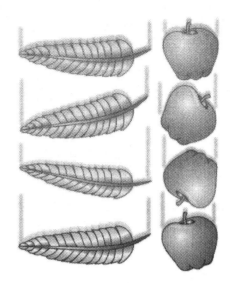

Figure 5–5: Without air resistance, feathers and apples fall at the same rate.

locity at any moment in time, by using a modified version of the equation, $d = \frac{1}{2}at^2$. When making calculations for free-falling objects above the Earth's surface, it is common practice to substitute gravitational acceleration, g, for the acceleration term. This results in:

$$d = \frac{1}{2}gt^2$$

If a visitor dropped a rock into the Grand Canyon and found that it took 17.6 seconds for it to reach the river below, the calculation for the depth of the canyon at that point, assuming no air resistance, is:

$$d = \frac{1}{2}gt^2$$

$$d = \frac{1}{2}(9.81 \text{ m/s}^2)(17.6 \text{ s})^2$$

$$d = 1,519 \text{ m}$$

Law of Universal Gravitation

Newton's **law of universal gravitation** states that all objects that have mass attract each other with a force that is directly proportional to the product of their mass and inversely proportional to the square of the distance between them. In the formula:

$$F_g = \frac{Gm_1m_2}{r^2}$$

F_g is the gravitational force on each object in newtons; m_1 and m_2 are their respective masses in kg; r is the distance between them in meters; and G is the **gravitational constant**, whose value is dependent on the units used. In the SI System, the value of G is 6.67×10^{-11} N · m2/kg^2 or, substituting the equivalent SI units for N results in 6.67 $\times 10^{-11}$ m^3/kg · s^2

From the law of universal gravitation, the direct relationship between mass and gravitational force can be seen. The greater the mass, the greater the gravitational force, which means that if the mass of Earth were to double, then the acceleration of gravity would be twice as great.

> ### Special Topics Box 5-1 ■ How did Newton figure out the law of gravitational acceleration?
>
> According to Carl Sagan in *Cosmos*, Newton was one of the greatest theoretical physicists in history. His reply to those who would question how he was able to make such amazing discoveries was simple. "By thinking on them." This is, in fact, what theoretical physicists most often do – think. Then they use data and mathematics to verify the things they think.
>
> During a break from college, Newton spent a lot of time thinking. While lounging outdoors he watched an apple fall from a tree with the moon in the distance. At that moment, he realized that a falling apple and the moon in its orbit around the Earth are operating under the same universal force. In short, the equation for the law of universal gravitation is not limited to Earth-bound phenomena. Anywhere in the universe, every object attracts every other.
>
> Newton credited Kepler with providing clues to the relationship between the gravitational force and the distance between the objects. Kepler had stated that the planets revolving around the Sun moved faster when closer to the Sun and slower when farther away. This is Kepler's third law of planetary motion. The data both sparked Newton's idea, and ultimately supported his conclusion that the force of gravity is mathematically related to the inverse of the distance squared. This is the essence of science. Not only do scientists attempt to predict systems mathematically, but sometimes it takes inspiration, creativity, and a guess.

The effect of distance on gravitational acceleration is the opposite because it fades by the square of the distance. As the distance between an object and the Earth increases, the acceleration of gravity decreases. Furthermore, since the term is squared, if you double the distance, the acceleration becomes only one-fourth as much.

The value of the acceleration of gravity and weight also varies slightly depending on the depth from the surface of the Earth to its core. Gravitational acceleration, therefore, is a little stronger at the slightly flattened poles (9.8321863685 m/s^2) and weaker at the bulging equator (9.7803267715 m/s^2). This difference may seem unimportant unless you are buying gold in Singapore that was weighed in Alaska.

Because of the mass and size (radius) of our moon, it has about one-sixth the gravitational acceleration (1.6 m/s^2) of the Earth. This fact was clearly demonstrated when the astronauts walked on the moon. It looked like they were bouncing in slow motion, and when they jumped they seemed to stay aloft too long.

Check Your Understanding

1. Explain the effects of gravity in your own words. What is gravitational acceleration?

2. Consider two objects that are near to each other, but far from anything else in space. How does an increase in mass, of either, affect the strength of the gravitational force between them?

3. What is the terminal velocity of an object falling toward the Earth, and what factors have an effect on it?

4. Describe the motion of a golf ball that is dropped from the top of a house.

5. What is the law of universal gravitation and what does it describe?

5-3   Newton's Laws of Motion

Objectives

Upon completion of this section, you will be able to:

■   Describe a force.

■   State Newton's first, second, and third laws.

■   Describe the effects of the three laws of motion.

■   Calculate force, mass, or acceleration given any two of these parameters.

■   Describe a force pair.

Newton's First Law (Inertia)

A **force** is defined as a push or pull. It is a vector quantity, so it has both magnitude and direction. To keep something in motion on Earth we have to continually apply force, or it will eventually come to a stop. That's because of friction – it's everywhere.

Newton recognized that friction confuses the issue of motion. If friction didn't interfere, he reasoned, then moving objects would keep moving until they ran into something. Even then, the object would just bounce off the obstacle and head in another direction. Try to imagine what a world without friction would be like. In that world, striking a cue ball on a billiard table (with no pockets), would result in the balls rolling around, bumping into each other and off the rails indefinitely. This is not reality on planet Earth. However, it is exactly what you need to imagine for **Newton's first law** of motion to make sense. This law states that an object at rest will stay at rest, and an object in motion will stay in motion with a constant velocity, until acted on by a force. Sometimes called the **law of inertia**, Newton's first law of motion is also the definition of **inertia**.

When a problem in physics states there is no net force in any direction acting on an object, you are being asked to visualize an ideal situation. The condition of no force at all is a special case; the result of several forces that combine in such a way that they cancel each other out. For example, imagine driving on a sheet of ice. The light ahead suddenly turns red. Can you stop? Generally, stopping isn't a problem because friction is there to help. When working with objects in motion, friction is often – but not always – the unseen force that causes motion to end. When there is no friction, the idea of mass having inertia makes more sense. We will talk more about friction after introducing the other laws of motion.

Newton's Second Law (Force, Mass, and Acceleration)

Experience has taught us that when something has more mass, it is more difficult to change its state of motion. The attempt to quantify this simple observation, however, was not easy. Newton knew that objects could be moved if they were subjected to enough force. Mass is a measure of inertia, which is a measure of the body's resistance to acceleration.

Experimentally, it can be shown that there is a simple relationship among force, mass, and acceleration. For example, an object with twice the mass takes twice the force to give it the same amount of acceleration. Likewise, if the mass is constant and we double the force, the object will accelerate twice as fast.

Newton's second law of motion states that when an unbalanced force is applied to an object, the object accelerates in the direction of the resultant force. The acceleration is proportional to the magnitude of the resultant force and inversely proportional to the mass being moved. This can be expressed mathematically by:

$$\text{acceleration} = \frac{\text{force}}{\text{mass}}$$

$$\text{force} = \text{mass} \times \text{acceleration}$$

$$F = m\,a$$

In the SI System, when mass, m, is expressed in kg and acceleration, a, in m/s^2, the unit for force, F, is the newton. A newton, therefore, is the amount of force required to accelerate a 1 kg mass by 1 m/s^2. In the English system, force is expressed in pounds-force (lb$_f$) mass in slugs, and acceleration in ft/sec^2. A pound-force is defined as the force required to accelerate a 1 slug mass 1 ft/sec^2.

Earlier it was noted that when an object is allowed to fall freely near the surface of the Earth, the accelerating force is gravity. We then made a substitution in the equation, $d = \frac{1}{2}at^2$, that converted it to $d = \frac{1}{2}gt^2$, where the constant acceleration term, a, was replaced by the constant acceleration of gravity, g.

In the same way, Newton's second law can be transformed from $F = ma$ to $F = mg$. The equation ($F = mg$) indicates that the force acting on an object close to the surface of the Earth depends on only two parameters – its mass and the acceleration of gravity. This force acting on the object is commonly referred to as its weight, W, and is given by the formula, $W = mg$.

Weight and mass are not the same. Mass is the constant property of any piece of matter, regardless of its location. To have weight, the matter must be under the influence of a gravitational field. For this reason, if an object with constant mass, m, is moved to the moon, its weight becomes ($W = mg/6$) or one-sixth its weight on Earth. In space, where gravitational acceleration is nearly zero ($W = m \times 0$), it is easy to see why objects become weightless.

Newton's Third Law (Action and Reaction)

Newton's third law of motion is summarized in the frequently heard statement: "For every action there is a reaction." Let's look at what this law really means. Forces are the pushes or pulls that cause objects to accelerate. Because objects have mass, however, they offer resistance to how much they accelerate. As stated in Newton's second law, the more mass something has, the more it resists acceleration.

Consider the forces at work on a baseball. A pitcher's hand must push on the ball to move it toward the batter, but the ball also pushes back on the pitcher's hand. As the ball flies toward the batter, the gravitational the force of gravity is pulling the ball toward the Earth, but the gravitation field of the ball is also pulling the Earth toward the ball. Because of the vast difference in their masses, the ball moves toward the Earth more than the Earth moves toward the ball.

When the fast-moving ball crashes into the more massive bat, they both experience the same force, but in opposite directions. The ball, being lighter, is the object that abruptly changes direction while the bat finishes its swing, but at a slower velocity.

When walking, feet exert a force (backward) against the ground and the ground simultaneously exerting a force (forward) against the feet. In the same way, bicycle tires – again simultaneously – exert a force against the street and the street exerts a force against the tires. These are examples of **action-reaction pairs**. Neither object is the "action" agent nor the "reaction" responder. The forces are simultaneous, equal, opposite in direction, and are not acting on the same body. The Earth propels us forward because we are applying forces backward with feet or tires. The forward force by the Earth is only possible because of the presence of friction between the surface of the Earth and the feet, tires, or other bodies involved. Forces never act alone, but must always come in pairs. Interacting bodies exert forces each on the other – never on the same body. Since they act on different bodies they can never cancel or balance each other.

Propulsion

If forces come in equal and opposite pairs, how can anything ever get moving? Acceleration given to an object depends on the sum of all of the individual forces acting on that body. We commonly speak of the **net force** to represent this concept. Each material body possesses mass, a measure of its resistance to acceleration by force of any kind. The greater the mass of the body, the smaller its acceleration will be, whatever net forces apply.

If you have ever shot a gun, you know about kickback or recoil. As the bullet is shot out of one end, the other end of the gun moves in the opposite direction resulting in recoil. The action-reaction pair is the bullet's force on the gun and the gun's force on the bullet. This is an obvious case of equal and opposing forces acting on two different objects. Due to its smaller mass, the bullet exits the gun much faster than the gun recoils.

So what about jet propulsion and rocket engines? When a balloon full of air is released, it flies wildly around the room. The forces acting in this situation are the balloon's elastic walls pushing the air out, while the exiting air pushes back against the balloon propelling it forward.

In a rocket engine, the volume and force of the gases being released by the burning fuel push against the rocket and the rocket pushes on the hot gases. Due to the difference in mass, the acceleration of the rocket is far less than the acceleration of the hot gases, but it can be controlled by how much fuel is being burned.

Balanced Forces

The idea of balanced forces acting on a single object is often misrepresented as an action-reaction pair to illustrate Newton's third law. For example, consider a book on a table. The book is acted upon downward by the force of gravity, and upward by the force of the table. These forces of gravity and the table are equal and opposite; therefore, the book does not move. These forces are not an action-reaction pair, but each is one-half of an action-reaction pair acting on the same body – the book. Even if two forces are equal and opposite in direction, as the weight of the book and the supporting force of the table are, they do not constitute a third-law pair because they act on the same body – the book. This illustrates Newton's first law, not the third.

Check Your Understanding

1. What two conditions must be known to describe a force?

2. Sudden acceleration in a car gives one the feeling of being thrown to the side, pushed back, or pulled forward. Using Newton's first and second laws, explain in your own words what is happening.

3. Restate, in your own words, Newton's first, second, and third laws.

4. Which of Newton's laws explains why a gun recoils? Explain the source of the recoil.

> What if everyone in the world (6 billion people) faced east and took a step simultaneously? Would the total force exerted by the people be enough to slow the Earth's rotation?

5-4   Friction

Objectives

Upon completion of this section, you will be able to:

- Define friction.

- Explain the difference between static friction and kinetic friction.

- Use the coefficient of friction to calculate the force of friction.

A force known as **friction** provides resistance whenever the surface of one solid body slides over another. Looking again at Newton's second law, $F = ma$, this mathematical statement implies that acceleration has no limit. Ideally, even the tiniest force is enough to make an object accelerate. In reality, we know this does not happen.

Imagine pushing a heavy wooden box on a freshly waxed floor. You must push hard at first, to get the box to move. Once it is moving, however, you don't have to continue pushing as hard to keep it moving. You might even be tempted to give the box a big shove to see if it will slide to where you want it to go. If that destination is very far, friction will probably slow the box to a stop before it arrives. In a perfectly frictionless world, the box would continue to slide forever.

Two kinds of friction are at work in this example: static and kinetic friction. **Static friction** is the force of resistance between objects that are not moving relative to one another. In the case of the box, it is the force that must be overcome to get it to move. **Kinetic friction** is the force of resistance between objects that are moving relative to one another. Once the box is in motion, it is the force that can be felt pushing against it (see Figure 5-6). Static friction is usually greater than kinetic friction. Both, however, provide resistance to an object's motion.

Eventually, as you push less and less on the box, you reach a point where it stops accelerating. At this point, you have reached an equilibrium in which the pushing force is balanced by the kinetic force of friction. (Equal and opposite forces acting on the same body.) Mathematically, this situation is expressed this way:

$$F_{push} = F_{friction}$$

Stated another way, when there is no net force, there is no further acceleration. Motion may continue, but it will have a constant velocity.

$$\text{no excess force} = \text{no acceleration}$$

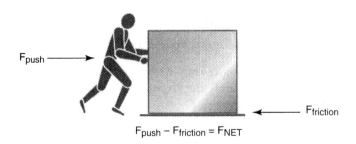

$$F_{push} - F_{friction} = F_{NET}$$

Figure 5–6: The net force acting on the box is the difference between the push force and the static friction force.

Now imagine pushing the box over a rocky driveway. On such a surface, the friction might be so great that you can't move the box at all. The physical characteristics of the surfaces of the objects affect how great the forces of friction will be. Rougher surfaces have a greater ability to resist motion. The measure of surface roughness is known as the **coefficient of friction**. It is represented by the Greek lower case letter mu, μ and typically falls between 0 and 1. The rougher the surface, the closer μ is to one.

On flat (level) surfaces, the force of friction is directly related to an object's weight as well as the roughness of the surfaces it contacts and exists only when a tangential force is applied. Experimentally, it can be shown that the force of friction, whether static or kinetic, is equal to the product of the coefficient of friction and the weight of the object.

$$\text{friction} = \text{coefficient of friction} \times \text{weight}$$
$$F_{\text{friction}} = \mu\ W$$
$$F_{\text{f}} = \mu\ W$$

The static frictional force balances the applied force until the object begins to move ($F_{\text{f}} \leq \mu W$). The concept of friction is commonly used in accident reconstruction. Given the weight of the car, its coefficient of friction, and the skid marks, a vehicle's speed at the time of impact can be calculated. The ability to determine this information can be of great economic value to the insurance companies who must determine the responsible party.

Check Your Understanding

1. What is friction?

2. What is the difference between static friction and kinetic friction?

3. If an object is being pushed across a surface, at what point will acceleration cease?

4. What are the two factors that must be used to determine either static or kinetic friction?

5-5   Momentum

Objectives

Upon completion of this section, you will be able to:

1. Define momentum.

2. Describe the effect of an impulse on momentum and velocity.

3. Calculate the momentum of a system.

4. Use the conservation of momentum to analyze collisions.

Newton's laws have helped us to understand much about the motion of objects. We will now bring the concepts of these laws together to discuss momentum and other aspects of motion. Objects in motion are said to have **momentum**, p. Sports announcers sometimes use the term to describe a situation in which a team seems to be unstopable. Momentum is also used to describe the difficulty of stopping a heavy vehicle.

The product of the mass of an object and its velocity is equal to its momentum. The change in the momentum of an object equals its mass times its change in velocity ($\Delta p = m\Delta v$). A unit used to express momentum is kilogram-meters per second, also known as the Newton-second.

The momentum of an object can be changed only by a force acting on it for a time interval. The product of the force, F, and the time interval, Δt, is called an **impulse** (impulse = $F\Delta t$). It can, therefore, be said that an impulse is required to change the momentum of an object. This relationship can be derived mathematically, starting with Newton's second law, $F = ma$, and substituting $\Delta v/\Delta t$ for the acceleration, a.

$$F = ma$$

$$F = m\frac{\Delta v}{\Delta t}$$

$$F\Delta t = m\Delta v$$

This equation states that the force times the time interval over which it acts equals the mass times the resultant change in velocity. As shown above, we see that the force times the time interval ($F\Delta t$) is called the impulse and the mass times the change in velocity ($m\Delta v$) is the change in momentum. Therefore, the equation ($F\Delta t = m\Delta v$) shows that

$$\text{impulse} = \text{change in momentum}$$

If an applied impulse changes the momentum of an object, then they must have the same units – kilogram-meters per second, $kg \cdot m/s$.

An impulse can come from any object that has the ability to apply force over a time interval. If the force is small, but the time is long, a significant impulse may result. For example, if a car's brakes are applied lightly for a long time period the car can slowly and safely come to a stop. On the other hand, if the brakes are slammed on for a very short time period, the car can be brought to a screeching stop. The car was subjected to the same impulse and experienced the same changes in both their velocity and momentum.

To illustrate further the importance of both the force magnitude and time interval, consider the following example. When a person jumps from a building, the fall isn't what injures – it's the sudden stop! If the stopping time is short, the force must be great. By use of a stuntman's air bag or a firefighter's net, the impact time can be lengthened. Regardless of how the person is stopped, the momentum change is the same – dependent only on their mass and velocity change. However, by increasing the impact time, the force must be smaller and the person is less likely to suffer injury. Other practical applications of this principle include automobile bumpers, foam packing peanuts, and baseball gloves.

It can also be shown that momentum is conserved during many real-world collisions. The game of billiards, for example, depends on this physical principle. The **law of conservation of momentum** states that the momentum of a body, or system of bodies, does not change except when an external force is applied. The momentum of a system is simply the sum of all the individual momentum values, including direction, before and after a collision. This can be mathematically stated by the following equation:

$$\sum P_{\text{before}} = \sum P_{\text{after}}$$

A moment is commonly defined as an infinitesimally brief period of time or an instant. Scientists theoretically stop the motion of all the objects in a system and calculate the momentum of each object at that instant. Then they repeat the analysis after the objects have collided. Therefore, we can use the equality of momentum before and after a collision to calculate either the mass or the velocity of all the objects involved. Our confidence in the law of conservation of momentum, when applied to the deterioration of atoms, for example, explains why physicists believe in neutrinos, which they cannot see. (See the Beta Radiation section of Chapter 11.)

To develop a better understanding of the conservation of momentum, consider the following example as shown in Figure 5-7. The couple – we'll call them John and Tina – is skating on a frictionless ice rink. They stop for an instant, put their hands together, and push off moving away from each other in a perfect line. If John's mass is 80 kg (about 175 lbs) and Tina's mass is 32 kg (about 70 lbs), and he travels 1.5 m/s (about 3.5 mph) after they push off, how fast will she be going?

The fact that they stop for an instant is significant. They start out with their total momentum (sometimes referred to as the momentum of the system) equal to zero ($p = 0$). If we know that the momentum at the beginning equals zero, then the momentum at the end must also be zero and we can solve for Tina's velocity. How can a system with two moving people have a momentum equal to zero? If one moves in exactly the opposite direction as the other with the same amount of momentum, then the sum of the total momentum equals zero. This is due to the vector nature of both velocity and momentum.

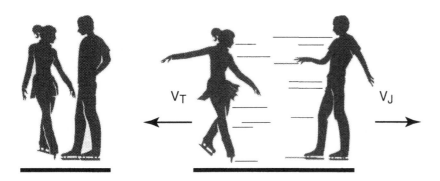

Figure 5–7: Initially, the skaters are at rest and, therefore, have no momentum. After pushing away from each other, one goes left and one goes right, but the sum of their momenta must still equal zero.

Special Topics Box 5–2 ■ How is momentum involved in a game of pool?

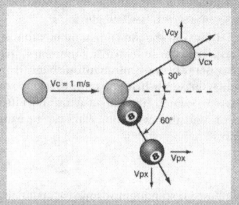

Consider this shot in which the cue ball collides with another ball of equal mass. The initial velocity of the cue ball is 1 m/s. The pool ball travels away from the collision at an angle of 60° below the cue ball's initial direction. The cue ball leaves the collision traveling at an angle of 30° above its original direction. Assuming the conservation of momentum, how fast must each ball be traveling after the collision?

First we will consider the x-direction component of motion. Let the masses of the balls be represented by m, and their velocities be represented by v_c, for the cue ball, and v_p for the pool ball.

$$p_{before} = m \, v$$
$$p_{before} = m \, (1 \text{ m/s})$$
$$p_{after} = m \, v_{px} + m \, v_{cx}$$
$$p_{after} = m \, (v_p \cos 60°) + m \, (v_c \cos 30°)$$
$$\sum P_{before} = \sum P_{after}$$
$$m \, (1 \text{ m/s}) = m \, (v_f \, 0.5000) + m \, (v_c \, 0.8660)$$

Dividing the equation through by m, yields:

$$1 \text{ m/s} = 0.5000 \, v_p + 0.8660 \, v_c$$

We now use the information from the y-direction to finish solving the problem. If we use the y-direction information to solve for either ball's velocity, we can substitute this into the x-direction equation and solve for the other velocity.

$$p_{before} = 0$$
$$p_{after} = m \, v_{cy} - m \, v_{py}$$
$$p_{after} = m \, (v_c \sin 30°) - m \, (v_p \sin 60°)$$
$$\sum P_{before} = \sum P_{after}$$
$$0 = m \, (v_c \, 0.5000) - m \, (v_p \, 0.8660)$$
$$m \, (0.5000 \, v_c) = m \, (0.8660 \, v_p)$$
$$v_c = 1.7320 \, v_p$$

Now that we have the velocity of the cue ball, v_c, in terms of the velocity of pool ball, v_p, we can substitute this value into the x-direction equation.

$$1 \text{ m/s} = 0.5000 \, v_p + 0.8660 \, v_c$$
$$1 \text{ m/s} = 0.5000 \, v_p + 0.8660 \, (1.7320 \, v_p)$$
$$1 \text{ m/s} = 0.5000 \, v_p + 1.4999 \, v_p$$
$$1 \text{ m/s} = 2.0 \, v_p$$
$$v_p = 0.50 \text{ m/s}$$
$$v_c = 0.87 \text{ m/s}$$

This example illustrates the vector nature of momentum, that is the x- and y-components of momentum are both conserved.

$$\sum P_{\text{before}} = \sum P_{\text{after}}$$

$$\sum P_{\text{b}} = \sum P_{\text{a}}$$

$$P_{\text{b}} = 0$$

$$\therefore P_{\text{a}} = 0$$

The total momentum after the skaters push off must be zero, since it was zero the instant before the skaters pushed off.

We know John's mass, Tina's mass, and John's velocity. The total momentum, both before and after the push-off, is zero. We must also decide which skater will be assigned the positive and which will be assigned the negative direction. Movement to the right is usually considered the positive direction, so John has been put on the right and Tina on the left. Therefore,

$$p_{\text{a}} = m_{\text{J}}v_{\text{J}} + m_{\text{T}}v_{\text{T}}$$

$$0 = (32 \text{ kg})v_{\text{T}} + (80 \text{ kg})(1.5 \text{ m/s})$$

$$32 \text{ kg } v_{\text{T}} = -120 \text{ kg} \cdot \text{m/s}$$

$$v_{\text{T}} = \frac{-120 \text{ kg} \cdot \text{m/s}}{32 \text{ kg}}$$

$$v_{\text{T}} = -3.75 \text{ m/s}$$

The negative sign simply means Tina is traveling in the opposite direction (left) from John. Her speed then is 3.75 m/s or about 8.5 mph.

A few aspects of this problem deserve additional comment. Note that Tina's speed is 2.5 times faster than John's speed and that John weighs 2.5 times more than Tina. In all cases where two objects start out at rest and go in exactly opposite directions, the resulting speeds are exactly (inversely) proportional to the masses of the objects involved. This relationship can be shown mathematically by the following equations:

$$m_{\text{J}}v_{\text{J}} = m_{\text{T}}v_{\text{T}}$$

$$\frac{m_{\text{J}}}{m_{\text{T}}} = \frac{v_{\text{T}}}{v_{\text{J}}}$$

Let's return to our two skaters, who now perform a different kind of maneuver. John skates with a velocity of 4.2 m/s toward Tina, who is stopped. When he reaches her, he scoops her up and continues in the same direction. Applying conservation of momentum, how fast will they be going after the pick-up?

We approach this problem in much the same way as before. First, we need to calculate the momentum before the pick-up. Initially, John had all the momentum in the system because Tina was stationary. After the pick-up, they move as a single unit with the same velocity. The momentum before the collision is given by:

$$p_{\text{b}} = m_{\text{J}}v_{\text{J}}$$

$$p_{\text{b}} = 80 \text{ kg} \times 4.2 \text{ m/s}$$

$$p_{\text{b}} = 336 \text{ kg} \cdot \text{m/s}$$

Since momentum is conserved, it must be the same after the pick-up.

$$p_{\text{a}} = 336 \text{ kg} \cdot \text{m/s}$$

$$p_{\text{a}} = (m_{\text{J}} + m_{\text{T}})v$$

$$p_a = (80 \text{ kg} + 32 \text{ kg})v$$

$$p_a = (112 \text{ kg})v$$

Substituting 336 kg · m/s for p_a, allows you to solve for v.

$$336 \text{ kg} \cdot \text{m/s} = (112 \text{ kg})v$$

$$v = \frac{336 \text{ kg} \cdot \text{m/s}}{112 \text{ kg}}$$

$$v = 3 \text{ m/s}$$

Notice that after John picks up Tina, their combined masses are traveling slower than John's original velocity. Does that seem reasonable? It is always a good problem solving technique to ask yourself if the answer seems reasonable.

Check Your Understanding

1. How is momentum defined in physics?

2. How does the total amount of momentum before and after a collision compare? Explain.

3. Explain, in terms of impulse and momentum, the reason for sitting as far away from a car's air bag as possible.

4. If one skater is moving and a second is stopped, what happens to the moving skater's speed after picking up the stopped skater? What determines their combined speed?

Angular Motion

Objectives

Upon completion of this section, you will be able to:

■ Define angular motion.

■ Calculate the velocity of a particle in angular motion.

■ Calculate the acceleration of a particle due to angular motion.

■ Describe the difference between centrifugal and centripetal acceleration.

For the purposes of this discussion, **angular motion** will be limited to just **circular motion**. There are many common devices that use circular motion to perform their function, including gears, wheels, CD players, pumps, and merry-go-rounds.

Have you ever noticed that you need to hold on tighter as a ride spins faster? The force required to keep you from flying off the ride is obviously related to the velocity of the circular motion. But do you think that this force would be greater or less if you selected a seat closer to the center (axis) of the ride?

Consider the example depicted in Figure 5-8. Mary and Larry each weigh 20 kg. They are both on a merry-go-round that is turning at the rate of one revolution every 6 seconds. The pony Mary is riding is 3.5 m from the center (axis) and the pony Larry is riding is 3 m from the center of the ride. Which child will require the greater force to stay on the ride?

First, we must find each child's velocity. To find velocity we must know both how far they travel and how long it takes. To calculate how far Mary and Larry travel in one revolution, we use the formula for the circumference of a circle, $d = 2\pi r$.

$$d = vt$$

$$d = 2\pi r$$

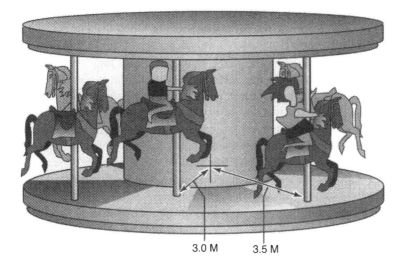

3.0 M   3.5 M

Figure 5–8: Children riding on a merry-go-round at different distances from the axis of rotation experience different amounts of acceleration.

$$vt = 2\pi r$$

$$v = \frac{2\pi r}{t}$$

For Larry	For Mary
$v = \dfrac{2\pi(3\text{ m})}{6\text{ s}} =$	$v = \dfrac{2\pi(3.5\text{ m})}{6\text{ s}} =$
$v \cong 3.14\text{ m/s}$	$v \cong 3.66\text{ m/s}$

One way to calculate force involves using Newton's second law, $F = ma$. For circular motion, acceleration, a, is due to rotational velocity, which can be found by using the equation:

$$a = \frac{v^2}{r}$$

Substituting this into $F = m\,a$ results in the equation:

$$F = \frac{m\,v^2}{r}$$

For Larry	For Mary
$F = \dfrac{20\text{ kg }(3.14\text{ m/s})^2}{3\text{ m}}$	$F = \dfrac{20\text{ kg }(3.66\text{ m/s})^2}{3.5\text{ m}}$
$F \cong 65.7\text{ N}$	$F \cong 76.5\text{ N}$

It can be seen from these calculations that Mary requires a greater force, because she is a greater distance from the center of the ride and must, therefore, travel farther to complete one revolution. Converted into familiar units, Larry requires a force of about 15 pounds while Mary requires a force of about 17.5 pounds.

To avoid being thrown off a circular ride like a merry-go-round requires that the direction of the rider's velocity be constantly deflected inward by a **centripetal force**. To the rider, it feels like there is a **centrifugal force** pulling outward while they are going around. This is considered a **fictitious force**, since there is really no outward force – there is only inertia.

For a rider to keep up with their accelerating surroundings, they must pull themself inward, providing inward acceleration to their body that equals the acceleration of the merry-go-round. From the rider's perspective, this need to pull themself inward results from the fact that it feels like a force is pulling them outward.

This circular acceleration is similar to the linear acceleration, which results from slamming on a car's brakes. The rider feels like a force is pushing them forward off the seat. In reality, it is inertia that causes them to continue moving forward.

So, what happens if the rider on a merry-go-round can't hold on? Which way will they go if they fly off the ride? As shown in Figure 5-9 (left), they will continue traveling in the direction of their velocity vector at the moment they leave the ride.

Perhaps we can now understand how a centrifuge or the spin cycle of a washing machine removes water. As shown in Figure 5-9 (right), water has only adhesion to provide the centripetal force necessary to offset the outward centrifugal force! The basket applies centripetal force to the clothes to hold them in, but lets the water lose its grip and fly out through the holes in the side of the basket.

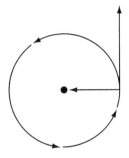
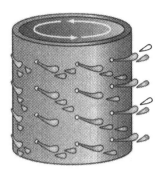

Figure 5-9: A person's grip provides the inner centripetal force (left) that must be applied to stay with the merry-go-round. If they lose that grip, the person will be thrown off the ride in the direction shown. The clothes are held in a circular path (right) by the force of the basket pressing upon them. The water is not and is thrown out of the holes in the basket.

Check Your Understanding

1. What is angular motion?

2. Define centrifugal and centripetal acceleration and give an example.

3. What is a fictitious force?

5-7   Conservation of Angular Momentum

Objectives

Upon completion of this section, you will be able to:

■ Describe acceleration due to a change indirection.

■ Define moment of inertia and angular momentum.

■ Describe how a change in the radius of an object affects its moment of inertia.

■ Use conservation of angular momentum to calculate an object's moment of inertia or angular velocity.

When an object is moving in a circular path with constant speed, the direction of its velocity is continually changing. The change in direction produces acceleration, but no change in the speed. Due to the definition of acceleration, any change in velocity constitutes acceleration. When measuring circular motion in the laboratory, it is found that the acceleration due to a change in direction is directly related to the velocity squared and inversely related to the radius of the turn.

$$\text{circular acceleration} = \frac{\text{velocity squared}}{\text{radius of the turn}}$$

$$a = \frac{v^2}{r}$$

For example, a car whose speed is 88 ft/s (60 miles/hour) rounds a curve whose radius of curvature is 484 ft. What is its acceleration?

$$a = \frac{v^2}{r} = \frac{(88 \text{ ft/s})^2}{484 \text{ ft}} = 16 \text{ ft/sec}^2$$

Most of the time we recognize acceleration while riding in a car by the feeling of being pushed back or pulled forward. When corners are turned quickly and acceleration or deceleration is sudden, then the inertia of items in your car may overcome friction and slide around.

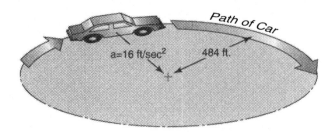

Figure 5-10: A car traveling in a curved path is accelerating. The arrow shows that the direction of the acceleration is always directed toward the center of the turn.

In the previous discussion of the rider on a merry-go-round, the distance of the rider from the center of rotation served as the radius of the mass, and the mass was treated as a single point in the plane of rotation. In the world of no friction, the merry-go-round would spin forever once it was set into motion. In reality, of course, we need a force to keep it turning by overcoming friction. We will simplify our discussion by assuming anything in motion remains in motion until we stop it.

Momentum in a straight line, known as **linear momentum**, is related to an object's inertia. Once something is in motion, it will not stop until something stops it. Inertia, as described by Newton's first law, is the tendency of an object to keep moving in a straight line at the same velocity unless acted upon by an outside force. Remember, mass is the measure of inertia.

Spinning objects experience the same kind of phenomenon. They continue to spin at the same speed and direction until changed by a torque. **Torque** is a force that produces or tends to produce rotation. Therefore, torque is to angular motion what force is to linear motion. The amount of torque is the product of the force, F, and its distance from the center of rotation, known as the **moment arm** or **lever arm**. Mathematically stated:

$$\text{torque} = \text{force} \times \text{lever arm}$$
$$\tau = Fr$$

Let's consider torque as it applies to a mother and child at play on a seesaw. As shown in Figure 5-11, the seesaw consists of a long board that rotates about a central pivot point. From experience, we know that if mother and child sit at the opposite ends of the board the mother's side will be on the ground and the child's will be in the air. However, if mother starts shortening her lever arm by moving toward the pivot point, a position can be found where both mother and child are in balance. With their torques in balance, mother and child can effortlessly bob up and down for as long as they want.

At the balance point, the torque of each person must be equal and opposite in direction. The torque produced can be found by multiplying their weight (force) by their distance from the center of rotation (lever arm). For example, if a child has a mass of 25 kg and is sitting 3.0 m from the center of rotation, how far from the center of rotation will a 60 kg mother have to sit to be in balance?

$$\tau_M = \tau_C$$
$$F_M r_M = F_C r_C$$
$$60 \text{ kg } r_M = 25 \text{ kg} \times 3.0 \text{ m}$$
$$r_M = 1.25 \text{ m}$$

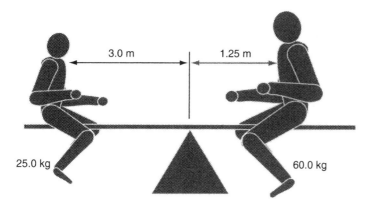

Figure 5-11: The product of force and length of the lever arm determines torque. An adult, therefore, must sit closer to the central pivot point to balance with a child.

As a second application, maybe you have tried loosening a bolt by using a short wrench and applying a lot of force. An experienced mechanic will quickly explain that if you use a longer wrench you won't have to use as much force. The torque equation shows that with a longer lever arm, the same force will produce a greater torque. It is important to understand, however, that only the force acting perpendicular to the lever arm contributes to the torque. Torque is a vector quantity. So to remove that stubborn bolt, use the longest wrench available and apply force in a direction perpendicular to the wrench, as shown in Figure 5-12.

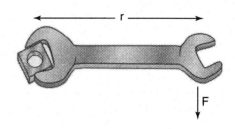

Figure 5–12: To remove a bolt, force must be applied in a direction perpendicular to the wrench. The longer the wrench, r, the more torque an equal force will apply.

The tendency for a rotating object to keep the same speed and direction of rotation is measured by its **moment of inertia**, I. As shown in Table 5-3, the numerical value for the moment of inertia of an object is dependent on its mass, shape, and axis. For a very small mass (**point mass**) revolving about a center (thin ring) it is simply the product of its mass, m, times the distance from the center of rotation squared, r^2. Using this relationship, the moment of inertia can be calculated for Mary on the merry-go-round. This calculation must assume, however, that Mary is a point mass.

$$\text{moment of inertia} = \text{mass} \times \text{radius}^2$$

$$I = mr^2$$

$$I = (20 \text{ kg})(3.5 \text{ m})^2$$

$$I = 180 \text{ kg} \cdot \text{m}^2$$

Measuring how fast something is rotating, or its **angular velocity**, ω, can be done in several different ways. One is to simply measure how many times an object completes a full revolution in a given length of time and report it in rotations per second. Another way is to measure the degrees of rotation in a given amount of time and report degrees per second. The preferred way to measure rotation, however, is to measure the radians of rotation (see Figure 5-3) in a given length of time and report radians per second.

One full rotation is 2π radians, or aproximately 6.28 radians. By expressing angular velocity in radians per second, the product of angular velocity and radius (ωr) becomes the distance traveled by the object per second. In other words (ωr) is the linear velocity, v.

If an object's moment of inertia is known, it can be used to calculate its angular momentum. **Angular momentum**, L, is the product of an object's moment of inertia and its angular velocity. If we continue the previous problem by calculating Mary's angular velocity, we must remember that the merry-go-round was rotating once every six seconds. For unit agreement, the rotation of the merry-go-round must be expressed in radians, where $360° = \text{one revolution} = 2\pi$ radians.

$$\text{angular momentum} = \text{moment of inertia} \times \text{angular velocity}$$

$$L = I\omega$$

$$L = (245 \text{ kg} \cdot \text{m}^2)(2\pi \text{ radians}/6 \text{ s})$$

$$L \cong (81.7 \, \pi \text{ kg} \cdot \text{m}^2/\text{s})$$

$$L \cong 257 \text{ kg} \cdot \text{m}^2/\text{s}$$

Body	Axis	Moment of Inertia
Cylinder (solid) of radius, r	Axis of the cylinder	$\frac{1}{2} mr^2$
Disk of Radius, r	Through center, perpendicular to disk	$\frac{1}{2} mr^2$
Disk of Radius, r	Along any diameter	$\frac{1}{4} mr^2$
Sphere of Radius, r	Any diameter	$\frac{2}{5} mr^2$
Thin ring of radius, r	Through center, perpendicular to ring plane	mr^2
Thin ring of radius, r	Along any diameter	$\frac{1}{2} mr^2$
Uniform thin rod of length, L	Perpendicular to rod at one end	$\frac{1}{3} mL^2$
Uniform thin rod of length, L	Perpendicular to rod at the center	$\frac{1}{12} mL^2$

Table 5–3: Moment of inertia of regular bodies.

Conservation of Momentum

Conservation of momentum and **conservation of angular momentum** have endured centuries of experimental verification. They are recognized as laws that predict the behavior of a system and the objects in that system. For angular momentum to be conserved, it is again necessary to consider the system completely isolated. When we say this, it means that it is frictionless, has no added pushes, pulls, or torques, and its mass does not change. A simple example involving a figure skater and some aspects of the conservation of angular momentum follows. The question is how much faster a skater will rotate if she moves her out-stretched arms closer to her body after she establishes her spin, as shown in Figure 5-13.

Figure 5–13: Skaters increase their speed of rotation by pulling their arms closer to their bodies after they establish a spin.

If the skater has an angular velocity of one revolution in 2 s, with her arms horizontally extended, what will be her angular velocity, ω_f, when she brings her arms close to her body? Assume that her arms have a mass of 3 kg each. You may assume her body's moment of inertia, I_{skater}, is constant and equal to 5 kg m^2. The original distance of her arm's center of mass was 0.5 m and their final distance is 0.2 m from her vertical axis.

To simplify the calculations, the skater can be thought of as two components: her body (a constant) and her arms. Her total moment of inertia, I, would be:

$$I = I_{skater} + I_{arms}$$

Since only the position of her arms change during the spin, her moment of inertia is given by:

$$I = I_{skater} + mr^2$$

At the start of the spin the moment of inertia is:

$$I_i = 5 \text{ kg m}^2 + 2 (3\text{kg}) (0.5\text{m})^2 = 6.50 \text{ kg m}^2$$

When her arms are close to her body the moment of inertia becomes:

$$I_f = 5 \text{ kg m}^2 + 2 (3\text{kg}) (0.2\text{m})^2 = 5.24 \text{ kg m}^2$$

The initial angular velocity, ω_i, is given by:

$$\omega_i = \frac{2\pi \text{ rad}}{2 \text{ s}} = \frac{\pi \text{ rad}}{s} = \frac{3.1415 \text{ rad}}{s}$$

Knowing that angular momentum, L, is conserved,

$$L_i = L_f$$

$$I_i\omega_i = I_f\omega_f$$

the final angular velocity, ω_f, is

$$\omega_f = \frac{\omega_i I_i}{I_f} = \frac{\pi \text{ rad} (6.50 \text{ kg m}^2)}{s(5.24 \text{ kg m}^2)}$$

$$\omega_f \cong \frac{3.9 \text{ rad}}{s}$$

Because angular momentum is conserved, this example shows that as the radius of the mass decreases, the angular velocity of the skater's spin must increase.

Check Your Understanding

1. Describe a situation in which speed remains constant but an object is accelerating.

2. Define moment of inertia and angular momentum.

3. What is linear momentum?

4. What is a point mass and what is the formula for finding its moment of inertia?

5. What are the preferred units for expressing angular velocity?

6. Describe conservation of angular momentum in your own words.

5-8 Waves

Objectives

Upon completion of this section, you will be able to:

- Define wave motion.

- Define amplitude, wavelength, velocity of a wave, period, and frequency.

- Calculate either wavelength, velocity, or frequency, if the other two quantities are known.

A **wave** is a vibration that travels through space and time. Strings on a guitar and jump ropes can be made to vibrate. In fact, both of these have similar vibrational patterns. When someone plucks the string on a guitar or gives a jump rope a jerk, a wave of energy is sent down its length. The string or rope will continue to vibrate until it runs out of energy. For simplicity's sake, let's focus on a guitar string. Once a plucked string stops vibrating, it is in the same relative position to the guitar as it was before it was plucked. Therefore, the energy wave did not move the string out of place. The wave that traveled the length of the string was just an energy form. In this example, the guitar string has elastic potential energy that causes it to vibrate when it is energized by plucking.

If you drop a pebble in a pan of water, you see a wave travel outward from the disturbance caused by the pebble. The surface of undisturbed water, is smooth and level at rest. When the water is disturbed, a wave pattern is created. As the wave travels across the surface, the water level first goes up, then back to level, then down, and finally back to its resting level. As shown in Figure 5-14, the distance the wave rises above or falls below it's resting level is known as its **amplitude**.

Figure 5-14 also illustrates that the **wavelength**, λ, is the distance from one wave crest to the next. In physics, it is usually expressed in meters. Amplitude and wavelength provide a two-dimensional description of a wave, but remember that the wave is also traveling away from the point of disturbance. Given the length of time it takes for a wave to travel between two fixed points, it is possible to calculate the wave's velocity. The measurement of how many waves pass a fixed point in a given length of time is known as the wave **frequency**, *f*. When waves are big and slow, such as in the ocean, we can actually count the number of waves in a given time period and divide by the time interval to find the number of waves per second.

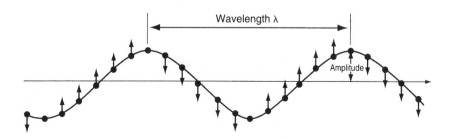

Figure 5–14: Cross-section of a wave showing its motion, amplitude, and wavelength, λ.

The symbol used for wavelength is the Greek lambda, λ. If you know the wavelength and frequency, it is possible to calculate the velocity of a wave using the following equation:

$$\text{velocity} = \text{wavelength} \times \text{frequency}$$

$$v = \lambda f$$

For example, if wave crests on a lake are separated by 0.4 meters, and 5 waves pass by a given point in 10 seconds, then the wave velocity is:

$$v = \lambda f$$

$$v = 0.4 \text{ m} \times \frac{5 \text{ waves}}{10 \text{ s}} = 0.2 \text{ m/s}$$

It should be noted that if wavelength is expressed in meters and frequency in waves per second, the units for the wave's velocity is meters per second. When multiplying waves per second by meters, the term "wave" is dropped, because it is a measurement without units.

Frequency can also be measured in waves per second, it can also be measured in vibrations per second, cycles per second, or any other periodic measurement where the length is not specified. The unit most often used to express frequency is the **Hertz**, Hz, (1 Hz = 1/sec).

Sometimes the **period**, T, of a wave, rather than its frequency, is known. Whereas the frequency is the number of waves passing a given point per second, the period is the number of seconds between waves. As shown by the equation below, mathematically, the period is the reciprocal of the frequency.

$$\text{period} = \frac{1}{\text{frequency}}$$

$$T = \frac{1}{f} \ \text{ or } \ f = \frac{1}{T}$$

Because of this relationship, it is also possible to use the period and wavelength to calculate the velocity of a wave. The equation can be derived by combining two of the equations shown above:

$$v = \lambda f$$

$$v = \lambda \frac{1}{T}$$

$$v = \frac{\lambda}{T}$$

$$\text{velocity} = \frac{\text{wavelength}}{\text{period}}$$

If a wave in the ocean has a wavelength of 15.0 m and a period of 5 seconds, how fast is a surfer on the crest of the wave traveling?

$$v = \frac{15.00 \text{ m}}{5 \text{ s/wave}} = 3.0 \text{ m/s} \ (\cong 6.7 \text{ mph})$$

Constant Velocity of Waves in Specific Media

If the measurements are taken carefully, it can be shown that even a wave on a jump rope has a fixed velocity, regardless of amplitude, frequency, or wavelength. The wave velocity is characteristic of the medium in which it travels. You may have noted that amplitude is not even a part of the equation for finding the speed of a wave in most cases. The only quantities necessary to calculate wave speed are frequency and wavelength. Because they are inversely proportional, the wavelength decreases as the frequency of the wave increases.

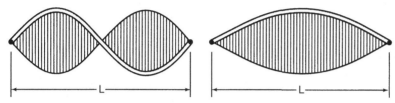

Figure 5–15: A full standing wave of wavelength, L, (left) and a half standing wave of wavelength, L/2 (right). The heavier line indicates the position of the rope when at maximum displacement. The shaded areas show the regions covered by the motion of the rope.

You can test this relationship by getting a friend, 3 or 4 meters of rope, and a stopwatch. Tie the rope to something fixed. Have your friend shake their end of the rope up and down with a frequency that forms a full **standing wave** by reflecting from the barrier and traveling back in sync. A standing wave is formed by the combining of the two traveling waves, e.g., the wave and its reflection, that are of identical amplitude, wavelength, and passing in opposite directions. As shown in Figure 5-15 (left), the pattern of a one wavelength standing wave consists of a node (the point in the center where their amplitudes cancel) and two antinodes (the points where the two amplitudes add). Now count the number of vibrations the rope makes in ten seconds. Be sure to divide this measurement by ten to get the frequency. Calculate the velocity of the wave by multiplying the frequency and the wavelength.

If the above procedure is repeated after establishing a half wavelength standing wave as shown in Figure 5-15 (right), the two velocities should compare quite well. Remember that since the standing wave created in this way is only half a wave length, the distance between the knot and the person shaking the rope will need to be multiplied by two.

Medium	m/s	ft/s	Medium	m/s	ft/s
Gases			Solids		
Air, 0°C (dry)	331.5	1,088	Ivory	3,013	9,885
Air, 20°C	344	1,129	Iron	5,150	16,902
Helium	965	3,166	Glass (pyrex)	5,640	18,511
Liquids			Steel	5,960	19,561
Water. 0°C	1,402	4,600	Aluminum	6,420	21,070
Water, 20°C	1,482	4,862			
Sea water	1,522	4,993			

Table 5–4: Speed of sound in various mediums. Source: Robert C. Weast, *Handbook of Chemistry & Physics*, CRC Press, 1974–75 ppE–47.

Table 5-4 shows that mechanical waves in elastic media have much different velocities. For example, sound waves in 20°C water move 4.3 times faster than they do in 20°C air. The speed of sound is 331.5 m/s in dry 0°C air; also indicated in Table 5-4. This is approximately one-fifth of a mile per second. In other words, it takes 5 seconds for the sound to travel one mile. By measuring the time difference between the flash of lightning and the clap of thunder, it is possible to estimate the distance to a storm. This, of course, assumes that the light from the lightning reaches you instantly.

Check Your Understanding

1. What is a wave?

2. Diagram a wave and label its amplitude and wavelength.

3. What is the period of a wave? What is the frequency of a wave? How are they related?

4. What is a standing wave?

5. Express the velocity of a wave in terms of its wavelength and frequency.

5–9 **Transverse and Longitudinal Waves**

Objectives

Upon completion of this section, you will be able to:

■ Explain the difference between transverse and longitudinal waves.

■ Give examples of each type of wave.

■ Describe the amplitude, wavelength, and frequency for transverse and longitudinal waves.

In the previous section, waves that cause the medium to move up and down (perpendicular) relative to the wave direction (velocity) were discussed. These are termed **transverse waves** and the movement of the medium is always perpendicular to the movement of the wave. Some waves require no medium, e.g., electromagnetic waves. Other waves cause the medium to move in the same direction as the vector. These are known as **longitudinal waves**. When the medium and the wave travel in the same direction, it results in the medium being compressed and decompressed. Both types of waves continue to move through the medium until they lose so much energy through friction and heating of the particles rubbing against one another that they fade out.

As shown in Figure 5-16, sound waves are a good example of longitudinal waves. Sound is compressed and decompressed air that is pushed along in a wave motion and that travels in the same direction as the wave. To see what this might look like, try tying a slinky between two fixed points, bunch the slinky toward one end, and then let go. You will see the compression wave travel the length of the slinky.

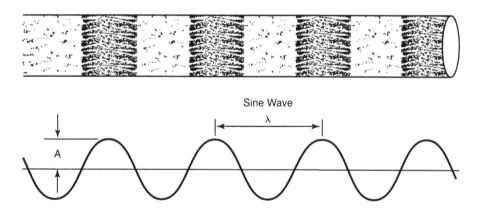

Figure 5–16: Longitudinal sound waves in a tube are composed of air compressions and decompressions. When graphed as a transverse wave they become peaks and valleys.

For longitudinal waves, the definitions for wavelength and frequency are comparable to that of transverse waves. The wavelength is the distance between two maximum compression points or, alternatively, between two minimum compression points. When longitudinal waves are drawn, they are frequently represented in the same way as transverse waves, because most of us are not patient or artistic enough to draw them realistically. When doing so, maximum compression is represented as the highest part of the wave and maximum rarefaction is the lowest part of the wave. Very often, longitudinal waves are pressure waves, e.g., sound waves.

Although the speed of longitudinal waves is constant in a given medium, it may vary due to other factors. For example, the speed of sound in air varies, depending on both its temperature and humidity. As shown in Table 5-4, sound waves travel slower in cold than warm air. The average change in speed is about six tenths of a meter per second per degree (0.6 m/s/degree).

Sound also travels faster in humid than dry air. Careful examination of Table 5-4 shows that sound travels fastest in solids and slowest in gases. In general, as the density of the medium increases, so does the speed of sound. There are stories that people used to put their ear to the rail before attempting to walk across a long railroad bridge. Since sound travels faster in steel than the air, they could more accurately determine if a train was coming.

Technology Box 5–1 ■ How is the epicenter of an earthquake determined?

The task of figuring the exact center of an earthquake (**epicenter**) is not so different from calculating how far away a lightning strike occurred. When locating the epicenter of an earthquake, three observers are needed. As shown in the figure below, each of the three seismic stations must first determine how far they were from the epicenter. Next, each seismic station draws a circle around their location that represents their distance from the epicenter. There can be only one point where the three circles overlap. This process is called triangulation and it allows a quake's epicenter to be pinpointed.

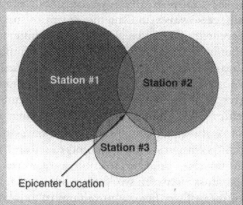

Three overlapping circles representing the distance of each station from the epicenter of the earthquake can be used to pinpoint the epicenter of an earthquake.

Next let's consider the bigger question. How do each of the seismic stations determine their distance from the epicenter? Over the years, seismologists have learned that earthquakes generate two kinds of waves in the Earth's crust: longitudinal waves and transverse waves.

The longitudinal waves, which travel faster, are compression waves. These are called the **P-waves**. The transverse or shear waves travel through the Earth's crust more slowly. They are called the **S-waves**. By comparing the difference in time between the arrival of the S- and P-waves, it is possible to calculate the distance to the epicenter.

For example, the velocity of the P-wave is approximately 6,400 m/s compared to 3,200 m/s for the S-wave in German basalt, a type of igneous rock. Graphs are available that allow seismic stations to calculate how far they are from the epicenter based upon the lag in time between the arrival of the P- and the S-waves.

In actual practice, this process is not quite so simple, since the velocity of waves change as the composition of the Earth's crust varies. As more seismic stations report their data, the location of the earthquake can be reported more accurately. It is obvious that the closest station will have the least amount of lag time between the S- and P-waves, known as the S-P lag. The closest station will obviously be the first to record the earthquake. By assembling and analyzing all this information, the exact location of an earthquake can be pinpointed.

Check Your Understanding

1. How does the movement of the medium differ when disturbed by a transverse wave as opposed to a longitudinal wave?

2. Give an example of a longitudinal wave and a transverse wave.

3. Describe how the amplitude of a longitudinal wave differs from the amplitude of a transverse wave.

Summary

The motions of large (planets) and small (atoms) pieces of matter have been observed, recorded, and analyzed for years. Sometimes matter is at rest and at others it is in motion. When in motion, it may be accelerating, decelerating, curving, rotating, or vibrating. In general, the various motions of matter have been found to be predictable and can be represented by various mathematical formulas.

When matter is in motion, it has a velocity. Velocity is a vector quantity, so it encompasses both speed and direction. In the same way, matter that is in motion is either moving closer or farther from a fixed point. It is possible to measure the distance or displacement of this movement. Distance is a measure of how far something has traveled, but displacement is a vector quantity, so it must encompass both distance and direction.

Matter in motion often experiences forces that cause it to speed up, slow down, or change direction. When matter changes its velocity (speed or direction) it is said to be accelerating. Objects free-falling toward the Earth experience the uniform acceleration of gravity.

Newton's law of universal gravitation states that all objects that have mass attract each other with a force that is directly proportional to the product of their masses and inversely proportional to the square of their distances. Newton also described several other important relationships about matter in motion. These are now generally referred to as Newton's laws of motion.

Newton's first law states that an object at rest will stay at rest and an object in motion will stay in motion with a constant velocity until acted on by a force. This is also the definition of inertia and the law is sometimes called the law of inertia.

Newton's second law states that when a force is applied to an object, it accelerates in the direction of the force. $F = ma$ is a mathematical summarization of this statement.

Newton's third law states that for every action there is a reaction. This means that forces must always come in pairs.

When the surfaces of two pieces of matter move with respect to one another, a force called friction results. If the matter is initially at rest, static friction must be overcome to start one object moving with respect to the other. If the matter is moving, then enough force must be applied to overcome kinetic friction in order for the object to continue in motion. The nature of the two surfaces in contact determines the coefficient of friction, which typically ranges from zero to one.

Matter in motion possesses momentum, which is the product of the mass and velocity of an object at any instant in time, and is always conserved. Matter may also be in circular motion. In this instance, the acceleration consists of a continuous change in the direction of the object's velocity.

Rotating matter, in a manner similar to matter moving in a straight line, tends to continue rotating at the same speed and direction until acted upon by a torque. The moment of inertia depends on both the mass and shape of an object, as well as the axis chosen for rotation.

When rhythmic or repetitive energy is added to matter, it may result in a wave pattern. The amplitude of a wave is its height above the medium's resting position. The distance between wave crests is its wavelength. The number of waves passing a fixed point per second is known as its frequency or alternatively, the time between waves is know as its period. Wave frequency and period are, therefore, the reciprocal of one another. The velocity of a wave is the product of the wavelength and frequency ($v = \lambda f$) or the wavelength divided by its period ($v = \lambda/T$). It is common to express frequencies such as cycles per second or vibrations per second in Hertz.

When a specific type of wave travels through a specific medium, it does so with a constant velocity. Waves may be of two different types; transverse and longitudinal. Transverse waves cause the movement of the medium to be perpendicular to the velocity of the wave. Longitudinal waves travel in the same direction as the medium, resulting in it being compressed and decompressed.

Chapter Review

1. Find the average velocity of an airliner if it flew (11° South of East) from Los Angeles to Dallas, a distance of 1,450 miles, in 2 hours and 10 minutes. Express the velocity in miles/hr, feet/s and m/s.

2. If $360° = 2\pi$ radians, how many radians would be equal to 140°?

3. Using the Pythagorean theorem ($a^2 + b^2 = c^2$), what distance would someone be from their starting point if they drove 300 km east and 100 km north?

4. Using the information in question 3 and graphing, what would be their displacement?

5. Use the relationship between speed, time, and distance to calculate the average speed that one must travel in order to cover 550 miles in 8 hours. If the speed limit is 65 mph, can you do this without breaking the law?

6. What is the acceleration of a car traveling at 27 m/s, when it makes a turn with a 100 m radius?

7. Assuming no air resistance and starting from rest, how far will a steel ball free-fall in 6 seconds? (g = 9.81 m/s)

8. According to Newton's law of universal gravitation, what is the gravitational force between two 10,000 kg spheres that are 10 m apart?

9. What force would be required to accelerate a 100 g sphere by 3.5 m/s^2?

10. Use Newton's second law to find the force necessary to accelerate a 1,500 kg car from 0 to 80 km/hour in 10 seconds. Remember to convert hours to seconds in order to find the answer in Newtons.

11. Calculate the kinetic friction of a 150 N object that has a coefficient of kinetic friction of 0.3.

12. A box is being pushed around a warehouse at a constant velocity. If the coefficient of static friction is 0.6 and the box weighs 500 Newtons, how much force does it

take to just get the box moving? If the coefficient of kinetic friction is 0.25, how much force does it take to keep the box moving?

13. How much force, in pounds, must you use to get your 1,000 pound piano moving if the coefficient of static friction between the piano and the floor is 0.72?

14. What is the momentum of a billiard ball with a mass of 235 g that is traveling at 2 m/s?

15. If Stacey is traveling north 80 mph in her 1,200 pound sports car and Debbie is traveling south 50 mph in her 3,500 pound SUV, which one has more momentum and by how much?

16. If a 10 kg ball traveling 4 m/s comes to a stop when it hits a stationary 5 kg ball, what should the velocity of the 5 kg ball be after the collision? Hint: momentum is conserved.

17. What is the velocity (in meters per second) of a car traveling around a circular track with a radius of 500 meters if the car makes 1 revolution in 50 seconds?

18. What is the acceleration of the car in Question 17 if the speed remains constant?

19. Both James and Justin are on skateboards facing each other when Justin pushes James. Assume both Justin and James have frictionless skateboards. James and his board have a mass of 60 kilograms and Justin and his board have a mass of 90 kilograms. How fast will James go if Justin ends up traveling 2 m/s after the push?

20. Which will have a greater moment of inertia; a 1 kg disc with a radius of 4 cm, or a 1 kg disc with a radius of 10 cm? Prove your answer.

21. Two identical spaceships are orbiting Earth. NASA calculates that both have the same angular momentum. Spaceship A, however, is farther from Earth than spaceship B. Which one must have the greater velocity?

22. What is wavelength of a wave, if the frequency is 10 waves per second and the velocity is 20 m/s?

23. If the wave in Question 22 has a wavelength of 4 meters, what will be its frequency?

24. What would be the torque on a bolt, if a force of 15 N is applied to the end of a one meter pipe wrench? Express the answer in N-meters and in foot-pounds.

25. Consider a roulette table with a marble rotating about the center, completing 3 rotations every 5 seconds. If the marble has a mass of 2 grams and is rotating 0.25 meters from the center of the table, find its moment of inertia. Then find its angular momentum. Finally, determine how many rotations it will make in 5 seconds, when it has moved to 0.10 meters from the center. Treat the marble as a point mass rotating about the center. Neglect the rolling motion of the marble.

26. If a wave has a frequency of 2 Hz and a wavelength of 0.75 meters, what is its velocity?

27. What is the wavelength of a light wave, if its frequency is 3×10^{15} Hz?

What is energy?

Energy

As early as 400,000 BC, fire was kindled in the caves of Peking Man. Revered as a deity and the basis of many myths, fire has been an essential element in the technologies on which civilized societies are founded. Engines driven by fossil fuels have replaced human and animal muscle, precipitating the rise of industrialized societies. Today cities, industrial facilities, and transportation networks could not function without regular supplies of energy.[1]

6-1   The Definition of Energy

Objectives

Upon completion of this section, you will be able to:

■ Explain the nature of energy and work.

■ Calculate the kinetic and gravitational potential energy of a mass.

■ Describe the various forms and types of energy.

■ Understand the many energy transformations that take place in nature.

Energy can be neither created nor destroyed, but is transformed through many processes that take place on Earth and elsewhere in the universe. It powers our world and provides the fuel for all processes to occur, including the development and sustenance of life. Energy is the currency, the driving force of the universe. It is fundamental to the structure and evolution of our physical world.

As you learned in Chapter 4, even the ancient Greeks recognized energy, in the form of fire and included as one of their four basic elements. The energy needs of the earliest humans were limited to heating, cooking, and lighting. However, the amount of energy that humans consume each day, on average, has steadily grown throughout history. As new machines and processes were developed, the world's population became increasingly dependent on energy. The availability of convenient, economical,

---

1.   Davis, G. R.: *Scientific American*, vol. 263, no. 3, p. 55, 1990.

and reliable sources of energy has become one of the driving forces that shape our way of life.

How often do you think about energy? Perhaps, only when you are hungry or the electricity bill arrives. Energy can be found in the far reaches of the universe and it is like our childhood superhero who could be invisible one moment and very much apparent the next. **Energy** is defined as the capacity for doing work.

Work

Work takes on many different connotations in our daily lives, but it is clearly defined in physics as the product of force and the displacement through which the force acts, as given below:

$$\text{work} = \text{force} \times \text{displacement}$$

For example, in order to lift a mass a vertical distance, you would apply an upward force to that mass. Assuming the force you apply is of greater magnitude than the downward force of gravity, the mass will move upward. The product of the force you apply and the distance it moves (displacement) represents the work done on the mass.

The unit for work is the joule, J, where 1 J = 1 newton × 1 meter. The joule, newton, and meter are the SI units for energy, force, and distance, respectively. Note that the units for energy and work are identical. Other units that may

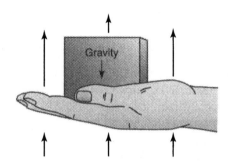

Figure 6–1: To lift a mass, an upward force must be applied by the hand that is greater than the downward gravitational force acting on the mass.

be used are the British thermal unit, Btu, which is often used when describing the cooling capacity of air conditioners; horse-power, used to describe motors and engines; and kilowatt-hours, used to calculate your electric bill. These units, and others relevant to energy, will be discussed later in the chapter.

To do work there must be a force and a displacement. According to this definition, each of us is constantly doing work as shown in Figure 6-2. For example, you in-

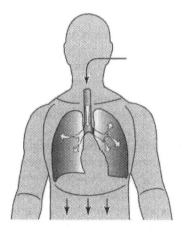
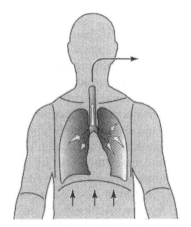

Figure 6–2: An individual breathes by contracting the diaphragm (left), which creates a pressure differential between the lungs and their surroundings.

hale by contracting your diaphragm downward, which results in a pressure decrease around your lungs. The difference in pressure between your lungs and the atmosphere constitutes a driving force that acts on the air, causing it to rush into your lungs. The atmospheric pressure does work expanding your lungs and then when exhaling, you do work on the atmosphere.

Since energy is everywhere, we need to carefully describe what energy we are considering. When we refer to a **system** in chemistry and physics, we simply mean a region in space, a substance, a machine, or any confined, bounded space. The region just outside a system is the **surroundings** and the **boundary** of the system is the real or imaginary surface that separates the system from its surroundings.

Every system possesses a particular amount of energy. The reason that energy is considered an abstract quantity is that we cannot measure or observe it directly. However, in certain cases, we can determine the amount of energy a system possesses by using some other easily measurable quantity, such as its temperature and volume or velocity and mass. Energy is simply one of many quantities, properties, or characteristics that can describe the condition or state of a system.

Conservation of Energy

The **first law of thermodynamics** states, in effect, that the total energy in the universe is constant; energy is neither created nor destroyed. This statement is also commonly known as the **law of conservation of energy**. Conservation of energy is not limited to any particular mechanism by which energy will be transformed. It simply states that the total amount of energy must remain constant throughout the process.

You are probably familiar with Einstein's hypothesis stating that energy and mass are equivalent as described by the equation, $E = mc^2$. According to this equation, energy, E, in joules, is proportional to the amount of mass, m, in kilograms, that is destroyed. The velocity of light, c, is a constant that is equal to 2.998×10^8 m/s. At first, the creation of energy may appear to be a violation of the law of conservation of energy. However, for it to happen an equivalent amount of mass must be destroyed. In this instance, matter can merely be thought of as a form of stored energy. During most energy transformations, this mass-to-energy conversion is negligible and can be disregarded. Therefore, we will focus on conserving energy only, rather than the sum of energy and mass converted to energy.

Still, it offers an interesting observation. If a very small amount of mass were destroyed, it would produce a huge amount of energy. Consider, for example, that a power plant designed to produce 750,000 kW of electricity per year, would consume about 275 tons of coal per hour or about 2,500,000 tons of coal per year. Conversely, a 750,000 kW nuclear power plant would consume only about 1.25 tons of uranium fuel per year. However, if there existed the means to generate 750,000 kW of electricity by the complete conversion of mass into energy, Einstein's equation indicates that only about 0.5 lb_m (0.260 kg) would be needed. Combustion and nuclear fission reactions that are being used in commercial power plants, however, produce a large quantity of undesirable waste products. The mass of the waste generated is still roughly equal to the amount of fuel consumed.

The equation below shows how to calculate the amount of fuel that would be required to produce 750,000 kW (7.50×10^8 J/s) of power, if the mass were completely converted to energy. Note that the unit of energy, the joule, equals 1 kg · m²/s². The total energy produced, E, is equal to the power, P, multiplied by time, t, since power is the rate at which energy is produced. It is also helpful to know that there are 3.15576×10^7 seconds in a year.

$$E = mc^2 \text{ or } m = \frac{E}{c^2}$$

$$E = Pt, \text{ therefore, } m = \frac{Pt}{c^2}$$

$$m = \frac{(7.5 \times 10^8 \text{J/s})(1 \text{ year})}{(2.998 \times 10^8 \text{m/s})^2}$$

$$m = \frac{(7.5 \times 10^8 \text{J/s})(1 \text{ year})(3.17556 \times 10^7 \text{s/year})}{8.988 \times 10^{16} \text{m}^2/\text{s}^2}$$

$$m = \frac{0.265 \text{ J}}{\text{m}^2/\text{s}^2} = \frac{0.265 \text{ kg} \cdot \text{m}^2 \cdot \text{s}^2}{\text{m}^2 \cdot \text{s}^2}$$

$$m = 0.265 \text{ kg}$$

When considering the conservation of energy, our main concern is keeping the books straight, that is, accounting for the energy entering and leaving a system during a process. We will soon apply the mathematical representation of the first law of thermodynamics, which allows us to perform this energy balance.

Like most people, you are probably sensitive to the environmental impact of burning fossil fuels and recognize the virtues of reducing our energy consumption. But are you actually conserving energy by reducing the amount of energy you consume? The often-used phrase "to conserve energy" is misleading – we always conserve energy! You cannot "use up" energy. Rather, you can convert it from one form to another, e.g., converting the chemical potential energy in gasoline to the mechanical energy of a rotating shaft. Ultimately it will be converted into thermal energy and dissipated into the environment.

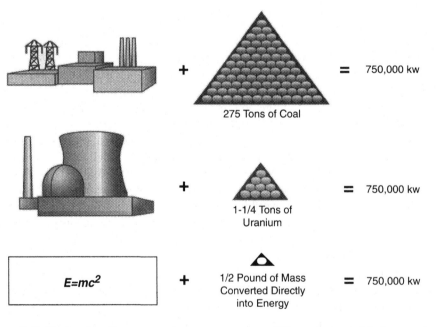

Figure 6–3: Coal and nuclear power plants consume very different masses of fuel to produce the same amount of electricity.

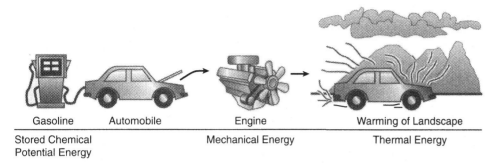

Gasoline	Automobile	Engine	Warming of Landscape
Stored Chemical Potential Energy		Mechanical Energy	Thermal Energy

Figure 6–4: Energy changes form as it passes through an automobile. Chemical potential energy, in the form of gasoline, is combusted and converted to mechanical energy. The mechanical energy moves the car and is ultimately converted to thermal energy that warms the surroundings.

After you have returned from a trip and your car cools, it is accurate to say that all of the energy stored in the gasoline was converted into thermal energy and dispersed into the environment. None of the energy was lost, but the energy consumed has been converted to a form that is of a much lower quality and can no longer provide a useful effect. In fact, all the energy you consume on a daily basis ultimately returns to the universe with nothing more than a little additional long-wavelength radiation, a form of energy to be discussed further in Chapter 9.

Forms of Energy

Humans have many basic energy needs. Some are obvious and have existed for thousands of years. Others are more recent and include cooling, manufacturing, communications, transportation, and agricultural applications. Have you ever considered what form of energy best serves a specific need?

In this text, energy will be subdivided into six major forms – mechanical, electrical, electromagnetic, chemical, nuclear, and thermal. Each of these forms, in turn, may assume one of two types. Energy that flows from one place to another is **transitional energy**. As it flows across a system boundary and does work as it is transformed. **Stored energy** resides in some form such as the chemical energy in gasoline, the electrical energy in a battery, or the mechanical energy in a compressed gas that can be made available to do work.

Mechanical

You are probably most familiar with **mechanical energy**, often defined as that which can be used to raise a weight and sometimes called the **energy of motion.** The transitional type of mechanical energy is called kinetic energy and does work. The stored type is called potential energy because it has the potential to do work. If you have ever set a mousetrap or inflated a tire, then you have established a reserve of potential energy, E_p. When you lift a mass, m, to a higher position, h, the work you do on the mass is transformed into **gravitational potential energy**. This is described by the following equation, in which the constant g is the acceleration due to gravity and is equal to 9.8 m/s².

$$E_p = m g h$$

Potential energy can also take the form of a compressed fluid or a stretched rubber band. In either case, a force acts on a system that is able to cause a displacement and thus there is the potential to do work. Even the kinetic energy, E_K, related to the relative velocity, v, of a mass, m, as expressed by the following equation can be considered a form of potential energy because it can be used to do work. A spinning flywheel is an example of a device that stores energy in the form of a moving mass.

$$E_K = \tfrac{1}{2} mv^2$$

We can use the two equations above to determine the amount of gravitational potential energy or kinetic energy a mass gains or loses as either its height above some surface or its speed changes. For example, if you raise a 3 kg block by 2.5 meters, the gravitational potential energy it gains is given by:

$$E_p = mgh$$

$$E_p = (3 \text{ kg}) (9.8 \text{ m/s}^2) (2.5 \text{ m})$$

$$E_p = 73.5 \text{ J}$$

If the block is then allowed to fall freely, assuming no air friction, then all the gravitational potential energy is lost as gravity does work (force over a distance) on the block. At the instant the block strikes the surface from where it began its journey, its potential energy is again zero, since all of the gravitational potential energy has now been converted into the kinetic energy of its motion. Equating these energies, we can calculate the speed of the block, just as it strikes the surface as:

$$E_P = E_K$$

$$73.5 \text{ J} = \tfrac{1}{2} mv^2$$

$$v = \sqrt{\frac{(2)(73.5 \text{ N} \cdot \text{m})}{3 \text{ kg}}}$$

$$v = \sqrt{\frac{(2)(73.5 \text{ kg} \cdot \text{m})\text{m}}{3 \text{ kg} \cdot \text{s}^2}}$$

$$v = 7.0 \text{ m/s}$$

Electrical

Electric current is the flow of electrons through a conductor. This is analogous to a flow of particles, such as water molecules, through a pipe. It is rather easy to measure the current flowing if one splices into the conductor a meter designed to make such a measurement. An ideal current meter does not resist the flow of electrons and thus does not interfere with the circuit in any way. It effectively counts the number of electrons passing by a point in the conductor. These instruments are called ammeters, and the unit of electric current is the Ampere (amp).

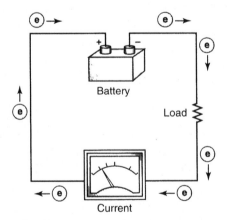

Figure 6–5: Individual electrons are "counted" as they pass through a current meter.

Like water, electrons only flow when they are "pushed" by an "electrical pressure" (voltage). This "electrical pressure," measured in volts, gives the electrons energy that can be used to do work. The energy of a single electron is measured in electron volts (eV's) and depends on how much of a voltage is driving the flow. When a voltage of 1 volt is driving the flow, each electron carries with it 1 eV or 1.6×10^{-19} J of energy. The total energy carried along a conductor over some time interval depends on the total number of electrons passing a fixed point along the conductor, and how much "electrical pressure" (voltage) each electron is under. The flow of electrons constitutes a transitional type of electrical energy.

Electrical energy may also be stored, such as when charges (electrons) accumulate on a surface (electrostatic) or when a magnetic field is established by the flow of electrons through a wound coil (electromagnetic). Like mechanical energy, electrical energy is a very useful form because it can be converted into other energy forms easily and efficiently

Electromagnetic

The only form of energy that is not associated with mass and cannot exist as stored energy is called electromagnetic energy. You have heard of visible light, ultraviolet radiation, microwaves, X rays, and radio waves. All are manifestations of electromagnetic radiation, a form of transitional energy that travels at the speed of light. Figure 6-6 shows several different divisions of electromagnetic radiation, each of which differs only in their wavelengths.

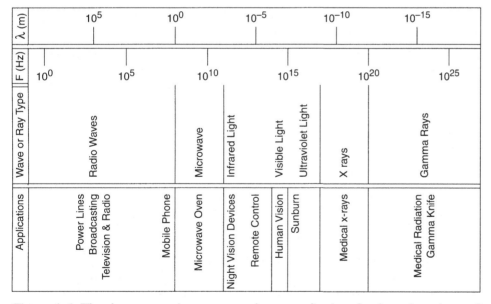

Figure 6–6: The electromagnetic spectrum and some applications for the various classes of radiation.

Note in Figure 6-6 that the visible range of wavelengths is very narrow. However, numerous technologies have developed ways to harness and utilize the energy of many parts of the spectrum. For example, we generate long-wave radiation to broadcast navigation information and television, and microwave radiation for cooking and cellular communications. Infrared radiation can be detected by special types of semiconductor devices and used for temperature measurements and surveillance. The medical profession utilizes X rays to study teeth, bones, and to diagnose diseases.

Even higher energy, shorter wavelength (nuclear) radiation is used to treat tumors. These examples are just a few of the possible applications of electromagnetic energy. More examples will be given in Chapter 9.

Chemical

Chemical energy exists only as a stored type of energy. It is released during exothermic reactions in which molecules with high-energy chemical bonds are converted into compounds with lower bond energies. Your body, like any living organism, requires a constant source of energy in order to survive. The energy is released by respiration, one of many processes through which stored chemical potential energy is converted to other forms.

The most important exothermic chemical reaction, in terms of modern development of human civilization, is the combustion of fossil fuels. This reaction provides a convenient energy form that drives many industrial processes, including manufacturing, electricity generation, and cooling. The combustion of fossil fuels entails an oxidation reaction in which the three principal combustible elements found in most fossil fuels, carbon, hydrogen, and sulfur, are converted into carbon dioxide, CO_2, water, H_2O, and sulfur dioxide, SO_2, respectively.

Nuclear

Nuclear energy can be released or absorbed whenever a nucleus is changed or formed. As with chemical energy, nuclear energy is a form of stored energy until it is released by radioactive decay, fission, or fusion reactions (see Figure 6-7). The radioactive decay process occurs when one unstable nucleus, commonly called a radioisotope, randomly decays into a more stable configuration, releasing particles and energy in the process. This process provides the high-energy electromagnetic (gamma) radiation needed to treat tumors, as described above.

The fission reaction drives most nuclear reactors used in power plants around the world. **Fission** occurs when a heavy-mass nucleus, such as uranium-235, absorbs a neutron and the resulting excited nucleus splits into two or more light-mass nuclei. Energy release occurs because the smaller fission product nuclei are more tightly bound and, therefore, more stable, than the original nucleus. This reaction is illustrated in Figure 6-8.

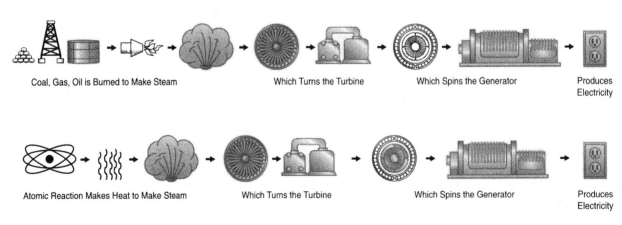

Figure 6–7: Electrical energy generated in a power plant derives from either the combustion of fossil fuels (top) or fission of nuclear materials (bottom).

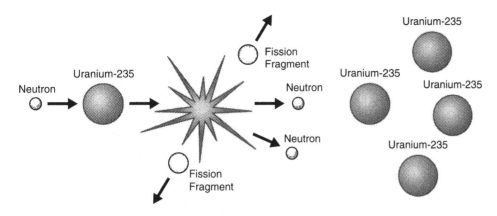

Figure 6-8: The fission reaction of uranium-235 is typically triggered by the capture of a free neutron. The fission of the uranium nucleus, in turn, produces several fission fragments and two or three free neutrons, which can trigger other fissions.

When two light-mass nuclei combine to form a heavier nucleus, the process is called nuclear **fusion**. This is the reverse of nuclear fission. The mass of the final nucleus is less than the combined masses of the original nuclei and the loss of mass corresponds with a release of energy according to the equation, $E = mc^2$.

Nuclear energy is converted to other forms of energy continuously within the Sun. These reactions provide us a steady stream of electromagnetic energy that sustains life forms and literally powers all other aspects of our environment.

Thermal

If you could monitor the total amount of thermal energy in a closed system, you would find that it is constantly increasing at the expense of all other energy forms. The first law of thermodynamics requires that the total quantity of energy in the universe remain constant. However, the **second law of thermodynamics** states, among other things, that thermal energy is a basic energy form. Nature permits complete conversion of other energy forms into thermal energy, but severely limits the conversion of thermal energy back into other energy forms. A more detailed explanation for this paradox will be provided later in the chapter.

As shown in Figure 6-9, **thermal energy** is the energy associated with the random, uncoordinated motion of atoms or molecules. It is also related to the mass of

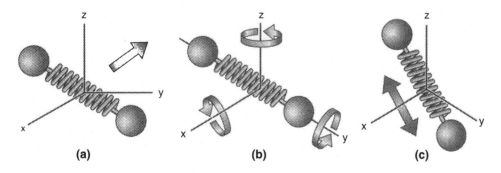

Figure 6-9: A diatomic molecule can have (a) translational motion, (b) rotational motion about the various axes, and (c) vibrational motion along the molecular axis, all of which contribute to thermal energy.

the material and to its ability to conduct heat. In fact, the temperature of a substance is actually a direct measure of the average kinetic energy of the atoms and molecules of that substance. Heat is the end result of any chain of energy conversions involved in any process. Recall, for instance, the energy chain shown in Figure 6-4 associated with the operation of an automobile. Ultimately, the stored chemical potential energy of the fuel, through a series of conversions, becomes thermal energy, resulting in a slight warming of the environment.

There is a difference between heat and thermal energy. We may say, "You had better wear a hat to avoid losing too much body heat." We define this quantity lost by a person as thermal energy. The thermal energy may increase or decrease as heat flows into or out of the person. The term "heat" then refers to a quantity of thermal energy as it passes into or out of a system or we could say it represents the transitional type of thermal energy. This subtle distinction between heat and thermal energy is important when defining, describing, or analyzing problems. The stored type of thermal energy results in either a change of temperature or a change in phase (melting or boiling) of a material at constant temperature.

Special Topics Box 6–1 ■ How often would you have to buy fuel if you had a nuclear-powered car?

The next time you fill your car's fuel tank, record your mileage and the date. When the gasoline is almost gone, fill-up again, and note the number of miles traveled, the volume of gasoline required, and the date. By performing the following calculation, you can then determine your car's average fuel economy:

average fuel economy (mpg) = miles/gallons

Now, divide the number of gallons of gasoline by the number of days between fill-ups. Then multiply this value by the energy released by each gallon of gasoline, which is approximately 1.18×10^5 Btu/gal. This calculation will give you, on average, how much energy is required to operate your car each day.

Energy required per day

= (gallons/day) $(1.18 \times 10^5$ Btu/gallon) = _____ Btu/day

(An individual who burns 3.5 gallons per day would use approximately 4.13×10^5 Btu/day.)

Fission of one pound-mass of uranium-235 will release 2.89×10^{10} Btu of energy. If you could build a nuclear-powered car with only 0.25 lb_m of uranium-235 fuel, you would have available $(0.25 lb_m)(2.89 \times 10^{10}$ Btu/$lb_m) = 7.2 \times 10^9$ Btu of energy. Let's calculate the lifetime of a 0.25 lb_m of nuclear fuel for the individual who, on average, burns 3.5 gallons of gasoline each day:

Lifetime of fuel = (total energy available)/(energy consumed per day)

= $(7.2 \times 10^9$ Btu)/$(4.13 \times 10^5$ Btu / day)

= 17,000 days ≈ 47 years

In other words, a nuclear fuel element with the mass of a hamburger patty would power your car for almost 50 years!

Try calculating how long one ounce of nuclear fuel would last in your nuclear-powered car, based on your daily fuel consumption.

Energy Transformations

You are now aware that energy exists in many different forms. The first law of thermodynamics states that energy is neither created nor destroyed. Energy changes, however, from one form into another. All human activities, mechanical processes, and physical interactions involve the conversion of energy from one form into another. For instance, when you consume food that contains chemical energy it is stored in your body. Through biochemical processes within muscles and nerves, this energy is converted into work, the transitional form of mechanical energy, and ultimately to heat, the transitional form of thermal energy. You may say that you have consumed an amount of energy, but your body is simply changing one form of energy into other lower forms of energy. Your next meal will provide you the opportunity to convert even more stored chemical energy into even more mechanical and thermal energy. Even when you are asleep, you continue to use energy to maintain body functions and a stable body temperature.

In general, a **reversible process** is defined as a process that can be reversed without leaving any trace on the surroundings. This implies that the system and its surroundings could be returned to their original state at the end of the process. This is never the case in the real world. A reversible process is simply an idealization of an actual process. For instance, if you allow a battery to power a heater, it will eventually discharge. The stored chemical potential energy converts first to electrical energy and finally to thermal energy. It is possible to re-charge the battery, thus returning it to its original state, but this will require that electricity be provided to the battery. That is the catch! The process of electricity generation is needed to re-charge the battery. This leaves a trace on the surroundings!

We have already recognized that all forms of energy are ultimately converted into thermal energy during any real or **irreversible process**. The conversion of mechanical energy into thermal energy occurs because of the presence of friction, an undesirable phenomenon in most thermodynamic or lubrication processes. Friction is sometimes a desirable phenomenon. For instance, were it not for the friction between the soles of your shoes and the floor, it would be impossible to walk.

From:	To: Mechanical Energy	Electrical Energy	Electromagnetic Energy	Chemical Energy	Nuclear Energy	Thermal Energy
Mechanical Energy	Gear box	Generator				Refrigerator
Electrical Energy	Motor		Lasers, Radio transmitter	Electrolysis, Battery charging		Resistance heating
Electromagnetic Energy	Spinning radiometer	Photovoltaic solar cell		Photosynthesis	Gamma-ray reaction	Absorption reaction
Chemical Energy	Engine, Animal muscle	Battery, Fuel cell	Candle, Chemiluminescence (fireflies)		Phosphors	Combustion, Digestion, Other exothermic reactions
Nuclear Energy	Particle emission	Nuclear battery	X ray and gamma ray emission	Ionization		Radioactive decay, fission and fusion reactions
Thermal Energy	Steam turbine	Thermo-couple	Thermal radiation	Endothermic reactions		Heat exchanger

Table 6–1: Energy transformation matrix. Some terms in the matrix may be unfamiliar, but are described in subsequent chapters.

Technology Box 6-1 ■ Is a fuel cell like a dry cell?

In 1839, the British physicist William R. Grove, discovered a unique way to combine hydrogen and oxygen molecules that produces electricity in addition to water. It wasn't until the 1960s, however, that fuel cells, based on this concept, evolved into a practical technology. The development was spurred by the deployment of lightweight – and expensive – versions of these devices by the National Aeronautics and Space Administration (NASA).[1]

A fuel cell is really a simple device, consisting of an electrolyte – a material that passes ions, but blocks electrons sandwiched between two electrodes. Hydrogen gas flows to the anode where the electrons are freed, leaving behind positively charged hydrogen ions, H^+, that readily diffuse through the electrolyte to the cathode. Meanwhile the freed electrons travel across an electrical circuit to the cathode where they react with the hydrogen ions and oxygen to produce water. This passage of electrons constitutes an electric current that can serve as a power source.

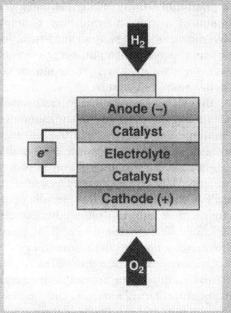

Fuel cells and batteries are similar in that both rely on electrochemistry. The difference lies in the reactants – fuel cells rely on hydrogen and an oxidizer (the atmosphere often serves this purpose), whereas the reactants in a battery are the materials used in the electrodes such as nickel oxyhydroxide and cadmium.

Today's fuel cells are still too expensive for widespread use. However, they represent a clean energy technology that provides energy in sufficient quantities to power cars and trucks, buses, homes, even cellular telephones and laptop computers! Various research groups will continue their efforts to reduce the costs and improve the technology to make the widespread application of fuel cells a reality.

---

[1]*Scientific American*, vol. 281, no. 1, pp. 72–73, 1999.

Electrical energy can also be converted into thermal energy. For example, when an electrical current passes through a conductor, its temperature increases because of its resistance. Electrical power is sacrificed at exactly the rate thermal energy is produced, as measured in watts.

Table 6-1 shows an energy transformation matrix, illustrating some typical energy forms and means for conversion. These are only a few of the many different energy transformations that exist and it would be a difficult task to account for every one. If you pay special attention, you will note that throughout this text energy is a concept that permeates all aspects of physics, chemistry, and technology.

Units and Dimensions

Today the United States remains a dual-system society. The aspiring technology professional, scientist, or engineer will quickly learn that various sectors of technology may embrace either the SI or the English system of units. To further complicate matters, the derived units from either system may also vary from technology to technology. A single industry may even adopt two different derived units to quantify the same dimension! An example lies in the power industry where the thermal power derived from fossil fuel is measured in British thermal units, Btus, (English system) and the electrical power produced by the plant is measured in kilowatt-hours, kWh, (SI system). Given the seemingly haphazard adoption of units throughout the various sectors of technology, it is critical that today's technology professional become familiar with both systems of units. Some common energy units and their conversions are provided in Table 6-2.

	J	ft·lb$_f$	cal	Btu	kWh
J	1	0.73756	0.239	9.488×10^{-4}	2.78×10^{-7}
ft·lb$_f$	1.35582	1	0.324	1.285×10^{-3}	3.77×10^{-7}
cal	4.186	3.09	1	3.968×10^{-3}	8.60×10^{5}
Btu	1.054×10^{3}	778.17	252	1	2.93×10^{-4}
kWh	3.60×10^{6}	2.65×10^{6}	1.16×10^{-6}	3.42×10^{3}	1

Table 6–2: Common energy units and conversion factors. Each number gives the unit value of energy named at the left in terms of the unit named at the top.

Check Your Understanding

1. What is the definition of work?

2. What is a system, as defined in science?

3. What is the first law of thermodynamics?

4. What are the six forms of energy?

5. What is the difference between transitional and stored energy?

6. Which kind of chemical reactions release energy?

7. What does the second law of thermodynamics severely limit?

8. What is a reversible process?

Objectives

Upon completion of this section, you will be able to:

■ Understand the properties of a thermodynamic system.

■ Convert between the various units of temperature and thermal energy.

■ Calculate the energy change by a system due to heat flow or work.

■ Balance the energy of a system with respect to heat and work.

■ Describe the first and second law of thermodynamics.

The word **thermodynamics** derives from the Greek words *therme* and *dynamis*, which mean heat and force, respectively. The science of thermodynamics emerged in the early part of the nineteenth century, based on the principle of conservation of energy, and can best be described as the study of processes involving energy transformations and heat. It provides an accounting system for keeping track of the amount of energy within a system and energy flow into and out of the system. Thermodynamics and other engineering sciences provide tools for the analysis and design of mechanical systems intended to meet human needs. They produce improved designs that increase output of work, reduce consumption of natural resources, and lessen environmental impact.

During the industrial revolution, there was considerable interest in building machines to do mechanical work. The first commercially successful steam engine for this purpose was patented by Thomas Savery in 1698. Many improvements were made to its design during the eighteenth century. In 1711, Thomas Newcomen introduced his engine to pump water in mines. In this design, steam, under pressure, was admitted from the boiler into a cylinder to raise a piston. As seen in Figure 6-10, a counterweight on the other end of the beam also helped lift the piston. The steam valve was then closed and the steam in the cylinder condensed by a jet of cold water. This resulted in a large pressure drop in the cylinder. The piston was forced back into the cylinder by atmospheric pressure, thus producing work to operate the pump. The water in the cylinder was then expelled through an escapement valve by the entry of more steam.[2] Later in the eighteenth century, the great inventor James Watt made many significant improvements in the design of the steam engine.

The early steam engines were extremely inefficient. As designs improved and efficiency increased, scientists began to wonder if these machines had an ultimate level of efficiency. This led to the development of thermodynamic principles that allowed scientists to quantify the maximum theoretical performance of such machines.

Properties

The many characteristics of a system are known as **properties**. The more common ones include temperature, pressure, volume, and mass. Other properties include internal energy, enthalpy, entropy, thermal conductivity, and electric resistivity. Of this second set, internal energy will be considered in this chapter. The **internal energy** represents all the energy belonging to a system while it is stationary, including nuclear, chemical, and thermal.

---

2. *Steam / Its Generation and Use*, Babcock & Wilcox Company, New York, 1975.

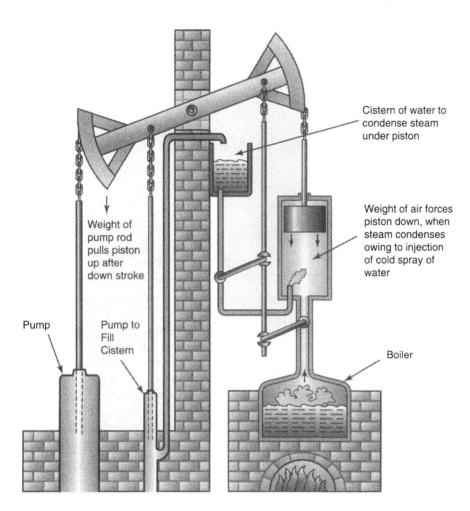

Figure 6–10: Newcomen's steam engine used steam pressure and a counterweight to raise a piston. The steam valve was then closed and the steam condensed by a jet of cold water. The resulting pressure drop allowed the atmosphere to force the piston back into the cylinder doing work.

Properties are considered either intensive or extensive. **Intensive properties** are those that are independent of the size of a system and **extensive properties** are those that depend on the size of the system. Some examples of extensive properties are mass, m, volume, V, and total energy, E. Examples of intensive properties include specific volume, v, $(v = V/m)$, and specific total energy, e, $(e = E/m)$. It should be noted that these intensive properties are equivalent to their related extensive properties, divided by the mass of the system. Generally, uppercase letters are used to denote extensive properties, while lowercase letters are reserved for intensive properties. The three major exceptions are for the extensive property mass, m, and the intensive properties pressure, P, and temperature, T.

A straightforward method to determine whether a property is intensive or extensive is to assume a simple system comprising a fixed mass of gas. Divide the system into two parts as shown in Figure 6-11. The intensive properties in each half will retain the same values as in the whole system. The values of each extensive property will take half its original value. For example, each half of the system shown in Figure 6-11 has half the original mass, volume, and total energy of the original system. But the temperature, pressure, and density of each half are the same as that of the original system.

Gas Properties	Gas System		Property
	Part 1	**Part 2**	
Mass, m	$\frac{1}{2}$ m	$\frac{1}{2}$ m	
Volume, V	$\frac{1}{2}$ V	$\frac{1}{2}$ V	Extensive
Total energy, E	$\frac{1}{2}$ E	$\frac{1}{2}$ E	
Temperature, T	T	T	
Pressure, P	P	P	Intensive
Density, ρ	ρ	ρ	

Figure 6-11: Criteria to differentiate between intensive and extensive properties.

When a system is in equilibrium, each of its properties takes on a constant value and the total set of property values completely describes the state of that system. In fact, for simple systems, we only need to determine the value of two independent, intensive properties to completely define the state of the system. As an example, assume a mass of H_2O, at equilibrium, atmospheric pressure, and a temperature of $-2°C$. Knowing temperature and pressure, we can determine the value of every other property and predict the phase of the system. In this case, it would be a solid, since temperature and pressure are both intensive properties and are independent of one another at this state.

Temperature

It was explained earlier that the thermal energy in a substance is simply the kinetic energy associated with the random motion of its atoms and molecules. Random motion takes place among the atoms and molecules, but doesn't happen on a scale that we can see.

The motion of each molecule changes billions of times each second because of collisions with other molecules. A single molecule may move faster or slower after any single collision, depending upon the nature of the collision. However, if we were to examine the average motion of any single molecule for a second or more, we would find that the average motion of each water molecule is about the same.

Compare this to one hundred cars, all traveling to the same destination along the same route. Each driver may start, stop, accelerate, or slow, at his or her own discretion. If all drivers manage to arrive at the final destination in the same amount of time, then they all have driven at the same average velocity. This is in spite of the fact that at any instant during the trip they were each likely driving at different velocities.

Back to our cup of water. It should be clear then that as the temperature of water increases, the water molecules translate (move) from one place to another faster, vibrate with greater amplitude, and rotate faster. In other words, the average energy associated with each type of motion increases. The opposite occurs as the water cools.

Temperature is an indicator of the average thermal energy of the atoms or molecules in a substance. Direct temperature measurement quantifies the energy or energy change of a system more specifically than subjective terms like hot, cold, lukewarm, warmer, and colder. While you may have noticed a cold drink grows warmer in a warm

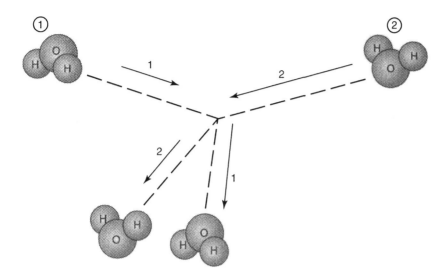

Figure 6–12: Two water molecules before and after collision. The molecule entering from the upper-left is moving slower before the collision, but faster afterward. The molecule entering from the upper-right is moving faster prior to the collision, but slower afterward.

room, this change can be quantified with great accuracy and precision using a thermometer.

It is important to note that temperature is a measure of the average thermal (kinetic) energy per molecule or atom, whereas thermal energy represents the total random kinetic energy of all atoms or molecules in a system. To illustrate this point, consider two systems, a pint of water and a gallon of water. Both are at the same temperature. The average thermal energy of any molecule of water in either system is, therefore, the same. However, the total thermal energy in a gallon of water is eight times greater than that of the pint of water, because it weighs eight times as much.

Temperature can be measured using a variety of different instruments, each based on the principle that an observable or measurable property of a substance changes in a predictable manner with respect to temperature. Such a property is called a **thermometric property**. You are familiar with one of the most common instruments for temperature measurement – the thermometer – in which a narrow (capillary) tube is attached to a bulb containing a liquid. As the liquid in the bulb warms, it expands into the narrow tube. Marks can then be inscribed on the tube at points that equal the liquid level when placed in a bath of known temperature.

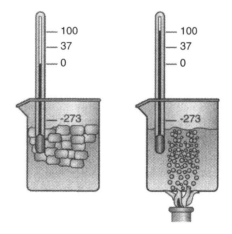

Figure 6–13: A mixture of water and ice have a temperature of 0° Celsius, while boiling water at sea-level pressure has a temperature of 100° Celsius.

The common temperature scales – Fahrenheit and Celsius – are based on the freezing and boiling points of water. On the Celsius temperature scale, 0° and 100° are assigned to these points respectively and the space between divided into 100 equal divisions or degree intervals. On the Fahrenheit temperature scale, 32° and 212° are assigned to the freezing and boiling

---

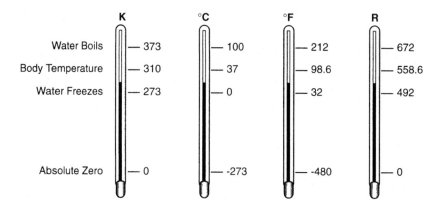

	K	°C	°F	R
Water Boils	— 373	— 100	— 212	— 672
Body Temperature	— 310	— 37	— 98.6	— 558.6
Water Freezes	— 273	— 0	— 32	— 492
Absolute Zero	— 0	— -273	— -480	— 0

Figure 6-14: The freezing point of water, as measured on the four common temperature scales.

points of water, with the space between these marks divided into 180 equal degree divisions or intervals.

The numbers 32 and 212 may seem arbitrary. In fact, the Fahrenheit scale was created by the German physicist Gabriel Fahrenheit during the early eighteenth century. Fahrenheit wanted a scale in which all temperatures commonly measured would be positive. At the time, the coldest temperature he was able to obtain was that of a freezing mixture of salt and water. This, and the lowest outdoor temperature he expected to measure in Holland where he lived, were both greater than 0°F. Thus, he assigned a value of 32°F to the freezing point of water, certain that all temperatures he would ever need to measure would be positive numbers. Fahrenheit chose 212°F as the boiling point of water so that normal human body temperature would be 100°F. Clearly, he was mistaken about the human body temperature and the range of temperatures that people would someday need to measure! For example, the coldest outdoor temperature ever recorded was –127°F on August 24, 1960 at Vostok Station, Antarctica.

As discussed in Chapter 2, it is believed that the coldest any object can possibly become is –459.67°F or –273.15°C, known as absolute zero. At this temperature, the random motion of atoms and molecules is at a minimum. From the point of view of the atoms and molecules, all of the kinetic energy has been removed, so they don't move or collide. They may not even vibrate. This state has never been reached and some quantum theories suggest that it is impossible to achieve. There is, however, no known upper limit to temperature.

The equations for converting between Fahrenheit, T_F, and Celsius, T_C, and Kelvin and Rankine temperature scales were given in Chapter 2. Remember that temperatures expressed in absolute units are abbreviated K and R, not °K and °R, according to convention. Temperature intervals are the same on the Celsius and Kelvin scales and on the Fahrenheit and Rankine scales, as shown in Figure 6-14.

Heat

We previously defined heat and work as the transitional forms of thermal and mechanical energy, respectively. We also learned that thermodynamics is about energy transformation and heat. It is not surprising, therefore, that heat and work have the same units. Heat and work are forms of energy, moreover both are energy in motion – energy flowing across the boundary of a system.

Substances change temperatures at different rates, when they are exposed to the same temperature difference. Why would a camper put hot water bottles in his or her sleeping bag on a cold night? Why not a hot aluminum shovel – it takes less time to raise the temperature of a mass of aluminum than it does an equivalent mass of water. So, why waste time? This time difference, however is a good clue about the characteristics of water. It indicates that water is absorbing more thermal energy than an equivalent mass of aluminum. In fact, water is able to store more thermal energy, per unit mass, per degree, than most other substances. It also cools slowly, providing warmth longer to the camper.

The driving force that promotes energy transfer, in the form of heat, between a system and its surroundings is the presence of a temperature difference. If two objects are in **thermal contact**, heat will flow from the hotter to the cooler object.

The quantity of heat required to raise the temperature of a given mass of substance by a specific amount varies from substance to substance. For example, the energy required to raise the temperature of 1 kg of water by 1°C is 4,186 J, but it takes only 900 J to raise the temperature of 1 kg of aluminum by the same amount. We define this property as the **specific heat**, c, of a substance. It is the thermal energy required to raise the temperature of 1 kg of a substance by one degree Celsius. From this, we can conclude that the specific heat of water is 4,186 J/kg · °C and for aluminum is 900 J/kg · °C. We can then calculate the thermal energy, Q, transferred to or from a mass, m, as its temperature changes by an amount ΔT, where ΔT equals final temperature minus initial temperature according to the following equation:

$$Q = m\,c\,\Delta T$$

This equation basically states that the larger the mass, specific heat, or change in temperature, the more thermal energy is transferred. Consider the winter camping scenario again. If 2 one-quart water bottles containing approximately 2 kg of water at 95°C, were allowed to cool to body temperature (37°C), how many joules of thermal energy will they provide?

$$Q = m\,c\,\Delta T$$

$$Q = (2\ \text{kg})\,(4{,}186\ \text{J/kg} \cdot {}^\circ\text{C})\,(T_{\text{final}} - T_{\text{initial}})$$

$$Q = (8{,}372\ \text{J/}{}^\circ\text{C})\,(37^\circ\text{C} - 95^\circ\text{C}) = -485{,}576\ \text{J}$$

$$Q = -485.6\ \text{kJ}$$

The result is negative since the temperature decreased, thus indicating that the water gives up energy. Recall that power is the rate of energy production or consumption over time. If we assume it takes six hours for the water bottles to reach body temperature, then we can determine the average power, P, delivered by the bottles over that span of time.

$$P = \frac{Q}{t} = \frac{485{,}576\ \text{J}}{(6\ \text{hr})\,(3{,}600\ \text{s/hr})} = \frac{22.5\ \text{J}}{s} = 22.5\ \text{W}$$

Note that we have dropped the negative sign, which indicates that heat was lost from the water bottles. The interesting result is that the average power provided is significant! In a well-insulated sleeping bag, two quarts of hot water will provide a very cozy environment for much of the night!

Another term we use when discussing thermal energy is the extensive property of **heat capacity**, C. The heat capacity of a particular system is the amount of heat required to raise the temperature of the entire system by one degree Celsius. Specific heat, c, can be thought of as the heat capacity, C, of a system divided by the mass, m, of the material, $c = C/m$, or

$$C = mc$$

Physical Science: What the Technology Professional Needs to Know

In general, specific heats vary with temperature and depend on whether the measurements are made at constant pressure or at constant volume. For solids and liquids, however, these differences are usually less than a few percent and are often neglected. Table 6-3 shows the specific heats, expressed in SI units, for some common substances at various states.

Liquids	Temperature	$c, \dfrac{kJ}{kg \cdot K}$	Solids	Temperature	$c, \dfrac{kJ}{kg \cdot K}$
Ammonia	liquid at −20°C	4.52	Aluminum	300 K	0.90
Ammonia	liquid at 10°C	4.67	Copper	300 K	0.38
Ammonia	liquid at 50°C	5.10	Copper	400 K	0.39
Ethyl alcohol	at 1 atm, 25°C	2.43	Ice	200 K	1.56
Glycerin	at 1 atm, 10°C	2.32	Ice	240 K	1.86
Glycerin	at 1 atm, 50°C	2.58	Ice	273 K	2.11
Mercury	at 1 atm, 10°C	0.14	Iron	300 K	0.45
Mercury	at 1 atm, 315°C	0.13	Lead	300 K	0.13
Refrigerant 12	liquid at −20°C	0.90	Silver	300 K	0.24
Refrigerant 12	liquid at 0°C	0.96			
Water	at 1 atm, 0°C	4.22			
Water	at 1 atm, 27°C	4.18			
Water	at 1 atm, 100°C	4.22			

Table 6–3: Specific heats of some common substances at various states.

First Law of Thermodynamics

If you have ever balanced a checkbook, then you are ready for the first law of thermodynamics. While energy is the currency that drives the universe, conservation of energy is the accounting method that keeps track of the flow of that currency (energy) into and out of a system. The balance of energy in a system varies only by the amount of energy added to or subtracted from that system.

When we discuss a **thermodynamic system**, we mean a quantity of matter or a region of space, chosen for study. As previously noted, the region outside the system is the surroundings and the surface that separates the system from its surroundings, whether real or imaginary, is the boundary of the system. An example of a real boundary is the aluminum can that encloses a beverage and a gas. The aluminum physically prevents the movement of mass across the boundary. Now let us imagine that we open the can. There becomes a region of the boundary across which mass may flow. We can assume that the boundary has not changed, it remains a closed surface, but that area that was cut open by a can opener now represents the imaginary part of the boundary.

Within a **closed system,** the mass remains constant. For example, the liquid and gas inside the sealed aluminum can (see Figure 6-15). When the can is opened, an **open system** or **control volume** is created where mass may flow across the boundary. Since liquid flows into and out of a pump, for example, it would be modeled as an open system. Energy, in the form of heat or work, may flow across the boundary of either a closed or an open system. Finally, a closed system that has neither heat nor work interactions is called an isolated system. An **isolated system** is one that has a fixed mass and constant internal energy.

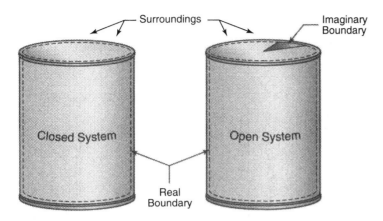

Figure 6–15: A sealed and an opened aluminum can, representing closed and open systems respectively. The dashed lines represent the system boundaries.

Remember that the first law of thermodynamics is simply an accounting system. For a closed system, the first law can be expressed mathematically as:

$$\Delta E = Q_{net} + W_{net}$$

In this equation, E represents total energy, $Q_{net} = Q_{in} - Q_{out}$, and $W_{net} = W_{in} - W_{out}$. In other words, the change in the total amount of energy contained within a closed system during some time interval is exactly equal to the net amount of energy added, minus the net amount of energy leaving the system. All the heat flowing into and work done on the system ($Q_{in} + W_{in}$) and all the heat flowing out of and work being done by the system ($Q_{out} + W_{out}$) must be compared. Consider your checking account as an analogy. The change in your balance is equal to all the deposits minus all the withdrawals.

Earlier in this chapter, it was mentioned that there are many different forms of energy. However, for the most common devices that involve heat and work interactions, we need only consider changes in three forms of energy – thermal, kinetic (mechanical), and gravitational potential energy. The change in total energy, ΔE, of the system can then be expressed as:

Energy change = Thermal energy change + Kinetic energy change + Potential energy change

$$\Delta E \quad = \quad \Delta U \quad + \quad \Delta KE \quad + \quad \Delta PE$$

$$\Delta E \quad = \quad U_2 - U_1 \quad + \quad \tfrac{1}{2} m(v_2^2 - v_1^2) \quad + \quad mg(h_2 - h_1)$$

In these equations, U is the internal energy of the system. **Internal energy** is all the energy belonging to a system while it is stationary, including nuclear, chemical, and thermal energy.

For a simple process – in the absence of any nuclear or chemical reaction – any change in internal energy simply equals the change in thermal energy. To calculate the internal energy, we need to know two independent properties, one of which must be extensive. So, we need only consider changes in temperature and any other independent extensive property. Kinetic energy is determined by changes in velocity and potential energy is determined by changes in elevation. All must be accounted for to determine ΔE. The first law of thermodynamics, therefore, can be simplified to:

$$\Delta U + \Delta KE + \Delta PE = Q_{net} + W_{net}$$

For example, consider process a that takes a system from an initial state, i, to final state, f, by the system absorbing 16 kJ of heat (energy in) and doing 12 kJ of work (energy out). Such a process might occur in a piston-cylinder device such as is depict-

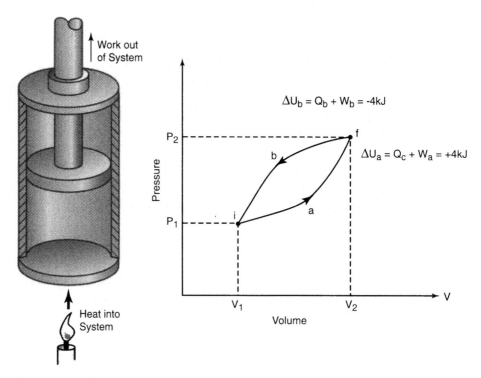

Figure 6–16: The internal energy increases during process *a*, heating, and decreases by the same amount during process *b*. The internal energy, *U*, depends only on the state of the system, not on the path it takes to attain that state. The process path is determined by work and heat interactions.

ed in Figure 6-16. Process *b* returns the system from state *f* back to its original state, *i*, as 18 kJ of heat is rejected (energy out) by the system.

Neglecting ΔKE and ΔPE by assuming that neither velocity nor elevation of the system significantly changed during either process, we can apply the equation to determine the change in internal energy of the system.

$$\Delta U_a = U_f - U_i$$

$$\Delta U_a = Q_a + W_a$$

$$\Delta U_a = 16 \text{ kJ} + (-12 \text{ kJ}) = 4 \text{ kJ}$$

In words, this reads that the change in internal energy of the system, ΔU_a, during process *a*, equals the internal energy at its final state, U_f, minus its internal energy at its initial state, U_i. This difference is equal to the energy added due to heat in, Q_a, minus the energy lost by work out, W_a. Since 16 kJ of energy was added and 12 kJ of energy was removed, the balance or net change in internal energy of the system, is 4 kJ. It is very important to note that heat and work occur during the process and that the internal energy changes during the process. Internal energy is a property of the system, but heat and work are not. Rather, they represent energy transfer across the boundary of the system.

Now assume that during process *b*, the heat transfer Q_b equals −18 kJ (negative because heat is rejected). We also know that during process *b* the system returns to its original state, so it loses the same amount of internal energy that it gained during process *a*.

$$\Delta U_b = U_i - U_f$$

$$\Delta U_b = -\Delta U_a = -4 \text{ kJ}$$

From the first law, we can determine the work done during process *b*.

$$\Delta U = Q_{net} + W_{net}$$

$$W_b = \Delta U_b - Q_b$$

$$W_b = -4kJ - (-18 \text{ kJ})$$

$$W_b = 14 \text{ kJ}$$

Since the work is positive, energy is coming into the system, or we say that work is being done on the system. The system is being compressed during process *b*.

The kinds of thermodynamic processes to be considered in this text are assumed to occur slowly enough that the properties of the system, such as temperature and pressure, are the same everywhere at any given time during the process. This is a reasonable assumption for many systems and simplifies the analysis.

Second Law of Thermodynamics

Consider again a cold beverage sitting in a warm room. From experience, you know that the beverage will get warmer as it gains heat from the room, it will not get colder by delivering heat to the room. However, the first law of thermodynamics would allow this to happen provided that the energy lost by the beverage is equal to the energy gained by the air in the room. It is the second law of thermodynamics that governs the direction in which a process must occur. It states that whenever two objects at different temperatures are placed in thermal contact with one another, heat must flow from the warmer to the cooler object. Thus, a cold beverage receives heat from the warm air in a room until the beverage and air are at **thermal equilibrium**.

According to the first law of thermodynamics, the internal (thermal) energy of a system may be increased or decreased by allowing heat to flow, or work to be done, across the boundary of the system. The first law pays no attention to the difference between heat and work. The second law of thermodynamics, however, places restrictions on which processes can and cannot occur in nature and recognizes that heat and work are very different mechanisms for energy transfer.

Furthermore, the second law of thermodynamics explains why it is easy to build a machine that converts mechanical energy completely into thermal energy. For instance, when you rub your hands together, all of the mechanical energy you generate is converted to thermal energy, thus warming your palms and muscles. However, the second law places restriction on conversions in the other direction. For instance, a machine designed to convert thermal energy into mechanical energy, such as an automobile engine, is limited. Only a fraction of the thermal energy produced by burning the fuel can become mechanical energy. The rest remains thermal energy and is lost purposefully through the exhaust system of the vehicle, which elevates the temperature of the exhaust system and its surroundings.

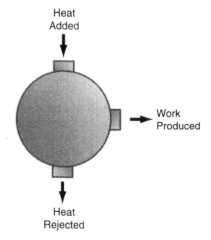

Heat Added

Work Produced

Heat Rejected

Figure 6–17: A heat engine converts a portion of each unit of thermal energy absorbed into work, and rejects the rest as heat.

In summary, the second law of thermodynamics states:

— You can never convert all of the energy in a system into work. There will always be some energy lost.

— You can never have a 100% efficient engine. Losses are inevitable.

— Systems tend to move from order to disorder.

— Heat flows in only one direction – from hot to cold.

A **heat engine** is a device that converts thermal energy into other useful forms such as mechanical or electrical energy. For instance, in a power plant a fossil fuel such as coal or natural gas is burned to produce steam at high temperature and pressure – a reserve of thermal energy. As shown in Figure 6-18, the steam is directed at the blades of a turbine, which causes it to rotate and convert much of its thermal energy into kinetic energy. The mechanical energy of the turbine is then converted into electrical energy as the shaft driven by the turbine is used to rotate a generator. It is interesting to note, however, that only a fraction of the energy released through the combustion process is ever converted into electricity! The remainder of it leaves through the stack, the cooling water, or is lost because of irreversible phenomena such as friction. Researchers of turbine and engine design are investigating ways of reducing these losses by using superconductive materials and components with reduced friction.

The **thermal efficiency**, ε, of a heat engine is defined as the ratio of the net work done by the engine, W, to the heat added or thermal energy absorbed by the engine, Q_h.

$$\varepsilon = \frac{W}{Q_h}$$

A heat engine with perfect efficiency (100%) would convert all thermal energy absorbed into mechanical work, but because of the second law of thermodynamics we

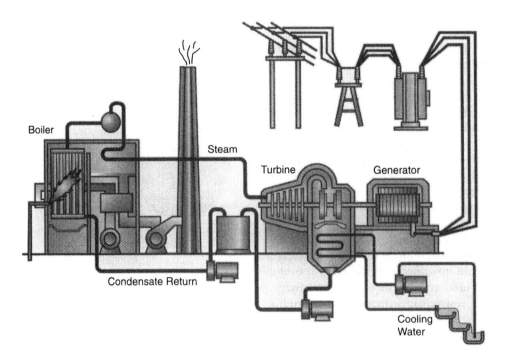

Figure 6–18: Features of a typical power plant used to produce electricity.

know this cannot happen. In practice, all heat engines convert only a fraction of the thermal energy absorbed into mechanical work. For instance, a well-designed automobile engine has an efficiency of about 20% and diesel engines range between 35% and 40%.[3]

Consider a small engine that absorbs 2,000 J of heat and produces 300 J of mechanical work in the form of a rotating shaft each time the engines turns one revolution. What is the thermal efficiency of this engine? By applying the above equation, we find that:

$$\varepsilon = \frac{W}{Q_h}$$

$$\varepsilon = \frac{300 \text{ J}}{2,000 \text{ J}} = 0.15 = 15\%$$

The remaining 1,700 J of thermal energy, Q_c, is exhausted to the environment.

Check Your Understanding

1. What is the meaning of the term thermodynamics?

2. What is the difference between an intensive and extensive property?

3. What property is an indicator of the average thermal energy of the atoms or molecules in a substance?

4. What is absolute zero?

5. What is the definition of specific heat?

6. How are specific heat and heat capacity related?

7. What must remain constant in a closed system?

8. What is governed by the second law of thermodynamics?

9. What is a heat engine?

10. What is the definition of thermal efficiency?

3.   Serway, R. A., *Physics for Scientists and Engineers*, Saunders College Publishing, Philadelphia, 1997.

6-3 Gas Laws

Objectives

Upon completion of this section, you will be able to:

- Describe and distinguish between absolute pressure and gauge pressure.

- Understand the relationship between temperature and average kinetic energy of molecules.

- Apply the ideal gas equation of state.

- Determine the work done on or by a system.

- Apply the first law of thermodynamics.

Imagine that you are suddenly transformed into a molecule in the air and reduced to about $1/10,000,000,000$ (10^{-10}) your normal size; a size comparable to that of all the other molecules around you. In this state, you are separated from adjacent molecules by an average distance of about ten times your own size. For the sake of comparison, this is roughly the distance between several people spread randomly across the floor of a large gymnasium.

You and your fellow molecules would feel little effect from each other were you not traveling at about 1,000 mph and constantly crashing into each other. These collisions are violent and occur at a rate of about 300 billion per minute, each causing a change in the direction of both molecules. Because of the random nature of these collisions, there is little net displacement during that one minute. You would simply be playing a game of random energy and momentum exchange, finding yourselves at the end of each minute still within a few centimeters of your respective starting positions. Over time, however, this rapid crashing of molecules may result in a gradual drift away from the original starting point. This net movement of one gas through another was introduced in Chapter 4 and called diffusion.

Pressure

The pressure, P, of a system is equal to the force, F, exerted by the substance per unit area, A.

$$P = \frac{F}{A}$$

The area – or system – can be a container, like a balloon, or the atmosphere of Earth. Each molecule in any system is in constant motion, crashing into other molecules and physical barriers, such as the walls of a balloon or the people on the Earth (see Figure 6-19). The sum of all the individual forces that the molecules apply during these collisions, regardless of the size of the system, constitutes pressure. During each sec-

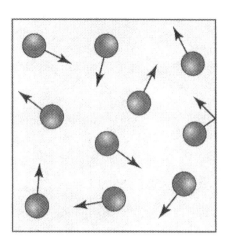

Figure 6-19: Gas molecules moving about within a container.

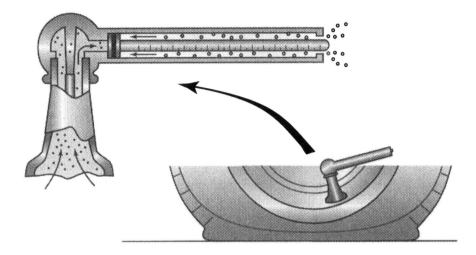

Figure 6–20: Gauge pressure is the difference between the pressure inside and outside a tire. Although the tire is flat and the gauge pressure is zero it still contains air, but the pressure of the air striking either side of the plunger in the gauge is equal.

ond, about 10^{23} molecules in the air strike each square centimeter of your skin[4]! You do not feel the force of each collision, but the net effect is a constant force of nature called atmospheric pressure. The mass of air that makes up the atmosphere of the Earth is also subject to the force of gravity, like any other mass. The air above rests on and squeezes the air below, contributing to atmospheric pressure and explaining why the pressure at sea level is greater than the pressure at higher elevations.

At sea level, the particles that constitute our atmosphere cause an average force of 1.01×10^5 N on each square meter of surface (1.01×10^5 N/m^2), where 1 N/m^2 = 1 Pascal, Pa. In English units, atmospheric pressure equals 14.7 lb$_f$/in^2. Pressure is a convenient property with which to determine the state of a system because it is relatively easy to measure.

Measurements taken with most pressure-measuring devices, or gauges, are different from what is known as absolute pressure. **Absolute pressure** is measured relative to a vacuum or zero pressure. Most pressure-measuring devices are calibrated to read zero at atmospheric pressure. This means that any non-zero reading would actually represent the difference between absolute pressure and the local atmospheric pressure. This difference is called **gauge pressure**. As shown in Figure 6-20, when a tire is flat, the tire gauge reads zero. This gauge pressure reading of zero means that the air pressure inside the tire is the same as the air pressure outside the tire. The tire is flat because it can't hold up the weight of the vehicle. Though the gauge reads zero, the actual, or absolute, pressure in the tire would be the same one atmosphere of pressure that is outside the tire: 14.7 lb$_f$ / in^2 or 1.01×10^5 Pa or 760 mm Hg.

If absolute pressure is less than local atmospheric pressure, it is referred to as **vacuum pressure**. Absolute, gauge, and vacuum pressures are all positive quantities. Absolute pressures greater than atmospheric pressure are related to each other by the following equation:

$$P_{gauge} = P_{abs} - P_{atm}$$

4.   Van Heuvelen, A., *Physics – A General Introduction*, Little, Brown and Company, Boston, 1982.

Technology Box 6–2 ■ The Barometer: How do meteorologists predict changes in the weather?

Whether you watch the weather channel or listen to the news, you will hear about high and low pressure systems that may affect what you will want to wear on a particular day. Meteorologists detect the presence of these systems using an instrument called a barometer. A barometer is a sealed-end manometer, with the open end placed into a container of mercury that is in contact with the atmosphere.

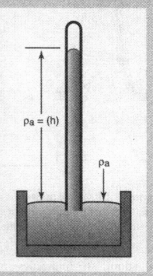

As shown in the figure, atmospheric pressure pushes down on the surface of the pool of mercury. Changes in this pressure cause the mercury in the tube to rise or fall. By measuring the height of mercury in the tube, the absolute air pressure can be measured. Average sea-level air pressure may be reported in a variety of units ranging from 760 mm of Hg, 29.92 inches of Hg, 1 atmosphere, 1,013.25 millibars, or 14.7 pounds per square inch (psi).

Look in your newspaper or on the Internet and track the barometric pressure in your area for a few days. Does there seem to be a correlation between changes in the barometric pressure and local weather conditions?

Technology Box 6–3 ■ How can you raise a heavy object using only a small force?

The hydraulic lift is a simple device composed of a piston with a small cross-sectional area, A_1, that exerts a force, F_1, on the surface of a liquid such as oil. The applied pressure, $P = F_1/A_1$, is transmitted uniformly throughout the fluid according to Pascal's law. Therefore, $F_1/A_1 = F_2/A_2$, which leads to

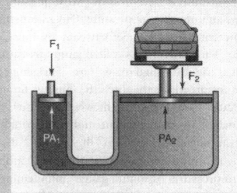

$$F_2 = F_1 (A_2 / A_1).$$

Since piston A_2 is much larger than piston A_1, it follows that force F_2 will be proportionally greater than force F_1. The hydraulic lift is a force-multiplying device with a multiplication factor equal to the ratio of the areas of the pistons. This principle is applied in a variety of applications, including car lifts and hydraulic jacks, dentists' chairs, automotive brakes, and some elevators.

Absolute pressures less than atmospheric pressure related to each other by the this equation:

$$P_{vacuum} = P_{atm} - P_{abs}$$

When the tire is repaired, a new gauge pressure reading indicates 32.0 pounds per square inch (lb_f/in^2). Assume the atmospheric pressure is known to be 14.7 pounds per square inch. (Common pressure units and conversions factors are provided in Table 6-4.) The equation for pressures that are greater than atmospheric can be applied to determine the absolute pressure of the air inside the tire.

$$P_{abs} = P_{gauge} + P_{atm}$$

$$P_{abs} = \frac{32.0 \; lb_f}{in^2} + \frac{14.7 \; lb_f}{in^2}$$

$$P_{abs} = \frac{46.7 \; lb_f}{in^2}$$

	N/m² (pascal)	lb_f/in²	atm	bar	mm Hg (torr)
N/m² (pascal)	1	1.45×10^{-4}	9.87×10^{-6}	1×10^{-5}	7.50×10^{-3}
lb_f/in²	6.89×10^3	1	6.80×10^{-2}	6.89×10^{-2}	51.7
atm	1.01×10^5	14.7	1	1.01	760
bar	1×10^5	14.5	0.987	1	750
mm Hg (torr)	133	1.93×10^{-2}	1.32×10^{-3}	1.33×10^{-3}	1

Table 6–4: Common pressure units and conversion factors.

There are many different ways to measure pressure. One of the simplest pressure-measurement devices is the open-ended **manometer**, a plastic or glass U-tube containing a fluid such as water, or oil. It works by comparing the pressure on one side to atmospheric pressure on the other. If the pressures are equal, the fluid levels will be the same on both sides of the U-tube, as shown in Figure 6-21a. When a person blows

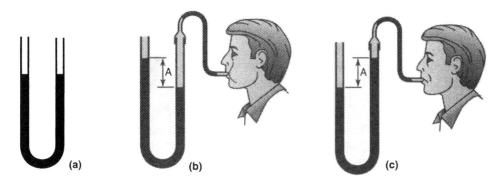

(a)   (b)   (c)

Figure 6–21: A open-ended manometer with pressure (a) equal to, (b) greater than, and (c) less than atmospheric pressure.

into the right side of the manometer (Figure 6-21b), the extra pressure on that side forces the liquid down and up into the left side. When a person sucks on the right side (Figure 6-21c), the reduced pressure causes the liquid to rise on that side. The pressure difference, or gauge pressure, equals $\rho g h$ where ρ is the density of the fluid, g represents the acceleration of gravity, and h is the height difference between the fluid levels in the two sides of the U-tube.

A useful extension of the equation $P = F/A$ is **Pascal's law**, which states that the pressure applied to an enclosed fluid is transmitted, undiminished, to every portion of the fluid and the walls of the containing vessel. A practical application of Pascal's law is discussed in Technology Box 6-3.

Equations of State

The properties that define a gaseous system can be related to each other in a mathematical relationship called the **equation of state**. This relationship can be used to calculate the changes expected to occur in one property because of changes in other properties. As shown in Figure 6-22, if a helium-filled balloon is allowed to float away, the volume of the balloon will increase as it gains altitude. Since it is the weight of the air that is providing the external pressure, and since air pressure lessens with altitude, there is less pressure acting on the balloon as it rises. Thus, the flexible skin of the balloon allows it to equalize the internal and external pressures by expanding – until it bursts.

Figure 6–22: As the altitude of a balloon increases, its volume expands due to decreasing atmospheric pressure.

In the seventeenth century, Robert Boyle discovered an equation of state that related the pressure and volume of a confined gas at constant temperature. By using a column of mercury to trap air on the closed side of a J-tube, as shown in Figure 6-23, the pressure on the air could be increased by adding mercury to the open column. Boyle found that if the temperature and amount of air in the tube were held constant, the volume of air trapped would vary inversely to the pressure acting on it. Thus, if the pressure were doubled, the volume would be decreased to one-half its original value. Boyle described this relationship mathematically by the following expression, commonly known as **Boyle's law:**

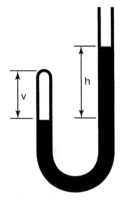

Figure 6–23: Boyle's J-tube experiment lead to Boyle's law, $PV = k$, at constant temperature.

$$PV = \text{constant (at a constant temperature)}$$

Therefore, if the pressure or volume changes, the new pressure and volume are related to the previous conditions by the following relationship:

$$P_1 V_1 = P_2 V_2$$

where, P_1 and V_1, are the pressure and volume of the gas initially and P_2 and V_2 are the pressure and volume of the gas finally. In this equation, absolute pressure must be used and the temperature of the system must remain constant.

Boyle's is only one of many empirical equations that provides a means to calculate expected changes in one property value given changes in others. Another simplified relation is known as **Charles' law**, which states that when the pressure and amount of a gas are held constant, the volume is directly proportional to the absolute temperature. This relation is described mathematically by the following equation, in which Kelvin or Rankine absolute temperature must be used:

$$\frac{V}{T} = \text{constant} \quad \text{(where pressure is constant)}$$

$$\frac{V_1}{T_1} = \frac{V_2}{T_2}$$

Ideal Gas Law

The best known equation of state combines the separate laws, relating the properties of pressure, P, temperature, T, volume, V, and the amount of gas, expressed in moles, n. It derives primarily from work done early in the nineteenth century by Jaques Charles and J. Gay-Lussac of France. They experimentally determined that at low pressures, the volume of a gas is proportional to the number of moles present and its absolute temperature, according to the following relationship:

$$PV = nRT$$

This is known as the **ideal gas equation** or **ideal gas law**. The constant of proportionality, or **universal gas constant**, R, varies with the units used. Some of the more commonly used values of R are 8.314 J/mole · K; 0.08206 liters atm/mole · K; and 1.987 cal/mole · K. Put another way, R is a measure of the energy needed to change the temperature of one mole of any gas by one K.

The ideal gas equation provides a very good approximation for most real gases at low density and at temperatures well above their condensation point. It is also important to remember that temperature and pressure must be expressed in absolute units, whenever the ideal gas equation is applied. Consider an example in which we wish to determine the volume occupied by two moles of helium gas, at 25.00°C and a pressure of 1.00 atm. The first step in completing the calculation is to convert the temperature from degrees Celsius to Kelvin. For SI unit agreement, 1.013×10^5 Pa is substituted for the pressure of 1.00 atmosphere. The resulting calculation in SI units is, therefore:

$$V = \frac{nRT}{P}$$

$$V = \frac{(2.000 \ \text{mol})(8.314 \text{J}/\text{mol} \cdot \text{K})(298 \text{K})}{1.013 \times 10^5 \text{Pa}}$$

$$V = 4.892 \times 10^{-2} \ \text{m}^3$$

In this example, the value used for the universal gas constant, R, was based on the fact that 1 Pa is equal to 1 N/m^2. Since the pressure is expressed as a force, we used the value of 8.314 $J/mol \cdot K$, where 1 J = 1 m $\cdot$ N.

The ideal gas law can also be used to determine the volume occupied by 55.0 g or 1.25 moles CO_2 gas, measured at one atmosphere (absolute pressure) and 25°C. First, we need to convert 25°C to absolute temperature units of Kelvin (25 + 273 = 298 K). Next, 0.08206 liters $\cdot$ atm/mole $\cdot$ K is selected for the universal gas constant, R.

$$V = \frac{1.25 \text{ moles } CO_2 (0.082 \text{ liters} \cdot \text{atm})(298 \text{ K})}{1 \text{ atm} \qquad \qquad \text{mole} \cdot \text{K}}$$

$$V \cong 30.6 \text{ liters } CO_2$$

Note that because of the universal gas constant, R, used in this example, the volume of gas is expressed in the non-SI units of liters rather than cubic meters, m^3, as in the previous example.

Kinetic-Molecular Theory

The **kinetic-molecular theory** is based on a number of assumptions. One is that gases are composed of particles in continuous and random motion, which causes them to collide with one another and the walls of their container. Another assumption is that the collisions are completely elastic (no energy is lost), and that the particles are extremely small compared to their average separation distances. Because of these and other assumptions, the average energy of gas particles is proportional only to absolute temperature.

In 1843, James Joule applied what was, at the time, a hypothesis of the intense motions of molecules in gases. He assumed that all molecules in a sample had the same mass, m, the same speed, v, and that an experimental volume of gas, V, contained N molecules. He also assumed that the molecules struck the walls of the container in a direction perpendicular to the wall with speed, v, and momentum, mv, and rebounded elastically. By application of Newton's second law and the simplifying assumption that exactly one-third of the particles travel in each direction (x, y, and z) within a container, Joule derived an equation for determining the pressure on the walls:

$$P = \frac{1}{3}\left(\frac{Nmv^2}{V}\right)$$

In the last 150 years, this equation has been extended to include assumptions that are more realistic. Surprisingly, the results changed little. Joule's highly simplified assumptions led to correct or nearly correct results. The kinetic-molecular theory also predicts that the total thermal energy, E_{th}, in a gas at temperature, T, and containing, N, particles is given by the following expression,

$$E_{th} = \frac{3}{2}kNT$$

where k, is the average kinetic energy of the particles in the gas and known as **Boltzman's constant** (1.38×10^{-23} joules/K). From this equation, we can see that the absolute temperature, T, is proportional to the thermal energy, E_{th}. We must be clear, however, that temperature is not the same as energy. Temperature is a way of detecting the movement of energy, but internal energy is also present in a system that cannot be quantified by temperature alone.

First Law Revisited

Recall that the first law of thermodynamics for a closed system assumes changes in only three forms of energy – internal, kinetic (mechanical), and gravitational potential. For most problems, only some of the terms in the equation need to be applied. Consider the piston-cylinder assembly shown in Figure 6-24 in which there is no friction between the piston and cylinder. Heat is added to the system (Q_{in}) causing the gas in the cylinder to expand, and subsequently work is done by the system (W_{out}).

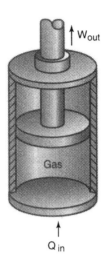

Figure 6–24: Heat added to the system, Q_{in} causes the gas in the piston to slowly expand, resulting in work being done by the system, W_{out}.

Any time a gas occupies a space, it exerts a force on any solid boundaries and the force the gas exerts is related to both its pressure and to the area over which it acts. If the piston is moving slowly at a constant speed, Newton's third law requires that the upward force on the piston and atmosphere be balanced by the downward force due to the weight of the piston and the atmosphere. The weight of the piston and atmosphere are both constant. Therefore, the pressure of the gas in the cylinder must also be constant! As the piston moves up a distance, Δh, the work, W, done by the gas on the piston (and, therefore, on the surroundings) is given by:

$$W = F\Delta h$$

$$W = PA\Delta h$$

$$W = P\Delta V$$

A change in volume of a cylinder, ΔV, equals the product of the cross-sectional area A and a change in height, Δh. In this instance, the first law (change in internal energy, ΔU) can be expressed as:

$$\Delta U = Q_{in} - W_{out}$$

$$\Delta U = Q_{in} - P\Delta V$$

As the gas expands, ΔV and the work term are both positive. The change in the internal energy of the system depends upon the amount of energy entering the system as heat, and the amount of energy leaving as work. Clearly, if the work done on the piston equals the heat added to the system, the internal energy of the system remains constant ($\Delta U = 0$). This does not happen in practice, however. The second law of thermodynamics addresses the loss of heat in the real world. The energy associated with changes in temperature and pressure enables heat engines to power machines, vehicles, and airplanes that provide us with the conveniences that shape our lives.

Let's consider a piston-cylinder device with a diameter of 10 cm. As 65 J of heat is added, the piston rises by one-half centimeter and the pressure within the cylinder remains constant at 540 kPa throughout the process. How much work is done and what is the change in the internal energy of the system? The above equation can be applied to calculate a solution.

$$\Delta U = Q_{in} - P\Delta V$$

$$\Delta U = 65 \text{ J} - (5.4 \times 10^5 \text{ Pa}) \, A\Delta h$$

$$\Delta U = 65 \text{ J} - (5.4 \times 10^5 \text{ Pa}) \, (\pi r^2)(0.5 \text{ cm})$$

$$\Delta U = 65 \text{ J} - (5.4 \times 10^5 \text{ Pa}) \, \pi \, (5 \text{ cm})^2 \, (0.5 \text{ cm})$$

$$\Delta U = 65 \text{ J} - (5.4 \times 10^5 \text{ N/m}^2)(39.27 \text{cm}^3)(1 \text{ m}^3/10^6 \text{ cm}^3)$$

$$\Delta U = 65 \text{ J} - 21.2 \text{ J} \cong 44 \text{ J}$$

We see that 21.2 J of work is done by the system on the surroundings and the internal energy of the system increases by about 44 J during this process. Internal energy is stored by the system and ready to do more work. Note that we neglected changes in kinetic and potential energy in this example, since neither the velocity nor the height of the gas within the piston changes significantly during the process.

Consider for a moment that in a jet engine a volume of air (the system) passes through the engine (see Figure 6-25). In such a case, four terms from the first law equation would apply, according to the following four statements:

— The engine does work on the system by compressing the volume of air at the engine inlet.

— The engine adds heat to the system during the combustion process.

— The system does some work on the engine as it expands through the turbine.

— The system does additional work on the engine as it expands through the nozzle of the engine and accelerates.

The velocity of the volume of air (the system) increases as it passes through the engine. The force that the jet engine applies to accelerate the air must be balanced by the force that the air applies back on the engine according to Newton's third law. This force constitutes the thrust developed by the engine.

It is important to understand that all systems and machines are different and, therefore, each problem requires a unique treatment of, and approach to, the first law

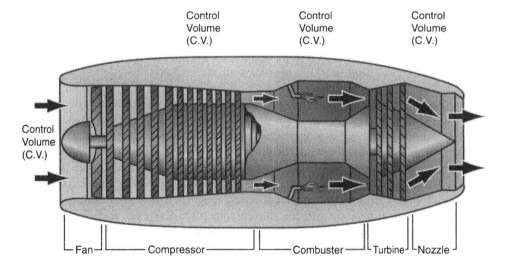

Figure 6–25: A control volume, comprised of a fixed mass of air, is compressed, heated, expanded, and accelerated as it passes through a jet engine.

of thermodynamics. If we were to analyze each of the four processes that pertain to a jet engine, we would have to consider how the first law applies to each process. Clearly, in a simplified analysis, the air is involved in either a heating process or a work process, but not both.

Check Your Understanding

1. What is the difference between absolute, gauge, and vacuum pressures?

2. If the pressure is equal on both sides of a manometer, how will the fluid levels in the two tubes compare?

3. What does the kinetic-molecular theory (kinetic theory) of gases state?

4. What relationship is mathematically described by Boyle's law?

5. What relationship is mathematically described by Charles' law?

6. What is the ideal gas equation?

7. What determines the numerical value of the gas constant, R, used in the ideal gas equation?

Heat Transfer

Objectives

Upon completion of this section, you will be able to:

■ Describe each mode of heat transfer.

■ Calculate the heat flux associated with each mode of heat transfer.

■ Identify the dominant mode of heat transfer for a given problem.

■ Determine when a system is at steady state.

It has long been observed that wherever a temperature difference exists, heat flows from a region or body of higher temperature to another of lower temperature. The rate at which energy flows or heat is transferred is called the **heat flux**, Q. As you might think, heat flux depends on temperature difference, and temperature difference is the driving force behind the heat transfer.

Every area of science and technology involves the flow of heat. In fact, some components are built specifically for the purpose of transferring heat from one medium to another. Power plants have cooling towers to dissipate waste heat into the atmosphere. Cars have radiators. Furnaces have heat exchangers, and your refrigerator has a condensing coil. The length, diameter, and material of the condensing coil on the back of your refrigerator are specified precisely to transfer the heat absorbed by the refrigerant (system) from the refrigerator space and the compressor to the air in the room (surroundings). (See Technology Box 6-4.)

Virtually any industrial process, from manufacturing to computer hardware, biotechnology, and aerospace, makes use of a careful analysis of heat transfer. Electronic components that operate above design temperatures fail prematurely. The strength of metal joints can be compromised if the welding process occurs at too high or too low a temperature. The growth rates of bacteria are highly sensitive to temperature. We know that temperature is proportional to the thermal energy contained within a system. We also know that heat transfer is the flow of thermal energy across the bound-

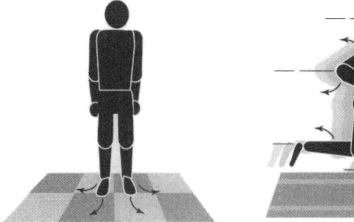

Figure 6–26: A person loses heat from their body by both conduction and convection. The barefooted person (left) is loosing heat to the cold floor by conduction. The person jogging (right) is loosing heat to the air passing across his body by convection.

ary of a system. We realize, therefore, that characterization and control of heat transfer is critical in processes that are sensitive to temperature.

Earlier in this chapter, it was explained that two bodies must be in thermal contact for heat to flow from one to the other, but the modes by which heat transfer may occur were not discussed. It is customary to consider three – conduction, convection, and thermal radiation.

Conduction is the movement of thermal energy from one place to another inside a single body or between two bodies in direct contact. This movement of thermal energy comes from the transfer of kinetic energy between interacting particles. As collisions and other interactions occur between these particles, kinetic energy is transferred from the more energetic (hotter) ones to less energetic (cooler) ones. As shown in Figure 6-26 (left), your feet transfer heat to a cold floor by conduction, since they are in direct contact.

Technology Box 6–4 ■ How do refrigerators and air conditioners work?

Air conditioners and refrigerators make use of the same process. It is based on cycling a working fluid, called a refrigerant, through a series of compression, cooling, expansion, and heat transfer cycles. When a gas is compressed, its molecules are pushed closer together and it becomes hotter. Likewise, when the compressed fluid expands, it absorbs energy from its surroundings to give the molecules the energy they need to move farther apart.

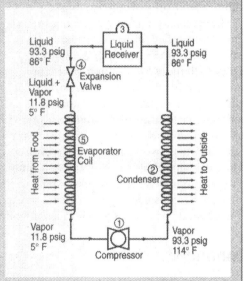

These steps in a typical refrigeration cycle are shown in the figure. Starting at the compressor, the refrigerant (usually one of the family of Freons or ammonia) is compressed and the hot vapor sent to the condenser on the back of the refrigerator. The temperature of the vapor is above room temperature so heat flows from the vapor to the room. As the vapor cools it becomes a liquid that is still under high pressure. The high-pressure liquid is then pumped to the expansion valve where it suddenly experiences a pressure drop. This causes some of the liquid to vaporize, cooling of the liquid/vapor mixture occurs. The liquid/vapor mixture then passes through the evaporator coil, where air from the food storage area transfers heat absorbed from the food to the refrigerant and more refrigerant vaporizes. The cool low-pressure vapor then returns to the compressor, ready for another trip through the cycle.

Freon is an ideal refrigerant because it is chemically inert and will not corrode the compressor or coils. It is stable and will not decompose, even after many, many cycles through the system. Due to its small size, it can escape from the system. Once in the atmosphere, freon rises to the upper edge where it interacts with ozone in the presence of ultraviolet light, reducing the amount of our protective ozone layer.

Convection occurs as a consequence of the motion of a fluid (liquid or gas) relative to a surface. When a fluid at a lower temperature flows over a surface at a higher temperature, heat transfers from the surface to the fluid. As is shown in Figure 6-26 (right), a runner's body transfers more heat when moving than if they were running on a treadmill because more air passes over the runner at a greater velocity.

Radiation, unlike conduction and convection, requires neither direct contact nor the presence of a fluid for heat transfer to occur. In fact, **thermal radiation** is most efficient in the absence of any medium! All bodies continuously emit energy in the form of long wavelength electromagnetic radiation. The warmth you feel while sitting near a fire is due to the thermal energy radiating from the source and striking your body. An infrared image is shown in Figure 6-27. It was taken by a camera that is sensitive to long-wave radiation emitted by a human body or other warm object.

In real life, the temperature of a system is controlled by the combined effects of all three modes. A rigorous analysis of the flow of thermal energy into and out of a system, therefore, must consider conduction, convection, and

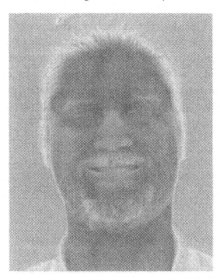

Figure 6–27: An infrared thermogram or "heat picture" of one of the authors.

radiation heat transfers simultaneously. Fortunately, such an analysis is often unnecessary to obtain useful results, since it is typical that one mode is much more dominant than the other two in any given circumstance. For instance, the outside surface of a hot water pipe feels hot due to heat conduction through the material. In order to study the transfer of heat from the pipe to the air flowing across it, only the effects of convective heat transfer need to be considered. In the following sections, each heat transfer mode will be discussed in detail and equations for quantifying it provided.

Conduction

Heat transfer by conduction occurs as a result of the motion of atoms and electrons. Consider Figure 6-28 in which an individual is holding a metal rod over an open flame. As the flame heats the rod on one end, the metal atoms and electrons near that end begin to vibrate with larger and larger amplitudes. These, in turn, collide with nearby atoms and electrons transferring some energy to them. As more energy (heat) is absorbed, a temperature profile (see Figure 6-28) of the rod develops, creating a hot, T_h, end and a cooler, T_c end. It can be demonstrated that there is a uniform temperature decrease from the hot to the cooler end of the rod.

For a length, L, of rod with a cross-sectional area, A, the rate of heat conduction, $\dot{Q}_{cond}$, is defined as the flow of thermal energy past any point on the rod per unit time. This rate is expressed in units of power, such as watts or Btu per hour. It is calculated according to the follow equation:

$$\dot{Q}_{cond} = \frac{Q}{t} = \frac{-kA\Delta T}{\Delta x}$$

in which, $\Delta T/\Delta x$, is the **temperature gradient** or the change in temperature along the length of the rod. Here, k represents the **thermal conductivity**, or how easily a

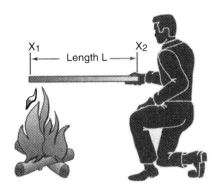
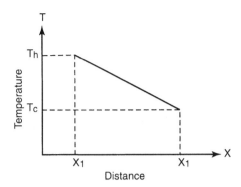

Figure 6–28: One end of a metal rod of length, *L*, is held over an open flame. The temperature profile shows the variation in rod temperature with respect to the distance from the flame.

material allows heat to flow, and has units such as watts per meter per degree Celsius, $W/m \cdot °C$. The equation is negative because heat is lost as one moves from the hot to the cooler end of the rod. In other words, heat flows away from increasing temperature.

To simplify the analysis, we assume that the rod is uniform in size from end to end insulated on all sides so that the heat flows in one direction only at **steady state**. The temperature gradient can then be expressed as the temperature difference at either end of the rod divided by the length of the rod:

$$\dot{Q}_{cond} = \frac{-kA(T_h - T_c)}{L}$$

The degree of thermal conductivity is dependent on the nature of the material. It is influenced by its molecular structure, density, and the number of free electrons. Metals have a large number of free electrons and are generally good thermal conductors. At the other end of the scale, as shown in Table 6-5, gases are poor thermal conductors because they have a disordered structure, low density, and fewer free electrons.

Substance	Thermal Conductivity W/m · °C	Substance	Thermal Conductivity W/m · °C	Substance	Thermal Conductivity W/m · °C
Aluminum	238	Air	0.0234	Asbestos	0.08
Copper	397	Helium	0.138	Concrete	0.8
Gold	314	Hydrogen	0.172	Diamond	2300
Iron	79.5	Nitrogen	0.0234	Glass	0.8
Lead	34.7	Oxygen	0.0238	Ice	2
Silver	427			Rubber	0.2
				Water	0.6
				Wood	0.08

Metals at 25°C / *Gases at 20°C* / *Nonmetals (approx. values)*

Table 6–5: Thermal conductivities for some substances. Serway, R. A.: *Physics for Scientists and Engineers*, Saunders College Publishing, Philadelphia, 1997.

Technology Box 6–5 ■ What is the R-value of your home?

If you have ever been involved in constructing or remodeling a house, then you have probably heard the term "R-value." Simply put, the R-value is a measure of how well a building component – a wall, a roof, a window – resists the flow of heat. The R-value equals the thickness of the building material, such as drywall or insulation, *l*, divided by the thermal conductivity of the material, *k*. When using R-values, the thermal conductivity equation becomes:

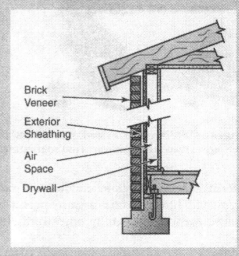

Brick Veneer

Exterior Sheathing

Air Space

Drywall

$$\dot{Q} = \frac{A \Delta T}{R_{tot}}$$

The convenient aspect of R-values is that you can add them together, if your system involves multiple layers of different materials. The diagram shows an exterior wall of a house containing several different layers. Assume that this wall is composed a layer of brick, a layer of exterior sheathing, an air space, and an inner layer of drywall. These layers have R-values of 4.00, 1.32, 1.01, and 0.45 hr · °F · ft²/Btu, respectively. This wall then has a total R-value, R_{tot}, equal to 6.78 hr · °F · ft²/Btu.

If you were to fill the air space with fiberglass insulation that has an R-value of 9.88 hr · °F · ft²/Btu, how does it change the total R-value for the wall? The rate of heat lost will decrease by whatever factor the R-value increased, and so will your energy bills!

Imagine two rods of the same length and diameter – one copper and one glass – both resting over an open fire. Which one would you rather pick up with your bare hand? Would the length of the rods make a difference? How would you define the parameters in the above equation, such that heat conduction from the fire to your hand is minimized? Keep in mind that, as shown in Table 6-5, copper is hundreds of times more conductive than glass!

Convection

Consider why people frequently blow across the surface of their hot beverage before sipping. Without thinking about the physics of this act, they are tying to hurry its cooling. Although our breath is close to body temperature, it is far cooler than the beverage and it readily absorbs thermal energy as it moves along its surface.

When a fluid at one temperature flows over a surface at a different temperature, heat transfer between the fluid and the surface occurs. Your body, for example, transfers heat more efficiently to the air while running outdoors than while running indoors on a treadmill because of the relative velocity between the cooling air and your body.

If the fluid motion is induced by a machine, say a pump or fan, then the heat transfer is said to be by **forced convection**. If the fluid motion results from differences in density – for example, the cool air drawn into the bottom as the warmer air rises above a fire – it is referred to as **natural convection**. This process also occurs when cool air in a room is heated by a hot radiator. The rising warmer air is replaced by cooler falling air, setting up an air current as shown in Figure 6-29.

It is a complicated matter to determine the temperature distribution and heat transfer where convection is the principle mode of heat transfer. The temperatures are different at different

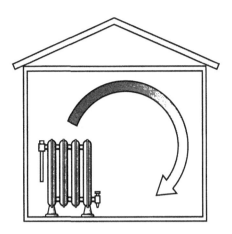

Figure 6–29: Radiators cause natural convection currents in the room being heated.

points within the fluid passing over a surface. These temperatures are influenced by the fluid motion and by the temperature difference between the fluid and the surface. We can create an equation that looks like the one for conduction. There are, however, a couple of important differences. First, we don't have a hot and cold end. We have a surface temperature, T_s, and an overall temperature for the fluid, T_f. Secondly, the thermal conductivity constant, k, is replaced by the **heat transfer coefficient**, h, of the fluid. The surface area being cooled or warmed is represented in the equation as A. As before, $\dot{Q}_{cond}$ represents heat flux and the heat transfer coefficient may have units of watts per square meter per degree Celsius, $W/m^2 \cdot °C$.

$$\dot{Q}_{cond} = hA(T_s - T_f)$$

Alternately, for heat transfer from a hot fluid to a cool surface, the equation would be modified slightly, producing:

$$\dot{Q}_{cond} = hA(T_f - T_s)$$

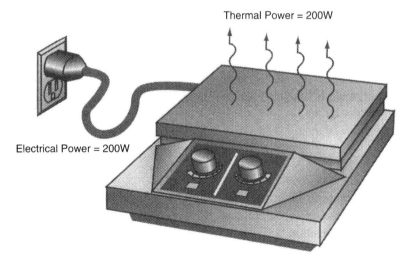

Thermal Power = 200W

Electrical Power = 200W

Figure 6–30: A hot plate operating at steady state has a balance of electrical power entering and thermal power delivered to the surroundings.

The heat transfer coefficient, h, varies according to whether the fluid flow is smooth or turbulent, the shape of the surface, the properties of the fluid, even the temperatures of the fluid and the surface. It also depends on whether the convection is forced or induced naturally. Typical values of h are shown in Table 6-6.

Think about the following questions: If you were in a hurry to eat a bowl of hot soup, how could you cool it most quickly? Would it speed the process, to blow on each spoonful separately? Would it cool faster in a large bowl or a polystyrene foam cup?

Consider an electrically powered hot plate, as shown in Figure 6-30. Assume that it is rated at 200 W and the surface of the plate is 15×15 cm. When the hot plate is operating, the heat transfer coefficient along the surface is 80 W/m$^2 \cdot$ °C). If the room temperature is 23°C, what is the temperature of the hot plate?

We assume that the hot plate converts all electrical energy to thermal energy and by application of the above equation, we are therefore able to determine its surface temperature, T_s.

$$T_s = \frac{\dot{Q}}{hA} + T_f = \frac{200\,W}{(80\,W/m^2 \cdot °C)(0.15\,m)^2} + 23°C = 134.1°C$$

Type of flow	Convective Heat Transfer Coefficient (W/m² · °C)
Natural convection, ΔT = 25°C	
0.25 m vertical plate in:	
– atmospheric air	5
– engine oil	37
– water	440
0.02 m diameter sphere in:	
– atmospheric air	9
– engine oil	60
– water	606
Forced convection	
Atmospheric air at 25°C with 10 m/s flow over a flat plate with:	
– plate length = 0.1 m	39
– plate length = 0.5 m	17
Condensation of steam at 1 atm	
– film condensation on horizontal tubes	9,000 – 25,000
– film condensation on vertical surfaces	4,000 – 11,000
– drop-wise condensation	60,000 – 120,000

Table 6–6: Typical values of the convective heat transfer coefficient encountered in some applications.

Radiation

Consider all the forms and types of energy that run our machines, light our streets, and heat our homes. The underlying source of this energy is the Sun. Energy from the Sun drives the weather, promotes photosynthesis, and provides us with both light and warmth. Petroleum, natural gas, and coal deposits that provide a convenient energy source today are actually forms of stored ancient sunlight. It is fair to say that the Sun is the major source of all the energy we have been discussing.

There are no direct mechanical linkages to conduct energy between the Sun and the Earth and no atmosphere that would permit heat transfer by convection. So, how does the Sun's energy get to the Earth? Through thermal radiation a miniscule fraction of the energy released by the Sun reaches the Earth, Moon, and all of the other planets.

Approximately 1,340 joules of thermal energy from the Sun strike each square meter of the top of the Earth's atmosphere each second. Some of the energy is reflected into space; the remainder is transmitted through the atmosphere, as shown in Figure 6-31. In fact, enough energy arrives at the surface of the Earth to meet all human needs hundreds of times over – if only the technology to harness it efficiently and inexpensively were available!

Recall from the thermogram in Figure 6-27, that all objects continuously emit energy in the form of thermal radiation. The rate of emission is proportional to the fourth power of the temperature of the object, expressed in absolute units.

$$\dot{Q}_{rad} = \varepsilon \sigma A T^4$$

In this equation, known as **Stefan's law**, $\dot{Q}_{rad}$ represents the power radiated by a body as measured in watts. The value ε is a property of the radiating surface known as **emissivity**. It is a dimensionless value between 0 and 1. A body that reflects all incident radiation and emits nothing has an emissivity equal to zero. One that absorbs all incident radiation and emits the maximum power possible given its temperature, has an emissivity of one and is known as a **blackbody**. Table 6-7 contains some typical emissivity values for surfaces of various substances. The constant σ is the Stefan-Bolt-

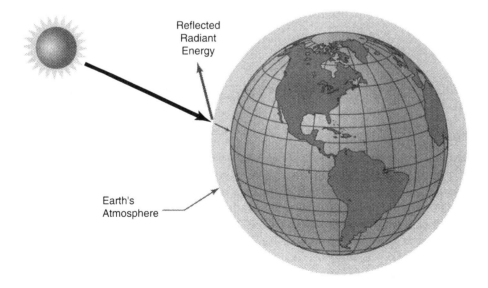

Figure 6–31: The Earth intercepts a small fraction of the energy emitted by the Sun, a portion of which is transmitted through the atmosphere and the remainder is reflected back into space.

zmann constant and equals 5.6696×10^{-8} watts per square meter per Kelvin to the fourth power, $W/m^2 \cdot K^4$, and A represents the area of the emitting surface in square meters.

Material	Emissivity (T = 300K)
Aluminum	
– polished	0.04
– rough plate	0.06
– oxidized	0.15
Cast iron	0.50
Sheet steel	0.70
Wood, black lacquer, white enamel, plaster, roofing paper	0.90
Porcelain, marble, brick, glass, rubber, water	0.94

Table 6–7:   Typical emissivity values for surfaces of various substances.

You might wonder, how much thermal energy does a typical human body radiate? Human skin has an emissivity of nearly one, regardless of its color. In fact, color is hardly related to emissivity for any material! If we assume that a typical human body has a surface area of $2.5m^2$ and body temperature of $37°C$, we can readily estimate the rate of radiation heat loss from a naked body.

$$\dot{Q}_{rad} = (1)\,(5.6696 \times 10^{-8}\ W/m^2 \cdot K^4)\,(2.5\ m^2)\,(310\ K)^4 = 1{,}309\ W$$

A naked human will radiate as much energy as thirteen 100 W light bulbs! This provides at least one good reason for wearing clothing – to capture much of the thermal energy we would otherwise lose.

As objects radiate energy, they also absorb thermal radiation from their surroundings that are also radiating. Otherwise any object would eventually radiate away all its energy and its temperature would reach absolute zero! If an object is at some temperature, T, and its surroundings are at a temperature, T_s, the net rate of energy gain or loss by the object by thermal radiation can be determined by:

$$\dot{Q}_{net} = \varepsilon\,\sigma\,A\,(T^4 - T_s^4)$$

$$\dot{Q}_{net} = \frac{(1)(5.6696 \times 10^{-8}\ W)}{M^2 \cdot K^4}\,(2.5m^2)[(310\ K)^4 - (298\ K)^4] = 191\ W$$

An object is said to be in equilibrium with its surroundings when it radiates and absorbs energy at the same rate; in other words, when $\dot{Q}_{net}$ equals zero. This assumes that we are ignoring heat transfer by conduction and/or convection.

Now reconsider the human body discussed above. Assume it is receiving thermal radiation from the surroundings at a temperature of $25°C$. The above equation shows that the net rate of heat loss is only 191 W because, although there is a loss of 1,309 W, there is also a power gain of 1,118 W. We rely on our metabolism to provide the energy that is constantly lost by heat transfer to our surroundings.

The temperature of a system is controlled by the combined effects of all three modes of heat transfer. Therefore, a rigorous analysis that considers conduction, convection, and radiation heat transfer simultaneously is required to determine exactly the flow of thermal energy into and out of a system. Fortunately, such an analysis is often unnecessary in order to obtain useful results, since it is typical that one mode of heat transfer is much more dominant than the other two in any given circumstance?

Technology Box 6–6 ■ How can the same container keep hot things hot and cold things cold?

The thermos bottle, called a Dewar flask by scientists, is a marvelous example of a device designed specifically to minimize, if not eliminate, all three modes of heat transfer. Although the materials used to construct the vessel conduct some heat, selecting certain materials and constructing the system in a special way can reduce heat transfer by conduction, convection, and radiation. Since the glass walls are surrounded by an evacuated space there is no medium across which heat can conduct or convect in either direction. Furthermore, the inner walls are silvered to reflect infrared radiation and, therefore, minimize any absorption of heat into the vessel.

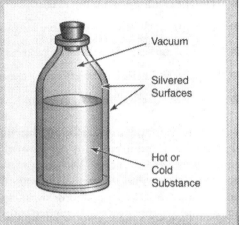

As with any real storage device, the thermos bottle is not an ideally isolated system. Some thermal energy will be transferred at the neck of the bottle, but this is small since glass is a poor conductor of heat and the cross-section is small. Since the flask reduces the rate of heat transfer, it is equally good at preventing heat gain to cold contents or heat loss to hot contents.

Check Your Understanding

1. What is the definition of the term "heat flux"?
2. What are the three distinct modes of heat transfer? Describe each one.
3. What is the difference between forced and natural convection?
4. What is the primary source of energy on Earth?
5. What is emissivity?

Simple Machines

Objectives

Upon completion of this section, you will be able to:

- Describe and calculate the mechanical advantage of using a lever arm.

- Describe and calculate the mechanical advantage of using pulleys and gears.

- Explain and calculate the force vectors involved in the use of an inclined plane.

- Explain and calculate the effect of the use of an inclined plane on friction.

- Apply the normal force to problems involving highway curve banking.

A **machine** is simply a device for applying energy to do work. It must receive energy from a source, and it cannot do more work than the energy it receives. In short, machines do not create energy. The energy a machine receives may be in the form of heat, mechanical, electrical, or chemical energy. We will consider only those machines that employ mechanical energy and do work against mechanical forces. Simple machines are supplied energy by a single applied force and do useful work against a single resisting force. The frictional force that every machine encounters during use is responsible for some waste of energy, which we will ignore for reasons of simplicity. Most machines, no matter how complex, are actually little more than combinations of levers and inclined planes.

Levers

Machines make work easier by decreasing the force necessary to do it. The use of a lever, for example, always requires a greater input distance, but less force. Since most of us have limits on how much force we capable of applying, there is often no choice other than using a simple machine.

Whenever you open a can of paint, change a tire, pull a nail, or use a furniture dolly to move a heavy object you have used a lever. The seesaw, as shown in Figure 5-14, is also a lever. The two riders can equalize their torques (force times length of the lever arm) by adjusting their distances from the fulcrum, and when their torques are equal can move up and down nearly effortlessly. Like a seesaw, a lever uses the position of the fulcrum to change the length of the lever arm over which the force is applied.

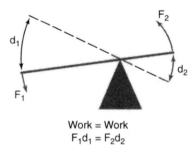

Work = Work
$$F_1 d_1 = F_2 d_2$$

Figure 6–32: The work done ($W = Fd$) on each side of the fulcrum must be equal.

As shown in Figure 6-32, the **fulcrum** is the point about which everything else rotates.

We can analyze the situation in Figure 6-32 by comparing the force arrows (vectors) and distances on the two sides. On the left-hand side, the lever arm is longer, but the force arrow (vector) is smaller. On the right-hand side, the lever arm is shorter, but the force arrow is longer. Work is defined as force times displacement ($W = Fd$) and we know that the work accomplished by each side must be equal, therefore:

Work on the left side = Work on the right side

$$W_1 = W_2$$

$$F_1 d_1 = F_2 d_2$$

The force and the displacement must also be in the same direction. Any portion of the force that is not acting in the same direction as the displacement, does not contribute to the work done on the object.

If a mechanic, for example, needs to lift a 2,000 lb car 10 inches to change a tire, he must move the jack handle up and down through a long distance to lift the car only a short distance. If each full pump moves the jack handle 20 inches, but only moves the car up by π inch, what force is being applied to the jack handle? Over what distance must the force be applied to the lift the car 10 inches?

Work of car jack = Work on car

$$W_j = W_c$$

$$F_j d_j = F_c d_c$$

To answer the first question, we are looking for F_j, the force applied to the jack.

$$F_j = \frac{F_c d_c}{d_j}$$

$$F_j = \frac{2{,}000 \text{ lbs}(1/4 \text{ in})}{20 \text{ in}}$$

$$F_j = 25 \text{ lbs}$$

To lift the 2,000 lb car, a force of only 25 lbs must be applied to the jack handle. The jack is acting as a force-multiplying device.

The total distance can be found by several different methods. To illustrate another use for the above equation, however, we will also use it to solve for the total distance over which the force must be applied. This time we solve the equation for d_j, the distance through which the end of the jack handle moves to raise the car 10 inches.

$$d_j = \frac{F_c d_c}{F_j}$$

$$d_j = \frac{2{,}000 \text{ lbs} \times 10 \text{ in}}{25 \text{ lbs}}$$

$$d_j = 800 \text{ in}$$

The force of the jack handle must be applied over a total distance of 800 inches to lift the car 10 inches. This would equate to 40 full pumps (up and down) of the jack handle. The force used by the person pumping the jack is 1/80 the weight of the car. The advantage of the lever to the person using it is clear. In fact, the advantage of using a machine is called the **mechanical advantage,** and is defined as the ratio of the input force to the output force.

The mechanical advantage can be calculated in either of two ways. Assuming there is no friction, the force of the load can be divided by the force applied or the distance applied can be divided by the distance the load moves.

$$\text{Mechanical Advantage} = \frac{\text{Force of load}}{\text{Force applied}} = \frac{2{,}000 \text{ lbs}}{25 \text{ lbs}} = 80$$

$$\text{Mechanical Advantage} = \frac{\text{Distance applied}}{\text{Distance load moves}} = \frac{800 \text{ in}}{10 \text{ in}} = 80$$

Both methods show that in the car and jack example, the jack presented a mechanical advantage of 80.

Pulleys and Gears

Why do you change gears when pedaling a bicycle up hill? Pulleys, sprockets, and gears operate under the same principle as a lever arm, but have circular shapes. In this case, the larger distance is obtained by using differing radii. Generally, the pulley, sprocket, or gear with the larger radius is connected to the pull (input) and the one with the smaller radius is connected to the load (output), as shown in Figure 6-33.

Due to the longer distance over which the pull is applied, the force necessary to lift a load will be smaller. Given a large enough difference in radii between the two, extremely heavy objects can be lifted. If the radius of the shaft turned by the pulley is 5 inches and the pulley has a radius of 25 inches, a 500-pound engine can be lifted with a pulley of 100 pounds. The distance pulled can also be compared to the distance the engine is lifted by one rotation of the pulley.

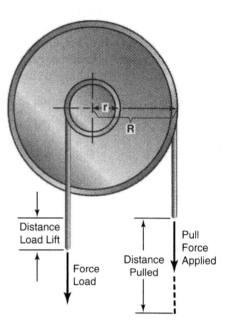

Figure 6–33: Pulleys, sprockets, and gears operate under the same principle as a lever arm but are circular in shape.

Work In = Work Out

$$W_I = W_o$$

$$F_I \times d_I = F_o \times d_o \text{ (where } d = 2\pi r)$$

$$F_I 2\pi r_I = F_o 2\pi r_o$$

Solve for the radius of the pull pulley, r_I.

$$r_I = \frac{F_o \cancel{2\pi} r_o}{F_I \cancel{2\pi}}$$

$$r_I = \frac{(500 \text{ lbs})(5 \text{ in})}{100 \text{ lbs}}$$

$$r_I = 25 \text{ in}$$

The radius of the pulley would need to be at least 25 inches to allow the mechanic to lift the engine with a force of no more than 100 pounds.

To compare the distance pulled to the distance the engine moves, use the equation for the circumference of the circle ($c = 2\pi r$) for each radius.

distance of the pull = d_I	distance engine is lifted = d_o
$d_I = 2\pi r_I$	$d_o = 2\pi r_o$
$d_I = 2\pi(25 \text{ in})$	$d_o = 2\pi(5 \text{ in})$
$d_I \cong 157$ in or 13 feet	$d_o \cong 31.4$ in or 2.6 feet

From this, it can be seen that for every 13 feet of pull, the engine is lifted only about 2.6 feet. Remembering that mechanical advantage is the distance applied over the distance the load moves, the mechanical advantage of this hoist would be, 13 ft/2.6 ft = 5.

On a 10-speed bicycle, when the going gets tough, the tough shift gears. As shown in Figure 6-34, shifting gears means changing the ratio of sizes between the larger pedal and smaller rear wheel sprockets. This, in turn, changes the ratio of the distance applied to the distance the load moves, changing the mechanical advantage.

As you turn the pedal one full rotation, it moves the chain forward, let's say 10 links, which turns the rear sprocket by a certain amount. When you shift to a higher gear ratio, you access a smaller sprocket on the rear wheel. Since the smaller sprocket has fewer teeth, moving 10 links across it makes the rear wheel turn more. Put another way, one turn of the pedals causes the back wheel to turn more times, but it requires more pedaling force.

The opposite is true when shifting to a lower gear (larger rear sprocket). The mechanical advantage is decreased when you shift down (a lower gear ratio). One turn of the pedals results in fewer turns of the back wheel, but requires less pedaling force. Since a reduction in the pedaling force is the desired effect, it is attained by shifting down.

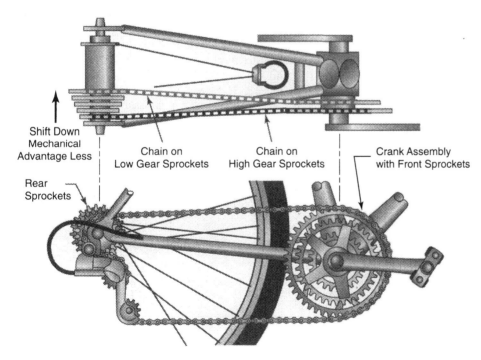

Figure 6-34: The combination of sprockets on a 10-speed bicycle makes many different gear ratios possible. In general, the larger the wheel sprocket (lower gear), the less pedaling force required, but the more peddling revolutions required.

A final thought concerning mechanical advantage. Our equation assumes that 100% efficiency is attainable – it isn't. In the real world, the total energy input can never equal the total energy output once work has been done. The losses in energy from friction, heat, and sound prevent us from making a perfect machine. Some energy will always be lost as the work is extracted.

Forces on an Inclined Plane

Forces are vectors. They act in specific directions. In Figure 6-35 (left) a force diagram is shown for a wooden crate being pulled along a flat surface. The weight force of the crate acts straight down. The force of the floor pushes up on the crate and is equal and opposite to the weight force. The pulling force acts to move the crate to the right. Finally, the force of friction acts to oppose the movement of the crate and is opposite in direction to the force of the pull. The formula for calculating the force of friction on a flat surface was given in Chapter 5 as $F_f = \mu w$, or the product of the coefficient of friction and weight.

If the surface is inclined through an angle θ, as shown in Figure 6-35 (right) the forces that act on the crate are oriented differently. The weight force, w, always acts in a downward direction (toward the center of the Earth), but now the force of the inclined surface on the crate does not act along the same line as the weight force. When an object sits on an inclined plane, its weight is effectively broken into two components. One component of the weight acts parallel to the surface of the inclined plane, w_x, and is often called the x-component. The other component of the weight acts perpendicular to the surface of the inclined plane, w_y, and is known as the y-component. The magnitude, or size, of these components is determined by two variables, the angle of the incline, θ, and the weight of the object on the incline.

By choosing to orient the components of the weight force parallel (w_x) and perpendicular (w_y) to the inclined plane, as shown in Figure 6-35 (right), they make a right triangle. The weight vector, therefore, becomes the hypotenuse (longer side) of the right triangle and the use of geometry shows that the angle θ formed will always be the same as that of the inclined plane. Since the hypotenuse of a right triangle is always the longest side, it is easy to see that each of the components of the weight force is less than the total weight of the crate.

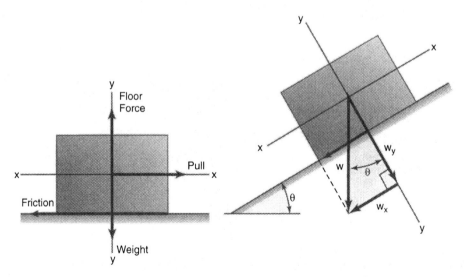

Figure 6–35: Force vectors acting on a crate being pulled on a flat surface (left) and on an inclined plane (right).

Trigonometry is the study of angles and trigonometric functions. The two trigonometric functions needed here are the sine (sin) and cosine (cos) of angle θ. By using these functions and a calculator, it is possible to determine the weight force components for each side of the right triangle, w_x and w_y.

The sine of an angle is found by dividing the side opposite, w_x, by the hypotenuse, w, as shown in Figure 6-36, or $\sin\theta = w_x/w$. Rearrangement of these terms results in $w_x = w\sin\theta$, for the x-component of the weight force. The cosine of an angle is found by dividing the side adjacent, w_y, by the hypotenuse, w, or $\cos\theta = w_y/w$. Rearrangement of these terms results in $w_y = w\cos\theta$ for the y-component of the weight force.

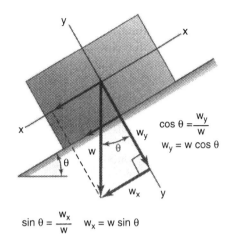

$$\cos\theta = \frac{w_y}{w}$$
$$w_y = w\cos\theta$$

$$\sin\theta = \frac{w_x}{w} \qquad w_x = w\sin\theta$$

Figure 6–36: The sine and cosine of the angle θ can be used to determine the value of the weight force for the x- and y-components, respectively.

Using the relationships developed above, consider their application when pulling a 1,000 lb piano up a ramp of 20°. What are the forces acting parallel (x-component) and perpendicular (y-component) to the ramp? The relationships for the x- and y-components would be the following:

$$w_x = w\sin\theta \qquad\qquad w_y = w\cos\theta$$

$$w_x = 1{,}000\ \text{lb}_m\sin 20° \qquad\qquad w_y = 1{,}000\ \text{lb}_m\cos 20°$$

With the aid of a calculator that includes trigonometric functions, we can solve each of these equations by first finding the numerical values for sin 20° and cos 20°, which are 0.3420 and 0.9397 respectively. (Depending on the type of calculator used, you may have to experiment, check your instruction manual, or ask a friend. Calculators using algebraic logic perform this calculation by first entering the angle (20) and then pressing the desired trigonometric function key.)

$$w_x = 1{,}000\ \text{lb}_m\ (0.3420) \qquad\qquad w_y = 1{,}000\ \text{lb}_m\ (0.9397)$$

$$w_x \cong 342\ \text{lb}_m \qquad\qquad w_y \cong 940\ \text{lb}_m$$

Furniture movers use ramps to reduce the force necessary to move objects. As you can see from the above example, each of the weight force components is smaller than the full weight of the piano. Logic tells us that if the movers had elected to lift the piano straight up, the distance the piano would travel would be shorter, but the weight force would be equal to the full weight of the piano. Application of the above relationships shows this is true. For example, when the ramp is lying flat, the angle θ is zero. If you use an angle of 0 degrees, it produces the following results:

$$w_x = 1{,}000\ \text{lb}_m\sin 0° \qquad\qquad w_y = 1{,}000\ \text{lb}_m\cos 0°$$

$$w_x = 1{,}000\ \text{lb}_m\ (0.0000) \qquad\qquad w_y = 1{,}000\ \text{lb}_m\ (1.0000)$$

$$w_x = 0\ \text{lb}_m \qquad\qquad w_y = 1{,}000\ \text{lb}_m$$

The results of these calculations confirm that our logic was correct. When the ramp is flat, the total weight force is in the *y*-component, indicating that the movers will have to lift the full weight of the piano straight up. In a similar way, let's assume that the movers placed their ramp straight up, making a 90° angle.

$$w_x = 1,000 \ \text{lb}_m \ \sin 90° \qquad\qquad w_y = 1,000 \ \text{lb}_m \ \cos 90°$$

$$w_x = 1,000 \ \text{lb}_m \ (1.0000) \qquad\qquad w_y = 1,000 \ \text{lb}_m \ (0.0000)$$

$$w_x = 1,000 \ \text{lb}_m \qquad\qquad\qquad w_y = 0 \ \text{lb}_m$$

The *x*-component of the weight force – that is, the parallel component – will equal the weight because the incline is at a 90° angle. Typically, the incline used will have an angle somewhere between these two extremes. Between the extreme values of 0 and 90 degrees, the sine and the cosine of the angle θ are always less than one. Therefore, the *x*- and *y*-components of the weight are less than the total weight.

An inclined plane is another kind of simple machine that reduces the force required to do work. It introduces more friction, and thus requires more energy and work, but the force required is less. Clearly, from the above examples, the force necessary to move a heavy object into a truck can be decreased by using an inclined plane with a small angle.

Friction on an Inclined Plane

Since we're looking at the forces acting on an object on an incline, let's examine further the effect that the incline has on the force of friction. A crate resting on a flat surface won't slide unless it is acted on by a force. Our experience tells us that the crate shown in Figure 6-37 will slide down the incline if the angle, θ, becomes too large. The reason an object will move down the incline is that there is a force pulling it. What is that force? None other than w_x, the component of the weight force acting parallel to the incline.

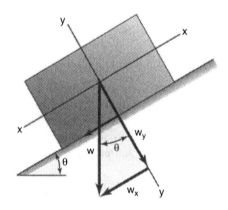

Figure 6–37: There is a force, w_x, acting on the crate that tends to pull it down the incline.

The formula for the calculation of either static or kinetic friction of an object on a flat surface is the product of the coefficient of friction times the weight. Friction occurs whenever two surfaces rub against one another, and it opposes movement. Therefore, if the force of friction is great enough, it will prevent the crate from sliding. If the force of friction is not great enough, an additional force must be applied to keep the crate in place.

When someone tries to slide something up an incline, they must overcome two things that resist its movement – inertia and the force of friction. When an object rests on an incline, the force of friction is no longer the product of the coefficient of friction times its weight. It is now equal to the coefficient of friction times only the component of the weight force that is perpendicular to the surface, w_y. So, even though the coefficient of friction remains the same, the component of the weight force is smaller. A diagram of the forces acting on a crate is shown in Figure 6-38. It includes the **normal force**, $\mathcal{n}$, which is the force equal in magnitude and opposite in direction to w_y.

If the crate shown in Figure 6-38 weighs 250 newtons and is on an incline with an angle of 37°, what would be the normal force and the force that acts to pull it down the incline?

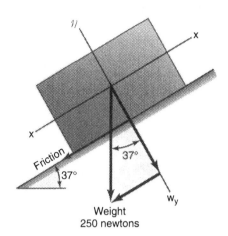

Normal Force

$n = w \cos \theta$

$n = 250 \text{ N} \cos 37°$

$n = 250 \text{ N} (0.7986)$

$n \cong 200 \text{ N}$

Down Incline Force

$W_x = w \sin \theta$

$W_x = 250 \text{ N} \sin 37°$

$W_x = 250 \text{ N} (0.6018)$

$W_p \cong 150 \text{ N}$

Figure 6–38: The normal force is equal in magnitude and opposite in direction to w_y, the component of the weight acting perpendicular to the inclined plane.

If the coefficient of static friction for the crate is 0.60, the force of friction can be found by multiplying the normal force (200 newtons) by the coefficient of static friction.

$$F_f = \mu n$$

$$F_f = (0.60)(200 \text{ N})$$

$$F_f = 120 \text{ N}$$

The x-component of the weight force (pulling the crate down the incline) is 150 N. The force of friction, however, is pushing in the opposite direction with a force of 120 N. From this, it can be seen that the weight force is 30 N larger than the frictional force and, therefore, the crate will slide down the incline, if left unattended. Put another way, a force of 30 N must be applied to prevent the crate from sliding down the ramp.

Check Your Understanding

1. Explain how a lever provides a mechanical advantage.

2. What is a fulcrum?

3. How does placing a crate on an incline effect the weight force and the frictional force?

4. Is it the x- or y-component of the weight force that acts perpendicular to the incline?

5. To what force is the normal force equal in magnitude, but opposite in direction?

Special Topics Box 6–2 ■ Why are curves on the highway banked?

The factors that answer the above question are similar in concept to the crate on the incline example. There is one big exception, however – the road is curved. Our incline was a straight, flat ramp. The variables used to determine the safest velocity of a vehicle are the degree of banking and the radius of the turn. Surprisingly, if friction is not considered, then the mass of the vehicle does not matter. The force of friction can also help keep a vehicle from sliding on a curve. However, it is best to build a turn so friction will serve only as a back-up. That is, friction will help if someone takes the turn faster or slower than its design indicates.

In order to deal with this problem, we again need work with the normal force, which operates perpendicular to the surface of the road. As shown in the figure, its vertical component is equal to the weight of the car, $w = mg$, times $\cos \theta$.

Let's look at all the forces as if they act at the center of the car. When the normal force is broken into its components, the vertical component $n = w_y$. But $w_y = w \cos \theta$, and since the weight of the car is equal to its mass times the gravitational force, mg, then $n = mg \cos \theta$.

As the car executes the curve, it is accelerating because its velocity is changing direction. This requires an inward force of $F = mv^2/r$ to produce this acceleration and keep curving. This inward force can be provided by

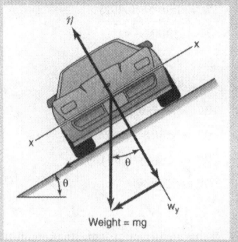

For a car in a turn, the normal force, n, has a vertical component $n \cos \theta$ toward the center, which provides the centripetal force.

the horizontal component of the normal force, $F = n \sin \theta$). Otherwise, the car would have to depend on friction between the tires and the road for this inward force, which could involve some dangerous screeching of tires. Substitution of the above relationship for n, gives $F = (mg \cos \theta) \sin \theta$.

The two force relationships, $F = m v^2/r$ and $F = mg \cos \theta \sin \theta$, can be set equal, resulting in the equation:

$$\frac{mv^2}{r} = mg \cos \theta (\sin \theta)$$

$$v^2 = rg \cos \theta \sin \theta$$

The above equation can be further simplified using the trigonometric double angle identity, $\sin 2\theta = 2\sin \theta \cos \theta$. Multiplying each side of the equation by two, we can then make the substitution, which simplifies the equation to:

$$v^2 = \tfrac{1}{2} gr \sin 2\theta$$

Using this equation, the safe velocity for any curve can now be found if its banking angle and radius are known.

$$v = \sqrt{(\tfrac{1}{2} gr \sin 2)\theta}$$

> ## Special Topics Box 6–2 ■ Why are curves on the high-way banked? (Continued)
>
> For example, the safest velocity a car may travel on a curve that has a radius of 0.5 miles (2640 feet) and a banking angle of 7°, would be:
>
> $$v = \sqrt{\tfrac{1}{2}\, 32.2 \text{ ft/sec}^2 (2{,}640 \text{ ft}) \sin 2(7°)}$$
>
> $$v = \sqrt{10{,}282.6 \text{ ft}^2/\text{sec}^2}$$
>
> $$v = 101.4 \text{ ft/sec}$$
>
> The velocity can be changed to miles/hour, by performing a final unit conversion.
>
> $$v = \frac{101.4 \text{ ft}}{\text{sec}} \times \frac{1 \text{ mile}}{5{,}280 \text{ ft}} \times \frac{3{,}600 \text{ ft}}{1 \text{ hr}} \times = \frac{69.1 \text{ miles}}{\text{hr}}$$

Summary

The first law of thermodynamics, also known as the law of conservation of energy, states that the total energy in the universe is constant and that it can neither be created nor destroyed. It is the driving force of the universe and comes in one of six major forms – mechanical, electrical, electromagnetic, chemical, nuclear, and thermal. Energy is critical to processes that occur throughout our universe as either a measurable quantity of heat or some amount of work. It can be stored or transferred between bodies.

Kinetic energy is the energy of motion. When energy is transferred by the application of a force with a resulting displacement, it is called work. Whenever an applied force acts to change the height of an object, it changes its gravitational potential energy. Electrical energy is often transferred from place to place by electric currents, which consist of moving electrons inside a conductor. In gases, liquids, or plasmas, the electric current includes moving ions. Stored electrical energy may be in the form of either an electrostatic or a magnetic field. Chemical and nuclear energy exists only as a stored type of potential energy in chemical or nuclear bonds. When thermal energy is transferred to or from a body, it results in either a change of temperature or a change in the phase of a material, at constant temperature. Heat is energy that transfers from one body to another because of temperature differences, by means of conduction, convection, or radiation. Energy can be transformed from one form to another.

Thermodynamics is the study of heat and internal energy. Internal energy represents all the energy belonging to a system. The properties included in the internal energy may be intensive (independent of the size of the system) or extensive. Typical extensive properties are mass, volume, and total energy. Intensive properties include temperature, pressure, and density.

Temperature is an indicator of the average thermal energy of the atoms or molecules in a substance. It is measured using the Fahrenheit, Celsius, Kelvin, or Rankine temperature scale. Both the Kelvin and Rankine temperature scales are based on absolute zero; the coldest any object can possibly be. The heat capacity of a system is the amount of heat required to raise the temperature of the entire system by one degree

Celsius. The second law of thermodynamics states that on its own, heat must always flow from the warmer to the cooler object.

The kinetic-molecular theory assumes that gases are composed of particles in continuous and random motion, which causes them to collide with one another and the walls of their containers. These collisions are responsible for the pressure of the gas. Absolute pressure is measured relative to a vacuum. Gauge pressure is the difference between the pressure of the gas and local atmospheric pressure. Various equations of state have been developed, but of these the ideal gas equation is best known ($PV = nRT$). The value of the universal gas constant, R, varies with the units used.

Heat flux is the rate at which energy is transferred. There are three distinct modes of heat transfer: conduction, convection, and thermal radiation. Conduction is by direct contact, convection is by the motion of a fluid relative to a surface, and radiation is by long wavelength electromagnetic radiation. Of the three, only radiation does not require a transfer medium.

Machines are devices for applying energy, received from some other source to do work. Many simple machines, ignoring frictional losses, make work easier by decreasing the force necessary to do it. In these cases, the distance over which force must be applied is always greater than the distance the load moves. This ratio of the applied distance divided by the distance the load moves is known as the mechanical advantage.

The weight forces that act on an object on an inclined plane are subdivided into two components. One component acts parallel to the surface and the other (normal force) perpendicular to the surface of the inclined plane. The magnitude of these components is determined by the angle of the incline and the weight of the object.

Chapter Review

1. Explain the difference between energy and work.

2. Consider a ball thrown straight up into the air. Explain how its gravitational potential energy and kinetic energy vary as the speed of the ball slows to zero. Compare the gravitational potential energy of the ball at its highest, middle, and lowest points.

3. Assume that a speeding locomotive has a mass of 4.45×10^5 kg and a speed of 65 km/hr. How fast must Superman, with a 90 kg mass, be moving to have the same kinetic energy as the locomotive?

4. Convert the values given in questions 3 into English units (i.e. kg to lb_m) and rework the problem, verifying that your new answer, after conversion, is equivalent to your original answer.

5. If you add 15 kJ of energy to a closed system and during this process the system does 12 kJ of work on its surroundings, then what is the net change the system has undergone and is it an increase or decrease? Show your work.

6. The second law of thermodynamics states that an ideal heat engine would convert all heat absorbed into work. Explain why such an engine does not exist.

7. If for every 45 Btu of heat added to a heat engine, the engine produces 10 Btu of work, calculate the efficiency of the engine.

8. The carbon dioxide that we exhale freezes at −109°F. Determine the equivalent temperature in degrees Celsius, in Kelvin, and in Rankine.

9. An archeologist working out in the field pours her morning coffee into an aluminum cup of mass 120 g and with an initial temperature of 20.0°C. She pours pre-

cisely 300 g of coffee that is initially at 70.0°C. What is the final temperature of the coffee and the cup when they have attained thermal equilibrium? (You may assume that coffee has the same specific heat as water.)

10. If the coffee were to cool from 70° to 20°C in thirty minutes, what would be the average power dissipated by the coffee as heat during those thirty minutes?

11. A 5-pound-mass lead projectile is shot out of a cannon. At one instant during its flight its velocity is 70 mph, it is 45 feet higher than the cannon barrel, and its temperature is 27 Fahrenheit degrees hotter (due to the charge and friction) than before it was shot out. How much greater is the total energy of the projectile at this instant than before it was shot?

12. A certain engine absorbs 4.5 Btu of heat from a combustion source, but discards 3.1 Btu of that energy to the surroundings. How much work can this engine do? What is the efficiency of the engine?

13. If the absolute pressure inside a closed tank is 6 psi and the atmospheric pressure is 14.6 psi, then what is the pressure inside the tank and what kind of pressure is it called?

14. If the internal energy of a system decreases by 14 J as 21 J of work is done on the system (W_{in}), how much heat must have entered the system during the process? Show your work.

15. Write a brief explanation for the three modes of heat transfer. Which one does not require a medium?

16. An empty automobile tire is inflated using air originally at 10°C and 1 atmosphere pressure. As the tire is pumped, the incoming air is compressed to 28% of its original volume, and its temperature rises to 40°C. What is the tire pressure in pascals and in lb_f/in^2?

17. In an automobile engine, a mixture of fuel and air is compressed in the cylinders before it is ignited by the spark plug. A typical engine has a compression ratio of 9.0 to 1, meaning that the gas in each cylinder is compressed to 1/9.0 of its original volume. If the original pressure of the fuel-air mixture is 1 atm, and its initial temperature is 31°C, determine the temperature of the compressed gas assuming that its pressure is at 22 atm.

18. A 20-liter tank contains 0.23 kg of helium at 24°C. The molecular mass of helium is 4.00 g/mol. How many moles of helium are in the tank? What is the pressure in the tank in pascals and in lb_f/in^2?

19. In a certain process, 4.25×10^5 J of heat is rejected by a system and at the same time the system contracts under a constant external pressure of 7.20×10^5 Pa. The internal energy of the system does not change during the process. Find the change in volume of the system. (Do not treat the system as an ideal gas.)

20. Nitrogen gas filling an expandable container is cooled from 80° to 10°C at a constant pressure of 3×10^5 Pa. The total heat removed from the gas during this process is 6.5×10^4 J.
    a. Calculate the number of moles of gas in the container.
    b. Determine the change in internal energy of the gas.
    c. Find the work done by the gas.
    d. How much heat would be removed from the gas over the same temperature change if the volume were held constant?

21. Explain why roads tend to wind up steep hills, rather than going directly up the slope.

22. A steel bar 10 ft long is placed over a wooden block and under the edge of a large safe. If the block is 1.5 ft from the safe, how much force must be applied to the bar to lift a 500 lb_m safe?

23. A pulley system is used to lift a 1,000 lb_m block of stone a distance of 10 ft by the application of a force for a distance of 80 ft. What is the mechanical advantage of the system?

24. What would the x- and y-components of the weight force be for an 800 lb_m rock on a 25° ramp. Draw a diagram and show your calculations.

25. Consider the rock used in question 24. Explain why its force of friction would be less on the ramp than when it is on a flat surface.

Physical Science: What the Technology Professional Needs to Know

How do scientists predict chemical reactions?

Chemical Reactions and Solutions

Humans have always manipulated the substances that surround them. Through trial and error coupled with careful observation, ancient civilizations learned to make an impressive array of useful substances like medicines, building materials, cosmetics, and dyes. Many of these products involved mixing, heating, and allowing chemical reactions to occur. Early discoveries were as likely the result of accidental as intentional manipulations. Either way, it was learned over the years that some things burn, while others do not. For example, some rocks are changed into a powder by heating. Foods undergo a change in both color and texture during heating, but will char and burn if the process is continued for too long. The valuable processes learned by early humans were passed along to each succeeding generation. A rich lore slowly developed that allowed emerging civilizations to ferment juices and produce alcoholic drinks, make glass, free various metals from their ores, and make alloys.

Although many of the early people involved in manipulating substances might be called chemists today, it is only a recent term for those who specialize in the study of these transformations. In Chapter 4, the different ways atoms rearrange or share their electrons to form stable chemical compounds were presented. In this chapter, you will learn about some of the discoveries that have been made and how to predict the results of various chemical combinations.

7-1   Chemical Reactions

Objectives

Upon completion of this section, you will be able to:

■ Balance simple chemical equations.

■ Use general terminology to describe groups of chemicals.

■ Discuss the four types of chemical reactions.

■ Predict and write a correctly balanced chemical equation.

■ Explain oxidation and reduction reactions.

■ Describe the parts of a simple electrochemical cell.

Over time, two important principles have been identified and serve as the foundation for our understanding of chemical interaction. The **law of conservation of matter** states that matter can be neither created nor destroyed. The sum of the masses of the materials reacting (**reactants**) must equal the sum of the masses of the materials produced (**products**); simply put, what goes in must come out! The **law of conservation of energy** states that during chemical reactions energy cannot be created or destroyed. It is possible for it to change its form, such as chemical energy becoming heat, light, or electrical energy.

Chemical reactions requiring the constant input of energy are called **endothermic** reactions. Photosynthesis is a familiar example. As soon as the light source is removed, photosynthesis stops. As shown in Figure 7-1, endothermic reactions can be thought of as rolling a rock up an energy hill. Chemical substances (A) with lower amounts of energy are elevated to a higher energy product (C). Chemical reactions that release energy are called **exothermic** reactions. Burning a piece of paper is an example. These reactions are like a rock rolling down an energy hill, because the products contain less energy than the starting materials. In both endothermic and exothermic reactions, an energy barrier (B) must be overcome before a reaction can occur. This barrier, called the **activation energy**, is the reason a match must be used to start the paper burning.

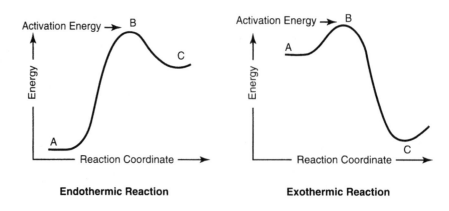

Figure 7-1: Energy diagram for endothermic and exothermic reactions.

During the last century, a connection between matter and energy was discovered and demonstrated ($E = mc^2$). This resulted in a combining of the two laws, now known as the **law of conservation of mass and energy**. The combined law allows for the possibility of a conversion between mass and energy. The conversion of mass to energy during ordinary chemical reactions, however, is so slight that it is undetectable under normal laboratory conditions.

Writing and Balancing Chemical Equations

Chemical reactions are frequently written in a shorthand form known as a **chemical equation**. Examination of Figure 7-2, shows that chemical equations are similar to mathematical equations with the reactants written on the left-hand side and the products on the right-hand side. When there is more than one reactant or product, a plus sign (+) is used between them. An arrow (→), read as "yields" or "produces," is used between the reactants and products to show the direction of the reaction. It conveys the same meaning as an equal sign (=) in a mathematical equation. If the chemical equation includes the amount of energy involved in the reaction, it may be shown on the left-hand side if endothermic or on the right-hand side if exothermic.

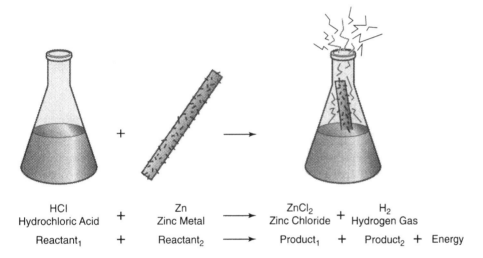

HCl Hydrochloric Acid	+	Zn Zinc Metal	→	$ZnCl_2$ Zinc Chloride	+	H_2 Hydrogen Gas	
Reactant₁	+	Reactant₂	→	Product₁	+	Product₂	+ Energy

Figure 7–2: Parts of an exothermic chemical equation.

The law of conservation of matter requires that atoms be conserved. This means that each of the atoms appearing on the left-hand side must also be accounted for on the right-hand side of the equation. If more or less atoms of a particular element are present on one side of the equation, it is not balanced. A chemical equation is not correctly written unless it is balanced. For example, if the diatomic element hydrogen, H_2 is allowed to react with the diatomic element nitrogen, N_2, they combine to produce the substance ammonia, NH_3. When this information is put into a **word equation** and then into an **unbalanced chemical equation**, it looks like this:

$$\text{Hydrogen} \quad + \quad \text{Nitrogen} \quad \rightarrow \quad \text{Ammonia}$$
$$H_2 \quad + \quad N_2 \quad \rightarrow \quad NH_3$$

A careful inventory of the atoms shows that two atoms of hydrogen have magically become three atoms on the right-hand side, while one atom of the nitrogen has disappeared. This is an unbalanced equation. Balancing an equation involves counting the atoms, one kind at a time, then inserting numbers – called **coefficients** – until the total number of atoms of reactants on the left is the same as the total number of atoms of the product on the right. In the equation below, the whole numbers needed have been added. This process produces a balanced chemical equation.

$$3\,H_2 \quad + \quad N_2 \quad \rightarrow \quad 2\,NH_3$$

Inspection of this equation shows that six atoms of hydrogen were used and the six atoms of hydrogen are now present in the two molecules of ammonia. The one molecule of nitrogen used contained two nitrogen atoms. Those two atoms separated and each is now incorporated into one ammonia molecule. Although it was not included in the equation, this is an endothermic reaction that requires 11.0 kcal/mol NH_3.

As a second example, consider the reaction of the ingredients in a common antacid, calcium carbonate, $CaCO_3$, with stomach acid, HCl. This reaction produces three products, calcium chloride, $CaCl_2$, water, H_2O, and carbon dioxide, CO_2.

$$CaCO_3 \quad + \quad HCl \quad \rightarrow \quad CaCl_2 \quad + \quad H_2O \quad + \quad CO_2$$

By counting each of the elements on the left-hand and right-hand sides of the equation, you will find that this is not a balanced equation. Before reading any further, what coefficients must be added to balance it?

$$CaCO_3 \quad + \quad 2\,HCl \quad \rightarrow \quad CaCl_2 \quad + \quad H_2O \quad + \quad CO_2$$

All that needs to be added in this example, is a two in front of the hydrochloric acid.

Chemical Formulas and Terminology

Before moving on to a discussion of chemical reactions, it is necessary to understand more about the general terminology used by chemists. Compare these descriptions of two animals: 1) a St. Bernard and a Siamese cat, 2) a dog and cat, and 3) two pets. Each of these statements conveys a different amount of information. The first statement provides very specific information; the second is more generic, covering all breeds of dogs and cats, and the third description is even more general and covers all household animals. Similarly, varying degrees of specificity are used when discussing chemicals.

One category of chemical compounds, including CaO, K_2O, and Al_2O_3, are the metal oxides (MO), which represent a combining of any metal (M) with the nonmetal element oxygen (O). CO_2, SO_2, and P_2O_5 are examples of nonmetal oxides (NmO) because they represent a combining of any nonmetal (Nm) with oxygen (O). Other general groups are acids – hydrogen with nonmetals or with nonmetal oxides (HNm, HNmO); bases or metal hydroxides (MOH); and salts (MNm or MNmO). Within the category of salts, if the positive ion is sodium, Na^+, then regardless of the negative ion these substances can be grouped and referred to as sodium salts. In a similar way, if all the negative ions are iodide, I^-, then the salts are all iodide salts.

Table 7-1 provides the names of a number of groups, the general formula for the group, and several examples. This is not meant to be an exhaustive list, but it illustrates that chemical substances may be grouped according to common ions. For example, one of the most common chemical reactions involves an acid reacting with a base to produce a salt and water. It is not necessary to memorize every acid/base reaction, just to recognize the reactants as belonging to these groups and know what they always produce. Biology teaches that crossing a horse with a donkey produces a hybrid known as a mule. It is not necessary to know whether the horse is a pinto or a palomino to predict the offspring!

General Term	General Formula	Examples
Acid or oxyacid	HNm or HNmO	HCl, HF, H_2SO_4, H_3PO_4
Base	MOH	$NaOH$, KOH, $Ca(OH)_2$
Metal	M	Na, Co, Cu, Ag, Au
Metal oxide	MO	Na_2O, CaO, Al_2O_3, Fe_2O_3
Nonmetal	Nm	S, P, H_2, O_2, Br_2
Nonmetal oxide	NmO	CO, CO_2, SO_3, N_2O
Salt	MNm or MNmO	$NaCl$, $MgBr_2$, $Al_2(CO_3)_3$
calcium salt	CaNm or CaNmO	$CaCl_2$, $Ca(NO_3)_2$, $CaSO_4$
chloride salt	MCl	$NaCl$, $BaCl_2$, $AlCl_3$, NH_4Cl
carbonate salt	MCO_3	Na_2CO_3, $CaCO_3$, $Fe_2(CO_3)_3$
Peroxide	MO_2 or NmO_2	Na_2O_2, Li_2O_2, H_2O_2, BaO_2

Table 7–1: Chemical groupings by general terms and formulas.

Types of Chemical Reactions

It is possible to predict what will be produced when two chemicals are combined. Careful study has shown that most inorganic reactions can be put into one of four types: combination, decomposition, single replacement, and double replacement.

Perhaps the easiest type of chemical reaction to understand is the **combination reaction**. In this kind of reaction two elements or molecules produce a more complex substance, $(A + B \rightarrow AB)$. There is no absolute method for correctly predicting the product of a chemical combination. The information presented in Table 7-2, however, and the ability to recognize chemical groups will enable many accurate predictions.

Combination Reactions

To predict the product of a reaction between magnesium and oxygen it is necessary to recognize that magnesium is a metal and that both substances are elements. Because they are both elements, their reaction must produce a more complex substance. Table 7-2 indicates that all metals will react with oxygen to produce a metal oxide. With this much information, the following partial equation can be written:

$$Mg + O_2 \rightarrow MgO$$

Before proceeding, two points need clarification. First, oxygen is written as O_2 because it is one of the seven diatomic elements (H_2, N_2, O_2, F_2, Cl_2, Br_2, and I_2). When diatomic elements are uncombined, they must be written with the subscript "2." Second, magnesium oxide is MgO and not some other combination because the charge on the magnesium ion, Mg^{2+}, and oxide ion, O^{2-}, are equal and opposite, so it takes only a 1:1 ratio to balance their charges. Now consider whether the equation is balanced. The number of oxygen atoms on each side of the arrow is not equal, so coefficients must be added.

$$2\ Mg + O_2 \rightarrow 2\ MgO$$

The equation is now balanced. It predicts the combination of magnesium metal and oxygen will result in the formation of the metal oxide, magnesium oxide. Identify each of the following examples as one of the reactants in Table 7-2.

$$2\ Al + 3\ Br_2 \rightarrow 2\ AlBr_3$$
$$Na_2O + H_2O \rightarrow 2\ NaOH$$
$$NH_3 + HBr \rightarrow NH_4Br$$

Answers: 3, 4, & 8

Combination:	A + B	$\longrightarrow$	AB
1. metal	+ oxygen	$\longrightarrow$	metal oxide
2. nonmetal	+ oxygen	$\longrightarrow$	nonmetal oxide
3. metal	+ nonmetal	$\longrightarrow$	salt
4. a soluble metal oxide	+ water	$\longrightarrow$	base
5. nonmetal oxide	+ water	$\longrightarrow$	an oxyacid
6. metal oxide	+ nonmetal oxide	$\longrightarrow$	salt
7. anhydrous salt	+ water	$\longrightarrow$	hydrate
8. ammonia	+ binary acid	$\longrightarrow$	ammonium salt

Table 7–2: Combination reactions.

Decomposition Reactions

Decomposition reactions are the reverse of combination reactions, and are easy to spot since they have only one reactant. In these reactions, a complicated substance is broken into simpler substances (AB $\rightarrow$ A + B). In Table 7-3, a few general reactions of this type are given.

Fireworks are enjoyed at many celebrations. One of the essential ingredients in fireworks is an oxidizer – a chemical that will supply large amounts of oxygen to sustain the explosion. One of the oxidizers commonly used is potassium chlorate, $KClO_3$. The equation below shows potassium chlorate releasing oxygen. In this equation you will notice the inclusion of a delta (Δ) under the arrow. This may be written either above or below the arrow and does not have to be included, but when it is, it indicates that heat is required to make the reaction occur.

$$2\ KClO_3 \xrightarrow{\Delta} 2\ KCl + 3\ O_2$$

Decomposition:	AB	$\longrightarrow$ A	+ B
1. heavy metal oxide	$\longrightarrow$	metal	+ oxygen
2. metal chlorate	$\longrightarrow$	metal chloride	+ oxygen
3. metallic nitrates	$\longrightarrow$	metal nitrites	+ oxygen
4. electrolysis of compounds	$\longrightarrow$	element	+ element
5. hydrates	$\longrightarrow$	anhydrous salt	+ water
6. metallic carbonate	$\longrightarrow$	metallic oxide	+ CO_2
7. metallic bicarbonate	$\longrightarrow$	metallic oxide	+ $H_2O + CO_2$
8. peroxides	$\longrightarrow$	oxides	+ oxygen

Table 7–3: Decomposition reactions.

Electrolysis is a process that involves passing a direct electric current through the liquid form of a substance. The symbol that is used to indicate electrolysis is the same as for the electron, e^-. The electrolysis of water shows that water produces H_2 and O_2 and contains twice the amount of hydrogen as oxygen.

$$2\ H_2O \xrightarrow{e^-} 2\ H_2 + O_2$$

Additional examples of decomposition reactions follow. Try to match them to the correct general equation.

$$CuSO_4 \times 5\ H_2O \xrightarrow{\Delta} CuSO_4 + 5\ H_2O$$

$$2\ NaHCO_3 \xrightarrow{\Delta} Na_2O + H_2O + 2\ CO_2$$

Answers: 5 & 7

Single Replacement Reactions

In **single replacement reactions** (Table 7-4), an element is put with a compound. Then it becomes a shoving contest to determine which element gets to be in the preferred ionic form. This kind of reaction has two subtypes: those in which the challenger is a metal and those in which the challenger is a halogen (AB + C → CB + A or AB + D → AD + B). Predicting the winner in these competitions requires familiarity with the **electromotive series**, a portion of which is shown in Table 7-5.

Single Replacement:	AB + C or AB + D ⟶	AC + B DB + A	
1. salt$_1$ + free halogen$_1$	⟶	salt$_2$ + free halogen$_2$	(see EM Series)
2. salt$_1$ + more reactive metal	⟶	salt$_2$ + free metal	(see EM Series)
3. very active metals + water	⟶	metal hydroxide + hydrogen gas	
4. active metals + acid	⟶	salt + hydrogen gas	

Table 7–4: Single replacement reactions.

The position of each element in the table was determined by comparing it with hydrogen, which was arbitrarily assigned a value of zero. If the element gives its electron(s) to hydrogen, it is above it in the series and referred to as an **active metal**. Those elements that cannot give their electron(s) to hydrogen were placed below it in the series. The potassium ion (K^+) is the least likely to take an electron, and the element fluorine (F) is the most likely to accept an electron to become F^-.

Some confusing terminology is often used in this area. For example, when an element or ion loses electron(s) it is called **oxidation**. The substance doing so is called the **reducing agent**. The **oxidizing agent** is the substance gaining electron(s) and it is said to be undergoing **reduction**. Table 7-5 is a partial listing of elements and their values as reducing agents.

Use of Table 7-5 makes it possible to accurately determine if a particular reaction will occur. Let's suppose that a child accidentally swallows a small piece of aluminum foil. What will happen? Since stomach acid is primarily composed of hydrochloric acid, HCl, the reactants will be:

$$Al + HCl \rightarrow$$

By referring to Table 7-5, we find that aluminum is above hydrogen, which means that it will give its electrons to the hydrogen ions. The products of the reaction, therefore, will be a solution containing aluminum chloride, $AlCl_3$, and hydrogen gas, H_2. The following is the balanced equation for the reaction.

$$2\ Al + 6\ HCl \rightarrow 2\ AlCl_3 + 3\ H_2$$

What would be the result of placing elemental bromine, Br_2, in contact with sodium iodide, NaI? From the relative positions of bromine and iodine in the table, it can be determined that bromine will take electrons from iodide ions, causing it to become elemental, I_2. The balanced equation for the reaction is:

$$2\ NaI + Br_2 \rightarrow 2\ NaBr + I_2$$

Element	Standard Reduction Potential, volts
$K^+ + e^- \rightarrow K$	−2.93
$Ca^{2+} + 2\,e^- \rightarrow Ca$	−2.87
$Na^+ + e^- \rightarrow Na$	−2.71
$Mg^{2+} + 2\,e^- \rightarrow Mg$	−2.37
$Al^{3+} + 3\,e^- \rightarrow Al$	−1.66
$Zn^{2+} + 2\,e^- \rightarrow Zn$	−0.76
$Fe^{2+} + 2\,e^- \rightarrow Fe$	−0.44
$Sn^{2+} + 2\,e^- \rightarrow Sn$	−0.14
$Pb^{2+} + 2\,e^- \rightarrow Pb$	−0.13
$2\,H^+ + 2\,e^- \rightarrow H_2$	**0.00**
$Cu^{2+} + 2\,e^- \rightarrow Cu$	0.34
$I_2 + 2\,e^- \rightarrow 2\,I^-$	0.53
$Ag^+ + e^- \rightarrow Ag$	0.80
$Br_2 + 2\,e^- \rightarrow 2\,Br^-$	1.09
$Cl_2 + 2\,e^- \rightarrow 2\,Cl^-$	1.36
$Au^{3+} + 3\,e^- \rightarrow Au$	1.42
$F_2 + 2\,e^- \rightarrow 2\,F^-$	2.87

Table 7–5: Electromotive series.

What would be the result of placing elemental bromine, Br_2, in contact with sodium iodide, NaI? From the relative positions of bromine and iodine in the table, it can be determined that bromine will take electrons from iodide ions, causing it to become elemental, I_2. The balanced equation for the reaction is:

$$2\,NaI + Br_2 \rightarrow 2\,NaBr + I_2$$

There are two commercially important single replacement reactions. The process of corrosion and the way batteries generate electricity both involve the movement of electrons from one substance to another. Reactions of this type are commonly referred to as **oxidation-reduction** or **redox** reactions. Millions of dollars are spent each year trying to either prevent redox reactions (corrosion) or encourage them (batteries).

Corrosion is the term often used to describe an unwanted redox reaction. Metals are often plated, painted, or plastic coated to provide a barrier that will prevent an unwanted redox reaction from occurring. We all want the aluminum metal in our shower and tub enclosures to remain bright silvery metal, without a bumpy grayish-white accumulation. Unfortunately, aluminum has a very high position in the table, meaning that there are many elements below that accept its electrons. When aluminum atoms are exposed to these other substances, its atoms are converted to aluminum ions. If chlorine is one of those substances, it will result in the formation of white aluminum chloride.

$$2\,Al + 3\,Cl_2 \rightarrow 2\,AlCl_3$$

Corrosion is often found on metal car parts near the battery. Knowing that iron, Fe, and lead, Pb, are both above hydrogen, makes this easy to predict and explain. If these metals were replaced by silver or gold (both lower than hydrogen), then corrosion would not occur. Because of their resistance to corrosion, there are times when expensive metals are used for critical applications. Gold has always been valued because it keeps its metallic luster and resists corrosion.

Batteries are an application of redox reactions designed to produce a useful electric current. We use these batteries in many items of convenience, ranging from computers and cellular telephones to remote controls for our TVs and stereos. The major difference between corrosion and a battery is a matter of proximity. During corrosion, the reducing and oxidizing agents are in direct contact; therefore, the electrons move directly from one to the other. In a battery, however, the oxidizing and reducing agents are kept separate, causing the electrons to move through an external wire to reach the oxidizing agent.

The driving force behind redox reactions is measured as voltage and can be used to predict whether the reaction will occur spontaneously. Figure 7-3 is an example of a simple electrochemical cell (battery). The left beaker contains a strip of the metal zinc, Zn, in a solution containing zinc ions, Zn^{2+}. The beaker on the right contains a strip of copper, Cu, in a solution containing copper(II) ions, Cu^{2+}. The metal strips in each of the beakers are known as **electrodes**. By convention, the electrode at which oxidation (loss of electrons) takes place is known as the **anode**, and the electrode at which reduction (gain of electrons) occurs is called the cathode.

Connecting the two beakers is a **salt bridge** that allows negative ions to migrate from the solution in one beaker to the other. Since positive ions are being formed on the left-hand side and removed from solution on the right-hand side, the migration of negative ions is necessary to maintain the electrical neutrality of each solution.

Anode reaction:	$Zn \rightarrow Zn^{2+} + 2e^-$	$E° = +0.76$ volts
Cathode reaction:	$Cu^{2+} + 2e^- \rightarrow Cu$	$E° = +0.34$ volts
Overall reaction:	$Zn + Cu^{2+} \rightarrow Zn^{2+} + Cu$	$E_{cell} = +1.10$ volts

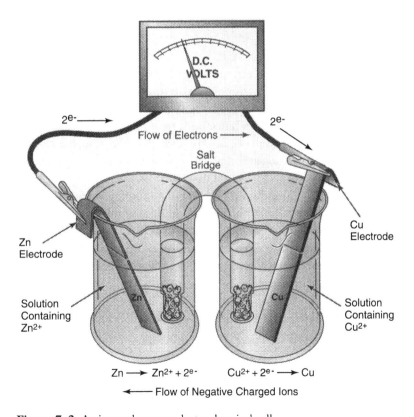

Figure 7–3: A zinc and copper electrochemical cell.

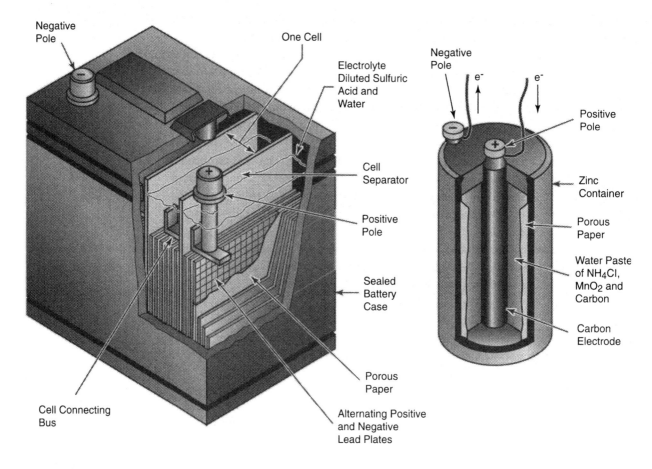

Figure 7–4: Cutaways of classic lead-acid storage and dry-cell batteries.

Half-cell reactions are frequently used to pinpoint what is happening on each side of a redox reaction. In this reaction, the **oxidation half-cell** has been written for the reducing agent (Zn) and the **reduction half-cell** for the oxidizing agent (Cu^{2+}). The voltages for each of these half-cell reactions have been taken from Table 7-5 and added to determine the cell total, E_{cell}. The fact that the sign for the combined half-reactions is positive (+) indicates that this reaction will occur spontaneously and that the total voltage produced by this cell is 1.10 volts.

One battery we all rely on is the 12-volt lead-acid storage battery used in cars and trucks. This battery is composed of six separate cells, each developing about 2 volts. By connecting these six cells in series, the overall voltage becomes the sum or 12 volts.

Although the chemistry of the redox reactions occurring within a lead-acid storage battery is more complicated (see Figure 7-4), the same overall processes are involved. The half-cell reactions are:

Anode reaction: $Pb + SO_4^{2-} \rightarrow PbSO_4 + 2\,e^- \quad E° = +0.351$ volts

Cathode reaction: $PbO_2 + 4\,H^+ + SO_4^{2-} + 2\,e^- \rightarrow PbSO_4 + 2\,H_2O \quad E° = +1.685$ volts

Overall reaction: $Pb + PbO_2 + 4\,H^+ + 2\,SO_4^{2-} \rightarrow 2\,PbSO_4 + 2\,H_2O \quad E_{cell} = +2.036$ volts

The traditional dry-cell or flashlight battery as shown in Figure 7-4 is a zinc-carbon battery. It derives its name from the fact that the liquid portion has been replaced by a moist paste of ammonium chloride (NH_4Cl), manganese dioxide

(MnO_2), and carbon. These components are the anode portion of the cell and the zinc container serves as the cathode. The half-cell reactions are:

Anode reaction: Zn $\rightarrow Zn^{2+} + 2\ e^-$ $E° = +0.76$ volts

Cathode reaction: $NH_4^+ + 2\ MnO_2 + 2\ e^- \rightarrow 2\ NH_3 + Mn_2O_3 + H_2O$ $E° = +0.74$ volts

Overall reaction: $Zn + 2\ NH_4^+ + 2\ MnO_2 \rightarrow Zn^{2+} + Mn_2O_3 + H_2O$ $E_{cell} = +1.50$ volts

In recent years, some of the applications where dry-cell batteries were used have been replaced by the more efficient Ni-Cd (nickel-cadmium) batteries. The reactions involved in these batteries are:

Anode reaction: $Cd + 2\ OH^-$ $\rightarrow Cd(OH)_2^+ + 2\ e^-$ $E° = +0.76$ volts

Cathode reaction: $NiO_2 + 2\ H_2O + 2\ e^-$ $\rightarrow Ni(OH)_2 + 2\ OH^-$ $E° = +0.49$ volts

Overall reaction: $Cd + NiO_2 + 2\ H_2O$ $\rightarrow Cd(OH)_2 + Ni(OH)_2$ $E_{cell} = +1.25$ volts

One of the great advantages that Ni-Cd batteries have over conventional dry-cell batteries is that they are rechargeable. Due to the increased use of consumer goods that require a dependable, rechargeable energy supply, battery research remains at an all-time high. This brief introduction to electrochemical reactions is only intended to stimulate your interest and encourage further study.

The last type of chemical reaction is the **double replacement**, in which two compounds are reacted. Each compound is composed of one positive and one negative ion. The positive and negative ions from the compounds exchange partners, forming two new compounds. A typical example would be the reaction of hydrochloric acid and sodium hydroxide to form the salt, sodium chloride, and water.

Double Replacement Reactions

$$HCl + NaOH \rightarrow NaCl + HOH$$

HOH is not the way water is typically written. The intent is to emphasize that water is the product of the combination of one H^+ and one OH^- ion. This reaction is also significant because it represents one of the most familiar of all double replacement reactions: acid + base $\rightarrow$ a salt + water. Table 7-6 lists some of the other double replacement reactions.

Double Replacement:	AB + CD	$\longrightarrow$	AD + CB
1. acid + base		$\longrightarrow$	salt + water
2. metal carbonate + acid		$\longrightarrow$	salt + water + CO_2
3. metal bicarbonate + acid		$\longrightarrow$	salt + water + CO_2
4. metal oxide + acid		$\longrightarrow$	salt + water
5. salt + base		$\longrightarrow$	insoluble base + salt
6. $salt_1$ + $salt_2$		$\longrightarrow$	insoluble $salt_3$ + $salt_4$
7. ammonium salt + base		$\longrightarrow$	salt + ammonia + water
8. metallic sulfite + acid		$\longrightarrow$	salt + SO_2 + water
9. salt + acid		$\longrightarrow$	insoluble salt + acid

Table 7-6: Double replacement reactions.

See if you can match the following reactions to the descriptions given in Table 7-6.

$$H_2SO_4 \quad + \quad 2\,NaOH \quad \rightarrow \quad Na_2SO_4 \quad + \quad 2\,H_2O$$

$$Pb(NO_3)_2 \quad + \quad 2\,KCl \quad \rightarrow \quad PbCl_2 \quad + \quad 2\,KNO_3$$

$$AgNO_3 \quad + \quad NaCl \quad \rightarrow \quad AgCl \quad + \quad NaNO_3$$

$$Na_2CO_3 \quad + \quad 2\,HCl \quad \rightarrow \quad 2\,NaCl \quad + \quad H_2O + CO_2$$

Answers: 1, 6, 6, & 2

Check Your Understanding

1. Explain what is accomplished by balancing a chemical reaction.

2. What is the difference between endothermic and exothermic reactions?

3. What are the four general reaction types?

4. What general type of chemical reaction always occurs with a single reactant?

5. Explain the difference between corrosion and what happens in a battery.

Chemical Calculations

Objectives

Upon completion of this section, you will be able to:

■ Calculate the formula mass of a pure substance.

■ Determine the percentage composition of a compound.

■ Explain the mole and use it in chemical calculations.

■ Solve simple stoichiometric problems.

In the previous section, the focus was on predicting the products of various chemical combinations and balancing the equation to account for all atoms. In each of these equations, the number of atoms or molecules was a small number like one, two, or three. It was not possible to weigh or even see single atoms or molecules in order to test predictions. Early chemists solved this problem by using larger amounts of each reactant in specific proportions. If one substance is twice as heavy as the other, then twice as much of that substance would be weighed out.

Let's examine this logic a bit more. If you were responsible for bagging the 300 small bolts used to assemble a lawn building, how would you do it? You could manually count them out each time, but that would be slow and probably inaccurate. Another option would be to count out and weigh 100 bolts, then calculate the weight for 300 bolts. By using an accurate balance or scales, you could quickly weigh out exactly 300 bolts time after time.

Here is another example. Suppose that all men weigh 200 lbs and all women weigh 100 lbs. A big party is being planned for couples only. Would there be the same number of individuals in 4 tons of men as in 2 tons women? Careful analysis reveals that as long as the 2:1 weight ratio is maintained, the number of individuals in each group will always be equal.

Carrying this idea a step further, suppose you react magnesium, Mg, and sulfur, S. The balanced equation for the reaction is:

$$Mg + S \rightarrow MgS$$

If you consult the periodic table, you will find that the average atomic mass of magnesium is 24 u and the average atomic mass of sulfur is 32 u This means that 24 grams of magnesium and 32 grams of sulfur each contain the same number of atoms. In fact, 24 tons of magnesium and 32 tons of sulfur would each contain the same number of atoms – it would just be a much larger number!

Chemists have found that using the formula mass of a substance, in grams, is a convenient way to measure equal quantities. This amount is called one mole. Chemists used this concept for more than 60 years without knowing how many atoms or molecules were actually involved. Although the number is now known as Avogadro's number, the Italian scientist Amadeo Avogadro (1776–1856) did not determine its numerical value.

Today, the mole is one of the seven SI units and is defined as the number of atoms present in exactly twelve grams (0.012 kg) of carbon-12 isotope. The accepted value for the mole is 6.0221367×10^{23} atoms of carbon, which is rounded to 6.02×10^{23} for most chemical calculations.

Calculation of Formula Mass

Based on the above discussion, the importance of knowing the formula mass, fm, of each substance involved in a chemical reaction can be seen. With only two common exceptions, determining the formula mass for an element is as simple as locating it on the periodic table and rounding its atomic mass to the nearest whole number. The exceptions are the elements copper and chlorine. Remember that the mass numbers appearing on the periodic table are typically a weighted average, based on the ratio of naturally occurring isotopes. For this reason, rounding the mass number is actually like selecting just one of the isotopes, which works well for some chemical calculations. For the elements copper and chlorine, rounding to just one of the isotopes introduces too much error, so they should never be rounded to less than 63.5 u and 35.5 u, respectively. The number of significant figures used for the atomic masses should always be determined by the number of significant figures in the calculation.

When a substance contains more than one atom or more than one kind of atom, then it is recommended that the formula mass be determined using the following procedure. 1) List each element present as shown below. 2) Determine the number of atoms of each element and place that number next to the symbol. 3) Round off the atomic mass listed on the periodic table for each of the elements and put that next to the number of atoms. 4) Multiply the number of atoms of each element times its atomic mass. 5) Total the contribution of each element to arrive at the formula mass for the substance. Example: Na_2CO_3

Na	$2 \times 23 =$	46
C	$1 \times 12 =$	12
O	$3 \times 16 =$	48
Formula Mass	$=$	106 g/mole

In this example, it can be seen that the units used to express the formula mass can be expressed in either u or grams/mole. Except when discussing a single molecule, the use of grams/mole is favored and will be used throughout the remainder of this discussion.

This method for determining formula mass makes it very easy to determine the contribution of each element. This approach can also be used to determine the percentage composition of elements in a compound. For example, the element mercury is often separated from the ore cinnabar, HgO. Using the above method provides an easy way to calculate the amount of mercury, by mass, that would be in one kilogram of ore. First, the formula mass must be determined. Example: HgO

Hg	$1 \times 201 =$	201
O	$1 \times 16 =$	16
Formula Mass	$=$	217 g/mole

To calculate the percentage composition, requires determining the contribution of each element to the formula mass.

Hg	$\frac{201}{217} \times 100 =$	92.6%
O	$\frac{16}{217} \times 100 =$	7.34%
		100.0%

From the calculation, it can be seen that 92.6%, by mass, of the 1 kg of ore is due to the mass of the mercury. This means that for each kilogram of ore, 926 g of mercury could be recovered. This process can be used to assay minerals (determine the elemental composition) or even to determine the amount of calcium present in a calcium supplement or antacid.

Numerous salts exist as **hydrates**. This means they have a definite mole-ratio of water incorporated in their crystal structure. This water can be removed by heating and results in the **anhydrous salt** and water. Such a substance is borax or sodium tetraborate decahydrate, $Na_2B_4O_7 \cdot 10\ H_2O$. Using the techniques developed above it is possible to calculate the percent of water contained within its crystal structure. Example: $Na_2B_4O_7 \cdot 10\ H_2O$

$$
\begin{array}{llll}
Na & 2 \times 23 & = & 46 \\
B & 4 \times 11 & = & 44 \\
O & 7 \times 16 & = & 112 \\
H_2O & 10 \times 18 & = & \underline{180} \\
& & & 382\ \text{g/mole}
\end{array}
$$

Note that water was left as a molecule in this calculation and, therefore, its formula mass (18 grams/mole) was used. To finish the calculation and determine the percent of water in this mineral, the following step would have to be completed.

$$\frac{180}{382} \times 100 = 47.12\%\ \text{water by mass}$$

It is interesting to consider that if you purchase borax, about one-half of it is just water!

Mole

Over the years, the use of the mole has become central to many different types of calculations. One of the very important relationships the mole provides is a connection between the number of atoms or molecules and their mass. As shown in Figure 7-5, 1 mole of an element always contains 6.02×10^{23} atoms or 1 mole of a compound always contains 6.02×10^{23} molecules. This is like saying 1 dozen peas always equals 12 peas or 1 dozen pea pods always equals 12 pea pods. So the quantity 1 mole always equals 6.02×10^{23}.

The second relationship the mole provides is between the number of moles and grams of a substance. The formula mass is always what relates these two amounts. No matter what the substance, 1 formula mass, in grams, equals 1 mole of the substance.

Use of the relationships in Figure 7-5, coupled with unit analysis provides a very powerful tool for performing many different calculations. For example, if 3 moles of water are in a container, how many water molecules does it contain? This question is asking for a number. It isn't much different than asking how many donuts are in three dozen. To solve the problem, you must first determine what is given (3 moles of water) and what you need to find (how many molecules). Once you have located these boxes in the diagram, examine what re-

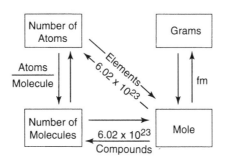

Figure 7–5: Relationships among grams, moles, and the number of particles.

lationship (conversion factor) relates the boxes. Since water is a compound, the relationship between the two boxes in the diagram is 1 mole = 6.02×10^{23} molecules. Using the unit analysis approach:

$$3 \text{ moles } H_2O \times \frac{6.02 \times 10^{23} \text{ } H_2O \text{ molecules}}{1 \text{ mole } H_2O} = 18.06 \times 10^{23}$$

or 1.81×10^{24} H_2O molecules

This calculation shows that 3 moles of water is the same as 1.81×10^{24} H_2O molecules. In a similar manner, the mass of 3 moles of water can also be determined.

$$3 \text{ moles } H_2O \times \frac{18 \text{ g } H_2O}{1 \text{ mole } H_2O} = 54 \text{ g } H_2O$$

This calculation indicates that the formula mass of water becomes the relationship (conversion factor) between the number of moles and grams of water.

The problems can be more complicated, but the method remains essentially the same. For example, how many molecules are present in 245 g of H_2SO_4? To move from the box containing grams to the box containing number of molecules, requires you to cross two sets of arrows or relationships. That means that you need to know two conversion factors. First, is the formula mass of sulfuric acid and second, the number of molecules in one mole.

$$245 \text{g } H_2SO_4 \times \frac{1 \text{ mole } H_2SO_4}{98 \text{ g } H_2SO_4} \times \frac{6.02 \times 10^{23} \text{ } H_2SO_4 \text{ molecules}}{1 \text{ mole } H_2SO_4}$$

$$= 1.50 \times 10^{24} \text{ molecules } H_2SO_4$$

Stoichiometry

In the previous problems, we learned how the amount of a substance can be converted from mass to moles to a number of particles. Chemists are often interested in determining the amounts of reactants or products involved in a chemical reaction. This area of chemistry, known as **stoichiometry**, can be defined as the area of science that deals with the quantities of substances involved in chemical reactions.

In Figure 7-6 a second half was added to the previous diagram making it possible to relate two different substances. The new conversion factor, however, that must be present is a balanced equation to relate the number of moles of each substance.

Stoichiometric calculations are subdivided into three different levels: mole-mole; mole-mass; and mass-mass. These subdivisions are based on what is known and what is requested by the problem. For example, suppose that 5 moles of sulfur are burned and we wanted to calculate the number of moles of sulfur trioxide that is produced. This would be an example of a mole-mole problem, since both the known and requested substance amounts are expressed in moles.

The first and essential step in solving all stoichiometric calculations is to write a balanced equation for the reaction. In this example, it would be:

$$2 \text{ S} + 3 \text{ O}_2 \rightarrow 2 \text{ SO}_3$$

2 moles   3 moles   2 moles

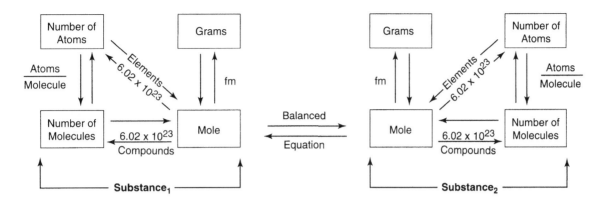

Figure 7-6: Stoichiometric relationships.

In the balanced equation, the coefficients not only account for all atoms, they also provide the mole-mole ratio between each of the substances. In this equation 2 moles of sulfur requires 3 moles of oxygen to produce 2 moles of sulfur trioxide. Referring to Figure 7-6, we find that the number of moles of one substance and the number of moles of a second substance, are separated by the relationship in the balanced equation.

The second step in solving stoichiometric problems, therefore, is to identify the box that represents what is given (5 moles S) and the box that represents what you wish to find (moles SO_3). Determine the relationship (conversion factor) between them and set up the mathematical relationship. The information given (5 moles S) is always written first, followed by the conversion factor relating the known and desired information. In this example it would be:

$$5 \text{ moles S} \times \frac{2 \text{ moles } SO_3}{2 \text{ moles S}} = 5 \text{ moles } SO_3$$

Notice that both the information given and the answer are expressed in moles. When the information given is in moles, but the requested information needs to be expressed in grams, it adds one additional step to the calculation. For example, suppose that the previous problem had asked you to calculate the number of grams of SO_3 that can be produced by the burning of 5 moles of sulfur. Again using Figure 7-6 as a guide, the information given (5 moles S) and the information requested (g SO_3) are separated by two sets of arrows. The first relationship, like last time, is the mole-mole ratio from the balanced equation. This time, however, we also need to know the formula mass of sulfur trioxide, SO_3. This information is not given in the problem statement, but it can be calculated using the atomic masses for each element from the periodic table. Its value is 32 + 3(16) g/mole. We now have values for each of the conversion factors separating the boxes. The set-up for the calculation would be:

$$5 \text{ moles S} \times \frac{2 \text{ moles } SO_3}{2 \text{ moles S}} \times \frac{80 \text{ g } SO_3}{1 \text{ mole } SO_3} = 400 \text{ g } SO_3$$

This time the information given is expressed in moles and the answer is in grams. This is, therefore, an example of a mole-mass problem.

Perhaps the most common type of stoichiometric problem is the mass-mass type. The reason for its popularity is that mass is the most common way of quantifying the amount of a substance used, especially for the non-scientist. Consider the following problem. How many grams of carbon dioxide, CO_2, would be produced by the complete combustion of 1,000 g of glucose? The balanced equation for the reaction is:

$$C_6H_{12}O_6 \;+\; 6\,O_2 \;\rightarrow\; 6\,CO_2 \;+\; 6\,H_2O$$

$$1\text{ mole} \qquad 6\text{ moles} \qquad 6\text{ moles} \qquad 6\text{ moles}$$

Again referring to Figure 7-6, we find that the gram-gram boxes for two different substances are separated by three sets of arrows; two of them involve formula masses and one of them is the mole-mole ratio between the two substances. The setup for the solution would be (1 mole of $C_6H_{12}O_6$ = 180 g and 1 mole CO_2 = 44 g):

$$1{,}000\text{ g }C_6H_{12}O_6 \times \frac{1\text{ mole }C_6H_{12}O_6}{180\text{ g }C_6H_{12}O_6} \times \frac{6\text{ moles }CO_2}{1\text{ mole }C_6H_{12}O_6} \times \frac{44\text{ g }CO_2}{1\text{ mole }CO_2} = 1{,}467\text{ g }CO_2$$

In a similar fashion, the amounts of each of the other substances can also be calculated.

$$1{,}000\text{ g }C_6H_{12}O_6 \times \frac{1\text{ mole }C_6H_{12}O_6}{180\text{ g }C_6H_{12}O_6} \times \frac{6\text{ moles }H_2O}{1\text{ mole }C_6H_{12}O_6} \times \frac{18\text{ g }H_2O}{1\text{ mole }H_2O} = 600\text{ g }H_2O$$

$$1{,}000\text{ g }C_6H_{12}O_6 \times \frac{1\text{ mole }C_6H_{12}O_6}{180\text{ g }C_6H_{12}O_6} \times \frac{6\text{ moles }O_2}{1\text{ mole }C_6H_{12}O_6} \times \frac{32\text{ g }O_2}{1\text{ mole }O_2} = 1{,}067\text{ g }O_2$$

The laws of conservation of matter and energy are important to all areas of science, but especially important to the area of stoichiometry. In this example, we have used the amount of just one reactant to calculate the number of grams of both products as well as the amount of the other reactant required. Addition shows that both the sum of the reactants (1,000 g of $C_6H_{12}O_6$ and 1,067 g of O_2) equals 2,067 g and sum of the products (1,467 g CO_2 and 600 g H_2O) equals 2,067 g. In other words, mass is conserved! The fact that this is always true during ordinary chemical reactions is a very powerful tool.

Limiting Reagents

What happens when the amounts of reactants are not present in stoichiometric proportions? The car provides a practical example. The reaction that makes a car go is the combustion of fuel, using oxygen from the air. So far, no car has ever run out of oxygen, but you have heard of someone running out of fuel! In this situation, the amount of fuel is the **limiting reagent** and determines how far you can travel.

Problems that involve a limiting reagent are typically easy to spot because the amounts of two reactants are given and the amount of a product is requested. For example, how many grams of hydrogen gas, H_2, can be produced by placing 50 g of zinc, Zn, into a solution containing 50 g of hydrochloric acid, HCl? The balanced equation for the reaction is: $Zn + 2\,HCl \rightarrow ZnCl_2 + H_2$.

At this point, there are two approaches that can be used to solve the problem. The simpler approach is to merely solve the problem twice. The first time, you use one reactant and the second time you use the other reactant. The smaller answer will *always* be the correct answer. This is like reading a cookie recipe and determining if you will run out of your supply of flour or sugar first. Whichever ingredient gets used up first determines the number of cookies you can make. The calculation based on using all of the zinc metal would give:

$$50 \text{ g Zn} \times \frac{1 \text{ mole Zn}}{65 \text{ g Zn}} \times \frac{1 \text{ mole H}_2}{1 \text{ mole Zn}} \times \frac{2 \text{ g H}_2}{1 \text{ mole H}_2} = 1.5 \text{ g H}_2$$

The calculation based on using all of the hydrochloric acid would give:

$$50 \text{ g HCl} \times \frac{1 \text{ mole HCl}}{36.5 \text{ g HCl}} \times \frac{1 \text{ mole H}_2}{2 \text{ moles HCl}} \times \frac{2 \text{ g H}_2}{1 \text{ mole H}_2} = 1.4 \text{ g H}_2$$

For this reaction the hydrochloric acid, HCl, is the limiting reagent and determines the amount of hydrogen gas, H_2, that will be produced. A similar calculation can be done to determine the amount of excess zinc metal present. The limiting reagent should be used in the calculation to determine the amount of excess reagent (Zn).

$$50 \text{ g HCl} \times \frac{1 \text{ mole HCl}}{36.5 \text{ g HCl}} \times \frac{1 \text{ mole Zn}}{2 \text{ moles HCl}} \times \frac{65 \text{ g Zn}}{1 \text{ mole Zn}} = 44.5 \text{ g Zn}$$

This calculation shows that only 44.5 g of zinc is required to use all of the hydrochloric acid; therefore, (50 g – 44.5 g) = 5.5 g of the zinc metal was in excess.

Theoretical Yield and Percentage Yields

One final use for stoichiometric calculations is to determine the theoretical yield or percent yield of a reaction. The **theoretical yield** is the maximum amount of product that can be produced when everything is ideal. The **percent yield** is the percent of the theoretical yield actually produced.

$$\text{percent yield} = \frac{\text{actual yield}}{\text{theoretical yield}} \times 100$$

For example, 340 g of $AgNO_3$ was reacted with NaCl and produced AgCl produces only 256 g of AgCl. Calculate both the theoretical and percent yields for this reaction. The equation for the reaction is:

$$AgNO_3 + NaCl \rightarrow AgCl + NaNO_3$$

$$340 \text{ g AgNO}_3 \times \frac{1 \text{ mole AgNO}_3}{170 \text{ g AgNO}_3} \times \frac{1 \text{ mole AgCl}}{1 \text{ mole AgNO}_3} \times \frac{143.5 \text{ g AgCl}}{1 \text{ mole AgCl}} = 287 \text{ g AgCl}$$

If, due to poor lab procedure, the actual yield was 256 g of AgCl, but the theoretical yield is 287 g, then the percent yield would be:

$$\text{percent yield} = \frac{256 \text{ g}}{287 \text{ g}} \times 100 = 89.2\%$$

An alternate way to represent experimental results is to report the **percent error**. The percent error is the difference between the theoretical and actual yields divided by the theoretical yield times one hundred. This is actually the difference between the percent yield and one hundred percent.

$$\text{percent error} = \frac{|\text{theoretical} - \text{actual yield}|}{\text{theoretical yield}} \times 100$$

When applied to the previous example:

$$\text{percent error} = \frac{|287\text{ g} - 256\text{ g}|}{287\text{ g}} \times 100 = 10.8\%$$

It should be noted that the numerator term is enclosed by, | |, indicating that it is the absolute difference that should be used. This means that it is not significant whether the difference between the theoretical and the actual yield is positive or negative.

Check Your Understanding

1. What is another name for Avogadro's number and its accepted (rounded) value?

2. Describe how the formula mass would be calculated for the compound, Na_2CO_3.

3. What is a hydrate?

4. What relationship exists between the number of grams and moles of any substance?

5. Define stoichiometry.

6. What item is essential for solving any type of stoichiometric problem?

7. What are the three subtypes of stoichiometric problems and how are they recognized?

8. How many conversion factors are required when solving a mass-mass problem?

9. How is the limiting reagent determined in a stoichiometric problem?

10. What relationships exist between theoretical yield, percentage yield, and percent error?

7–3   Solutions

Objectives

Upon completion of this section, you will be able to:

1. Discuss water as a solvent.

2. Identify the characteristics of true solutions, colloidal dispersions, and suspensions.

3. Calculate the percent concentration and molarity of a solution.

4. Discuss the colligative properties of a true solution.

Types of Solutions

A **solution** is a homogeneous mixture composed of two distinct parts: solute and solvent. The **solute** is the substance dissolved or present in the lesser amount. The **solvent** is the substance doing the dissolving or present in the greater amount. Salt water is a typical solution in which the salt is the solute and water is the solvent, and a martini is a solution in which the vermouth is the solute and the gin is the solvent. The coolant used in car radiators is also a solution composed of antifreeze and water. If the amount of antifreeze is less than the amount of water, then it is the solute and water is the solvent. When the amount of antifreeze exceeds the amount of water, then the roles reverse and water becomes the solute and the antifreeze becomes the solvent.

When a solute and solvent are put together, it may result in one of three different types of solutions: a true solution, a colloidal dispersion, and a suspension. A **true solution** is homogenous, transparent, stable, typically colorless, and can't be separated by filtering. It is composed of a solvent, typically water, and solute particles that are atoms, ions, or small molecules in the range of 5×10^{-9} to 1×10^{-7} cm. Solute particles in this size range tend to be small enough that light passes virtually unaffected by their presence and they can pass through the openings in filter paper. True solutions are most desirable and the type most often encountered in the chemistry laboratory. Two examples of true solutions would be saltwater and sugar water.

Colloidal dispersions are homogeneous, appear cloudy or milky, are stable and the solute cannot be separated by filtration, but can be separated by dialysis. They are composed of a solvent, typically water, and the solute particles range from large, to very large, to clusters of molecules in the range of 1×10^{-7} to 1×10^{-5} cm. Solute particles of this size are still small enough to pass through the openings in filter paper; however, they do interfere with the passage of light and cannot pass through a dialyzing membrane. This type of solution is very important in body chemistry, since blood and cells contain large molecules dispersed in water. Artificial dialysis is a process that is typically performed on persons whose kidneys are not functioning properly. By passing the blood through dialyzing tubing surrounded by a true solution, the small atoms, ions, and molecules that are waste products in the blood are removed, but the larger molecules, like the proteins called globins, cannot pass.

Suspensions are heterogeneous, opaque, unstable, and can be separated by filtration. The particles within a suspension are typically larger than 1×10^{-5} cm, ranging up to particles that are visible. Fast moving streams often form suspensions by picking up large amounts of soil, which may eventually be deposited in a delta. Many antibiotics carry directions to shake before dispensing. This is an indication that the antibiotic is in a suspension and must be re-mixed just before removal to deliver a uniform dose.

One of the most important characteristics of a solution is whether it is homogeneous or heterogeneous. The reason this is important is that solutions are often used as a convenient way to transfer known amounts of chemicals. Most students agree that it is easier to measure liquid volumes than to weigh out exact amounts on a balance. For this to be possible, however, the solutions must be homogeneous. If they are homogeneous, then identical volumes of the solution, whether taken from the top or bottom, will contain the identical amount of solute. This means true solutions and colloidal suspensions may be used for quantitative transfer, but suspensions cannot.

Before leaving this area, it should be noted that although our focus has been primarily on water-based (**aqueous**) solutions, a solution may be composed of any of the three states of matter. For example, did you ever wonder where the rubber from car tires goes? At least a part of it spends some time as a suspension in the air. In this case, the rubber would be the solute and the air would be the solvent. It is a suspension because these bits of rubber settle out, leaving a black residue on your house, car, or anything else outside.

Soda water is a true solution prepared by dissolving CO_2 gas into water. In this solution, a gas is the solute and a liquid is the solvent. Conversely, fog is considered a solution composed of a liquid (water), as the solute and a gas (air), as the solvent. The point here is that there are many other possible combinations of solids, liquids, and gases that result in solutions.

Solution Concentration

When a true solution or colloidal suspension is formed, the amount of solute placed in the solvent may vary. Salt water may contain only a few grains of salt in a large volume of water or it may be made to be very salty. Unlike pure substances that have constant composition, solutions are mixtures and are variable in their composition. Because they can vary in their composition, there is a need for a way to communicate its composition. This can be done by including information about its **concentration**. Many different methods of expressing concentration have been developed. They all have one thing in common. Each method expresses the number of things per a given amount.

$$\text{concentration} = \frac{\text{number of things}}{\text{amount}}$$

For example, 15 students in a classroom is a concentration. The things being counted are students and the amount under consideration is the volume in the classroom. If it were changed to 15 students in a telephone booth, the number of students remains the same, the volume becomes less, therefore the concentration becomes higher and they would have to get a lot friendlier!

Scientists have developed many different methods for expressing concentration that are particularly useful to them. Percent concentration and molarity are two that we will consider in more detail below. Regardless of the method used, concentration always expresses the number of things (grams, moles, parts) that are present in a given amount (milliliters, grams, liters, cubic meters, million parts).

Methods for Expressing Solution Concentrations

There are two kinds of concentration terms; non-quantitative and quantitative. Non-quantitative terms are in common use, but inexact. Suppose, for example, someone comments that a particular person is the tallest they have ever seen. Would you have any understanding of how tall the person is? In a similar way, the terms dilute and concentrated are in common usage. We often buy concentrated orange juice and then add water to dilute it. What do the terms dilute and concentrated mean? They do not transmit specific quantitative information, they are only a comparison. It can only be assumed that a **dilute solution** contains less solute per volume than one that is concentrated. The opposite is also true, a **concentrated solution** contains more solute particles per given volume than one that is dilute.

Only if there is prior agreement, can the terms dilute and concentrated be more informative. For example, it is common in many laboratories to fill stock reagent bottles with the product that comes from the manufacturer and label it as concentrated, e.g., Conc. H_2SO_4, Conc. HNO_3, Conc. NH_4OH, etc. Then these concentrated acids and bases are diluted 50:50 with water (reducing their concentrations by one-half) and put in bottles that are labeled dilute, e.g., Dil. H_2SO_4, Dil. HNO_3, Dil. NH_4OH, etc. One problem with this system, however, is that not all manufacturers produce the same concentration in the product labeled "concentrated." Therefore, two bottles labeled concentrated may have quite different concentrations.

Another set of non-quantitative terms is unsaturated, saturated, and supersaturated. These terms are slightly more valuable, if a solubility graph is available. Figure 7-7 is a graph showing the solubility of several different salts at different temperatures. The information for constructing this graph was experimentally determined. As you see, some salts ($BaCl_2$) are not very soluble in cold water, but become increasingly soluble as the temperature increases. Others (NaCl) maintain nearly the same solubility over a wide range of temperatures. The curve for each salt represents the maximum amount, in grams, that will dissolve in 100 g of water at each temperature. Each curve, therefore, represents a saturated solution over a range of temperatures.

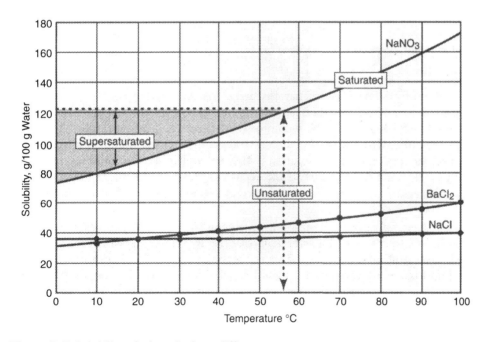

Figure 7–7: Solubility of selected salts at different temperatures.

A **saturated solution** is defined as one that has dissolved all the salt it can at a particular temperature. Conversely, an **unsaturated solution** is one that can still dissolve more solute at that temperature. As shown by the dotted line below the $NaNO_3$ curve, a solution that is anywhere below this curve is unsaturated. If the solution is saturated at 55°C, it will become unsaturated if the temperature is increased to 60°C. Raising the temperature to speed dissolving is a common cooking practice.

A supersaturated solution can only be produced by trickery. By definition, a **supersaturated solution** is holding more solute than it should at a given temperature. The only way to produce such a solution is by heating the solution until all the solution is dissolved, then covering and setting it aside to slowly cool. If all conditions are ideal, the solution may cool past the saturation point without crystals forming. As shown in Figure 7-7, if a $NaNO_3$ solution is cooled and it follows the dotted line above and to the left of the curve, it would then be an unstable supersaturated solution. The addition of a seed crystal or vibration would be all that is required for the solution to return to saturated state and to have the excess crystals form at the bottom of the solution. On occasion, clouds become supersaturated but it does not rain. Each raindrop needs a nucleating body (grain of pollen, dust, etc.) around which water molecules collect to form a large enough droplet of water to fall. Seeding the clouds is an attempt to provide these nucleating bodies and cause it to rain.

Percent Concentration

As strange as it may first sound, there are three different ways of calculating **percent concentrations**. The three formulas are shown below, but only one of the methods produces a true percentage.

Weight-volume percent
$$\frac{g\ solute}{ml\ solvent} \times 100 = \%_{(wt/vol)}$$

Weight-weight percent
$$\frac{g\ solute}{g\ solute + g\ solvent} \times 100 = \%_{(wt/wt)}$$

Volume-volume percent
$$\frac{ml\ solute}{ml\ solute + ml\ solvent} \times 100 = \%_{(vol/vol)}$$

Percent by definition is the part divided by the whole, times one hundred. In the weight-volume percent method, used in the medical profession, 6 grams of glucose dissolved into 100 mL of water would be labeled as a 6% $_{(wt/vol)}$ solution. The problem is that 100 mL of water has a mass of 100 grams, so when you add the mass of the glucose, the total mass of the solution is 106 grams. It, therefore, would not be a 6% solution, but a bit less. In comparison to its convenience, this slight deviation from the actual percentage is apparently not seen as important when administering 6% glucose intravenous (IV) solutions.

In a similar way, the volume-volume percent fails to be a true percent. For example, if 50 mL of ethyl alcohol and 50 mL of water were mixed, it would seem that you would have 100 mL of solution. This is not the case, however. Because of hydrogen bonding between the water and alcohol molecules, the actual volume is slightly less than 100 mL. This means that the solution is not a 50% solution, but would have a true percentage that is slightly greater.

From the previous discussion, it becomes obvious that the only calculation that results in a true percent is the weight-weight percent method. This is, therefore, the preferred method for most chemical work. In this method, the mass of the solute is divided by the combined masses of the solute and solvent and multiplied times one hundred. For example, if 34 grams of NaCl were dissolved into 200 grams of water the weight-weight percent would be calculated as follows:

$$\frac{34\ g}{34\ g + 200\ g} \times 100 = 14.5\%_{(wt/wt)}$$

Depending on the degree of accuracy required, it is customary to assume that one milliliter of water at room temperature has a mass of one gram (1 mL $H_2O = 1$ g H_2O). This is only strictly true, however, at $4°C$.

Molarity

In the section on stoichiometry, it was discussed that when chemicals react with one another, they always do so on the basis of a mole ratio. Chemists developed a concentration unit called molarity so that solution concentrations could be easily used in stoichiometric problems. **Molarity** (M) is the number of moles of solute per liter of solution. There are two formulas commonly used for calculating the molarity of solutions.

$$M = \frac{moles_{(solute)}}{liter_{(solution)}} \qquad M = \frac{g_{(solute)}}{fm \times liter_{(solution)}}$$

These two equations produce the same result. In the right-hand equation, dividing the number of grams of solute by its formula mass, fm, converts it to moles, this results in an answer in moles per liter. This equation, therefore, also gives the number of moles of solute per liter of solution.

Because this method is used for precise analytical work, it is also significant to note that it is the number of moles in a liter of solution, not the number of moles in a liter of solvent. A **standard solution** is one that has been carefully prepared to a known concentration and is used to determine the concentration of other solutions. To prepare a standard solution, the solute is first weighed out, then completely dissolved in a small amount of the solvent, before adding enough additional solvent to bring the total solution to an exact volume.

The simplest problems in this area express the amount of solute in moles and the volume in liters. For example, what would be the molarity of a solution made by dissolving 6 moles of NaOH into enough water to make 3 liters of solution?

$$\frac{6 \text{ moles}}{3 \text{ liters}} = \frac{2 \text{ moles}}{\text{liter}} \text{ or } 2 \text{ M}$$

Note that the unit moles/liter was replaced by the symbol M. In those problems where the volume is expressed in any other unit of volume, it must be converted to liters before solving the problem. If 4.50 moles HCl were dissolved into enough water to make 300 mL of solution, what would be the molarity of the solution?

$$\frac{4.50 \text{ moles}}{0.300 \text{ liter}} = \frac{15.0 \text{ moles}}{\text{liter}} \text{ or } 15.0 \text{ M}$$

If the statement of the problem uses the term mass or the amount of solute is expressed in grams, then the other formula for molarity can be used. For example, what would be the concentration of a sulfuric acid, H_2SO_4, solution prepared by dissolving 9.8 g of H_2SO_4 into enough water to make 800 mL of solution. One way to avoid making errors when completing these problems is to list all of the variables known, such as:

$$\begin{aligned} M &= ? \\ g &= 9.8\text{ g} \\ fm &= 98\text{ g/mole} \\ L &= 800\text{ mL} \quad = \quad 0.800\text{ L} \end{aligned}$$

The value of this tabular housekeeping is that it gives you an opportunity to collect all known data and provide items, such as formula mass, that are not provided. It also helps you to remember that the volume must be expressed in liters (L) rather than in milliliters (mL). You are now ready to enter the data into the right-hand equation.

$$M = \frac{9.8\ \cancel{g}}{98\ \cancel{g}/\text{mole} \times 0.800\text{ L}} = \frac{0.125\text{ moles}}{L} = 0.125\text{ M}$$

In a similar way, it is possible to calculate the amount of a 6.00 M HNO_3 solution that would be required to provide 34.5 g of HNO_3 for a reaction.

$$\begin{aligned} M &= 6.00\text{ M} \\ g &= 34.5\text{ g} \\ fm &= 63\text{ g/mole} \\ L &= ?\text{ mL} \quad = ?\text{ L} \end{aligned}$$

$$6.00\text{ M} = \frac{34.5\ \cancel{g}}{63\ \cancel{g}/\text{mole} \times L}$$

$$L = \frac{34.5\ \cancel{g}}{63\ \cancel{g}/\cancel{\text{mole}} \times 6.00\ \cancel{\text{moles}}/L} = 0.0913\text{ L} = 91.3\text{ mL}$$

For the units to cancel, it is necessary to change the molarity unit, M, back to its equivalent moles/liter. Again, it was necessary to calculate the formula mass for the nitric acid, HNO_3, since it was not provided in the problem statement.

Dilution

Perhaps the most common procedure employed when working with solutions is dilution. As the name implies, additional solvent is added allowing the solute particles to move further apart. The simple formula that can be used for all dilutions, regardless of the units used to express the concentration or the volume, is:

$$C_{(old)} \times Vol_{(old)} = C_{(new)} \times Vol_{(new)}$$

Here again, it is valuable to set up an accounting system for what you know and what you want to find. For example, what is the concentration of the resulting solution, if 300 mL of water is added to 500 mL of $6\%_{(wt/wt)}$ sugar water?

$$\begin{aligned} C_{(old)} &= 6\%_{(wt/wt)} \\ Vol_{(old)} &= 500\text{ mL} \\ C_{(new)} &= ? \\ Vol_{(new)} &= 500\text{ mL} + 300\text{ mL} = 800\text{ mL} \end{aligned}$$

$$C_{(old)} \times Vol_{(old)} = C_{(new)} \times Vol_{(new)}$$

$$6\%_{(wt/wt)} \times 500\text{ mL} = C_{(new)} \times 800\text{ mL}$$

$$C_{(new)} = \frac{6\%_{(wt/wt)} \times 500 \text{ mL}}{800 \text{ mL}} = 3.75\%_{(wt/wt)}$$

A slightly more involved problem would be, how much water must be added to 500 mL of a 4.00 M HCl solution to make it a 3.00 M HCl solution? It is necessary to first calculate the new volume.

$$C_{(old)} \quad = 4.0 \text{ M}$$
$$Vol_{(old)} \quad = 500 \text{ mL}$$
$$C_{(new)} \quad = 3.00 \text{ M}$$
$$Vol_{(new)} \quad = ?$$

$$C_{(old)} \times Vol_{(old)} = C_{(new)} \times Vol_{(new)}$$

$$4.00 \text{ M} \times 500 \text{ mL} = 3.00 \text{ M} \times Vol_{(new)}$$

$$Vol_{(new)} = \frac{4.00 \text{ M} \times 500 \text{ ml}}{3.00 \text{ M}} = 667 \text{ ml}$$

Once the new volume is known, it is possible to subtract the original volume to determine the amount of water added.

$$667 \text{ mL} - 500 \text{ mL} = 167 \text{ mL } H_2O$$

Colligative Properties

Those properties that are dependent on only the concentration of solute particles and not the chemical nature of the particles are called **colligative properties**. This means that the effect of an atom, ion, or molecule on the colligative property is equal – that is, only the concentration is important. Freezing point depression, boiling point elevation, vapor pressure lowering, viscosity increase, and osmotic pressure increase are considered colligative properties. For the purposes of this discussion, however, only a few of them will be considered. To further simplify the discussion, we will only consider molecular (covalently bonded) solutes, since the effect of ionic solutes may be double, triple, or even greater per mole.

When discussing colligative properties, yet another method is used for calculating solution concentration: **molality** (m). The only difference between molarity (M) and molality (m) is that molality is the number of moles of solute per 1 kilogram of solvent. The formula for molality, is:

$$m = \frac{g_{(solute)}}{fm_{(solute)} \times kg_{(solvent)}}$$

It is interesting to note that a single solute affects all of the colligative properties of a solution. You may have noticed in recent years that antifreeze for the radiator in a car is frequently referred to as antiboil. This means that the same ethylene glycol/propylene glycol mixture that protects your radiator from freezing in the winter prevents boil-over during the summer; two benefits for only one price!

The formula for calculating the colligative properties of freezing point depression or boiling point elevation for a solution is:

$$\Delta T = k\,m$$

where ΔT is the change in boiling or freezing point; k, which is solvent-dependent, expressed either as the molal boiling point constant, k_b, or molal freezing point constant, k_f, and m is the molality of the solution.

A typical problem is one involving the lowering of the freezing point. For example, what will be the freezing point if 30.00 g of table sugar, sucrose, $C_{12}H_{22}O_{11}$, is added to 30.00 g of water ($k_f = -1.86°C/m$). First determine the molality of the solution.

$$g_{(solute)} = 30.00 \text{ g}$$
$$g_{(solvent)} = 30.00 \text{ g} = 0.0300 \text{ kg}$$
$$fm = 342 \text{ g/mole}$$
$$\Delta T \quad ?$$
$$k_f = -1.86°C/m$$

$$m = \frac{30.00 \text{ g}}{342 \text{ g/mole} \times 0.0300 \text{ kg}} = 2.92 \text{ m}$$

Then calculate the freezing point depression.

$$\Delta T = -1.86°C/m \times 2.92 \text{ m} = -5.44°C$$

In this instance, ΔT is also the final freezing point of the solution, since the freezing point of water is 0°C. If a different solvent had given this result, then ΔT would need to be subtracted from the freezing point of the pure solvent.

A similar calculation could be performed to determine the effect of the $C_{12}H_{22}O_{11}$ on the boiling point of this solution.

$$g_{(solute)} = 30.00 \text{ g}$$
$$g_{(solvent)} = 30.00 \text{ g} = 0.0300 \text{ kg}$$
$$fm = 342 \text{ g/mole}$$
$$\Delta T \quad ?$$
$$k_b = +0.52°C/m$$

$$m = \frac{30.00 \text{ g}}{342 \text{ g/mole} \times 0.0300 \text{ kg}} = 2.92 \text{ m}$$

$$\Delta T = 0.52°C/m \times 2.92 \text{ m} = -1.52°C$$

This calculation indicates that the boiling point will be increased by 1.52°C. Since the normal boiling of water is 100°C, adding the sugar will make the boiling point of the sugar water 101.52°C.

A fluid's measured resistance to flow is called its viscosity. Viscosity is also dependent on the concentration of solute particles. The instrument used to measure viscosity is called a viscometer. It is sufficient to say here that the viscosity changes with concentration. A familiar example of this would be the reduction in viscosity that occurs when water is added to corn syrup.

Check Your Understanding

1. Identify the two components in all solutions and explain their roles.

2. What are the similarities and differences between a true solution, colloidal dispersion, and a suspension?

3. Why is it important that a solution be homogeneous?

4. Explain the difference between an unsaturated and saturated solution.

5. What is the most accurate method for calculating percent concentration of a solution?

6. What are the two formulas that may be used for calculating the molarity of a solution.

7. What is the formula for calculating the dilution of a solution?

8. Name five colligative properties of solutions.

9. Explain the difference between molarity and molality, if 4°C water is the solvent.

10. Write the formula for calculating dilution concentrations.

Water and Its Solutions

Objectives

Upon completion of this section, you will be able to:

- Discuss the ability of the water molecule to form hydrogen bonds.
- Explain the effect hydrogen bonding has on the melting and boiling points of water.
- Explain why ice floats in water and is slippery.
- Discuss the forces responsible for the surface tension of water.
- Explain water's unusually high vapor pressure compared to its formula mass.
- Identify the difference between dissociation and ionization and explain each term.
- Define an electrolyte.

Water

Water is without a doubt the most familiar and vital chemical compound on Earth. Not only is it essential for life, but its solutions are also among the most important. Before examining these solutions, let's first examine water, the chemical. As indicated by its formula, H_2O, water is composed of one oxygen and two hydrogen atoms. For reasons best explained by the quantum mechanical model (see Appendix A), it is not a linear (180°) molecule, but rather a bent molecule with its two hydrogen atoms at a 104.5° angle. This angle is apparent in the representation of water shown in Figures 7-8 and 7-9.

In the section on bonding, it was noted that when oxygen forms each of its covalent bonds with hydrogen, the electrons are pulled away from the less electronegative hydrogen nuclei and toward the more electronegative oxygen nucleus. This results in a polar covalent molecule, with one partially negative, δ–, and two partially positive, δ+, regions. In both the liquid and solid states, there is a tendency for water molecules to participate in hydrogen bonding with as many other water molecules as possible, as shown in Figure 7-8.

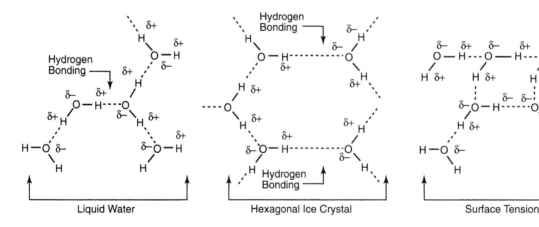

Figure 7–8: Hydrogen bonding in water, ice crystals, and surface tension.

Although hydrogen bonding is a weak attraction, it is responsible for some rather dramatic effects on the melting point, boiling point, vapor pressure, and surface tension of water. For example, most other substances with a formula mass of 18 g/mole are gases at room temperature. Without hydrogen bonding there would be no liquid water at Earthly temperatures. Based on the measured melting and boiling points of water, it appears that water would be better represented by the formula $(H_2O)_{10}$, thereby, giving it an effective formula mass of $(10 \times 18 \text{ g/mole})$ 180 g/mole. Many other substances with formula masses in this range do have 0°C melting and 100°C boiling points.

Solid water, ice, also has some unusual properties that are attributable to hydrogen bonding. Substances experience a slowing of the movement of their atoms or molecules as they are cooled. This results in them becoming more dense and eventually forming a crystal structure. Because the atoms or molecules are closer together in the crystal structure than they were in the liquid, the density of the solid state is always greater. For nearly all other substances, the solid form will sink in its liquid. This pattern is not true for water. As water is cooled below its boiling point its molecules move closer together. At 4°C, however, the effect of hydrogen bonding causes the molecules to move farther apart as they organize into hexagonal patterns, as shown in Figure 7-8. Within the center arrangement, a central void area forms. Thus, the density of water decreases between 4°C and 0°C. The resulting solid, therefore, has a lower density than its surrounding liquid. Water is an example of a substance in which its solid will float in its liquid. If ice were the typical solid, lakes would freeze from the bottom and a lemonade pitcher would not need to be designed to hold back the ice cubes!

There is yet another characteristic of ice that is unique – its slipperiness. People sometimes say that something is as slick as glass. Actually, we would have no trouble walking or driving on glass. When solids come into contact, friction typically prevents one solid from moving freely past the other. This creates the need for lubricants so the two parts will not wear out. This is not so when ice is one of the solids. Applying pressure to ice causes the surface molecules to move closer together. To be closer together, they must get out of the ice crystal and return to the liquid form. When you step on ice, your heel doesn't actually come into contact with ice, but rather with a thin film of water that forms between your heel and the ice. This water acts as a lubricant between the two solids and greatly reduces friction. Walking on ice is, therefore, similar to walking on a floor covered with small BBs!

Surface tension is defined as the attractive forces exerted by the molecules at and just below the surface of a liquid. This side to side and downward attraction causes the molecules on the surface to become more highly organized and move closer together as shown in Figure 7-8 (right). This results in a thin layer of tightly packed water molecules that is similar to a thin elastic sheet. The "V" you make with your hands when diving is to help separate this layer, which will otherwise cause a sharp slap to the top of your head!

Since nonpolar liquids have very little intermolecular attraction, they tend to have low surface tensions, while polar liquids tend to have high surface tensions. This is the reason water tends to bead on a non-wetting surface, such as the finish on your car, while nonpolar gasoline spreads over a large area.

Vapor pressure is a measure of the tendency for a liquid to change into its gaseous state at a given temperature. If molecules are attracted to each other through hydrogen bonding, they will have a lower vapor pressure than a nonpolar liquid; in other words, they require higher temperatures to evaporate at the same rate. On occasion, advantage is taken of this difference to separate nonpolar substances from water. The technology known as air-stripping, in which large volumes of air are blown through contaminated groundwater, is frequently used to separate nonpolar gasoline from polar water.

Dissociation and Ionization

Two other properties of water are directly attributable to its polar character: its ability to **dissociate** (separate) ionic solids and to **ionize** polar molecules. Ionic solids are composed of an orderly stack of oppositely charged particles, such as Na^+ and Cl^-. Within this structure, each ion is surrounded by a number of oppositely charged ions. Some ions must be on the surface and, therefore, cannot be surrounded by oppositely charged ions on all sides.

As long as the crystal remains dry, it is better for the surface ions to have a few oppositely charged neighbors, than none. The crystal is stable. When it is put into water, however, the polar water molecules offer an attractive alternative. By leaving their position on the surface of the crystal, each ion can become **solvated**, that is, surrounded by five or six doting water molecules. Depending on the charge of the ion, water either turns its partial positive or negative end toward the ion. Energy-wise this arrangement is more desirable than being on the surface of the crystal. Once the outer layer of ions is removed, then the next layer becomes the outer layer and the process is repeated until the crystal is completely dissolved. Figure 7-9 shows that when a sodium chloride crystal is placed in water, the surface ions are removed and surrounded by properly oriented water molecules. The resulting **hydrated** ions are denoted by $Na^+_{(aq)}$ and $Cl^-_{(aq)}$, where (aq) is from the Latin word for water, *aqua*.

Solutions resulting from the dissolving of ionic solids always contain ions that can conduct an electrical charge. Solutions that contain ions and conduct electricity are known as **electrolytes**. Most solutions in the human body are of this type. People occasionally develop a condition known as electrolyte imbalance. It is caused by the loss of a large amount of water and ions. If the condition is not treated, it can be fatal.

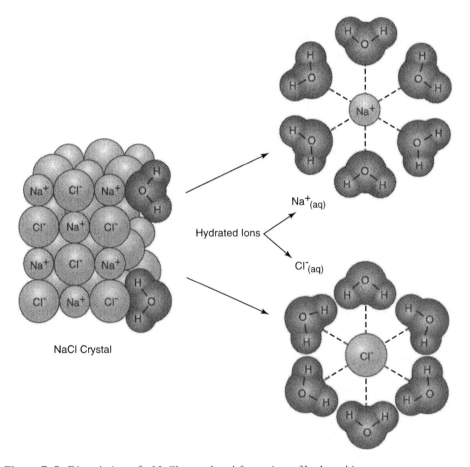

Hydrated Ions

$Na^+_{(aq)}$

$Cl^-_{(aq)}$

NaCl Crystal

Figure 7–9: Dissociation of a NaCl crystal and formation of hydrated ions.

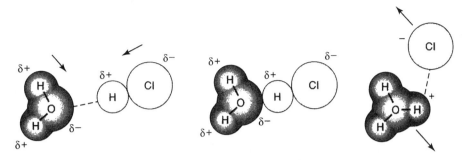

Figure 7–10: Water molecules ionize $HCl_{(g)}$ molecules.

The polar nature of water is also responsible for producing solutions containing ions, when only other *polar molecules* are present. Polar molecules attract each other by hydrogen bonding in much the same way that opposing poles of two magnets are attracted. In the helter-skelter world of molecules, they are constantly being jostled about, bumping into each other and into the walls of their container. Most of these molecular accidents simply result in a change in direction or momentum, but occasionally, with the assistance of an extra tug from hydrogen bonding, a molecule gets broken. When water causes other molecules to be broken into ions, the process is called **ionization**.

For example, Figure 7-10, shows that hydrogen chloride gas, $HCl_{(g)}$ is a polar molecule, with the more electronegative chlorine atom pulling the electrons toward it. When $HCl_{(g)}$ is bubbled into water, there is an attraction between the polar molecules. During chaotic molecular collisions between the water and HCl, this extra tug from the polar water causes most of the $HCl_{(g)}$ molecules to ionize forming $H_3O^+_{(aq)}$ and $Cl^-_{(aq)}$ ions. The resulting solution is called hydrochloric acid, with its formula written simply as HCl. In truth, hydrochloric acid solutions contain many water molecules and varying numbers of $H_3O^+_{(aq)}$ and $Cl^-_{(aq)}$ ions. It is the presence and number of **hydronium ion**, $H_3O^+_{(aq)}$, that gives it its acidic character. More will be said about this in the next section.

In review, there are two ways solutions containing ions can be formed: by the dissociation of an ionic salt and by the ionization of polar molecules. The presence and number of ions in a solution can be experimentally measured by passing an electric current through the solution. After years of experimentation, it was found that three types of compounds tend to produce electrolytes: acids, bases, and salts. In the following section, each of these will be examined.

Check Your Understanding

1. What is the bond angle between the hydrogen atoms in a water molecule?

2. Why are the melting and boiling points of water considered high?

3. What causes ice to float?

4. What is the cause of surface tension of water?

5. Is the vapor pressure of nonpolar substances high or low?

6. What is an electrolyte?

7. Describe the difference between dissociation and ionization.

8. What is a hydronium ion?

Acids, Bases, and Salts

Objectives

Upon completion of this section, you will be able to:

- Describe the characteristics of an acidic solution.

- Describe the characteristics of a basic solution.

- Describe the characteristics of salt solutions.

- Identify and explain the equilibrium constant of water.

- Calculate the pH, pOH, $[H^+]$ and $[OH^-]$ of a solution.

Long before acids were defined, scientists noted that some solutions always produced hydrogen gas when they were exposed to a metal, such as zinc. It was believed that there was some connection between this behavior and acids. It took years before the connection was made between acids, hydrogen gas, and metals.

Even today, some differences remain concerning how acids and bases should be defined. The oldest successful definition is that of Svante August Arrhenius, who stated in 1868 that an **acid** must produce hydrogen ions, H^+, in water and that a **base** must produce hydroxide ions, OH^- in water. Some years later it was argued that having water specifically mentioned made the definition too limiting, so in 1923 the Brønsted-Lowry definition was advanced. It stated that acids are proton donors (remember, a hydrogen atom minus its electron is just a proton, H^+) and bases are proton acceptors. Although it was more general, this definition still allows the OH^- to be considered as a base, because by accepting a hydrogen ion it becomes water ($H^+ + OH^- \rightarrow HOH$).

Organic chemistry provided more situations similar to acid-base interactions. In 1923, the Berkeley chemist G. N. Lewis proposed yet another definition to address these situations. In this definition, acids were considered electron-deficient and bases electron-rich substances. Although it was a much broader definition, it still included the hydrogen ion, H^+, (electron deficient) and the hydroxide ion, OH^-, (electron rich). Today, these definitions are still in use by various areas of chemistry. The Arrhenius definition is the one most often used in the area of inorganic chemistry. It will be our choice here.

Acids

From childhood, we have been repeatedly cautioned about the dangers of acids. Numerous labels, depicting the infamous skull and crossbones, warn us of their dangers. Just what are acids and what harmful characteristics do they have? To answer these questions, let's start with the word acid. It comes from the Latin word *acere*, which means sour. This is certainly one of their most familiar characteristics. Although tasting chemicals is never advised, you have tasted vinegar, lemon juice, and maybe even sour milk. It is now known that the hydrogen ion, H^+, or more accurately the hydronium ion, H_3O^+, is the only ion that causes our taste buds to register a sour response.

Acid solutions also cause some vegetable dyes to change colors. Most of you have experienced the effect vinegar or lemon juice has on red cabbage. In the chemistry laboratory, the most commonly used plant dye is litmus. It is red in acidic solutions and blue in basic solutions. So, acids taste sour and make blue litmus paper red, but what do they do to other substances? As a group, acids have a variety of chemical characteristics. The most familiar of these is that an acid reacts with a base to produce a salt and water (HNm + MOH → MNm + HOH), where MNm is the general formula for a salt and HOH is actually H_2O. This reaction is called **neutralization**. Acids and bases neutralize each other and always produce a salt and water. Acids also react with metal bicarbonates and carbonates, producing a salt, water, and carbon dioxide gas. The general equations for these reactions are HNm + $MHCO_3$ → MNm + HOH + CO_2 for the bicarbonate reaction and HNm + MCO_3 → MNm + HOH + CO_2 for the carbonate reaction. This latter reaction is used by geologists as a test for carbonate rocks. For example, limestone and marble are both composed of $CaCO_3$, so its reaction with hydrochloric acid, HCl, would be:

$$2\ HCl + CaCO_3 \rightarrow CaCl_2 + H_2O + CO_2$$

A second application of this reaction is in the removal of water spots from glassware. Water spots are composed primarily of calcium and magnesium carbonates. By placing the glassware in a vinegar solution for a few minutes, a few small bubbles form and the water spots magically disappear!

Acids also react with metal oxides to form a salt and water (HNm + MO → MNm + H_2O). For example, rust (Fe_2O_3) is commonly removed from pieces of iron using a **pickling** process. Pickling involves dipping the iron in a vat of acid. Sulfuric, H_2SO_4, and hydrochloric acids (HCl) are common choices.

$$Fe_2O_3 + 3\ H_2SO_4 \rightarrow Fe_2(SO_4)_3 + 3\ H_2O$$

To remove the $Fe_2(SO_4)_3$ from the rust-free metal, it is rinsed with water, dried, and is ready for welding, painting, or other manufacturing process.

As was noted in the discussion on single replacement reactions, acids will react with active metals (in Table 7-5, those above hydrogen on the electromotive force series) producing a salt and hydrogen gas. Even lead, Pb, which is just above hydrogen, will slowly react with the sulfuric acid from a car's battery and cause a white deposit ($PbSO_4$) to form on the battery terminals. The equation for the reaction is:

$$Pb + H_2SO_4 \rightarrow PbSO_4 + H_2$$

The higher a metal is found above hydrogen on the electromotive force series, the faster this reaction will occur.

In summary, acids produce hydrogen ions, taste sour, make litmus red, and react with bases, bicarbonates, carbonates, metal oxides, and active metals. That still leaves many other substances, e.g., plastics, glass, rubber, etc., that are generally unaffected by acidic solutions.

Bases

It has already been noted that bases neutralize acids. This is, in fact, the most important inorganic chemical reaction for bases. There are other reactions that bases undergo in the area of organic chemistry, but they will not be discussed here.

What are some other characteristics of basic or **alkaline** solutions? First, the word alkaline comes from the Arabic *al-qily*, which roughly translates to heating plants into ashes. Pouring water through the ashes of burned plant matter was the ancient way of making an alkaline solution. Even in the early West, pioneer women made the lye water needed for soap making by pouring water through ashes. The resulting solution is actually a mixture of sodium, NaOH, and potassium, KOH, hydroxides.

Basic solutions have a bitter taste. There aren't as many bitter tasting as sour tasting foods, although most of us have tasted baking soda. Unlike sour-tasting foods, which always contain H^+, most bitter-tasting foods do not contain OH^-. Chocolate is probably your favorite! Some people prefer milk chocolate, where sugar and milk have been added to mask the bitter taste of the nasty little alkaloid, theobromine, but for others, bittersweet chocolate is their favorite. Chayota squash, bitter melon, rhubarb, artichokes, and coffee are a few other foods that contain bitter tasting compounds.

Bases also have another characteristic that should not be intentionally tested. When the skin is exposed to a base, it feels slippery or soapy. Dermatologists agree that skin contact with bases can be more harmful than contact with acids. This is primarily due to the tendency of bases to form soaps with the lipid bi-layers that compose most biological membranes. Bases feel slippery on your skin because your skin is dissolving! In conclusion, bases produce hydroxide ions in water, turn red litmus blue, taste bitter, feel slippery, and neutralize acids.

Salts

The last category of compounds that produces ions in water is salts. Salts lack a common ion like acids and bases. **Salts** are defined, therefore, as substances that dissolve in water and produce a positive ion other than H^+ and a negative ion other than OH^-. Sodium chloride, NaCl, is the most familiar example and is commonly called – salt. It is important to understand, however, that it is just one of many salts!

You may not be surprised to learn that salts lack similar characteristics; for example, most do not taste salty! Some salts dissolve easily in water, while others don't. Four general guidelines will allow you to predict the solubility of many salts:

— All Li^+, Na^+, K^+, and NH_4^+ salts are soluble.

— All NO_3^- and $C_2H_3O_2^-$ salts are soluble.

— All Cl^-, Br^-, and I^- salts are soluble, unless the positive ion is Ag^+, Pb^{2+}, or Hg_2^{2+}.

— Most other salts are insoluble, meaning that they dissolve less than 1 g/100 g solution.

The above guidelines can be used to predict the results of the double replacement reaction between sodium chloride and silver nitrate solutions. This reaction is shown below in three different ways. The first equation is called a **molecular equation** because each of the substances is written as though it is composed of molecules rather than ions. In the second equation, called a **total ionic equation**, each substance is written in its ionic form, if it is soluble in water. A line has been drawn through each ion that is not involved in the reaction. Finally, in the last equation, called a **net ionic equation**, the **spectator ions** or non-participating ions have been removed leaving just those ions that are responsible for the formation of the new product. In this case, it is the formation of an insoluble salt or **precipitate**.

Molecular
Equation: $NaCl + AgNO_3 \rightarrow AgCl + NaNO_3$

Total Ionic
Equation: $\cancel{Na^+_{(aq)}} + Cl^-_{(aq)} + Ag^+_{(aq)} + \cancel{NO_3^-_{(aq)}} \rightarrow AgCl + \cancel{Na^+_{(aq)}} + \cancel{NO_3^-_{(aq)}}$

Net Ionic
Equation: $Cl^-_{(aq)} + Ag^+_{(aq)} \rightarrow AgCl\downarrow$

When you examine the above equations, you find that the sodium, $Na^+_{(aq)}$, and nitrate ions, $NO_3^-_{(aq)}$, are the spectator ions and were, therefore, eliminated from the net ionic equation. The method used to determine the participating and non-participating ions is based on the third guideline above: All Cl^-, Br^-, and I^- salts are soluble, unless the positive ion is Ag^+, Pb^{2+}, or Hg_2^{2+}. In the example, the positive ion is Ag^+. It will combine with Cl^-, to form a AgCl precipitate. The use of a downward pointing arrow next to the formula for the precipitate is optional.

Net ionic equations are frequently the choice of chemists. They simplify a chemical event by drawing attention to only those ions that actually participate, like a cartoonist who emphasizes some feature of a famous individual by exaggerating it.

We can use the example of the deadly effects of hydrogen sulfide gas (H_2S) to illustrate simplifying reactions. Hemoglobin carries oxygen though our body. Hemoglobin itself is a huge molecule containing four heme molecules, $C_{34}H_{32}FeN_4O_4$. At its center is a ferrous ion, Fe^{2+}, which binds the oxygen and carries it to the cells. Exposure to rotten egg gas, H_2S, allows the ferrous and sulfide ions to react, according to the following net ionic equation:

$$Fe^{2+} + S^{2-} \rightarrow FeS\downarrow$$

Once ferrous sulfide forms, heme can no longer transport vital oxygen and if this happens to enough hemoglobin molecules, death occurs.

The Equilibrium of Water

Polar molecules are sometimes ionized when placed in water, as mentioned earlier. Since water itself is a polar molecule, you might wonder if any water molecules ever become ionized. The answer is an emphatic, yes! The actual number at room temperature, however, is quite small. Careful conductivity experiments have shown it to be about 1.0×10^{-7} M. Each time a water molecule is ionized, however, it forms two ions: one hydronium ion, H_3O^+, and one hydroxide ion, $OH^-_{(aq)}$. To simplify this discussion we will refer to only hydrogen, H^+, and hydroxide, OH^-, ions, although it is understood that hydrogen ions are always hydrated. The equation for the ionization of water, then, is:

$$H_2O \leftrightarrow H^+ + OH^-$$

Notice that a double arrow, $\leftrightarrow$, has been used in this equation. That is because two reactions are simultaneously occurring. Due to bumping and jostling, some water molecules continue to be broken, while collisions between hydrogen and hydroxide ions continue to form new water molecules. Whenever two reactions are going in opposite directions, a point is eventually reached where the number of molecules and ions become constant. This point, where all concentrations remain steady, is called **equilibrium**.

At room temperature (25°C), the **equilibrium constant**, K_w, for water is given by the expression: $K_w = [H^+][OH^-]$. The use of square brackets, [ ], denotes that the concentrations are expressed in moles/liter, M. If the concentration of H^+ and OH^- at room temperature are substituted into the equilibrium constant expression, the value of the equilibrium constant of water, K_w, is 1.0×10^{-14}.

$$K_w = [1.0 \times 10^{-7}][1.0 \times 10^{-7}]$$

$$K_w = 1.0 \times 10^{-14}$$

This is an important number and will be used later in the discussion. What is necessary to understand here is that like any equation, whatever happens to one variable changes the value of the other variable proportionately. For instance, in the equation $xy = 24$, if $x = 3$, then y must equal 8; if $x = 6$, then $y = 4$. In this example, the product must always equal 24. In the same way, the product of the $[H^+] [OH^-]$ must always equal 1.0×10^{-14}. If an acid is added to water at equilibrium, it will increase the concentration of the hydrogen ions. The equilibrium of water must, therefore, shift to the left forming more water molecules, as in the above equation. This shift, however, means that the concentration of the hydroxide ions must decrease proportionally. The same would be true, if hydroxide ions (base) were added. The equilibrium of water would shift to the left forming more water molecules and reducing the concentration of hydrogen ions.

pH and pOH

The balance between hydrogen ions and hydroxide ions in a solution is of utmost chemical and biological importance. It is, however, difficult to use such small numbers. This problem was solved early in the last century by Sören Sörenson who developed a mathematical formula to convert small hydrogen ion concentrations into larger numbers that are easier to use and understand. These numbers are called **pH**. The use of pH is, therefore, just another way of expressing the concentration of hydrogen ions in a solution. The formula that makes the conversion is $pH = -\log_{10}[H^+]$. The part of this formula that may be new to you is the $\log_{10}$ part, which stands for the word **logarithm** to the base number 10. Stated another way, it means the power to which 10 must be raised to equal the number.

The logarithm for any number that is less than one is always a negative number. But how did we get these answers? The easiest way to determine logarithms is to use a simple scientific calculator and know something about using it. Scientific calculators can be identified by having keys marked with log (the common expression used for logarithms$_{10}$), 10^x, EXP or EE, INV, etc. Assuming your calculator is of this type, try the following example:

> To determine the log of 500, enter 500, then press the log key. Your display should read 2.69897, which may be rounded to 2.6990.

To determine the log for the number that was smaller than one, 0.00500, enter 0.00500, then press the log key. The display should read −2.3010 (rounded). Another way to enter this number is to first put it into scientific notational form. This means expressing the number in two parts. The first part is obtained by moving the decimal to either the left or right until you have a number that is between 1 and 10. The second part is to express the number of places that the decimal was moved left (plus) or right (minus) as the power of ten.

The number 0.00500 is the same as 5.00×10^{-3}. To enter this number in your calculator, push 5.00 then EE or EXP, followed by the +/− key, and finally the 3 key. Next, press the log key and the answer should read −2.3010. If this does not work for your calculator, refer to its owner's manual or ask for help.

Now that we have a basic idea of how to convert a number to its scientific notational form and find its logarithm, let's return to the equilibrium for water. We know that $[H^+]$ and $[OH^-]$ concentrations are 1.00×10^{-7} for pure water at room temperature. To convert this into the pH of water, simply plug it into the formula $pH = -\log [H^+]$.

$$pH = -\log[1.00 \times 10^{-7}]$$

$$pH = -(-7.0000)$$

$$pH = 7.00$$

The pH of pure water is seven (pH = 7.00) and it is neutral. From this example, it can be seen that only when the number of hydrogen and hydroxide ions are exactly equal is the solution neutral. When small amounts of acid are added to the water, the acid molecules add to the total $[H^+]$ and, therefore, change the pH of the solution. For practical purposes, the contribution of $[H^+]$ from the ionization of water is negligible, so the acid concentration can be used as the total $[H^+]$.

Example: To calculate the pH of a 0.100 M solution of hydrochloric acid, HCl, you would first need to write an equation for its ionization.

$$\underset{\text{HCl}}{\overset{0.100M}{}} \longrightarrow \underset{H^+}{\overset{0.100M}{}} + \underset{Cl^-}{\overset{0.100M}{}}$$

if $[H^+] = 0.100$, then

$[H^+] = 1.00 \times 10^{-1}$, and the pH would be

$$pH = -\log[1.00 \times 10^{-1}]$$

$$pH = -(-1.000) = 1.00 \text{ (rounded)}$$

In this example, it is assumed that every molecule of HCl underwent ionization. Acids that ionize nearly 100 percent are called **strong acids**, regardless of their molarity. **Weak acids** typically ionize to less than 1 percent, therefore, they produce fewer ions than the number of molecules present. The percent of acid molecules that ionizes can only be measured in the laboratory and must be included in the problem statement. For example, acetic acid, $HC_2H_3O_2$, is in vinegar. Only about 1 percent of its molecules becomes ions in a 0.100 M solution. The following steps, therefore, are required to calculate its pH.

$$\underset{HC_2H_3O_2}{\overset{0.100M}{}} \overset{1\% \times (0.1M) =}{\longrightarrow} \underset{H^+}{\overset{0.00100M}{}} + \underset{Cl^-}{\overset{0.00100M}{}}$$

if $[H^+] = 0.00100$, then

$[H^+] = 1.00 \times 10^{-3}$, and then pH would be

$$pH = -\log[1.00 \times 10^{-3}]$$

$$pH = -(-3.0000) = 3.00 \text{ (rounded)}$$

From this example, it can be seen that the pH of a weak acid is a larger number; therefore less acidic than for a strong acid of the same molarity. This is the reason why we can put vinegar on our salads and still enjoy them without damaging our mouth. By comparison, the pH of stomach acid is 1.0 or 10^2 times more acidic! Note, every pH unit means a ten-fold change in H^+ concentration.

In each of the above examples, a **monoprotic** acid was used. Monoprotic acids produce only one proton per molecule when they ionize. **Diprotic** acids, however, produce two. Sulfuric acid, H_2SO_4, is both a strong acid (100% ionized) and it is a diprotic acid. To determine the pH of a 0.100 sulfuric acid solution, the following steps would be needed:

$$\begin{array}{cccc} 0.100M & 100\% \times (0.100M) \times 2 = & 0.200M & 0.100M \\ H_2SO_4 & \xrightarrow{\hspace{2cm}} & 2\,H^+ & +\quad SO_4^{2-} \end{array}$$

if $\quad [H^+] = 0.200$, then

$[H^+] = 2.00 \times 10^{-1}$, and the pH would be

$$pH = -\log[2.00 \times 10^{-1}]$$

$$pH = -(-0.6990) = 0.70 \text{ (rounded)}$$

There are several important points to note about this example. First, since it was a diprotic acid, the concentration of hydrogen ions is twice the concentration of acid molecules. Second, the pH only changed from 1.00 to 0.70 although the concentration of hydrogen ions is twice that in the HCl example. The reason for the small change is that the pH scale is logarithmic. That is to say that the difference in hydrogen ion concentration must be ten times to change the pH number by just one!

In a fashion similar to pH, chemists have also devised an expression for $[OH^-]$, which is called the **pOH**. The mathematical formula is also similar:

$$pOH = -\log[OH^-]$$

Remember that pH and pOH scales are related to the ionization constant of water, K_w. Also remember that at 25°C, $K_w = [H^+][OH^-]$. In pure water both the $[H^+]$ and the $[OH^-]$ are 1.00×10^{-7}. As we saw earlier, when these values are placed in the equilibrium expression, $K_w = 1.00 \times 10^{-14}$.

$$[H^+][OH^-] = 1.00 \times 10^{-14}$$

By converting the above equation to its –log, we get:

$$pH + pOH = 14.$$

This useful relationship makes determining the pH of basic solutions much simpler.

For example, to calculate the pOH of a 0.002 M solution of the strong base, KOH, you would need to write an equation for its ionization.

$$\begin{array}{ccc} 0.002M & 0.002M & 0.002M \\ KOH & \xrightarrow{\hspace{2cm}} K^+ & +\quad OH^- \end{array}$$

if $\quad [OH^-] = 0.002$, then

$[OH^-] = 2.00 \times 10^{-3}$, and then pOH would be

$$pOH = -\log[2.00 \times 10^{-3}]$$

$$pOH = -(-2.69897) = 2.70 \text{ (rounded)}$$

It may be noted that these are exactly the same steps taken in the previous example for calculating pH, except this time it is for [OH] and the results are, therefore, pOH. To complete the calculation for pH, one step remains: if pH + pOH = 14, then when pOH = 2.70

$$pH = 14 - 2.70 = 11.30$$

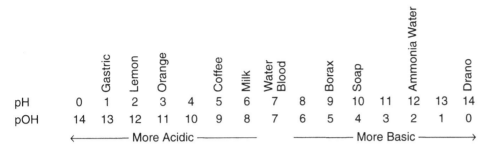

		Gastric	Lemon	Orange		Coffee	Milk	Water Blood		Borax	Soap		Ammonia Water		Drano
pH	0	1	2	3	4	5	6	7	8	9	10	11	12	13	14
pOH	14	13	12	11	10	9	8	7	6	5	4	3	2	1	0

←———— More Acidic ———— ———— More Basic ————→

Figure 7–11: Relationship between pH and pOH.

It is very important to stop and examine your answer. The pH of a solution is less than seven if it is acidic and more than seven if is basic. The relationship between pH and pOH scales is summarized in Figure 7-11.

From Figure 7-11 it can be seen that the numerical values on the pH and pOH scales run in opposite directions.

In 1935, Dr. Arnold O. Beckman began production of "acidimeters" (see Figure 7-12) to measure the pH of lemon juice in the Southern California citrus industry. These meters later became known as Beckman pH meters and their use revolutionized the industry. The pH meter measures the very small potential difference across a glass electrode that is caused by the hydrogen ions in the solution when compared to a reference electrode. This electrical difference is converted into a direct pH meter reading. Today, pH is defined as the reading obtained from a properly standardized pH meter.

Figure 7–12: The "acidimeter" later became known as the pH meter. Photo courtesy of Beckman Coulter, Inc.

If the pH of a solution can be obtained directly from a meter reading, then is it possible to convert it into its equivalent hydrogen or hydroxide ion concentration? The answer is yes and again the process is made simple by the use of a scientific calculator.

For example, if a pH meter determines that a solution has a pH = 6.50, what would be the [H^+] of the solution? The mathematical process is called finding the **antilog**. As its name suggests, this means finding the opposite of the log. The steps to do this are the following: First, enter the pH number 6.50; then push the +/− key; followed by the INV key; and the log key. On some calculators you will push the 10^x key, followed by 6.50 and the +/− key. In either case, the answer, 3.16×10^{-7} (rounded), is the [H^+] of the solution.

The sequence of keys used is the same whether you are converting from pH to [H^+] or from pOH to [OH^-]. If this has not worked on your calculator, consult your owner's manual or ask for help.

Figure 7-13 shows that the relationships between [H^+], [OH^-], pH, and pOH can all be put into a single diagram.

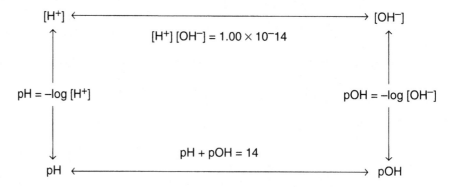

Figure 7–13: Relationship between $[H^+]$, $[OH^-]$, pH, and pOH.

As long as you have information about any one of the corners, it is possible to calculate the values for each of the others. For example, to calculate the $[H^+]$, $[OH^-]$, and pH of a solution with a pOH = 9.1.

$$pH \ + \ pOH \ = \ 14.0$$

$$pH \ = \ 14.0 - 9.1 \ = \ 4.9 \ (acidic)$$

For a solution with a pH = 4.9, the $\quad$ $[H^+] = 1.26 \times 10^{-5}$

For a solution with a pOH = 9.1, the $\quad$ $[OH^-] = 7.94 \times 10^{-10}$

It should be noted that by multiplying the $[H^+]$ times the $[OH^-]$, a value of 1.00×10^{-14} is obtained. This will always be true, because it is the ionization constant of water, K_w.

Hydrolysis of Salts

As was previously noted, not all salts taste salty. Some do, but others taste sour or bitter. A salt solution tasting sour or bitter indicates that a pH shift has occurred. How can a pH shift occur, if an acid or base has not been added? In the previous section we considered substances (acids and bases) that add ions to the solution, but what would happen if we were to add a substance that could remove hydrogen or hydroxide ions?

When a salt dissociates, it produces positive and negatively charged solvated ions. These ions encounter the small numbers of hydrogen and hydroxide ions resulting from the ionization of water. Because they are oppositely charged, there is an attraction between the ions from the salt and the ions from water. On occasion, the combination formed is that of a weak acid or base. What this means is that once formed, there is a tendency for them to remain a molecule, rather than ionizing back into ions. This tendency to capture hydrogen or hydroxide ions may soon result in a solution where the $[H^+]$ and $[OH^-]$ are no longer equal. As we learned above, solutions that do not have a balance of $[H^+]$ and $[OH^-]$ do not have pH = 7.

Most of us have experienced the bitter taste of baking soda, $NaHCO_3$. Baking soda is an example of a salt of a strong base and a weak acid. When it is placed in water it dissociates, forming a $Na^+_{(aq)}$ and $HCO_3^-_{(aq)}$. This means that the sodium ion will be attracted to hydroxide ions forming NaOH, but since NaOH is a strong base it splits back into ions. This process results in no change in the $[OH^-]$. When a $HCO_3^-_{(aq)}$ encounters a hydrogen ion, a molecule of the weak carbonic acid, H_2CO_3, is formed. Since it is a weak acid, it tends to stay as a molecule, thereby reducing the number of H^+ in the solution. Solutions that have an excess of OH^- are basic, hence the bitter taste. This process can be represented in the following way:

$$H_2O \rightarrow H^+ + OH^- \qquad\qquad NaOH \rightarrow Na+ + OH- \text{ (strong base)}$$

$$NaHCO_3 \rightarrow Na^+ + HCO_3^- \qquad \underline{H_2CO_3 \rightarrow \qquad\qquad \text{(weak acid)}}$$
$$\text{A basic solution with excess } [OH^-]$$

Acids		Bases	
Weak	Strong	Weak	Strong
Acetic, $HC_2H_3O_2$	Hydrochloric, HCl	Ammonium hydroxide, NH_4OH	Sodium Hydroxide, NaOH
Carbonic, H_2CO_3	Nitric, HNO_3	Other metal hydroxides	Potassium Hydroxide, KOH
Boric, H_3BO_3	Sulfuric, H_2SO_4		Other Column I and II hydroxides
Phosphoric, H_3PO_4			

Table 7–7: Common weak and strong acids and bases.

Using Table 7-7, it can be shown that when a salt of a weak base and strong acid, e.g., NH_4Cl, is put into water, an acidic solution results. Those salts of strong acids and strong bases, e.g., NaCl, or of weak acids and weak bases, e.g., $CuCO_3$, do not change the numbers of hydrogen and hydroxide ions; therefore, they remain neutral (pH = 7).

Buffers

Many substances, other than acids and bases, can change the pH of a solution. Fortunately, there are combinations of salts, called **buffers,** that can protect against these changes. For example, it is essential that our blood maintain a pH = 7.4. Yet, depending on our activity level, there are times when acidic products, such as lactic acid, must be moved from a working muscle to the liver. During those transfers, does the pH of the blood become more acidic?

There are two important pairs of buffering ions that are present in the blood: HCO_3^-/CO_3^{2-} and $HPO_4^{2-}/H_2PO_4^-$. Each set of ions can react with the addition of small amounts of either hydrogen or hydroxide ion, as shown by the following equations:

Addition of acid: $HCO_3^- + H^+ \rightarrow H_2CO_3$

Addition of base: $H_2CO_3 + OH^- \rightarrow HCO_3^- + H_2O$

From these equations, it should be noted that the bicarbonate ion captures excess hydrogen ions, forming weak carbonic acid. This is the reason why bicarbonate IVs are administered to heart attack victims. By flooding the blood with excess bicarbonate ions, it boosts the blood's capacity to buffer against a decrease in pH, caused by elevated CO_2 level before normal breathing resumes.

Buffered solutions do not necessarily have a neutral pH. Using different ion combinations, solutions with nearly any pH may be buffered.

Check Your Understanding

1. How did Arrhenius define acids and bases?

2. List seven characteristics of acidic solutions.

3. List four characteristics of basic solutions.

4. Write the four general solubility rules for salts.

5. What is the difference between the molecular, total ionic, and net ionic equation?

6. What is the numerical value of the equilibrium constant of water, K_w?

7. What is the concentration of $[H^+]$ and $[OH-]$ in water at 25°C?

8. What is the equation for the calculation of pH from a known $[H^+]$?

9. What does the sum of pH and pOH always equal?

10. Explain the difference between a strong acid or base and a weak acid or base.

11. Explain the effect of a salt hydrolyzing.

12. How do buffers work?

Summary

All chemical reactions must obey the law of conservation of matter. Reactions are either endothermic (produce energy) or exothermic (requiring energy). When chemical equations are written, the reactant(s) are usually written on the left-hand side and the product(s) on the right-hand side of the arrow. The arrow separating the reactant(s) and product(s) has the same meaning as an equal sign in an equation. The number of each kind of atoms on the left must equal those on the right. If after writing the correct formula for each reactant and product, atoms are not conserved, the equation must be balanced by inserting coefficients.

There are four basic types of chemical reactions: combination, decomposition, single replacement, and double replacement. Within each of these basic reaction types, there are specific reactions. Each reaction is described using general terminology, such as acids, bases, salts, metal oxides, metal chlorides, etc. Through an understanding of the basic reaction types and general terminology, one can accurately predict the products of thousands of combinations.

For many years, it has been the practice to count atoms by weighing. It is now known that the formula mass, in grams, of any pure substance contains one mole (Avogadro's number) of particles (atoms or molecules). Avogadro's number is 6.0221367×10^{23}, and is defined in the SI System of units as the number of atoms of carbon in exactly twelve grams of carbon-12. The sum of the individual atomic masses within a compound is its formula mass. Moles, therefore, can be changed into grams and visa versa, if the formula mass of the substance is known. It is equally possible to determine the number of atoms or molecules in a given mass of substance by applying Avogadro's number.

Stoichiometry is the branch of science that deals with the quantities of substances involved in chemical reactions. If the amount of reactant is known, then it is possible to apply stoichiometric principles and calculate the amount of product that will be produced. On occasion, the amount of reactants are not present in the correct stoichiometric ratios. In those instances one of the reactants becomes a limiting reagent for the reaction and determines the amount of product possible. The results from such calculations are considered to be theoretical yields. In actual laboratory practice, however, less product is always produced. By comparing the actual and theoretical yields, the percent yield can be calculated or, alternatively, the percent error calculated.

Solutions are composed of two parts: solute and solvent. There are three distinctive types of solutions: true solutions, colloidal dispersions, and suspensions. True solutions and colloidal dispersions are preferred, because they are homogenous. This means that repeated samples would contain the same solute to solvent ratio. Suspensions are heterogeneous and tend to settle after mixing.

Concentration is defined as the number of things in a given volume. Scientists have devised many different ways of expressing the concentration of solution. Three of the popular methods are weight-weight percent, molarity, and molality. Properties, such as freezing point, boiling point, vapor pressure, viscosity, etc. are dependent on the molality of a solution. These are colligative properties and explain the use of antifreeze in a car's radiator. All solutions can be diluted or concentrated by changing the solvent to solute ratio. The formula for calculating the final concentration is $C_{(old)} \times Vol_{(old)} = C_{(new)} \times Vol_{(new)}$.

Water, the most common solvent, is a unique substance. Due to hydrogen bonding, water has unusually high melting and boiling points, forms ice crystals that are less dense than the liquid, and exhibits surface tension, causing water to bead on a non-wetting surface. The polar nature of water also results in it being a powerful solvent for polar and ionic substances. When an ionic substance is dissolved in water, each ion becomes solvated, and is capable of carrying an electrical current. These solutions are known as electrolytes.

Water molecules can also be broken into hydrogen, H^+, and hydroxide, OH^-, ions. They also help other molecules to break apart and form ions. This process is known as ionization. Substances that produce hydrogen ions are called acids and those that produce hydroxide ions, bases. Acid solutions taste sour, turn blue litmus red, and react with many other substances. Basic solutions taste bitter, turn red litmus blue, feel soapy, and react with acids in a process known as neutralization. The other product of neutralization is a salt. Not all salts taste salty and some are soluble in water while others are not. Total and net ionic equations may be used to focus on the reacting ions, by eliminating the spectator ions from the net ionic equation.

To follow a neutralization process or to determine the ratio of hydrogen and hydroxide ions in a solution, Sörenson devised a mathematical formula to convert the small amounts of hydrogen ions into a larger number, called pH and can be calculated using the formula $pH = -\log[H^+]$. In general, pH values less than 7 indicate an acid solution and those that are more than 7 a basic solution. A solution is only said to be neutral (pH = 7) when: $[H^+] = [OH^-]$.

Salts, although they contribute neither hydrogen nor hydroxide ions, may also change the pH of a solution. This is accomplished by removing either hydrogen or hydroxide ions from the solution. Salts of strong acids and bases or weak acids and bases do not disturb the balance and result in solutions with a pH = 7. However, solutions containing salts of weak acids and strong bases or strong acids and weak bases do shift the pH, resulting in solutions with pH > 7 or pH < 7, respectively. Occasionally, it is important to protect the pH against small additions of acid or base. In those instances, a combination of salts, called a buffer, is added to the solution.

Chapter Review

1. Balance each of the following equations, using the smallest set of whole numbers.

    a. H_2 $\quad + \quad$ Br_2 $\quad \longrightarrow \quad$ HBr

    b. NH_3 $\quad + \quad$ HCl $\quad \longrightarrow \quad$ NH_4Cl

    c. $CaCO_3$ $\longrightarrow CaO$ $\quad + \quad$ CO_2

    d. H_3PO_4 $\quad + \quad$ $Ca(OH)_2$ $\quad \longrightarrow \quad$ $Ca_3(PO_4)_2 + H_2O$

    e. $AlCl_3$ $\quad + \quad$ $AgNO_3$ $\quad \longrightarrow \quad$ $Al(NO_3)_3 + AgCl$

2. Identify each of the equations in Question 1, as a combination, decomposition, single replacement, or double replacement reaction.

3. Calculate the formula mass and express your answers in g/mole for each of the following:

    a. $NaNO_3$

    b. $CuCl_2$

    c. K_2CO_3

    d. H_2O

    e. $BaCl_2 \times 2\,H_2O$

4. Calculate the percent composition of the first four compounds in Question 3. For the last item, calculate the percent water in the hydrate.

5. Calculate the number of molecules and grams in 5 moles of water, H_2O.

6. How many grams of silver would be required to get 1.2×10^{24} silver atoms?

7. According to the following balanced equation, how many moles of carbon monoxide will be produced by the incomplete combustion of 16 moles of carbon?

$$2\,C + O_2 \rightarrow 2\,CO$$

8. Based on the following balanced equation, how many grams of carbon dioxide will be produced by the complete metabolism of 18 grams of glucose?

$$C_6H_{12}O_6 + 6\,O_2 \rightarrow 6\,CO_2 + 6\,H_2O$$

9. The environmental impact of driving to school can be calculated by assuming a tank of gasoline is about 15 gallons and has a mass of 37.5 kg. If gasoline consists primarily of hexane (C_6H_{14}) and burns according to the following balanced equation, how many kilograms of carbon dioxide will be added to the atmosphere? How much water will be formed at the same time?

$$2\,C_6H_{14} + 19\,O_2 \rightarrow 12\,CO_2 + 14\,H_2O$$

If you burn one tank of gasoline each week, how many kilograms and pounds of carbon dioxide does it add to the air per year?

10. If 50.0 g of HCl and 50.0 g of NaOH are mixed, which of the reactants is the limiting reagent? Using the equation, $HCl + NaOH \rightarrow NaCl + H_2O$, justify your answer and calculate the number of grams of water that would be formed.

11. If in Problem 10, only 21.3 g of water was recovered, what would be the percent yield for the reaction? Show your work.

12. Select one solution type as an example, and write two paragraphs describing its characteristics.

13. Calculate the weight-weight percent concentration for a solution prepared by dissolving 35.6 g of $NaNO_3$ into 325 g of water.

14. Assuming 1.0 g = 1.0 mL for water, what would be the molarity of the solution described in Question 13?

15. If an additional 200 mL of water were added to the solution described in Question 13, what would be its new concentration? Express your answer in both $\%_{(wt-wt)}$ and M.

16. Calculate the freezing point of a mixture prepared by dissolving 10.0 g of $C_6H_{12}O_6$ into 2.00 kg of water. ($k_f = -1.86°C/m$)

17. Sodium chloride, NaCl, is frequently sprinkled on icy steps and sidewalks. Write a short explanation of why this is effective. Include terms such as dissociation, freezing point depression, hydration, etc. in your description.

18. Using the solubility of salt guidelines presented in Table 7-7, indicate which of the following salts would be soluble. If they are soluble, write them in ionic form on the space to the right.

    a. KCl   _____   _____

    b. $AlBr_3$ _____   _____

    c. $PbBr_2$ _____   _____

    d. $CaCO_3$ _____   _____

    e. $Mg(NO_3)_2$ _____   _____

19. Write the total ionic and net ionic equation for the following molecular reaction.

$$Na_2SO_4 \quad + \quad Ca(NO_3)_2 \quad \rightarrow \quad 2\,NaNO_3 \quad + \quad CaSO_4$$

____ + ____ + ____ + ____ → ____ + ____ + ____

____ + ____ → ____

20. Use a calculator to determine the logarithm of the following numbers.

    a. log 455   = _____

    b. log 0.00245   = _____

    c. log 0.0000000422 = _____

    d. log 6,560,000,000 = _____

    e. log 0.00000010   = _____

21. According to the following ionization equation, determine the pH of a 0.15M solution of hydrochloric acid, HCl.

$$HCl \xrightarrow{(100\% \text{ ionized})} H^+ + Cl^-$$

According to the following ionization equation, determine the $[H^+]$, pH and pOH of a 0.005M sulfuric acid, H_2SO_4, solution.

$$H_2SO_4 \xrightarrow{(100\% \text{ ionized})} 2\,H^+ + SO_4^{2-}$$

22. If a solution has a pH = 10.7, what would be the $[H^+]$ and $[OH^-]$?

23. Calculate the $[H+]$, $[OH^-]$, pH, and pOH of a 0.0015M solution of KOH.

$$KOH \xrightarrow{\text{(100\% ionized)}} K^+ + OH^-$$

24. Determine if the hydrolysis of aluminum sulfate, $Al_2(SO_4)_3$, would result in a solution that is acidic, basic, or neutral. Justify your answer.

25. Some aspirin products on the market say they are buffered. Do a little research to determine the buffering agent used and write a paragraph explaining why it works.

Who invented electricity?

Electricity and Magnetism

The use of electricity is, in a very important way, what defines the modern world. Think about the last time the power went off. If it was during the day, most likely your activities suddenly became restrained. The cash register at the mall, with its integrated scanner that automatically updates the store's inventory as it rings up your purchase, was useless. Traffic snarled as signal lights failed to operate. If it was at night, you may have lit some candles, used a flashlight, and perhaps listened to a battery-operated radio. If the blackout continued for very long, you probably just went to bed, since routine activities simply became too difficult or impossible.

It is ironic that we find it hard to function without electricity. For most of history, electrical energy was not available at the flick of a switch. In fact, it is still considered a luxury and is unavailable to a large portion of the population of the world. But the number of people in the United States who can even remember living without electricity is rapidly dwindling. It is interesting to talk to them and find out what life was like. It was certainly not dominated by television, CDs, movies, and the Internet.

Electricity is not a recent discovery, but the technology that allows us to use it is. You might believe that Benjamin Franklin discovered electricity, but that's not true. What he did do was to help establish that certain natural phenomena, most famously lightning, is electrical in nature. Many other scientists in the 1700s and 1800s provided additional clues that enabled the building of an understanding that has subsequently been harnessed to the advantage of humankind. Lightning may be nature's way of reminding us of the importance of electricity in all natural processes.

Chapter 4 presented the fact that electrical charge plays an important role in the composition of atoms and that it is what holds compounds together in the form of chemical bonds. In Chapter 5 we discussed how forces acting on objects cause them to change their motion (accelerate). Although it cannot be seen, the mechanism that transmits the force to accelerate an object is the electrical interaction between the electrons of the atoms of the object and those of the object applying the force.

Because we are far removed from the atomic-level interactions involved in pushing an object, we tend to call such interactions mechanical, although we know that electrical forces are involved. In Chapter 7, the oxidation-reduction reactions that are used to generate electricity in commercial batteries were introduced. When considering electricity, concepts such as current and voltage usually come to mind. In this chapter, we will concentrate on these, but it is still useful to remember that nature is not so easy to categorize. Electricity plays a major role in the understanding of many topics such as bonding, batteries, oxidation-reduction, mechanics, and magnetism.

8-1   Charge and Charge Carriers

Objectives

Upon completion of this section, you will be able to:

- Discuss the two kinds of charges in nature.

- Identify and discuss the nature of the particles that are charge carriers in the atom.

- Distinguish between the unit of charge and the elementary charge.

- Explain the importance of net and excess charge in electrical phenomena.

Static electricity is a popular, not a scientific, term. Nevertheless, it is useful to discuss it because everyone has experienced some phenomenon associated with it – clinging clothes, flying hair, or the shock received when touching a doorknob. What the term "static electricity" refers to is the buildup of electrical charges on an object or person. The electrical forces that bind compounds together are called chemical bonds. Clinging clothes demonstrate these same electrical forces, but on a much larger scale. Static electricity is a constant reminder that electrical charges are everywhere.

An understanding of the concept of charge is central to the discussion of electricity. Because everything physical is composed of matter, and matter is composed of atoms, and atoms are composed of charged particles, we are surrounded by countless numbers of electrical charges. This fact is often hidden because there are two kinds of charge.

This simple fact was introduced in the discussion of protons and electrons in Chapter 4. The proton carries a positive charge (+) and the electron carries a negative charge (–). The reason the charges were assigned a (+) and (–) is that under certain circumstances they cancel each other out. We will be more specific about this statement in a moment. First, take a look at Figure 8-1. You may have thought this was some kind of printing error. At first, you see only a simple gray area. It is worth a second look, because it is more complicated than it first appears. This book is printed with black ink on white paper. There is no gray ink used. Shades of gray are made by using halftones, which are comprised of a pattern of black dots and white areas. The halftone gray area in Figure 8-1 is known to printers as 50% black because half of the space is black and half is white. Take a very close look at the gray area – get right on top of it – and you will see that this is true.

Typically, matter – like the halftone or gray area depicted in Figure 8-1 – is composed of equal amounts of positive and negative charge. When equal amounts of opposite charges are brought together, the resulting matter is said to be electrically neutral or uncharged. Electrical phenomena do not affect uncharged matter. This gives us another principle: when equal amounts of positive and negative charge are close together, they cancel each other's effect. It is only when you get down to the scale of atoms that you can actually distinguish between the separate plus and minus charges.

Figure 8–1: Look very closely at this halftone. What appears from afar as a gray shade is actually made up of tiny areas of black and white.

Charge Carriers

Charge is a fundamental property of matter, just like size, mass, and temperature. If we say that an object has a mass of 45 kg, for example, we have provided some useful information. In the same way, describing an object as electrically charged or neutral is also useful information.

The flip side of the statement that charge is a fundamental property of matter, is that without matter there can be no charge. As previously noted, matter is made up of atoms and atoms are made up of electrons, protons, and neutrons. The electrons and protons are **charge carriers**: elementary particles that have the property of one type of charge or the other. The electrons carry the negative charge, while the protons carry the positive charge. It is impossible to separate the charge from these elementary particles.

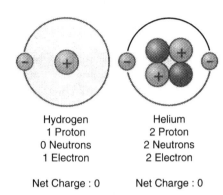

Hydrogen	Helium
1 Proton	2 Proton
0 Neutrons	2 Neutrons
1 Electron	2 Electron
Net Charge : 0	Net Charge : 0

Figure 8–2: Electrons and protons are called charge carriers. Because there are the same number of protons as electrons in an atom they appear to have no charge. The effects of the (+) and (−) charges cancel.

The Unit of Charge and the Elementary Charge

Since charge is a fundamental property of nature, in the SI system it has been given its own unit, the **coulomb,** C. Do not confuse the coulomb with the symbol used in equations to denote charge; usually q or Q.

The coulomb is large compared to the amount of charge carried by an electron or a proton. In a single grain of salt, for example, there are approximately 460 coulombs of positive and 460 coulombs of negative charge. The effect of these two charges is to cancel each other, so that the grain of salt is electrically neutral. Scientists have found that every proton and electron carries the same amount of charge as every other proton and electron. This amount of charge is called the **elementary charge** and been given the symbol, e. The elementary charge is small compared to the coulomb. In fact, e = 0.00000000000000000016 C. This is certainly an instance where the use of scientific notation (1.6×10^{-19} C) is helpful.

It is now possible to explain how the amount of charge in one grain (approximately 0.01 g) of salt is calculated. Sodium chloride, NaCl, has the formula mass of $(23 + 35.5)$ g/mole = 58.5 g/mole. Therefore, the number of ion pairs, N, in the grain of salt is found by:

$$N = \frac{0.01\,\text{g}}{58.5\,\text{g/mole}} \times 6.02 \times 10^{23} \text{ ion pairs/mole} = 1.03 \times 10^{20} \text{ ion pairs}$$

Each sodium ion contains 11 protons and each chloride ion has 17 protons. Although each proton carries only a tiny amount of charge, there are a lot of them. To calculate the total amount of positive charge, Q, requires the following calculation:

$$Q = (11 + 17)\frac{\text{protons}}{\text{ions}} \times (1.03 \times 10^{20}\,\text{ions}) \times (1.6 \times 10^{-19})\frac{\text{coulombs}}{\text{proton}}$$

Q = 461 coulombs

It can also be shown that the electrons carry 461 coulombs of negative charge that cancels the effect of the positive charges

Technology Box 8-1 ■ Why should you use a ground strap?

The simple act of walking across a carpet can transfer a large amount of charge. Picking up even a fraction of a microcoulomb, μC, (less than 10^{-6} C) may not sound like a big deal, until you realize that it is sufficient to give a person a charge of several thousand Volts. To an office worker, this phenomenon may be merely annoying – an occasional ouch upon reaching for the doorknob or touching someone else. To someone working with electronic components, this same phenomenon can be disastrous. The circuit elements in an integrated circuit (shown on the left below) actually occupy only a tiny fraction of the volume of the integrated circuit, IC, most of which is devoted to contacts and packaging. The energy contained in a small spark is enough to damage this structure, whose size is measured in microns (micrometers). Touching a sensitive component while you are charged can destroy it, even at charge levels so low that no noticeable spark can be seen or felt. Some particularly sensitive components are labeled as "ESD Sensitive." (ESD stands for electrostatic discharge.)

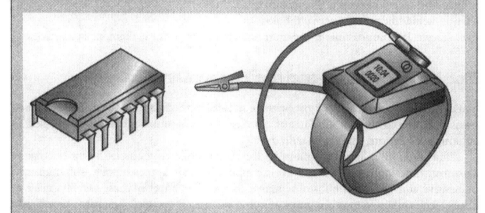

An integrated circuit or chip (left) can be damaged by a small, even invisible, electrical spark because its active components are built as microscopic structures. Wearing a grounding strap (right) will keep excess charge from building up on a person's skin, so that such chips can be safely handled.

The solution to this problem is to undo the charge imbalance between you and the components you are handling before touching them. In order to move, charge carriers must have material through which they can flow. Materials through which charges flow readily are called conductors. If you have ever installed a card or a disk in your personal computer, you were probably instructed to touch the metal frame of the computer and to avoid handling the card by its contacts. This action gives the charge carriers a chance to flow between your body and the frame of the computer, so that any charge imbalance is neutralized. If you were working on a complicated circuit, touching the frame periodically would be a waste of time. A better solution is to wear a grounding strap, as shown on the right-hand side above. Usually wrapped around your wrist, the strap is made of a conductive material and attached by a wire to a conductive mat on which you can place all of the parts you are working. This arrangement prevents net charge from building up and a damaging discharge from forming.

Net or Excess Charge

We now know that all matter contains electrical charges. Because the amounts of positive and negative charge are equal, most matter behaves as if it is uncharged – at least when viewed on a large scale and at long range. When dealing with electrical phenomena, we talk about the amount of charge on an object. However, what we are really talking about is the net or excess charge that is on the object.

By **net charge** or **excess charge,** we mean any charge that is not cancelled by an equal amount of the opposite type. For example, your body contains a enormous number of protons. It also contains the same number of electrons, so that under normal circumstances your body is electrically neutral. When you shuffle across the carpet, however, you may either pick up or leave behind some electrons. Now there is a charge imbalance on your body.

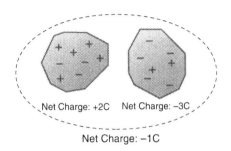

Net Charge: +2C   Net Charge: –3C

Net Charge: –1C

Figure 8–3: An object is considered to be charged if there is an excess of one or the other type of electrical charge. If an object with excess positive charge and one with excess negative charge are brought close together, the effect would be the same as the net charge on a single object.

There is also a reason why the two types of electrical charges were assigned plus and minus rather than red and blue. It is so you can calculate excess charges. Suppose, for example, you have an object with +2 C of excess positive charge and you add +3C more. The object will now have a total of: 2 C + (+3 C) = +5 C of (excess) charge. If, on the other hand, you add the same amount of charge, but with the opposite sign, you will have a total of: 2 C + (–3 C) = –1 C of (excess) charge.

Check Your Understanding

1. What are the two kinds of charge and what two particles are the charge carriers?

2. What is the elementary charge, e, and its relationship to the charge carriers?

3. What is the value, in coulombs, of the elementary charge?

4. What is meant by the term net or excess charge?

5. What is the reason for assigning (+) and (–) values to charge?

8-2 How Charges Affect Each Other

Objectives

Upon completion of this section, you will be able to:

■ Calculate the electrostatic force between two charges.

■ Explain the relationship between the electrostatic force and voltage.

■ Define what is meant by ground potential.

■ Describe the relationship between voltage and energy.

OK, so I collect a charge shuffling across the carpet, but why do I get a shock when I reach for the doorknob? If you think carefully about the last time you received a shock of static electricity, you actually felt the shock an instant before you touched the door-knob or other person. If you were in a darkened room, you may have even seen a spark jump the gap. This is one of the key aspects of electrical phenomena: electrical charges affect other electrical charges at a distance.

The spark you saw or felt resulted from the motion of the electrical charges; in this case, the charge carriers were electrons. Something caused the electrons to move. Electrons do not simply pick up and move spontaneously. They begin to move – accelerate – only when a force is acting on them.

You already know that like charges repel and unlike charges attract. This simple fact is the basis of all electrical phenomena. What it means is that like charges exert a repulsive force on each other and opposite charges exert an attractive force. Buried in this simple explanation is yet another important truth: electrical forces only act between electrically charged objects. Stated another way, neutral objects are unaffected by electrical forces.

When we say that the forces act between charged objects, we mean exactly that. As shown in Figure 8-4, two charged objects are either attracted to or repelled from each other. The force (attraction or repulsion) acts in a straight line between them and the forces are reciprocal. This means that if one object is attracted toward another, that one will also be attracted to the first. Under no circumstances will an object be attracted toward a charge that repels it. All forces describe interactions between bodies.

We can do better than just make general statements about these attracting and repelling forces. First, they are called **electrostatic forces** and have been given the symbol, F_E. Second, they can be quantified. In the late 1700s Charles Coulomb determined that the magnitude of the electrostatic force depends on both the magnitudes of the charges involved and the distance of their separation.

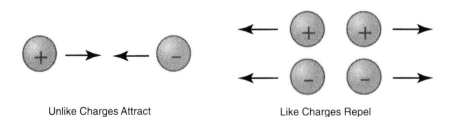

Unlike Charges Attract   Like Charges Repel

Figure 8–4: Like charges repel while unlike charges attract, so each object experiences a force due to the other.

The electrostatic force depends, in a precise way, on the charge and distance between the two charged objects; specifically on the square of the distance between
them. Summing this up, we have what is known as **Coulomb's law**:

$$F_E = \frac{k_c q_1 q_2}{r^2}$$

where the force, F_E, is given in newtons, the charges, q_1 and q_2, are measured in coulombs, and the distance, r, between the charged objects is measured in meters. The
proportionality constant $k_c = 8.9875 \times 10^9$ N · m^2/C^2.

A little dimensional analysis on the above equation will convince you that the
units and sign of the proportionality constant, k_c, are correct. If q_1 and q_2 are of opposite signs, their product is negative and F_E is attractive. Alternately, if they are of the
same sign, their product is positive and F_E is repulsive.

For example, according to the Bohr theory the hydrogen atom consists of an
electron revolving about a proton in a circular orbit. The radius of the Bohr electron
orbit hydrogen is 5.3×10^{-11} m. What is the force between the electron and proton?

$$F_E = \frac{k_c q_1 q_2}{r^2}$$

$$F_E = \frac{8.9875 \times 10^9 \text{N} \cdot \text{m}^2 (1.6 \times 10^{-19} \text{C})(-1.6 \times 10^{-19} \text{C})}{\text{C}^2 \quad (5.3 \times 10^{-11} \text{m})^2}$$

$$F_E \cong -8.2 \times 10^{-8} \text{ N}$$

Since the charge carriers (a proton and an electron) have equal but opposite
charges, the calculated electrostatic force, F_E, is negative. As stated above, a negative
electrostatic force indicates that the force is attractive and consistent with the understanding that opposite charges attract.

Force and Voltage

The force that one charge exerts on another charge is the fundamental force that
makes all modern electrical devices possible. For example, an incandescent light bulb
works because carriers of electrical charge – electrons – are pushed though the material
of the filament. During this process they transfer kinetic energy to the atoms of the filament, which in turn gains thermal energy, and emits a part of that energy in the form
of visible light. When you select a light bulb, you will find no mention of the electrostatic force involved. There will be only two numbers on the package: the power, in
watts, and the voltage at which it was designed to operate. For common household
lighting, it will be 120 V (AC voltage) and for your car it will be 12 V (DC voltage).
The power rating for the bulb will only be valid if it is operated at the stated voltage.

Voltage is closely related to the force that electrical charges exert on each other.
Consider the three situations shown in Figure 8-5. On the left-hand side of the figure
is a ball held above the ground. When it is released, it will fall toward the ground. This
is because the masses of the ball and the Earth attract each other. Another way to describe this behavior is to note that the gravitational potential of the ball becomes lower as the ball gets closer to the surface of the Earth. The natural tendency of the ball,

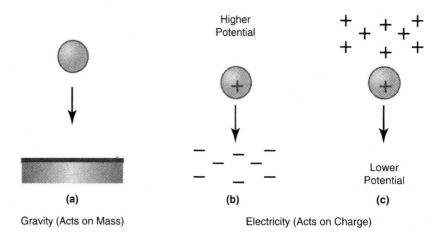

(a)

Gravity (Acts on Mass)

(b)   **(c)**

Electricity (Acts on Charge)

Figure 8–5: Masses experience a gravitational force that will cause them to move if they are free to do so (a). Similarly, charges experience an electrostatic force that will cause them to move if they are free to do so. Unlike charges (b) are attracted, while like charges (c) repel each other.

if it is free to move, is to move from a region with higher gravitational potential to a region with lower gravitational potential. This is true for a ball of any mass.

The two other situations shown in Figure 8-5 are of charged objects under the influence of many other charges. (We will assume gravitational effects are negligible, as in a weightless environment.) In both cases, the object has an excess of positive charge, so it is attracted to negative charge and repelled by positive charge. In either event, a statement can be made that is analogous to the mass under the influence of gravity. The natural tendency of a positively charged object is to move from a region with higher **electrostatic potential** to a region with lower electrostatic potential. The term **voltage** is another term used to describe electrostatic potential. (Voltage equals electric potential energy divided by unit charge.)

Voltage is usually given the symbol V (note the italics) and has SI units of the Volt, which also uses the symbol V (no italics). Voltage variations are the consequence of having an imbalance of charges in a region of space. The relationship between voltage and charge is that voltage is higher where there is excess positive charge. Voltage is lower where there is excess negative charge.

A positively charged object feels a force away from regions of higher potential and toward regions of lower potential because of the electrostatic force. But what about a negatively charged object? Figure 8-6 shows that the force a negatively charged object experiences is opposite to that of an object that has the same amount of positive charge. The

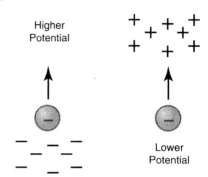

Figure 8–6: Charges may be repelled from like charges or attracted to unlike charges. In either case, negative charge moves from regions of lower electrostatic potential (or lower voltage) to regions of higher electrostatic potential (or higher voltage).

natural tendency of a negatively charged object is to move from a region with lower electrical potential to a region with higher electrical potential. This is one difference between how mass responds to gravitational potential and how charge responds to electrostatic potential – there is no kind of mass that falls up!

Technology Box 8-2 ■ What is a cathode ray tube or CRT?

Computer components are getting smaller – a trend that has continued since the early days of computing – but one component that still takes up a lot of space is the monitor. Although several promising flat screen technologies have recently been marketed, the majority of monitors are still essentially specialized television picture tubes, generically known as cathode ray tubes (CRTs). They are called cathode rays because they emanate from the hot cathode.

As shown below, a CRT is a specially shaped glass enclosure within which there is a vacuum. Within the tube, streams of electrons freed from a hot metal filament at the back of the tube, are accelerated toward the front screen. One reason the tube must contain a vacuum is that in air, the filament would quickly burn out. The other reason is that the freed electrons must not bump into other matter on their way to the screen. A voltage difference between the back of the tube and the screen accelerates the electrons from the back to the screen. Like any other accelerated object, they gain kinetic energy as their speed increases. When the electrons slam into the inside front surface of the tube, they lose this energy to the surface material. In a CRT, this surface material is a phosphorescent material that converts some of the energy into visible light. The pattern formed by the light emitted is what we see as the image. CRTs are still the most popular type of computer display because they are bright and easy to read in a well-lit area.

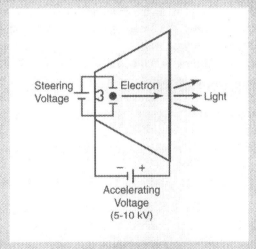

The image we see on the screen is formed by keeping the cathode rays as a tight beam as they are scanned across the phosphorescent materials on the back of the screen. A second set of voltages steer the electrons to the appropriate locations on the phosphorescent material on the back of the screen. The beam is also selectively turned on and off so only the portions of the screen that are needed to form the image are struck by the energetic electrons. The process, called a raster scan, is achieved by applying varying voltages to the two sets of metal plates that are located between the hot filament and the screen. The electrons in the beam never touch these plates, but the positive and negative charges that are collected there exert forces on the electrons that deflect them left and right, or up and down. These forces are what steer the beam to the correct location on the inside surface of the screen.

Ground Potential

It is a good bet that you are sitting in a chair. Imagine yourself standing on the chair. Now ask yourself how high you are. If you think about this for a moment, you will see that this is a nonsense question, because it is incomplete. You need to ask, "How high above what?" The floor? The ground? Sea level? Without a reference point, the question is meaningless.

The same thing holds for voltage. Unlike temperature, which has an absolute zero, voltage has no absolute zero. Therefore, all voltage measurements are relative. Because of this, descriptions such as "the voltage on the red wire is 15 V" are often used when analyzing circuits. Implicit in this statement is an assumption that somewhere in the circuit there is a reference voltage against which all other voltages are measured. This reference voltage is called the **ground potential** or simply **ground**. Numerically, the ground potential is the voltage that we choose to call 0 Volts.

Any point in an electrical circuit can serve as the ground. By convention, if there is a single voltage source, such as a battery, the ground point is usually taken to be the negative terminal of the battery. In some circuits, there are multiple voltage sources. The typical personal computer, for example, uses +5 V, +12 V, and –12 V relative to ground to operate its circuits.

The term "ground" comes from the fact that in most buildings, this voltage is literally the voltage of the ground surrounding the building. In modern buildings, there is a heavy copper wire leading to a metal stake in the ground. This wire is connected to the center round contact of a 3-prong socket. (In older buildings, the electrical outlets may have only two contacts. In this instance, the metal box in which the outlet is housed is the electrical ground.) Having a ground contact available means that each electrical appliance in the building can share a common voltage reference.

Knowing the local ground potential and connecting electrical equipment properly is an important part of electrical safety. In the early days of rock music, British bands playing in pubs with antiquated electrical wiring sometimes found that their electric guitars and microphones were at different voltages. Unfortunately, the band members usually discovered this when they leaned close to the microphone and received a nasty shock.

Voltage and Watts

Consider two houses in a new neighborhood. Both are wired for 120 V electrical service, so you can walk into either house and plug in an appliance and it will work. One house is occupied and its occupants are running several lights, a television, a radio, and some other electrical appliances. The house next door is still unoccupied. Which house will have the higher electric bill?

The obvious answer is that the occupied house will have the higher electric bill. Although both houses have the same voltage available, the unoccupied house is not using any electricity. The point of this comparison is that voltage by itself is only half the story. The utility company is in the business of selling electrical energy.

Look at a typical electric bill and you will see that electricity is metered in kilowatt-hours or kwh. A kilowatt, which is simply a thousand **watts**, is a unit of power. The product of kilowatts and time is a measure of energy. The two important facts to remember are that electric charges gain or lose energy when they move from one location to another where the voltage is different. Second, the amount of energy gained or lost depends on both the amount of charge moved and the size of the voltage change. Therefore, you must be moving electrical charges in order to be using electrical energy, which is exactly what you are doing when you run any electric appliance. The mathematical expression for this is:

$$\Delta U = Q \Delta V$$

where, ΔU is the change in potential energy that a charge, Q, experiences when it undergoes a change in voltage, ΔV.

Take, for example, the case of a single electron as it accelerates from the back of a computer monitor to the screen (see Technology Box 8-2). For safety's sake, the screen is at ground potential or $V = 0V$. The back of the monitor, where the electron originates, has to be more negative than ground, so the electron is repelled from the back and attracted toward the screen. Voltages of $-1,000$ V are typical in computer monitors. Therefore, the change in voltage that the electron experiences is $\Delta V = +1,000$ V. The electron has a charge of $Q = -1.6 \times 10^{-19}$ C, so that its change in potential energy, ΔU, would be given by:

$$\Delta U = (-1.6 \times 10^{-19} \text{ C}) (+ 1,000 \text{ V}) = -1.6 \times 10^{-16} \text{ J}$$

Since the change in potential energy is negative, the electron has lost potential energy, but gained kinetic energy. This means that the electron will be moving faster – right up to the point where it hits phosphors on the back of the screen. Some of that energy will be emitted in the form of visible light. (Remember that energy can neither be created nor destroyed.) In fact, the relationship, $\Delta U = Q \Delta V$ is the basis for the **electron-volt** or eV. The electron-volt is simply the energy that one electron gains or loses when it experiences a voltage change of 1 V. Therefore,

$$1 \text{ eV} = 1.6 \times 10^{-19} \text{ J}.$$

The electron-volt is not a unit in the SI system! To use energy in a calculation, use the above equality to convert energy in eV to joules, if you want the results to come out in SI units.

Check Your Understanding

1. What is the meaning of the statement "like charges repel and unlike charges attract?"

2. What is the name and symbol of the electrical forces that occur between charge carriers?

3. If the electrostatic force, F_E, between charges is negative, is it an attractive or repulsive force?

4. What is another name for electrostatic potential?

5. Is potential voltage higher or lower where there is excess negative charge?

6. What is the value of the ground potential voltage?

7. What are the two factors that determine amount of electrical energy?

8. Is the electron-volt an SI unit?

Simple Electrical Circuits

Objectives

Upon completion of this section, you will be able to:

■ Explain the difference between conductors and insulators.

■ Identify a conventional electrical current in a simple electrical circuit.

■ Calculate the amount of current flowing in a simple electrical circuit.

■ Calculate the amount of power dissipated in a simple electrical circuit.

Electrical equipment often warns: "Danger: High Voltage," but is it the current or the voltage that kills? The answer is that it's not one or the other. Let's look at the facts. The first fact is that it takes voltage to drive current, so high voltages (in excess of 20 Volts) can be dangerous because they can drive high currents through your body. There was a kitchen gadget developed that used this principle to quickly cook hot dogs. You put an electrode in each end of a hot dog and turned on the power. In short order, the hot dog was cooked. It's possible to do the same thing if you accidentally contact a large electrical voltage and the electrical current flows through your body. So, it's not the voltage by itself or the current by itself that does the damage, but the two working together. Voltage times current is power, and it is the product of power and time that is energy. It is the amount of energy deposited in your body that can be fatal.

There are times when high voltage alone can be dangerous. It can upset the electrical signals that operate the heart, causing it to stop or develop an irregular pattern called **fibrillation**. In such cases, medical technicians use a defibrillator (see Technology Box 8-3) to reestablish a normal rhythm.

Electrical Current

The concept of **electrical current** is straightforward – it is the movement of electrical charges from one location to another. Recall that electrical charges exist only on charge carriers. For charges to move from one place to another, therefore, the electrons or protons must move. Theoretically, the motion of either could constitute an electrical current; however, in the vast majority of materials used to build electrical devices, it is only the electrons that move and the protons remain stationary. In very few practical instances are electrons found moving anywhere other than within conducting material. A notable exception is the electrons within a CRT (see Technology Box 8-2).

Electrons are able to move very readily in some materials, while their movement is more difficult in others. Materials in which the electrons move easily are called **conductors**, while their movement is more difficult in **insulators**. All metals are conductors, as are other substances, such as graphite. Many plastics and fiberglass are used as insulators. A ceramic substance is used to connect power lines to the metal poles because it is an insulator. There is yet another important class of materials (discussed briefly later in the chapter) that lies between conductors and insulators. They are called **semiconductors** and are at the heart of the technology behind every computer in use today, as well as most other types of consumer electronics.

Technology Box 8–3 ■ Is a defibrillator really a reset switch for the heart?

The heart is a single muscle, but it is so large that a "heartbeat" is actually a series of contractions that is initiated in the atria followed by the ventricles. The pacemaker of a normal heart is the sinoatrial node (SA node), which is first chemically depolarized and initiates an action potential in the heart muscle fibers attached. This action potential is propagated through the atria and then to the atrioventricular node (AV node). A part of this specialized system, called the atrioventricular bundle, conducts the beating impulse from the atria and the AV node to the rest of the heart, which results in it contracting as a single unit. It is critical that the contractions and relaxations proceed in the proper sequence. The upper chambers (atria) contract first, then the lower (ventricular), etc. This is the mechanism by which blood is pumped throughout the body.

If you are ever unlucky enough to encounter a high voltage source, it could either stop your heart or interrupt its beating sequence. If your heart is stopped, you will need someone nearby to administer CPR immediately. They may be able to restart your heart and save your life. Equally dangerous, however, is fibrillation. Fibrillation means that the heart has its beating and relaxing sequence out of sync. In this case, the heart is working against itself and cannot pump blood. In this instance, it is necessary to defibrillate the heart and this is done by applying yet another electrical shock. The idea is to momentarily cause the entire heart muscle to contract, in the hope that when the voltage is removed the heart's contractions will return to normal.

The voltage is applied with a pair of paddle-shaped electrodes applied to the chest. This procedure is a staple of television medical dramas. The physicians using the defibrillator shout "Clear!" before applying the voltage, so that their colleagues know when to remove their hands from the patient. If they failed to do so, then the voltage applied might cause their own heart to fibrillate.

A familiar example that illustrates the role of conductors and insulators is the common extension or appliance power cord. The cord contains metal wire conductors surrounded by a rubber or plastic insulator. Copper is often used for the electrical wires because it is reasonably inexpensive and a good conductor. The rubber or plastic insulating material allows us to handle the cord safely.

The copper wire in a typical home electrical extension cord is approximately 2 mm in diameter. (This is a typical value, but thicker wires would be necessary for higher current applications, such as electric dryers and electric lawnmowers.) Let's consider a 10 cm length of this wire and treat it as though it were a cylinder, as shown in Figure 8-7. The volume V, of the wire would be given by:

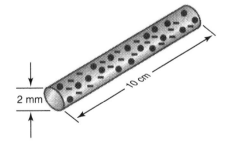

Figure 8–7: A copper wire 2 mm in diameter and 10 cm long contains 1.55×10^{24} separate electrical charges.

$$V = r^2 L$$
$$= \pi \, (0.1 \text{ cm})^2 \, (10 \text{ cm})$$
$$= 0.314 \text{ cm}^3$$

---

where r and L are the radius and length of the wire, respectively. Copper has a density of 8.96 g/cm³, so the mass, m, of copper is:

$$m = \rho V$$
$$= (8.96 \text{ g/cm}^3)(0.314 \text{ cm}^3)$$
$$= 2.81 \text{ g}$$

Since copper has a formula mass of 63.5 g/mole, the number of moles of copper in the wire is:

$$\text{mol} = \frac{2.81\,\text{g}}{63.5\,\text{g/mol}} = 0.0443 \text{ mol}$$

Each copper atom has 58 charge carriers (29 protons and 29 electrons), so the number of individual charge carriers, N, in this short length of wire would be:

$$N = (58 \text{ charges/atom})(6.02 \times 10^{23} \text{ atoms/mol})(0.0443 \text{ mol})$$
$$= 1.55 \times 10^{24} \text{ charges}$$

From this calculation, it can be seen that the total number of charges in this short wire is huge, but remember that half of these charges are positive, while the other half are negative. This means that the net electrical charge on the wire is zero. Even in a battery where one of the terminals is more positive, the electrical charges differ by only a tiny fraction. In this wire, there is no voltage difference, so the huge number of charges remains balanced.

If this wire were attached to the terminal of a battery, there would be a voltage difference between one end of the wire and the other. The voltage at the end attached to the (+) terminal of the battery would be higher than that attached to the (−) terminal. Each of the charges in the wire would experience a force. Positive charges (the protons in the copper atoms) would feel a force away from the (+) end and toward the (−) end, while the negative charges (electrons in the atoms) experience a force in the opposite direction.

There is a difference between the electrons and the protons however; the electrons can move, but the protons in the copper atoms cannot. This situation is similar to one that was discussed in Chapter 5, where the book on a table is affected by the downward force of gravity, but at the same time the table applies an upward force on the book. The copper atoms feel a force due to the electric field within the copper, but the rest of the copper atoms push back, so they cannot move.

Metals such as copper are good conductors because, the planetary model of the atom is replaced by one in which the outer valence electrons form a **sea of electrons** that is shared. Therefore, these electrons are able to wander from atom to atom. There are always the same number of electrons as there are protons, so that the (+) and (−) charges offset, but any single electron is free to move from one atom to another. When voltage is applied across the wire, these electrons experience a force that pushes them toward the positive end of the wire. When the atoms trade electrons, they do so with the electrons going primarily in one direction. This motion of the

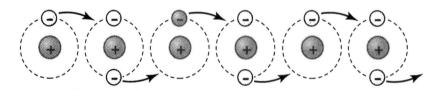

Figure 8–8: Electrical current flows in a wire when the electrons (charge carriers) are passed from one atom to another.

electrons is what is known as an electrical current. (See Figure 8-8.) It is possible to quantify the amount of electrical current flowing in a wire. The unit of current is the ampere, A, which is defined as 1 coulomb of charge flowing past any spot in the wire in 1 second.

In general terms, the current, I, is the change of charge, Δq, over a given interval of time, Δt, or

$$I = \frac{\Delta q}{\Delta t}$$

From previous discussions, we know that it takes a large number of charge carriers to move 1 C, since each electron carries only a single e ($e = 1.6 \times 10^{-19}$ C) of charge.

$$\frac{1\,C}{1.6 \times 10^{-19}\,C/e} = 6.25 \times 10^{18}\,e$$

1 coulomb = 6.25×10^{18} elementary charges

A current of 1 ampere (or amp for short) is equivalent to 6.25×10^{18} electrons flowing past any point in the wire each second. This is not an unusually high current. Typical circuits in a home are designed to carry 20–30 amps before a fuse or circuit breaker will stop the flow to prevent overheating of the wires.

In actual practice, we do not count each electron that goes by – we measure current by the effect that it has on other devices, like meter movements. Nevertheless, current is the result of the real motion of real particles in the conducting material.

Conventional Current

Which is correct: "Close the door, you're letting the heat out!" or, "Close the door, you're letting the cold in"? Actually, either is totally correct. What is happening when the door is open is that cold air is moving into the room, while hot air is moving out of the room.

In the case of electrical current, we have two possible types of charge carriers (protons and electrons) with different signs. Current is defined by a sign and a direction. Suppose we have a wire lying left to right on a table. The following situations would give us the same current in the wire (see Figure 8-9):

— 5 coulombs of positive charge every second, moving left to right

— 5 coulombs of negative charge every second, moving right to left

It turns out that electrons are what carry the current in a metal wire. Therefore, only the second case is physically possible. Nevertheless, when we describe current flow, it is conventional to talk as if there are positive charges moving in the conductor. This convention has caused endless grief for students of electricity and magnetism for years. Nevertheless, it has been well-entrenched since the days of Benjamin Franklin.

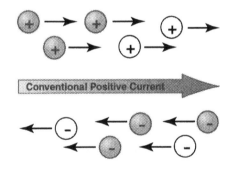

Figure 8–9: By convention, positive current in the right-hand direction can be either positive charges moving to the right or negative charges moving to the left.

Technology Box 8–4 ■ What are jumper cables?

Everyone should keep a set of jumper cables in the trunk of their car. It is almost inevitable that you will eventually have to jump-start your car or help someone else. A set of jumper cables consists of two thick wires with a set of heavy-duty clamps at each end. There is a correct, safe order in which to connect the cables. If you just try to remember the sequence, you have a 1 in 24 chance of getting it right. If you think about the physics behind the problem, you can't go wrong.

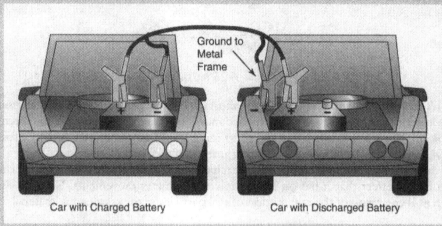

Car with Charged Battery   Car with Discharged Battery

When jump-starting a car, make sure the last connection is made to the frame of the car with the dead battery as shown in the figure. This keeps the sparks that will occur away from any hydrogen that may have accumulated in the battery.

The first important thing to remember is to connect positive of one battery to the positive of the other battery and negative to negative. The second important thing to remember is that the last connection must be made away from the dead battery to the frame of that car. This is to prevent a spark from occurring near the top of the dead battery. Hydrogen gas often accumulates within a battery. The spark that occurs when the last connection is made could ignite this gas and cause the battery to explode.

Why, you ask, do the previous three connections not share the same danger? It is because they do not complete the circuit through which a current can flow. No current, no spark! So, you can safely connect jumpers to both terminals of the good battery and the positive terminal of the dead battery. It is just the last connection that must be made away from the battery and to the frame of the car. The frame is electrically connected via a ground strap to the negative battery terminal. In other words, the frame is the ground potential of the car.

Does this mean that there can never be negative current? By no means! The following are also accurate descriptions of the current shown in Figure 8-9:

— 5 amps of current to the right

— –5 amps of current to the left

For the remainder of this chapter, we will employ the convention of positive current. We will talk about positive charges moving through an electrical circuit. We know that this is not actually correct; that the carriers of negative charge, the electrons, are actually moving. Nevertheless, when we adopt this convention, we get consistent and accurate, results, which in science is the bottom line. In solutions, gases, semiconductors, and plasmas, however, positive charges can also move.

Closed vs. Open Circuits

Attaching a wire to a battery makes the simplest electrical circuit possible. It is the circuit, for example, used in a flashlight. At one point in the circuit, the wire gets thin (the bulb) so it becomes hot and glows. We will discuss this phenomenon shortly. Nevertheless, we can trace a single path from one terminal of the battery to the other through a metal wire. The word circuit shares the same root as circle, and this is not a coincidence. For charge-carrying electrons to move through the wire, they must have some place to come from and some place to go. A single wire satisfies this basic requirement. The simplest geometry that neither gains nor loses charge carriers is the circle. This gives us our first requirement for electrical circuits. For a current to flow in an electrical circuit, there must be at least one unbroken closed-loop conducting path. The two key words in the preceding sentence are unbroken and conducting. As shown in Figure 8-10 (top) a circuit that contains a closed path is called a **closed circuit**.

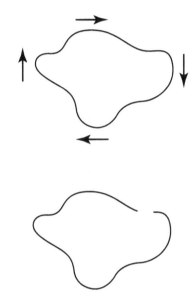

Figure 8–10: Current can flow in an electrical circuit only if there is a closed path of conducting material in which the charge carriers (electrons) may move (top). If this path is broken (bottom), then current cannot flow because the charge carriers must stay within the conducting material.

As shown in Figure 8-10 (bottom), a circuit that is not a closed path is called an **open circuit**. The function of the switch in a flashlight circuit is to break the closed path when the light is not needed. As we discussed earlier, without the movement of charge carriers, there can be no current. Charge carriers (electrons) are free to move in the conductor, but not through the insulator. Therefore, when we are tracing out the path of current flow, we follow the path laid out by the conductor. If this path is broken somewhere along the line, then current cannot flow.

When current flows in a simple circuit, such as the one shown in Figure 8-10 (top), it is important to realize that electrons are in motion everywhere in the circuit at once. When the circuit is broken, as in Figure 8-10 (bottom), the motion of the electrons stop, everywhere at once. Therefore, it does not matter where in the wire we look, because the same number of charge carries per second or equivalently, the same number of coulombs per second, is moving past each point in the circuit. Electrical current in a wire is similar to current in a river. If we stand on the bank of a river, the current is flowing, no matter if we are looking upstream or downstream. It is not the same water molecules, but they are all in motion at the same rate. Similarly, current is the rate of charge per second that is moving past any point in the wire.

There is a second requirement for current to flow in an electrical circuit – there must be a voltage source. An electric circuit has something in common with one of the desktop toys that are available in many varieties in stores (see Figure 8-11). In the toy a small-wheeled object is lifted to the top of a track, then it rolls down to the bottom, to be lifted again. In an electrical circuit, it is the battery that "lifts" charges from low to high potential. These charges then flow from high potential to low potential, to be lifted by the battery again.

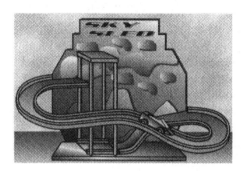

Figure 8–11: A mechanical analogue of the electrical circuit is the desk toy in which wheeled objects are lifted from the low end to the high end of a track, then roll down the track.

Resistance and Ohm's Law

Let's recap what we have learned about the effects that charges have on each other. An accumulation of excess positive charges results in a more positive voltage. An accumulation of excess negative charge results in a more negative voltage. Electrons and protons are charge carriers that feel a force due to these charge accumulations. The result is that positive charges, if they are free to move, move from regions of positive voltage to regions of negative voltage, and vice versa for electrons.

One of the basic equations in the study of physics, $F = ma$, relates the acceleration, a, of an object to its mass, m, and the force applied, F. This equation also applies to protons and electrons that carry electrical charge. It tells us that these charge carriers are accelerated by an electrostatic force. Indeed they are, but under one important circumstance: if they are free particles and essentially isolated from the influence of other matter. For example, because the electrons in a CRT are in a vacuum, they are accelerating all the way from the filament to the inside surface of the front screen (see Technology Box 8-2).

Most of the time electrons are in motion in an electronic device, they are not free to move outside the conductor. Therefore, they are continually running into the matter around them. The spacing between atoms in solid matter is on the order of 10^{-10} meters. What this means is that the farthest an electron can accelerate before running into an atom is about 10^{-10} meters. When the collision occurs, it is the lower mass electron that is most affected – like bouncing a ping-pong ball off a bowling ball. When the collision occurs, the electron may rebound straight back or be deflected to the side, as shown in Figure 8-12. It still feels an electrostatic force, so it starts to

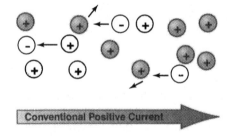

Figure 8–12: Electrons in a material that experience an electrostatic force "bounce off" the atoms in the material as they try to accelerate. Although the path of any one electron can be quite convoluted, overall the electrons reach a constant, average velocity in one direction.

accelerate all over again, but soon will undergo yet another collision. It never gets much of a chance to build up speed, but continues to drift in the general direction of the electrostatic force that is accelerating it.

The path of any single electron moving through a material can be very convoluted, but the electrical current is the sum of the motion of many, many electrons. At any one time, some may be moving faster and some more slowly or even backwards, but on the whole a single, average velocity describes their motion, and this average velocity is a constant value. This means that if we could sit in one spot and watch electrons move by, we would see a constant current.

The phenomenon just described is the **resistance** of the material to the flow of electric current. The symbol for resistance is R, and the SI unit of resistance is the ohm, for which we use the Greek letter Ω (omega). Resistance measures the difficulty of moving electrical charge carriers (in most cases, electrons) through a block of material. Resistance depends on both the shape of the block and the material from which it is made. The material property that describes how much the material impedes the flow of charge carriers is called **resistivity**. The symbol for resistivity is the Greek letter ρ (rho), and SI unit of resistivity is the ohm-meter (ohms times meters, usually symbolized as $\Omega \cdot m$).

Materials with low resistivity are called conductors, while those with high resistivity are called insulators. For example, the resistivity of copper, the most common material for electrical wires, is $1.678 \times 10^{-8}\ \Omega \cdot m$, a very tiny number, which means that copper is a good conductor. You may be familiar with expensive electrical connectors that are plated with gold, and think that gold is a better conductor than copper. Gold, however, has a resistivity of $2.24 \times 10^{-8}\ \Omega \cdot m$; still a good conductor, but not as good as copper. The advantage that gold has over copper is that it does not oxidize as readily. Copper oxide has a much higher resistivity than copper metal, so when a thin layer of oxide forms on the surface of a copper wire, it can make a big difference in how resistive the wire appears. As conductors go, silver is actually better than copper, with a resistivity of $1.586 \times 10^{-8}\ \Omega \cdot m$. Unfortunately, silver is not only much more expensive than copper, but it is also much more readily oxidized. Just for comparison, the resistivities of common plastics are in the range of $10^{12} - 10^{14}\ \Omega \cdot m$. That is why plastic insulation is frequently used on extension and power cords.

An electronic device whose primary characteristic is to impede the flow of electrical current is called a **resistor**. Of course, since resistivity is a property of all materials, some devices that are not designed with this purpose in mind inadvertently end up as resistors in electrical circuits. Physically, resistors come in many shapes and sizes, and you can even buy "chips" (integrated circuits) with many resistors in a single package, but the common individual or discrete resistor is shown in Figure 8-13. It has two connections, essentially an "in" and an

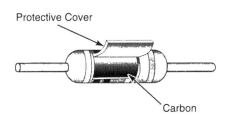

Figure 8–13: A physical resistor is composed of a resistive material, typically carbon, surrounded by an insulating shell or coating.

"out" for the flow of current. If we were to strip away the plastic insulating shell from the resistor, we would find a bit of material, typically carbon, to which the conducting metal leads of the resistor are joined.

Current flows in the resistor when the electrons in the resistor feel an electrostatic force. This happens when there is a difference in voltage (or an imbalance of charge) from one end of the resistor to another. We will call this voltage difference ΔV. Since it is the voltage difference that causes the current to flow, it should not be surprising that the larger the difference, the more current that flows. In fact, the current is directly proportional to the difference in voltage. The resistance of this simple electronic device, however, tries to impede the flow of current, so the greater the resistance, the smaller the current. The current turns out to be inversely proportional to resistance. Putting these together, we find that the current, I, depends on the voltage difference, ΔV, and the resistance, R, according to the following equation:

$$I = \frac{\Delta V}{R}$$

This relationship is known as **Ohm's law**, named after Georg Simon Ohm, whose name also graces the SI unit of resistance. In many textbooks, this equation is given as, $V = I R$. The fact that this equation is simply a rearrangement of the previous equation is not a problem, but it has also dropped the "Δ" from the voltage notation. This is unfortunate, because there is no absolute zero of voltage and all voltage measurements are relative. It is very important to realize that the voltage, V, that appears in this equation is the voltage difference between one end of the resistor and the other.

As we mentioned earlier, the resistance depends on the material and the shape of the resistor. The actual shape of the resistor is not the precise issue, but two quantities do matter: the length and cross-sectional area of the resistor. By length, we mean simply the length, L, as shown in Figure 8-14. The longer the physical length, the greater its resistance. This makes sense because there are simply more atoms in the way of the electrons trying to traverse the material. As for the cross-sectional area, it does not matter whether it is round, square, rectangular, or irregular. On the other hand, the greater the cross-sectional area, the lower the resistance. For the same reason, it is easier to move traffic down a four-lane than a one-lane road. When we lump together the material effects (resistivity, ρ) and the shape effects (length, L, and cross-sectional area, A), we have the expression for resistance, R:

$$R = \frac{\rho L}{A}$$

With this equation, it is possible to calculate how much current flows in the circuit shown in Figure 8-14. The battery is the voltage source. Its function, through chemical action, is to maintain a voltage difference of 12 V.

To calculate the voltage across the carbon block, it is first necessary to know the resistivity of the copper wire. Carbon's resistivity is $1.375 \times 10^{-5}\ \Omega \cdot m$, a small value, but over 1,000 times greater than that of copper. Based on this difference, copper can be considered a perfect conductor, with 0 resistivity. (A good approximation in this case. There are times, however, when such an assumption could get you in trouble –

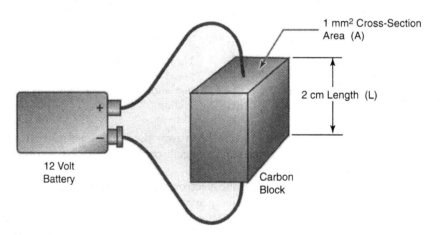

1 mm² Cross-Section Area (A)

2 cm Length (L)

12 Volt Battery

Carbon Block

Figure 8–14: A battery (voltage source) connected by copper wires to a block of carbon.

when you try to run too many appliances using only a thin extension cord, for example.) If copper is a perfect conductor, $R = 0$, the equation, $V = IR$, tells us that there will be no difference in voltage from one end of the wire to the other, regardless of how much current is flowing. We can now state that the voltage across the carbon block will also be 12 V.

The carbon block is 2 cm (0.02 m) long and has a cross sectional area of 1 mm^2 or 10^{-6} m^2. The resistance of the block can then be found by using the equation:

$$R = \frac{\rho L}{A}$$

$$R = \frac{(1.375 \times 10^{-5} \Omega \cdot m)(0.02\,m)}{10^{-6}\,m^2} = 0.275\Omega$$

Therefore, Ohm's law allows us to determine that the current flowing through this circuit is

$$I = \frac{\Delta V}{R}$$

$$I = \frac{12\,V}{0.275\Omega} = 43.6A$$

This is a lot of current through a very small piece of material, so the block will get very hot! It is possible to use the property of resistance to heat materials to temperatures high enough that they glow. This is how an incandescent light bulb works. The bulb is actually a small evacuated chamber in which the filament material is heated to a white-hot temperature. Because there is no oxygen inside the bulb, the filament material cannot "burn up." If the glass chamber is accidentally broken, the filament material will rapidly react with oxygen in the air and "burn" the filament out in a fraction of a second.

In the normal lifetime of a light bulb, the hot filament material gradually evaporates, so that the filament wire gets thinner and thinner, eventually breaking. Look carefully the next time you change a light bulb. The darkening you see on the inner glass envelope is due to the vaporized filament material condensed on the cooler inside of the glass.

Technology Box 8–5 ■ What is superconductivity?

Most, but not all, materials impede the flow of electrons through them. There is an important new class of materials, called **superconductors**, which do not. Their resistivity is zero!

The phenomenon of superconductivity is evidence of the quantum mechanical nature of matter. Superconductivity has been observed in materials when they are very cold (77 K and below), but the goal is to find a room temperature superconductor. Because superconductors have true zero resistance, the voltage necessary to maintain a current is zero. In addition, the power dissipated in a superconductor is zero. Therefore, one possible application of superconductors is energy storage, by using currents that circulate in a superconducting ring.

Research in the field of superconducting materials is still an exciting and important area of study, but it is not quite as frantic as it was at one time. One scientist described his past research as mix, patent, and test…in that order.

Power Dissipation in Resistive Circuits

Most people have accidentally touched a hot light bulb. Did you ever wonder why it gets hot? Let's compare two electrons – one is free in space and the other inside a wire – and see if it will provide an answer.

Assume that in Figure 8-15 the electron in the upper portion (a) of the drawing is inside a wire while the one in the lower portion (b) is in a vacuum. We will also assume that the left side is at a positive voltage compared to the right side. The electrons feel a force so they begin to move to the left. The lower electron can accelerate all the way across because it has nothing in its way. Because it is picking up speed, we know that it must be gaining kinetic energy. It gains this kinetic energy by trading potential energy for it. The potential energy lost is given by the equation,

$$\Delta U = (-e)(\Delta V)$$

where $(-e)$ is the charge. In this case, we know that the electron loses potential energy because although it goes to a region of higher voltage (ΔV is positive), it is a negatively charged particle, so that ΔU is negative.

The kinetic energy gained must be equal to the potential energy lost, so we can set them equal:

$$\Delta K = -\Delta U$$

$$\tfrac{1}{2} mv^2 = e \, \Delta V$$

By using the above equation, it is possible to calculate just how fast the electrons are moving before they slam into the phosphor screen inside your computer monitor as discussed in Technology Box 8-2, but we'll leave that for a homework problem.

The behavior of the lower electron in Figure 8-15 has not provided an answer to the question about heat. Perhaps examining what is happening to the upper electron will. Just like the lower electron (b), the upper electron (a) loses potential energy; in fact, the same amount since it experiences the same change in voltage. Unlike the lower electron, however, the upper electron never gets a chance to go very fast. It is constantly being stopped, and in some cases turned around, by the metal atoms in the wire. It is not exactly at rest, but its speed is nowhere near that of the lower electron. It is losing potential energy, but is not gaining as much kinetic energy. Now the question is, "Where did the energy go?"

The only place that the energy can go is into the atoms in the metal wire.

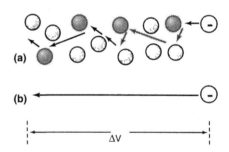

Figure 8–15: Two electrons falling through the same voltage difference lose the same potential energy. The bottom electron gains the equivalent amount of kinetic energy because it is accelerating in a vacuum. The top electron gains very little kinetic energy, but gives up energy to the surrounding atoms of the material in which it is moving.

Every time the electron slams into a metal atom, it shakes it up just a little. Vibrations as a form of energy were discussed in Chapter 5. In other words, the vibrational energy the electrons give to the metal atoms in the wire is a form of thermal energy. The amount of thermal energy that any single electron adds to the wire is small, but the effect of many electrons all moving at once can be quite large. Suppose that N electrons move through the metal in one second, each carrying one elemental charge, e. The potential energy they lose becomes thermal energy, ΔE, that the material gains and must dissipate as heat to its surroundings.

$$\Delta E = N e \Delta V$$

If this happens continuously, the rate of change of energy is given by $\Delta E/\Delta t$, which is called power, P. If the above equation is divided by Δt, we also get an expression for power:

$$P = \frac{\Delta E}{\Delta t}$$

$$P = \frac{Ne\Delta V}{\Delta t}$$

Now we need to recognize just one thing more. The total charge that these N electrons carry down the wire is $\Delta q = Ne$, but since we know that the rate of change of charge with respect to time, $\Delta q/\Delta t$, is current, I, we can arrive at the following relationship:

$$P = \frac{\Delta q \Delta V}{\Delta t}$$

$$P = I\Delta V$$

Power, as we have discussed, is a rate of energy used per unit time. Power, P, dissipated in the resistive material of the wire can be determined by the relationship $P = IV$. Again, you will usually see this equation written with the change in (Δ) voltage assumed.

What happens to this power? Initially, it goes into raising the temperature of the wire. As the wire gets hot, however, it both radiates and conducts heat to its environment. The temperature will rise until the power radiated and conducted away from the wire is equal to the electrical power dissipated. The faster the material loses the heat, the lower the temperature must be for this balance to be achieved. The cooling of electrical circuits has become an important part of the design of electrical devices, such as computers. (See Technology Box 8-6)

The SI unit of power is the watt, W. Most light bulbs used in the home consume between 60 and 100 watts of electrical power. A night light may use as little as 4 W, while a strong reading light may use as much as 150–200 W. Appliances such as hair dryers and electric ovens consume power in the thousands of watts or kilowatt (kW) range.

Electrical appliances dissipate power only when they are turned on. What the electric company sells is energy, not power. Power, P, times time, t, is energy, E, or stated mathematically:

$$E = Pt$$

Therefore, if we leave a 60 W light bulb burning for one hour, we have used:

$$E = 60 \text{ Watts}(1 \text{ hr})(60 \text{ min/hr})(60 \text{ sec/min}) = 216{,}000 \text{ Watt} \cdot \text{seconds}$$

$$E = 216{,}000 \text{ Joules}$$

The joule is the proper SI unit of energy. This is not the unit of energy that electric companies use to sell electricity to the consumer, however. Electrical utilities typically use the kilowatt-hour or kWh. It is easy to convert kWh to SI units, using the following equality:

$$1 \text{kWh} = 1{,}000\text{W}(1 \text{ hr})(60 \text{ min/hr})(60 \text{ sec/min}) = 3{,}600{,}000 \text{ Joules}$$

Technology Box 8–6 ■ What is that annoying noise in my computer?

How often have you found yourself in this situation – you have a paper to write, so you turn off the TV, sit down at your computer, and start to think about what to say. You are trying to concentrate, and now that the room is quiet, a sound that you may not usually notice seems very loud – the whining of your computer's cooling fan. The computer is an electronic device, so why does it have to have a fan to blow the air around? Actually, if you have ever looked inside a PC, you have seen something that looks like a little metal forest on the top of the microprocessor. As microprocessors have become faster their power use has increased, and the techniques used to remove the electrical energy that they dissipate as heat has also become more extensive.

Electrical energy is dissipated when it is turned into thermal energy inside the material of the microprocessor. If the energy has no outlet, it simply builds up and the temperature of the microprocessor increases. The thermal energy can be radiated away in the form of infrared electromagnetic radiation (heat). The energy that the microprocessor radiates per unit time increases rapidly as its temperature increases. Unfortunately, if radiation were the only way for the energy to escape, the temperature would soon become so high that the chip would be damaged. The other avenue for the energy to escape is by transfer to the surrounding air. The small forest of metal atop the microprocessor is a **heat sink** and provides the route for the energy to be moved from the interior of the chip to the air. It is designed to have a large surface area, because the larger the surface area, the more efficiently it can transfer energy to the surrounding air.

Usually, microprocessors draw current when their components switch states. This means that the faster the processor, the more current it draws, on average. Microprocessors historically have operated with +5V power supplies, but more modern computer logic is tending toward using lower voltages, so that the systems can operate faster with less power.

Photo courtesy ACK Technology, Inc.

The utilities probably think that it is easier to tell a customer that they have used 1 kilowatt-hour of energy than to say that they have used 3.6 megajoules. The cost of electricity varies from utility to utility, but averages about 10 cents per kilowatt-hour at this time. The cost of leaving a 60 W light bulb on from 8 p.m. to 8 a.m. can be calculated by the following:

$$60\,\cancel{W}(1kW/1{,}000\,\cancel{W})(12 \text{ hours})(\$0.10/kWh) = \$0.07$$

This seems like an inexpensive way to give your home that additional bit of security. Air conditioners, on the other hand, may consume 2 kW of power when running. If the system ran half of the time (called a 50% duty cycle), then to cool your house for 24 hours would cost:

$$2 \text{ kW } (24 \text{ hours} \times 50\%) \, (\$0.10/kWh) = \$2.40$$

When you consider that there are approximately 30 days per month, it is not difficult to understand why summer-time electric bills can be quite high.

Check Your Understanding

1. Is it the voltage or the current that is more likely to kill you? Explain.
2. What is the definition of current?
3. What is the difference between a conductor and an insulator?
4. In a material such as copper, what is the charge carrier?
5. What are the two factors that must be included to define current?
6. What is the difference between an open and a closed circuit?
7. What is the SI unit for resistance?
8. What is the difference between the terms resistance and resistivity?
9. What is the mathematical formula for Ohm's Law?
10. What is the mathematical formula for electrical power?

8–4 More Complex Resistive Circuits

Objectives

Upon completion of this section, you will be able to:

■ Read a schematic drawing of a simple electrical circuit.

■ Explain how Kirchoff's voltage and current laws apply to electrical circuits.

■ Identify parallel and series combinations of resistors.

■ Calculate the equivalent resistance of a combination of individual resistors.

■ Explain the difference between AC and DC voltage.

The impact of electric lighting on civilization cannot be overestimated. The simple circuit consisting of a voltage source and a light bulb has given us more productive and leisure hours per day, and has made our homes and communities more secure. Many of the electronic devices that we enjoy during these extended daylight hours, however, contain somewhat complicated electrical circuits. It is beyond the scope of this book to delve deeply into the realm of circuit design and operation, but electronic devices are an integral part of modern society, and it is useful to understand some basic circuit concepts.

Circuit Symbols

Discussing complicated electrical circuits requires knowledge of the graphic language of electrical schematics. Electrical engineers have developed a set of symbols that stand for the various components in an electrical circuit. There are different symbols for the many different types of components. Only the basic ones will be introduced here.

Wires in an electrical circuit are assumed to be perfect conductors. This means that they have no resistance and, therefore, there is no change in voltage anywhere along the wire. In point of fact, this is not completely true; even a good conductor such as copper has some resistance. In most cases, however, it is a good approximation. In situations where even the small resistance of a wire cannot be ignored, such as in circuits that draw very high currents, that resistance is indicated by a separate resistor symbol.

The symbol for a resistor, shown in Figure 8-17, is a zigzag line. Although three or four zigs and zags are used, the number is not significant. As discussed

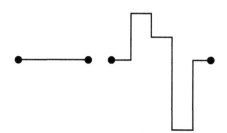

Figure 8–16: The simplest electrical symbol is the line, signifying wire. The line may be straight or have one or more right-angle turns. Curved lines are avoided in circuit diagrams.

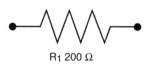

R₁ 200 Ω

Figure 8–17: The symbol for a resistor is a zigzag line, representing the path of the electrons through the material.

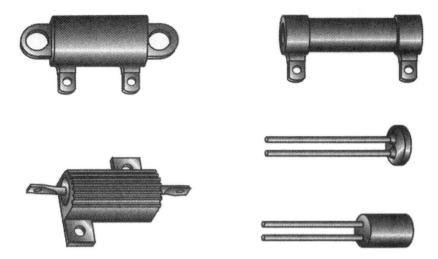

Figure 8–18: Resistors may be physically packaged in many different ways, but the same symbol is used in the circuit diagram for all.

in a previous section, the zigzag line represents the paths of the electrons as they move through the resistive material. The value of the resistor, in ohms, is indicated next to the resistor symbol. It is customary to give circuit elements letter designations, so that R_1 would be one resistor in a circuit, R_2 another, etc.

There are two important things to note about resistors. First, a resistor is a two-terminal device. This means there are two places where electrical connections (one in and one out) can be made. For small resistors, these are the wires at each end as seen in Figure 8-13. Larger resistors may have metal tabs with threaded screw holes. Some of the many ways that resistors can be packaged are shown in Figure 8-18. The second important thing to note about resistors is that the same symbol is used in the circuit diagram for a resistor, no matter how it is packaged. This is true for all electrical symbols on circuit diagrams. You may be able to order a particular electrical component in many different physical packages (metal cans, plastic chips, etc.), but its symbol and internal function remain the same.

The symbol for a battery is one or two sets of alternating short and long lines. These lines represent the components stacks that make up the electrochemical cell. By convention, longer lines represent the positive side and the shorter lines the negative side. Usually, the voltage is indicated next to the battery symbol. A couple of variations of the symbol for a battery are shown in Figure 8-19.

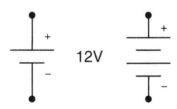

Figure 8–19: The symbol for a battery or any other voltage source is one or more pairs of long and short lines. The longer lines represent the positive side of the battery or voltage source.

Batteries use up their stored chemical energy during use and must be either recharged or replaced. Most electronic circuits that are designed for long periods of operation are plugged into an electrical outlet. The electrical outlet, in turn, provides electrical energy to a power supply inside the device. To provide electrical power, $P = V I$, the power supply must provide a certain amount of current while maintaining a voltage. Remember that the voltage and current are related by Ohm's law: $V = I R$.

In other words, if you connect the power supply to a resistor and know the voltage, *V*, you can determine the amount of current, *I*. Most power supplies set the voltage and then provide whatever current the circuit requires. These are called **voltage-regulated power supplies** and function just like batteries, except that they do not run out of energy.

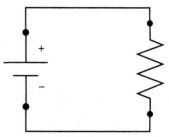

In electrical circuits, the same symbol is used for power supplies as for batteries. Like resistors, a battery or power supply is a two-terminal device.

We now need a way to connect all of these components symbolically. The simplest way to connect circuit elements is to draw lines (wires) between their terminals. Figure 8-20 shows what the simple resistive circuit we discussed earlier would look like, in the language of electrical schematics.

Figure 8–20: A simple resistive circuit, shown as a schematic diagram.

Figure 8-21a shows the conventional methods for diagraming connections. When lines (wires) simply cross there is no electrical connection. When they are connected, there is a dot. Four wires should never be connected in a cross, but this rule is commonly violated. In older circuit diagrams, it was customary to make one wire "jump over" the other, but the trend toward computer generated circuit diagrams has made this practice obsolete.

There is one more symbol frequently used in circuit diagrams – three lines, long, medium, and short, arranged in a triangle as seen Figure 8-21b. This is the symbol for the electrical ground. It is simply a symbol for a point that we will agree to call 0 Volts. As discussed earlier, the wires connecting the elements in a circuit diagram are considered perfect conductors, meaning that the negative end of the battery and the lower end of the resistor are all at 0 Volts.

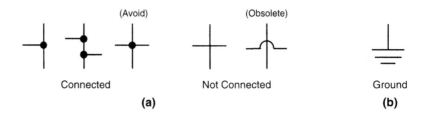

Figure 8–21: Examples of some wires that are and some wires that are not connected in an electrical circuit.

Kirchoff's Laws

When analyzing electrical circuits, there are two useful rules. One has to do with the current that is flowing and the other has to do with how the voltages add up. Together these are known as **Kirchoff's laws**, the first being known as **Kirchoff's current law** or KCL and the other being known as **Kirchoff's voltage law** or KVL.

Kirchoff's current law is perhaps the more intuitive of the two. It states that current flowing in a wire acts very much like current flowing in a river, especially when it branches. The point where the current branches is called a **node**. KCL states that the sum of all of the currents at a node must equal zero. This means that all of the current must go somewhere. In a wire, there is no place for the electrical charges to accumulate, so another way of stating the law is, "What goes in must come out."

Figure 8–22: When setting up a problem using Kirchoff's current law, pick directions for all of the currents. If you guess incorrectly, you will simply get a negative answer for the current.

As stated earlier, positive current in one direction is the equivalent to negative current in the other direction. Therefore, as shown in Figure 8-22, the currents into the node are $+I_1$, $-I_2$, and $-I_3$. Writing KCL mathematically, $I_1 + (-I_2) + (-I_3) = 0$. Rearranging this equation tells us that $I_1 = I_2 + I_3$. Kirchoff's current law guarantees that electric charge is conserved.

Take, for example, the situation described in Figure 8-22 (right), where $I_1 = 5A$ and $I_2 = -2A$.

$$I_1 + (-I_2) + (-I_3) = 0$$

$$5\,A + (-2A) + (-I_3) = 0$$

$$I = 3A$$

The beauty of KCL is that it is impossible to guess wrong when setting up the sign for the current. A negative answer simply means that the conventional positive current is flowing in the opposite direction of the arrow you have drawn.

Kirchoff's voltage law is somewhat less intuitive, but still uses a common-sense approach. It simply states that the sum of the voltages around any loop in a circuit is zero. Kirchoff's voltage law guarantees that electric energy is conserved.

When applying KVL, we move around the circuit loop. When we pass over any given circuit element, such as a battery or a resistor, we find ourselves moving uphill (toward a location with a more positive voltage) or downhill (toward more negative voltage). For a battery the situation is simple, as shown in Figure 8-23. Moving from the (−) to the (+) side, the voltage is positive; from the (+) to the (−) side, it is negative. It would actually be more accurate to say voltage change rather than voltage. Because the language of the electrical engineer is so pervasive in the field, we will adopt it even though it is technically incorrect.

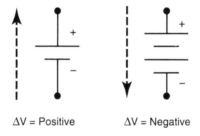

ΔV = Positive ΔV = Negative

Figure 8–23: Moving from the (+) to the (−) side of the battery is a negative change in voltage and visa-versa.

Positive Current, I

Figure 8–24: The (+) end of a resistor is the end from which positive current flows. The other end is the (−) end toward which positive current is flowing.

The factor that determines which end is positive in resistors is the way the current is flowing. The (+) end of a resistor is the end away from which the positive current is flowing. The (−) end is the end toward which positive current is flowing. Ohm's law ($V = I R$) tells us the magnitude of the voltage change. Figure 8-24 shows this in graphical form.

In the next example, KVL will be used to determine what happens if one of the three batteries in a flashlight is accidentally inserted backward (shown graphically in Figure 8-25). Assume that the current, I, through the circuit is flowing in a clockwise direction. (KCL is not used because there is only one path through which the current can flow.)

To apply KVL we simply make one complete circuit of this loop and add up the voltages. There is no connection between the direction of current flow and the direction we happen to select to go around the loop. If we arbitrarily choose to move around the loop in a counterclockwise direction, it will cause us to go from the (−) end to the (+) end of the resistor (the bulb filament). We will elect to call this voltage change negative. We will do the same thing when moving from the (−) to the (+) end of battery #3. For batteries #2 and #1, the situation is reversed, so the signs are also reversed. Applying KVL produces the following:

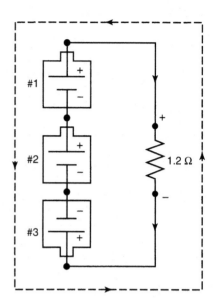

Figure 8–25: Kirchoff's voltage law can tell us about the behavior of a flashlight with one of three batteries placed in backwards. The dashed arrows indicate the direction that we will move around the loop to add up the voltages.

$$+V_{\text{filament}} - 1.5\,\text{V} - 1.5\,\text{V} + 1.5\,\text{V} = 0$$

$$V_{\text{filament}} - 1.5\,\text{V} = 0$$

$$V_{\text{filament}} = 1.5\,\text{V}$$

In other words, the reversed battery cancels the voltage of one of the other batteries. The net effect is as if there were only one 1.5 V battery in the flashlight. If it is known that the resistance of the filament of the light bulb is 1.2 Ω and the voltage across the filament is 1.5 V, the current through it and the power that it dissipates can be calculated.

Series and Parallel Circuits

Two resistors connected "in line," as shown in Figure 8-26, is known as a **series combination**. Imagine this combination of resistors placed inside a sealed box that has one "in" and one "out" electrical contact. If a battery is attached to the contacts, current will flow through the box. Since the box acts sim-

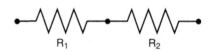

Figure 8–26: Two resistors connected end-to-end is known as a series combination.

ply as a two-terminal device, it would be logical to conclude that there is only a single resistor in the box. If the applied voltage, V, and current, I, that flows are measured, Ohm's law will allow us to determine the resistance, R, of whatever is inside the box according to the following equation:

$$R_e = \frac{V}{I}$$

Because the circuit inside the box contains multiple resistors, what has just been described is the **equivalent resistance** (hence the subscript "e"). The equivalent resistance is based on how much current flows when a voltage is connected. The circuit inside the box may be just a single resistor, two or more resistors connected end to end, or it may be a complicated electronic circuit. It may be of concern to you whether the electric heater or the air conditioner is running in your home, but the power company is only interested in how much current is being used, at what voltage, and at the end of the day, how much electrical energy (power × time) has been consumed. That is why the concept of equivalent resistance is useful.

Since we know what is inside the box, we will use Kirchoff's laws to analyze the circuit. There is no need to apply KCL since there is only one path for the current to take. We will pick a direction and call the current, I. Tracking the voltage through the circuit is also fairly straightforward if we keep track of the signs. We are going from (−) to (+) in the battery, so that adds $+V$ and we're going in the direction of the current through the resistors, so that we get $-IR$ for each of them. Together, Kirchoff's current law looks like

$$+V - IR_1 - IR_2 = 0$$

We can factor out the current, I, from the second and third terms, to get

$$+V - I(R_1 + R_2) = 0$$

$$-I(R_1 + R_2) = -V$$

$$(R_1 + R_2) = \frac{V}{I}$$

This looks like the equation for equivalent resistance! In fact, it is the same, if the sum of the resistances is called the equivalent resistance. The equivalent resistance of a series of two resistors is the sum of the individual resistances. This also works for more than two resistances. If there are three or more resistors, all in series, then the equivalent resistance is given by:

$$R_e = R_1 + R_2 + R_3 + \ldots$$

This is not the only way that two resistors can be connected, however. Suppose that, instead of being connected end-to-end, the resistors are connected as shown in Figure 8-27. This arrangement is known as a **parallel combination** of resistors.

In this case, Kirchoff's current law is needed. Designate the two currents through the individual resistors as I_1 and I_2, respectively. As previously discussed, these currents will be different unless R_1 and R_2 happen to be the same. Looking at the top node in the circuit, $+I$ is entering the node, and I_1 and I_2 leaving it, so that Kirchoff's current law becomes

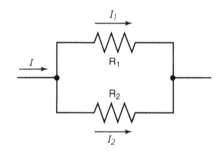

Figure 8–27: If two resistors are connected so they split the current path, the result is called a parallel combination of resistors.

$$+I - I_1 - I_2 = 0$$

or, more simply

$$I = I_1 + I_2$$

Looking at the circuit from the outside, Ohm's law can be applied to determine that $I = V/R_e$. The same thing applies to each resistor individually, $I_1 = V/R_1$ and $I_2 = V/R_2$. Making these substitutions,

$$\frac{V}{R_e} = \frac{V}{R_1} + \frac{V}{R_2}$$

Dividing both sides by V, produces:

$$\frac{1}{R_e} = \frac{1}{R_1} + \frac{1}{R_2}$$

It turns out that this same form of the relationship holds for any number of resistors in parallel, namely

$$\frac{1}{R_e} = \frac{1}{R_1} + \frac{1}{R_2} + \frac{1}{R_3} + \dots$$

One quick relationship that comes from this is in the case where $R_1 = R_2$, so that the equivalent resistance is given by:

$$R_e = \frac{R_1}{2}$$

The equivalent resistance of two identical resistors in parallel is half the resistance of either resistor individually. Another case to consider is if the two resistors are greatly different in resistance. Suppose, for example, a 10 Ω resistor is parallel with a 1,000 Ω resistor. Using the above equation, then,

$$\frac{1}{R_e} = \frac{1}{10\Omega} + \frac{1}{1000\Omega}$$

$$\frac{1}{R_e} = (0.1 + 0.001)\Omega^{-1} = 0.101\Omega^{-1}$$

$$R_e = 9.90\Omega$$

In parallel circuits, the equivalent resistance for the entire circuit is always smaller than that of the branch with the smallest resistance. Although the vast majority of the current flows through the lower-valued resistor, so the presence of the larger resistor in parallel only lowers the total resistance slightly.

AC Circuits

To this point, all of the voltage sources considered have delivered a constant voltage relative to the local ground potential. These sources are very good representations of batteries that power small electronic devices and the electrical systems in automobiles. They are not, however, a good representation of the most commonly used type of voltage: AC electrical sources. The term **AC** stands for alternating current, as opposed to **DC**, which stands for direct current. The fact that the current alternates in an AC circuit is due to the voltage supply. The difference in DC and AC voltage sources is shown in Figure 8-28. In the case of a DC source, the (+) terminal is always at a higher voltage than the (–) terminal. In the case of an AC source, the higher terminal alternates. It depends on where you are in the cycle at that moment.

One of the wires in an AC circuit must be a reference. In the case of the electrical wiring in a house, this reference wire is called the **neutral wire** and is one of the leads

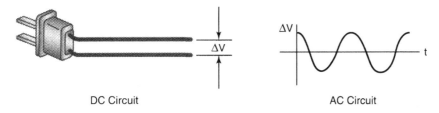

DC Circuit   AC Circuit

Figure 8–28: The difference between DC and AC circuits is that the voltage in an AC circuit is constantly changing.

in a two-pronged electrical plug. The voltage on this wire is not the same as the ground of the circuit, but it is usually close. If the voltage of the other wire is measured relative to the neutral wire, it would be seen to be constantly changing. Depending on when the measurement is made, the voltage might be larger or smaller, positive or negative, with respect to the voltage on the neutral wire. It is not changing randomly, however, but in a well-defined sinusoidal pattern, (see the AC circuit in Figure 8-28) so that the voltage is given as a function of time by the equation

$$V = V_0 \sin (\omega t)$$

In this case, the angular frequency, ω, is in radians per second. Since there are 2π radians for every complete cycle of this oscillation, the frequency of the oscillation is given by:

$$f = \frac{\omega}{2\pi}$$

In the United States the electrical power used is $\omega = 120\pi$ radians/second, so that the frequency at which the voltage on the AC line varies is $f = 60$ Hz. There is nothing magic about this value; it is just the agreed-upon frequency that power plants in this country use. In Europe, for example, 50 Hz AC power is common, while on aircraft, 400 Hz power is the standard for AC systems.

Look again at the equation $V = V_0 \sin (\omega \cdot t)$. It provides an indication of what voltages can be expected. The sine function never gets larger that +1, indicating that $+V_0$ is the greatest positive voltage that can be expected. The opposite extreme of the sine function is –1, so $-V_0$ is the most negative voltage that can be expected. Twice every cycle, the voltage between the two wires in the AC system will be zero.

Most residential power systems in the United States are 120 V AC. In heavier industrial applications, 208 V, 220 V, and 480 V systems are often used. Therefore, if the voltage between the two leads in the wall were measured, +120 V would be expected to be the greatest positive value, and –120 the greatest negative. This assumption would be reasonable, but wrong! The greatest voltage is actually almost +170 V because 120 V is the **root mean squared voltage** of the circuit.

The root mean squared voltage, or **RMS voltage**, is a type of average appropriate for sinusoidally-varying signals. The AC circuit in Figure 8-28 shows why a straight average would not be appropriate. Since the voltage is positive as often as it is negative, the average of the voltage is, therefore, zero. Calculating the root mean squared voltage, involves exactly what the name implies. First, each voltage is squared (making them all positive); second, the average of their sums is determined; and finally, the square root of the average is taken. This can be done for any repeating waveform, not just sine waves. The RMS result, for a sine wave in particular, is always the same relative to the peak value or amplitude of the signal. In this case the peak value, V_0, equals 170 volts.

$$V_{RMS} = \frac{V_0}{\sqrt{2}}$$

$$V_{RMS} = \frac{170 \text{ volts}}{\sqrt{2}} \cong 120 \text{ volts}$$

The RMS voltage is useful because the average power, $P_{average}$, consumed by a resistor connected to an AC voltage is given by:

$$P_{average} = \frac{V^2_{RMS}}{R}$$

This equation allows us to determine that a 60 W light bulb uses, on average, 60 W of power. If you followed its power consumption, you would find that there are times when it would be dissipating 120 W and other times when it would be 0 W. Since light bulbs are usually on for many cycles, the average power consumption is the measurement of interest.

Check Your Understanding

1. What assumption is made about the resistance of a wire in a circuit diagram?

2. What is the meaning of the dot and the zigzag line in a circuit diagram?

3. What is the symbol used for the electrical ground in a circuit diagram?

4. What is a node, and what is true about the sum of the currents into a node?

5. What is the name of the law that makes the statement that the sum of the voltages around any loop in a circuit is zero?

6. If you are moving from the (+) to the (−) side in a battery, is the voltage positive or negative?

7. In a series combination of resistors, are the resistors connected "in line" or "in parallel?"

8. What is the frequency, in cycles per second, of electrical power used in the United States?

8-5 Additional Circuit Components

Objectives

Upon completion of this section, you will be able to:

■ Identify resistors and capacitors in an electrical circuit.

■ Identify basic semiconductor devices in an electrical circuit.

Resistors and batteries are the most basic elements of electrical circuits. The resistors in a circuit are known as dissipative elements because they turn electrical energy into heat energy. In other words, it is the resistance of a circuit that determines how much power it uses. All circuit components have some amount of resistance, so that all circuits dissipate electrical energy at a certain rate. In an energy-hungry society, that aspect of electronic circuits is vital to understand. If you look inside your stereo amplifier or computer, however, you will see many components other than resistors. Some will be capacitors, there may be a few inductors and many will be semiconductor devices, such as integrated circuits (ICs) or chips. Capacitors and inductors are two devices that can be made out of metal, in the appropriate shape.

Capacitors

The **capacitors** on a circuit board look like small metal cans or flat disks with two wire leads. Their function is to store electrical charge. Capacitors are nothing more than two metal plates placed close together, as shown in Figure 8-29. These plates may be small or large. The plates inside the metal cans are made up of thin metal foil separated by thin plastic or paper. These thin metal foils are rolled up so that a large area can be packed into a small space.

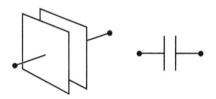

Figure 8–29: Any two plates of metal placed near each other will form an electronic device called a capacitor.

A capacitor is another two-terminal device. The **capacitance,** C, of a capacitor is measured in farads, F. A farad is a large unit, so capacitors in a typical electrical circuit are more likely to be measured in microfarads, μF. The capacitance depends on the area of the metal plates and the distance between them, according to the formula:

$$C = \varepsilon_r \varepsilon_0 \frac{A}{L}$$

where A is the area of the plates, L, is the distance between them, and ε_r is the relative dielectric constant, which is 1 for a vacuum and very close to 1 for air. The value of ε_0, known as the permeability of free space, is

$$\varepsilon_0 = 8.8542 \times 10^{-12} \text{ F/m}$$

The dimensions for both sides of the above equation for calculating capacitance, C, are in farads. This must be true if both sides of the equation are to balance. You can also see why microfarads are a common unit – for a reasonable set of dimensions for length and area – since the value of C will be small. To make for a high capacitance in a small package, the space between the plates is usually filled with a material like plastic that has a high relative **dielectric constant**; on the order of 2 to 4. Other possible dielectric materials are oils that are manufactured for this particular property.

A capacitor stores electrical charge by collecting an excess of positive or negative charge on the metal plates. As an excess of charge accumulates, the difference in voltage between the plates increases. The situation is identical to when we first started discussing the idea of different voltages, except that now the charges are not just hanging in space – they are located inside a physical material; the metal of the capacitor plates. The voltage difference between the plates depends on the amount of charge that is stored in the capacitor, and is given by

$$V = \frac{Q}{C}$$

Once again, the symbol for change, the Greek letter delta (Δ), is missing because all voltage measurements are relative. The charge, Q, stored by a capacitor must be provided by a flow of electrical charge or current onto the plates. An earlier equation illustrated that $I = \Delta Q/\Delta t$. This equation can be rearranged so that the charge, ΔQ, on a capacitor is given by $\Delta Q = I\Delta t$, if the current is constant. As the charge on the capacitor plates increases, so does the voltage between the plates, according to the equation $V = Q/C$. This indicates that it takes time to charge a capacitor. For this reason, capacitors are often used in circuits for timing elements. The reverse logic also holds – it takes time to discharge a capacitor. The small capacitors on the circuit board in a typical computer act as charge reservoirs. These reservoirs help smooth out variations in the voltages that are applied to the integrated circuits. In fact, some computer manufacturers have recently started using large-value capacitors (1–10 Farads) to store enough charge to take the place of batteries on motherboards. These capacitors eliminate the problem of diminishing storage capacity after many discharge/recharge cycles. Resistors are the opposite of capacitors because resistors use up, or dissipate, energy.

It is worth noting that large capacitors can store significant (lethal) amounts of electrical charge, even after the electrical power to a circuit has been removed. Even if the important safety rule of unplugging an electrical circuit before working on it is followed, caution must be exercised if it contains a large capacitor.

Inductors

The other circuit component that can be made out of metal is the **inductor**. An inductor is nothing more that a loop (usually many, many loops) of wire, as is shown in Figure 8-30.

Like capacitors, inductors store energy. However, capacitors store energy in the electric field between metal plates, whereas inductors store energy in the magnetic field that is created as the current flows through the loops of wire. (A magnetic field exists around any wire that is carrying a current.) In an induc-

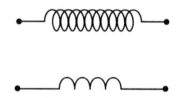

Figure 8–30: Any coil of wire (top) makes up a circuit element called an inductor. Usually, inductors are made up of many coils of wire. The circuit symbol for an inductor is shown (bottom).

tor, looping a long wire into a coil allows it to be packed into a small volume, greatly increasing its local magnetic field strength.

Inductors, like capacitors, are two-terminal devices. There is one critical difference between capacitors and inductors: there is a voltage between the terminals of an inductor only when the current through the inductor is changing. Whether there is no current or hundreds of amps, if the current through is constant, there is no voltage difference across the inductor.

The SI unit of **inductance** is the henry, H. Like the farad, the henry is a large unit, so for many applications the microhenry, μH, is a more common unit. Like capacitance, inductance depends on the geometry of the inductor, and is given by

$$L = \mu_r \mu_0 \frac{N^2 A}{L}$$

where A is the cross-sectional area of the wire loops, N is the number of loops, L is the length of the inductor, and μ_r is the relative permittivity, which – like the relative dielectric constant – is 1 for a vacuum and very close to 1 for air. The value of μ_0, known as the permittivity of free space, is

$$\mu_0 = 4\pi \times 10^{-7} \text{ H/m}$$

(The factor of 4π may seem unusual, but it is exact.) Because inductance increases as the square of the number of turns of wire, it is common for inductors to be made with hundreds or thousands of turns. Just as plastic is often placed between the plates of capacitors to increase the capacitance in a small package; metals like iron, with high relative permittivity, are often used in the center of inductors to increase their inductance. In fact, the simplest method is to wrap the wire directly around the so-called magnetic core. To prevent the wire from shorting out, special thin plastic coated or enamel insulation coated **magnet wire** is used. The increase in inductance between magnetic-core and air-core inductors can be substantial, with relative permittivities on the order of 1,000 to 100,000.

The voltage between the terminal of the inductor is given by the relationship

$$V = L\frac{\Delta I}{\Delta t}$$

There are a couple of useful applications of this fact. A transformer is essentially two, coupled inductors. When subjected to an AC power source, one inductor can induce a current in the other. Because this current is constantly changing, the voltage across the second inductor is changing, as well. By adjusting the number of turns in the wire in each inductor, it is possible to "step-up" or "step-down" the second voltage. Most electronic components have transformers built into them.

Electric motors use inductors. In some devices, such as refrigerators and air conditioners, the motor is turned on and off many times. Whenever the current in the motor is suddenly shut of, the voltage across the inductance of the motor increases for a short period of time.

For example, an inductance of 0.032 H would be typical for a 10 A motor running a refrigerator compressor. If the motor turned off in 1/1,000 s, then the voltage across the inductor would rise to:

$$V = 0.032 \text{ H} \times 10 \text{ A}/0.001 \text{ s} = 320 \text{ V}$$

This is almost three times the average (RMS) voltage of the AC line. Surge suppressors are devices that protect electrical components by only conducting when the voltage exceeds that normally present on the AC line, so that they short-circuit such voltage "transients" before they can do damage to sensitive electrical equipment like computers. (See Technology Box 8-7.)

Physical Science: What the Technology Professional Needs to Know

Technology Box 8–7 ■ How do you block inductive kicks?

The voltage defined by the equation, $V = L\,\Delta I/\Delta t$ is not merely a bit of mathematical trivia – it can be a real problem when the currents involved are large. This is the case in many industrial-scale motors and even the motors in household appliances. The trouble occurs when the switch tries to interrupt the current to the motor quickly. The voltage induced by this change in current can appear along the entire power line to which the motor is attached. Since it appears only for a short time, it is often called a "**spike**" in the line because of the way the transient voltage would appear on an oscilloscope trace. Although it is short-lived, it can be devastating to sensitive equipment like computers.

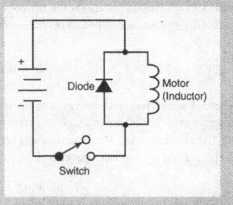

A snubber diode can help to keep the inductive voltage spike that occurs when a motor is shut off quickly from affecting other equipment.

Fortunately, the one-way switching characteristic of a diode provides a possible solution. Placing a diode in line with the motor as shown in the figure above does the trick. When the motor is running, the diode is reverse-biased (in other words, in backwards) so that it does not draw any current. When the switch opens up the circuit, the diode provides a current path that is local to the motor, so the voltage spike doesn't affect nearby equipment. Such diodes are often called **snubber diodes** and are highly recommended whenever it is necessary to switch an inductive load.

Semiconductor Devices

Many components in modern electronic devices are integrated circuits or ICs, so-called because they contain many individual functions in a single package. The variety of devices that are available is staggering. There are toys with built-in computing power today, that were only a dream to the *Apollo* engineers. Despite their wide variety of functions, these devices are all built with semiconducting materials based mostly on silicon, but increasingly using other materials such as gallium.

The use of semiconductors in electronics is a perfect example of an application in which the line between chemistry and physics blurs. The fact that a material is a poor conductor is not particularly interesting. Semiconductors become useful, however, when made with controlled imperfections. Figure 8-31 (left) shows a two-dimensional representation of a section of a perfect crystal. Each atom shares electrons with the atoms near it. Suppose that one of these atoms is replaced with an atom that has one less electron in its outer shell, as in Figure 8-31 (center). The three electrons can bind to surrounding atoms, but there is a "hole" where the fourth bond should be. The chemistry of the imperfect crystal has modified its electronic properties. Such a material with "missing" electrons is called **p-type** material. The absence of an electron is equivalent to a positive charge carrier. The non-silicon atoms that are introduced into a crystal are called **dopants**. It is also possible to introduce dopants with too many electrons in the outer shells producing **n-type** materials as in Figure 8-31 (right). The extra negatively charged electrons are not as tightly bound, so are more mobile.

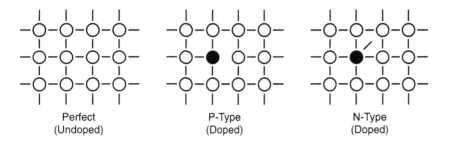

Perfect
(Undoped)

P-Type
(Doped)

N-Type
(Doped)

Figure 8–31: In a perfect crystal (left), each atom shares electrons with other atoms around it. In a p-type doped crystal (center), impurity atoms are introduced that have too few electrons in their outer shells creating "holes." In an n-type material (right), the impurity atoms form bonds with the surrounding atoms, but have "leftover" electrons that are free to move.

A common misconception about "doped" semiconductors is that the doping makes them electrically charged. This is not true. Materials of the n-type have an excess of free electrons, but not an overall excess of electrons. The dopant atoms, like the other atoms in the crystal, have as many electrons as protons, so the net charge is zero. These electrons can, however, move more easily than those being shared between atoms.

Doped semiconductors become useful when different types are stacked together. The simplest stack is one layer of n-type and one layer of p-type. This makes a device known as a **diode**, which is shown with its associated circuit symbol in Figure 8-32. In a diode, unlike a resistor, current can only flow in one direction; a property that is useful in circuit design.

A **transistor** is three layers of semiconductor material, stacked in one of two possible arrangements: an n-p-n stack or a p-n-p stack. The transistor, as shown in Figure 8-33, is the heart of the digital computer. The transistor has three terminals. The current that flows through the device, between the two outermost terminals, is controlled by the current into the third. If the base current is zero, then the current through the device is zero. In a computer these devices are used as switches and are either "on" or "off." Because these devices have two distinct states, they can be used to represent and manipulate the **binary logic** of the computer, in which the only two digits allowed are 0 and 1. The microprocessor, which is at the heart of not only the computer but the information revolution, is built by arranging literally millions of transistors on a single chip.

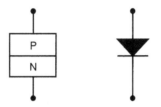

Figure 8–32: A stack of n-type and p-type doped semiconductor material forms a two-terminal device known as a diode. The symbol for a diode is shown to the right.

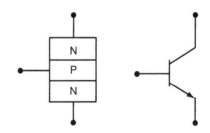

Figure 8–33: An n-p-n or p-n-p stack of n-type and p-type doped semiconductor material forms a three-terminal device known as a transistor. The symbol for a transistor is shown to the right.

A transistor can also be used with a base current that varies over a wide range. In this mode, it operates as an amplifier. Its output is called an **analog** signal because, unlike a digital signal that has a limited number of possible values, this output can vary continuously over a given range. Some transistors, such as those used in stereo amplifiers are physically large, because they are intended to handle a large amount of power. There are other applications, notably cellular telephones, where speed and low power usage are desirable characteristics. In these applications, new semiconductor materials, such as gallium, are replacing silicon.

The Future of Electronic Devices

The field of electronic devices is changing rapidly. Manufacturers of electronic components are making incremental as well as revolutionary improvements in devices ranging from the humble resistor to such cutting-edge technologies as **quantum dots** that rely on the quantum-mechanical properties of individual electrons. One promising area of research is the field of micro-electromechanical devices, or **MEMS** for short. Such devices use fabrication techniques that have been developed for electronic semiconductor devices to build miniaturized gears, levers, and other mechanical devices on a scale so small that these devices can only be seen under a microscope. Figure 8-34 shows a microscope picture of such a device. The next logical step is to combine the operation of mechanical and electrical components on a

Figure 8–34: A microscopic set of gears built with semiconductor fabrication technology is an example of a micro-electromechanical or MEMS device. Photo courtesy Sandia National Laboratories: Intelligent Micromachine Initiative; www.mems.sandia.gov.

single substrate. One of the first commercially available devices to make this step was an accelerometer (a device that measures acceleration) on a chip marketed by Analog Devices, Inc. This device measured acceleration by monitoring the bending of a set of microscopic beams that were built into the chip at the same time the semiconductor devices on the chip were manufactured.

To do more than scratch the surface of this exciting field is beyond the scope of this text. Nevertheless, the basic principles of voltage, current, and resistance play a role in even the most advanced device technologies. This is because these principles are fundamentally related to the energy that it takes to operate an electronic device. The desire to do more with less – specifically, to have available more computing speed and more communications bandwidth, while using less electrical power – is the driving force behind many of the advances in electronic devices.

Check Your Understanding

1. What is the composition of the interior of a capacitor and what is its function?

2. What is the general design of an inductor and what is its function?

3. What is the SI unit for inductance, and does it tend to be a large or small unit?

4. What is a dopant?

5. What is the composition of a diode and what function makes it useful?

6. What type of device is formed by semiconductor material stacked as either n-p-n or p-n-p?

8–6   **Magnetism**

Objectives

Upon completion of this section, you will be able to:

- Describe the nature of magnetic fields and forces.

- Calculate the magnitude and direction of a magnetic force acting upon a moving charge.

- Recognize the differences between electric and magnetic fields and forces.

- Calculate the magnitude and direction of a magnetic force acting upon a current-carrying conductor.

- Calculate the torque acting upon a current-carrying conductor.

- Apply Faraday's law of induction.

- Understand the nature of the magnetic field of the Earth.

Magnetism is among the oldest curiosities of science. More than 2,000 years ago it was known that a naturally occurring iron ore called magnetite (Fe_3O_4), or lodestone, possessed the mysterious property of attracting bits of iron. When a needle was rubbed with magnetite, it also became magnetic. In approximately 1,000 AD the Chinese found that when such a needle was freely suspended, it always took a position with its axis along a north and south line and that the same end always pointed north.

The resulting technology is the compass, and its use soon spread to Europe. Columbus used a compass when he crossed the Atlantic Ocean, noting that the needle deviated slightly from exact north, as determined by the stars. About 1600, William Gilbert, physician to England's Queen Elizabeth I, proposed that the Earth itself was a giant magnet. He further suggested that its magnetic poles were some distance from the Earth's axis of rotation.

Over the years, scientists have learned how to harness and control magnetism, resulting in many useful technologies that continue to enhance our lives. The way we generate and transmit electrical energy today became feasible only after the relationships between the laws of magnetism and the phenomena of electric currents were discovered. It is, therefore, appropriate to conclude our investigation of the electrical nature of matter with a study of magnetism.

"A wonder of this kind I experienced as a child of 4 or 5 years when my father showed me a compass. That this needle behaved in such a determined way did not at all fit into the kind of occurrences that could find a place in the unconscious world of concepts. . . . I can still remember – or at least believe I can remember – that this experience made a deep and lasting impression on me. Something deeply hidden had to be behind things."

Albert Einstein: Philosopher-Scientist, edited by Paul Arthur Schilpp (Library of Living Philosophers, Evanston, Illinois, 1949)

Magnetism is a physical phenomenon characterized by fields associated with moving electricity, and exhibited by both magnets and electric currents. Many of the conveniences of modern life operate only because of the presence of, and our ability to control, magnetic fields and magnetic forces. **Magnetic fields** are present in the portion of space near a magnetic or current-carrying conductor in which the magnetic forces can be detected. A **magnetic force** is due to the presence of a magnetic field produced by charges in motion. Electric motors, television and computer screens, microwave ovens, and stereo speakers are common examples of devices that are designed around the phenomenon we know as magnetism. The ability to generate a magnetic field, and to exert a magnetic force on an object or very small particle, makes these modern technologies possible.

The effect of the Earth's magnetic field on a compass needle is a familiar concept. If the compass is moved in a straight line, the needle holds steady, but if the direction of movement changes, the needle pivots to remain aligned with the Earth's magnetic field. Of concern at this time is the effect of the Earth's magnetic field rather than its source.

All materials develop some degree of magnetization when they are exposed to a magnetic field. Only those materials that develop enough magnetization to be used for engineering applications, however, are considered magnetic. Additionally, magnetic materials are often classified as either soft or hard, where a **soft magnetic material** can be magnetized and demagnetized easily, but a **hard magnetic material** resists changes in its magnetization. Hard magnetic materials are often used for permanent magnets. A material falling between hard and soft is classified as an **intermediate magnetic material** and may be used as recording media for analog and digital information such as audio, video, and computer storage.

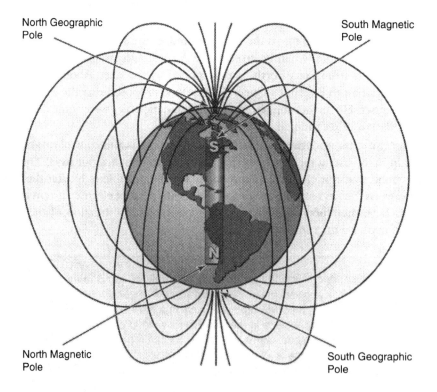

Figure 8–35: The magnetic south pole of the Earth is located in the northern hemisphere near the north geographic pole.

Magnets attract and repel each other like electric charges, but are they the same thing? Magnets have two poles, designated north (N) and south (S), from their use as compasses. The end of a compass needle that points northward is defined as the **north seeking pole (north pole)** of the needle and the opposite end is defined to as its **south seeking pole (south pole)**. Note in Figure 8-35 that the Earth's geographic and magnetic poles do not coincide – more about this later.

We also know that when two magnets are brought close together, the opposite poles attract and the like poles repel. It is an interesting bit of trivia to note that since the north pole of the compass needle points toward the Earth's geographic north pole, it must actually be the south magnetic. In other words, the areas of the globe we have always known as the north and south poles are, in fact, the magnetic south and north poles respectively!

In fact, the magnetic forces of attraction and repulsion between magnetic poles are similar to the electrical force interaction between positive and negative electrical charges. There are also striking differences between the two, as will soon be seen.

Magnetic Fields and Forces

An electric charge at rest creates an electric field in the surrounding space. The electric field, in turn, exerts a force on any other charge that is present in the field. A **field**, therefore, is a region in space where objects are influenced by an external force. For instance, the Earth's gravitational field applies a force to any mass within its reach. Since your body has mass, you are being held firmly to the Earth by gravitational force.

On the other hand, an electric current is a stream of charges in motion that creates a magnetic field in the surrounding space, as well as an electric field. The resultant magnetic field exerts a force on any other moving charge or current present in the field. Magnetic fields are vector fields. That is to say, a vector quantity is associated with each point in space that describes the magnitude and the direction of the field at that point. Magnetic fields are, therefore, represented by the symbol $\vec{B}$.

A convenient way to visualize a field is to consider how it acts on a test object. For example, consider how a moving, positively charged particle is affected by the presence of a magnetic field. In Figure 8-36 a positively charged particle, velocity $\vec{v}$, is moving out of the page in a uniform magnetic field, $\vec{B}$, that points to the right. An unexpected aspect of the magnetic force is that it acts on a moving charge in a direction that is perpendicular to both the velocity and the field vectors. The magnitude of this force is proportional to that of the charge, its velocity, and the field.

Consider a single charge moving through a magnetic field, for example. If you were to suddenly double the magnitude of the charge or its speed, or the strength of the magnetic field acting upon it – the magnitude of the magnetic force would also double.

Finally, the magnitude of magnetic force depends upon the angle between

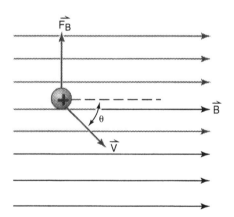

Figure 8–36: A positively charged particle with velocity $\vec{v}$ moving in uniform magnetic field $\vec{B}$ and experiencing force $\vec{F}_{B}$.

the magnetic field vector and the velocity vector. This angle is denoted by θ as shown in Figure 8-36. The mathematical relationship relating magnetic force to all of the above parameters is presented in the following equation:

$$F_B = |\vec{F_B}|$$

$$F_B = qvB \sin \theta$$

in which q, v, and B represent charge, speed, and magnetic field strength respectively. Note that the direction of $\vec{v}$ and $\vec{B}$ are not considered directly; the angle θ accounts for the orientation between these vectors. The term $B \sin \theta$ is the component of the magnetic field vector $\vec{B}$ that is perpendicular to the velocity and, therefore, the component of the field responsible for magnetic force.

The standard SI unit of magnetic field strength is the **tesla,** T, in honor of Nikola Tesla, the Serbian-American scientist and inventor. The tesla is related to other SI units in that 1 tesla equals 1 newton second per coulomb meter, or stated mathematically:

$$1.0 \text{ T} = 1.0 \frac{N \cdot s}{C \cdot m}$$

It can be easily shown that the product of the four terms on the right-hand side of the equation, $F_B = qvB \sin \theta$, results in the unit newton, the SI unit of force. A secondary unit of magnetic field strength is the **gauss,** G, named in honor of Carl Friedrich Gauss (1777–1855) which equals 1×10^{-4} tesla ($1 \text{ G} = 10^{-4}$ T). **Gaussmeters** are instruments used for measuring magnetic fields.

Consider the example illustrated in Figure 8-37 in which a proton moves with a speed of 8.0×10^6 m/s along the x-axis. It enters a region in which a uniform magnetic field of strength 2.5 T exists in the x-y plane, directed at an angle of 60° with respect to the x-axis. What is the initial magnetic force acting on the proton?

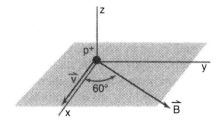

Figure 8–37: Magnetic force acting upon a proton in motion.

Using the above equation, the magnetic force can be found by making the following substitutions:

$$F_B = qvB \sin\theta$$

$$F_B = (+1.6 \times 10^{-19}C)(8.0 \times 10^6 \text{ m/s})(2.5T) \sin 60°$$

$$F_B = (+1.6 \times 10^{-19}C)(8.0 \times 10^6 \text{ m/s})(2.5T)(0.8660)$$

$$F_B = 2.8 \times 10^{-12} \text{ N}$$

Once the magnitude of $\vec{F_B}$ has been calculated, the direction of this vector can be determined by using the **right-hand rule**. When using this method you point the fingers of your right hand in the direction of the velocity vector, curl your fingers into the direction of the magnetic field vector, and observe the direction in which your thumb is pointing. This direction will coincide with the direction of the magnetic force acting upon a positive charge. Note that the direction of the force applied to a negative charge would be exactly the opposite.

Re-examine Figure 8-37. If you were standing behind the figure with your right hand extended as shown in Figure 8-38 (left) as if shaking someone's hand, your fingers would be pointing in the direction of the velocity. As you curl your fingers in the

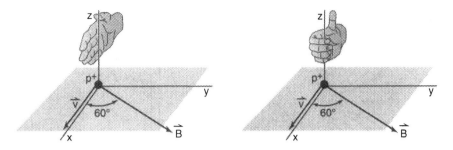

Figure 8–38: The right-hand rule. If you were standing behind the figure (left) with your right hand extended as if shaking someone's hand, your fingers would be pointing in the direction of the velocity, $\vec{v}$. As you curl your fingers in the direction of the field, $\vec{B}$, (right) your thumb would be pointing in the direction of the force vector.

direction of the field (right), in which direction would your thumb be pointing? It would point up, in the direction of the positive z-axis. The proton therefore experiences a force of 2.8×10^{-12} N upward, or in vector notation, $\vec{F}_B = (2.8 \times 10^{-12}$ N$)\,\underline{k}$, where $\underline{k}$ is a unit vector that signifies only direction. Again, if the answer calculated were negative, the direction indicated by the right-hand rule would be reversed.

It is important to observe the limiting cases with respect to magnetic fields and forces. The magnetic force depends upon the angle θ between the directions of the velocity and magnetic field vectors. Therefore, the magnitude of magnetic force, F_B, acting upon a moving charged particle or object equals zero when $\vec{v}$ and $\vec{B}$ are parallel ($\theta = 0°$ or $180°$) and F_B has a maximum value, qvB, when $\vec{v}$ and $\vec{B}$ are perpendicular ($\theta = 90°$).

We can now address the question posed at the beginning of this section and identify the very distinct differences between electric and magnetic forces that act upon charged particles or objects. The electric force is independent of the speed of a charged particle, but the magnetic force acts only on such a particle when it is in motion and is proportional to its speed. The electric force acts in, or opposite to, the direction of the electric field. The magnetic force, however, always acts in a direction perpendicular to the magnetic field.

Magnetic Field Lines and Flux

It is sometimes difficult to visualize a magnetic field, or to characterize the influence it exerts on a charged particle in motion. A simple visual tool and several analytical concepts follow to assist your understanding.

Magnetic fields are most easily visualized using **magnetic field lines**, such as those shown encircling the globe in Figure 8-35. These lines are drawn such that each line, as it passes through any point in space, is tangent (meets the curve) to the magnetic field vector, $\vec{B}$, at that point. Magnetic field lines also agree with the orientation of a compass needle placed at that point. This fact alone may help you better visualize the nature of magnetic field lines.

Typically, only a few representative field lines are drawn to avoid completely filling a space, but the convention is such that the strength or magnitude of a magnetic field is indicated by the density of field lines drawn. In other words, the closer together the lines, the stronger the field represented.

Magnetic field lines produced by several common sources of magnetic fields are shown in Figure 8-39. The first image is of a bar magnet; note that field lines always originate at the north pole and end at the south pole. The second image is of the mag-

Magnetic Field Lines

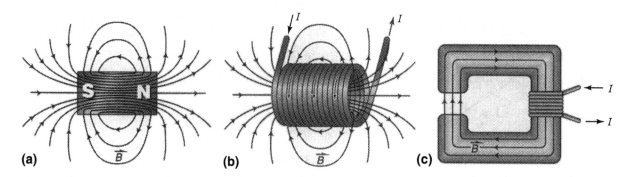

Figure 8–39: Magnetic field lines in a plane through the center of (a) a permanent magnet, (b) a cylindrical coil, (c) an iron-core electromagnet.

netic field lines generated by an electric current passing through a coil of wire. The last image is of an **electromagnet**. Note that between the poles of the electromagnet (Figure 8-39c) the field lines are nearly straight, parallel, and equally spaced. This indicates that the magnetic field in this region is nearly uniform.

Magnetic Flux

The term flux has many meanings. When discussing magnetic flux, however, it is necessary to think of it as a rate. For example, raindrops pass through an outdoor basketball hoop when it is raining. The number of drops passing through the hoop each second divided by the area of the hoop is its rain flux as shown in Figure 8-40.

Magnetic flux is the product of the surface area defined by a closed loop and the average component of the magnetic field that is perpendicular (normal) to that surface. A uniform magnetic field is shown in Figure 8-41. When the uniform magnetic field, $\vec{B}$, passes through an area, A, the magnetic flux, Φ_B, is given by

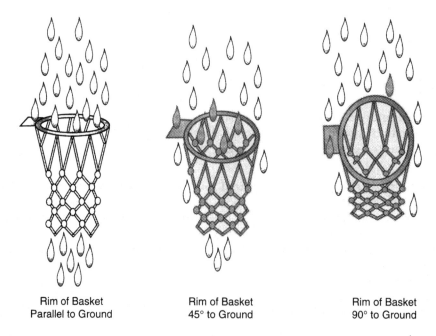

Rim of Basket
Parallel to Ground

Rim of Basket
45° to Ground

Rim of Basket
90° to Ground

Figure 8–40: The most rain drops fall through a basketball hoop the area of which is perpendicular to the rain fall. If the hoop is gradually rotated, the rain flux will decrease since there will be less area through which the rain drops may pass.

$$\Phi_B = BA \cos \phi$$

in which ϕ is the angle between the direction of the field and the normal to the surface of the area. The normal is simply the unit vector that is perpendicular to the area, and the cosine function accounts for the fact that the flux is largest when the field lines are parallel to the normal. To understand why this is true, try slowly turning a coin standing on its edge. Doesn't the coin also appear largest when its head or tail faces you directly?

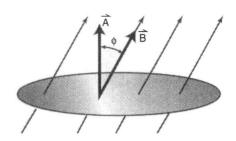

Figure 8-41: When a uniform magnetic field, $\vec{F}_B$, passes through a circular area, A, the magnetic flux, Φ_B, is equal to $BA \cos\phi$.

The SI unit of magnetic flux is the unit of magnetic field, T, times the area (m^2). This unit is called the **weber,** Wb, in honor of the German physicist Wilhelm Weber.

$$1 \text{ Wb} = 1 \text{ T} \cdot m^2$$

From the equation, $\Phi_B = BA \cos \phi$, it can also be shown that the maximum magnetic flux through an area occurs when the field is perpendicular to the area through which it passes. At this maximum, the magnetic flux, Φ_B, equals BA, because $\cos 0° = 1$. Again, consider the example of raindrops passing through the basketball hoop. The rate of raindrops passing through the hoop is greatest when the hoop is parallel to the ground. If the hoop is gradually rotated, the rain flux will decrease since there will be less area through which the raindrops may pass. When the hoop reaches a point where it is perpendicular to the ground the rain flux would be zero. In a similar way it can be shown that when $\phi = 90°$, $\cos 90° = 0$, therefore, the magnetic flux, Φ_B, would also be zero.

It is important to note that the calculation of magnetic flux is considerably more complicated when a non-uniform magnetic field is used. However, the techniques described here provide the means to analyze real problems in which the assumption of a uniform magnetic field is reasonable.

Magnetic Force on a Current-Carrying Conductor

Earlier in this chapter, it was explained that an electric current is simply a collection of charged particles in motion; that is, a stream of electrons moving through a wire. Above, it was explained that a charged particle, when passing through a magnetic field, experiences a force in a direction that is perpendicular (normal) to both the velocity of the particle and the direction of the field. It stands to reason, therefore, that a current-carrying wire also experiences a force when placed in a magnetic field. This is due to the net force acting upon all the electrons in motion that make up the current.

The relationship between magnetism and electricity was first observed in 1819 by the Danish scientist Hans Christian Oersted when, during a lecture demonstration, he found that an electric current in a wire deflected the needle of a nearby compass. As shown in Figure 8-42, compasses placed near a wire carrying no current are affected only by the magnetic field of the Earth. When a current is applied to the wire, however, it induces a magnetic field that is much stronger than the Earth's. Looking at the figure, it is apparent that the induced field lines must encircle the wire. Through the efforts of many scientists since that time, the intimate relationships between electricity and magnetism have been further developed and mathematical expressions relating the two derived.

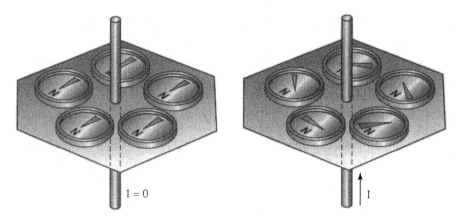

Figure 8–42: There is no effect on a compass placed near a wire that is not carrying a current (left). When a current is applied to the wire, it induces a magnetic field whose field lines encircle the wire (right).

Technology Box 8–8 ■ The Loudspeaker: Magnetic force in action

A common application of the magnetic force acting upon a current-carrying wire is found in a loudspeaker. The speaker is constructed of several components, but at its foundation is a permanent magnet that provides a steady magnetic field.

The speaker functions by having an amplifier deliver a varying electric current across a moveable coil of wire within the speaker. As the current alternates through the coil, the magnetic field created by the permanent magnet exerts a force on the coil that is proportional to the current in the coil. Therefore, the direction of the force is either to the left or to the right, depending upon the direction of the current.

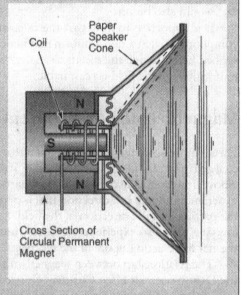

The signal from an amplifier causes the current to oscillate with varying frequency and amplitude, thus causing the coil to oscillate in a similar fashion. The coil is attached to the speaker cone, so it also oscillates in a similar way, thus generating pressure waves in the air that our ears detect as sound.

Most speakers can reproduce a wide range of loudness (amplitude) and pitch (frequency), not only of a single tone, but also for the harmonics of that tone. This capability is important so the loudspeaker can reproduce the qualities that distinguish the sounds of many different voices and instruments.

The following equation provides a straightforward means to determine the magnetic force acting upon a wire in a magnetic field through which an electric current flows.

$$F_{wire} = \left| \overrightarrow{F_{wire}} \right|$$

$$F_{wire} = ILB \sin \theta$$

In this equation, I represents the current and L the length of the wire. The angle θ defines the orientation between the direction of $\vec{L}$ and $\vec{B}$. Note that L only represents the length of the wire, but the vector $\vec{L}$ also incorporates a direction component that describes the orientation of the wire. In a fashion similar to what we saw earlier, the greatest magnetic force on a wire occurs when the wire is oriented perpendicular to the field, and no magnetic force occurs when the wire and the field are parallel. As before, the term $B \sin \theta$ represents the component of the magnetic field vector, $\vec{B}$, that is perpendicular to the wire and, therefore, the component of the field responsible for the magnetic force acting on the wire.

Note that in most textbooks, the symbol ⊗ implies a field line going into the page and a circle with a dot in the middle, ⊙, represents a field line coming out of the page. You can easily remember this by the × representing the tail of an arrow as it flies away from you and the dot, •, representing the tip of an arrow coming toward you.

Consider the example illustrated in Figure 8-43. A wire passes a current upward through a region in which a uniform magnetic field points into the page. The square region in which the magnetic force is present has an area equal to 1 m². The wire is carrying a current of 4.5 A. What is the magnitude and the direction of the magnetic force acting upon the wire?

It is possible to calculate the magnitude of the force, noting that the wire and field are perpendicular. This makes the angle θ equal to 90°. Note that the tesla unit was expanded in this solution to show that the primary SI units cancel, leaving only the newton as the unit of force.

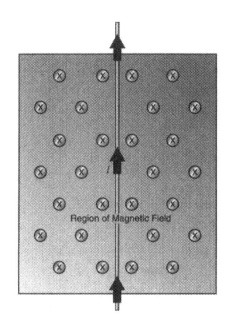

$F_{wire} = ILB \sin\theta$

$F_{wire} = (4.5\ A)\ (1\ m)(0.75\ T)(\sin 90°)$

$F_{wire} = (4.5\ A)\ (1\ m)(0.75\ T)(1.00)$

$F_{wire} = 3.38\ \dfrac{\cancel{C} \cdot N \cdot \cancel{m} \cdot \cancel{s}}{\cancel{s} \cdot \cancel{C} \cdot \cancel{m}}$

$F_{wire} = 3.38\ N$

Figure 8–43: A wire passes a current upward through a region in which a uniform magnetic field points into the page. The square region in which the magnetic force is present has an area equal to 1 m². The wire is carrying a current, I, of 4.5 A.

To determine the direction of the magnetic force, we again apply the right-hand rule. Simply point your fingers in the direction of the current, curl your fingers into the direction of the field, and the thumb will point in the direction of the magnetic force. In the example illustrated in Figure 8-43, you must place the fingers of your right hand pointing up and your palm must be facing the page so that you can curl your fingers toward the field. Consequently, your thumb will be pointing left, indicating that this is the direction of the magnetic force acting upon the wire.

Torque on a Current-Carrying Conductor

To this point, our discussion has moved from considering the magnetic force acting on a single moving charge, to that of a collection of moving charges or an electric current. Next, we will consider the effect exerted on a current loop when placed in a magnetic field. An understanding of this phenomenon will provide an explanation for several important technologies upon which we depend daily.

Consider the situation depicted by Figure 8-44a. A current, I, is passed through a rectangular loop of wire that has been placed in the presence of a magnetic field, $\vec{B}$, oriented parallel to the plane of the loop. The force on the horizontal wires is zero since they are parallel to the magnetic field. The magnitude of the forces acting on the vertical wires, however, equals IbB, as determined by the equation, $F_{wire} = \left|\overrightarrow{F_{wire}}\right| = ILB \sin \theta$, where b is the length of the vertical conductor, L. The right-hand rule shows that the force on the left side of the loop is acting out of the paper and the force on the right side of the loop is acting into the paper. A bottom view of this arrangement is shown in Figure 8-44b. If the loop is free to rotate about points O and O', then these forces produce a torque about points O and O' that causes the loop to rotate clockwise.

As presented in Chapter 5, the magnitude of torque, τ, equals the product of force and the length of the lever arm. For the situation presented in Figure 8-44, the maximum torque applied to the loop occurs when the plane of the loop is exactly parallel to the magnetic field. This is calculated as:

$$\tau_{max} = F_1 \frac{a}{2} + F_2 \frac{a}{2}$$

$$\tau_{max} = (IbB)\frac{a}{2} + (IbB)\frac{a}{2}$$

$$\tau_{max} = I\,ab\,B$$

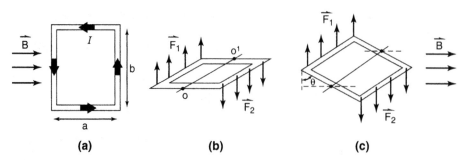

(a) **(b)** **(c)**

Figure 8–44: (a) Front view of a rectangular loop in a uniform magnetic field. (b) Bottom view of the rectangular loop showing the forces acting to create a torque. (c) Bottom view of the loop oriented at an angle θ from parallel to the field.

where the lever arm about the point O equals $a/2$ for each force. Since the product, ab, represents the area, A, of the loop, the above equation can be simplified producing:

$$\tau_{\text{max}} = IAB$$

The torque will actually vary in a sinusoidal fashion as the coil spins. In Figure 8-44b, the forces $\vec{F_1}$ and $\vec{F_2}$ are always directed up and down respectively. But torque depends not only on the magnitude of the force applied, it varies depending upon the component of that force that acts perpendicular to the lever arm across which it applies. This is shown in Figure 8-44c. The component of force acting perpendicular to either lever arm is $F \sin \theta$, and so the above equations can be modified as shown below to determine the torque applied to the loop in any given orientation θ:

$$\tau = F_1 \frac{a}{2} \sin \theta + F_2 \frac{a}{2} \sin \theta$$

$$\tau = IbB\left(\frac{a}{2}\sin\theta\right) + IbB\left(\frac{a}{2}\sin\theta\right)$$

$$\tau = I \, ab \, B \sin\theta$$

$$\tau = I \, AB \sin\theta$$

The torque that a magnetic field applies on a conducting loop is the basis on which an electric motor operates. Consider, for example, that a square wire loop measuring 0.70 m on each side and free to rotate about point O is placed in a uniform magnetic field as shown in Figure 8-45. A current, I, of 3.0 A flows clockwise around the loop and the magnetic field, $\vec{B}$, points to the right with a magnitude of 1.5 T. What would be the net force and torque acting upon the loop?

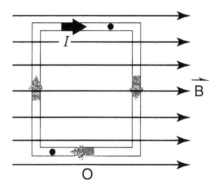

Figure 8–45: Clockwise torque exerted by a magnetic field on a conducting loop.

Since the horizontal segments of the loop are parallel to the magnetic field lines, no force is exerted on them. The vertical segments, however, experience a force that is equal in magnitude, but opposite in direction. Therefore, the left vertical segment experiences a force into the page and the right vertical segment experiences a force out of the page. When viewed from above, this means that the loop will experience a clockwise torque, which can be calculated as follows:

$$\tau = I \, AB \sin \theta$$

$$\tau = (3.0 \; A) \, (0.70 \text{ m})^2 (1.5 \text{ T}) \sin 90°$$

$$\tau \cong 2.2 \; \frac{\cancel{C} \cdot N \cdot m^{\cancel{2}} \cdot \cancel{s}}{\cancel{s} \cdot \cancel{C} \cdot \cancel{m}} \, (1)$$

$$\tau \cong 2.2 \; N \cdot m$$

Technology Box 8–9 ■ Direct-current Motors

The direct-current (DC) motor is one of the workhorses of modern industry. Its flexibility makes it the motor of choice for many different industrial applications. It can be accelerated and decelerated quickly and smoothly, and accurately controlled over a wide range of speeds. Since industry is supplied with alternating current (AC) power, solid-state rectifier diodes and **thyristors** are used to convert and provide DC power to the motors. (A thyristor is capable of producing large direct currents by rectification of alternating currents and can be automatically triggered off for specified time periods.)

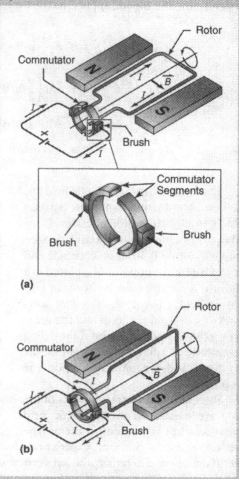

A schematic diagram of a simple DC motor is shown. The wire loop that is free to rotate is called the rotor and the rotor ends are attached to the two curved conductors that form the **commutator**. Note that the two commutator segments are insulated from each other. See (a) at right.

In the figure it can be seen that the brushes are aligned with the commutator segments, so the current flows into the left side of the rotor and out of the right side. The magnetic torque applied is down on the left side and up on the right side, which causes the rotor to spin counterclockwise. As the angle between the rotor and the field lines, which are fixed pointing left to right, increases from 0°, the net torque acting on the rotor decreases until the rotor has turned by 90° as shown in (b) at right.

At an orientation of 90° each brush is in contact with both commutator segments, so the current bypasses the rotor altogether. As the rotor continues to rotate beyond this point, current flow resumes through the commutators into the rotor, but the current has reversed its direction such that the torque continues to act in the same direction as before. This process continues so long as power is delivered to the coil. Naturally, the torque applied to the coil varies as the coil rotates, but at normal operating speeds the motor delivers mechanical energy smoothly to a load and returns it to the original position. See (a) above.

Look around and try to identify the various items that use electric motors. They are everywhere – driving the cooling fan in your computer, turning the compressor in your refrigerator, raising and lowering your garage door, etc. The electric motor is a critical component of many modern conveniences.

Faraday's Law of Induction

We now understand that electric power can be harnessed to create mechanical work in the form of a spinning rotor – the heart of an electric motor. These motors, in turn, drive everything from our food processors to the huge fans that deliver the air for combustion in power plants. The unit of power that represents the standard by which motors are rated – **horsepower** – reminds us of the less convenient source of work upon which our ancestors relied (1 horsepower $\cong$ 745.7 watts). The electric motor is arguably one of the most important innovations of the industrial age. How, then, is the electricity needed to run these motors and other appliances created? To explain this requires yet another concept – the induced electric field.

One of the most technologically significant discoveries of the nineteenth century was made by Michael Faraday. He was experimenting with magnets and wires when he noticed that as a wire is moved through a magnetic field, a voltage is created within the wire. In 1821, he reported in the *Quarterly Journal of Science* that whenever a magnetic field increases or decreases – called a **time-varying magnetic flux** – it produces an electric field. He also noted that the faster the magnetic flux increases and decreases, the greater the electric field induced. This electric field subsequently causes the electrons to move through the wire, a phenomenon recognized as an **induced current**. This is the essence of **Faraday's law of induction**. It explains, as shown in Figures 8-46a and 46b, that when a permanent magnet is held close to a loop of wire and moved either toward or away from the loop, an induced current is produced. The amount of current can be measured by a **galvanometer** connected in series with the loop. The direction of the current is clockwise as the flux decreases and counterclockwise as it increases. When the magnet is stationary with respect to the wire, there is no time-varying magnetic flux; therefore, no current is induced in the loop.

Now consider what happens if the galvanometer is replaced by a load; that is, any device (light bulb, motor, etc.) that offers resistance to the flow of the current. Whenever the loop is exposed to a time-varying magnetic flux and a current is induced, then an **induced voltage** will exist across the load. One might even say that this induced voltage simulates the effect that a battery might play on the circuit.

Faraday's law of induction also states that the voltage, also known as the **electromotive force,** or **emf**, is proportional to the time rate of change of magnetic flux, which is approximated by the expression $\Delta\Phi_B/\Delta t$, through the loop. Mathematically stated as:

$$emf = V$$

$$V = \frac{\Delta\Phi_B}{\Delta t}$$

$$V = \frac{\Delta(BA\cos\phi)}{\Delta t}$$

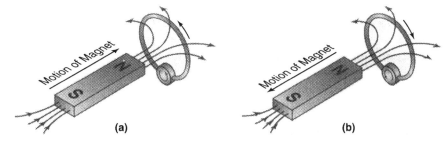

Figure 8–46: A permanent magnet moving toward a closed wire loop causes an increasing magnetic flux through the loop and therefore induces a current to flow counterclockwise (a). The same magnet when moving away from the loop causes a decreasing magnetic flux through the loop and subsequently a current to flow clockwise (b).

Technology Box 8–10 ■ The Alternating-current Generator

The electric generator is a device that converts work, or mechanical energy, into electrical energy, which is opposite to the way an electric motor works. The generator operates on the principle of electrical induction by which electricity is produced as a circular conductor or loop is rotated within a constant magnetic field. The figure at right shows this conversion in action. We see that the flux through the loop varies as the coil rotates since the angle ϕ is changing with respect to time.

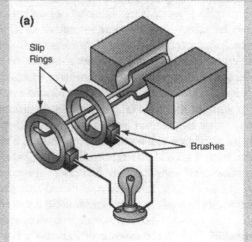

It is more common in large generators to keep the loop (armature) winding or stator (stationary) and rotate the magnetic field. In either case, the relative motion between winding and the magnetic field induces an alternating voltage. Note in the figure to the right (b) that the rotor, defined here as a rotating electromagnet, supplies a permanent magnetic field and spins with the stator. The electromagnet is powered by a relatively low-output direct current fed by means of slip rings and carbon or copper mesh brushes.

This form of generator is particularly suited for high voltages and is designed to provide a three-phase output as depicted in (c). This is accomplished by positioning three separate windings around the rotating field 120° apart, thus there is always current available for use since the three sine waves peak at different times and never reach zero simultaneously.

A three-phase power supply simply means that three separate alternating voltages, each out of phase by 120° with the other two, is provided on three separate lines. This is the form of power that most heavy machinery is designed to utilize.

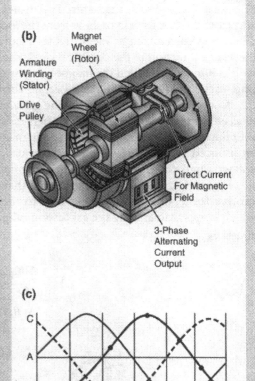

This equation reinforces the concept that voltage is induced only by *changing* the magnetic flux. The flux is a function of the area of the loop A, the strength of the magnetic field B, and the orientation between field lines and the normal to the loop area, ϕ. If any one or a combination of these parameters changes with respect to time, then the flux also changes with time, and a voltage is induced. In fact, in applications such as generators, many loops of wire in series are exposed to the varying magnetic flux to induce a voltage. In order to account for multiple loops, the above equation is revised as follows:

$$emf = V$$

$$V = -N\frac{\Delta\Phi_B}{\Delta t}$$

$$V = -N\frac{\Delta(BA\cos\phi)}{\Delta t}$$

where N represents the number of loops of wire that constitute a single coil. Note that N is left out of the expression, $\Delta\Phi_B/\Delta t$, since the number of loops would always be constant in any typical application of Faraday's law.

You may wonder why a negative sign appears in the *emf* equation. This is because the induced voltage varies inversely with respect to the change in flux. Explained another way, the negative sign accounts for the fact that the magnetic field generated by the induced current is oriented in a direction that opposes the magnetic field responsible for the induced current. In this way, energy is conserved during any application of Faraday's law. In many instances, such as in the next example, the *emf* is taken as the absolute value of the time rate of change of the magnetic flux, $\Delta\Phi_B/\Delta t$.

Consider a coil that is wrapped with 200 turns of wire on the boundary of a closed square frame 18 cm on each side. A uniform magnetic field perpendicular to the plane of the coil is applied. If the field changes linearly from 0 to 0.5 Wb/m^2 in 0.8 s, determine the magnitude of the induced *emf* in the coil while the field is changing. The solution is determined by applying the following equation:

$$emf = V$$

$$V = \left| -N\frac{\Delta(BA\cos\phi)}{\Delta t} \right|$$

$$V = \left| -(200)(0.18\,m)^2\frac{\Delta B}{\Delta t} \right|$$

$$V = \left| -(6.48\,m^2)\frac{(0.5-0)\,Wb/m^2}{(0.8-0)s} \right|$$

$$V = 4.05\,V$$

(It should be noted that the expression, $\Delta(BA\cos\phi)/\Delta t$, is approximated in this example by dividing the change in the magnitude of the field by the change in time. If a function of $|\vec{B}|$ with respect to t were given, then derivation of that function would represent the preferred approach.)

To summarize, a current is induced in a loop as long as a magnetic flux through the loop is changing with respect to time. The greater the magnetic flux change over time, the greater the current induced through and voltage induced across the loop. This warrants yet another observation: as the speed of the permanent magnet in Figure 8-46 is increased in either direction, the induced current and voltage are also increased.

Earth's Magnetic Field

The Earth has a permanent magnetic field of about 1 G or 10^{-4} T. A very crude model treats the Earth as one large bar magnet, as shown in Figure 8-47. The magnetic axis of the Earth is, in fact, inclined with respect to its spin axis by approximately 12° and the magnetic north and south poles are opposite of the geographic poles.

The origin of the Earth's magnetic field is a subject of much interest and research. One of the things of which we are certain, however, is that this field is not caused by a huge, embedded permanent magnet. In fact, the high temperatures in the interior of the Earth prohibit the formation of a permanent magnetic field. Despite the fact that the Earth is composed of a great deal of iron and other ferromagnetic material capable of sustaining a permanent magnetic charge separation (dipole moment) on a large scale.

It is considered likely by most experts that the magnetic field of the Earth is caused by convection currents of ionic material in the Earth's fluid core. These currents are influenced by the rotation of the Earth. It has been observed that the strength of the magnetic field of a planet is related to the speed at which it rotates. Realize, however, that a detailed understanding of all the mechanisms associated with magnetic fields on a planetary scale is not yet available.

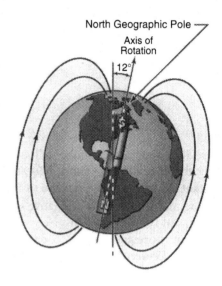

Figure 8–47: The magnetic axis of the Earth is in fact inclined with respect to its spin axis by approximately 12° and the magnetic north and south poles are opposite of the geographic poles.

A rather peculiar phenomenon associated with the Earth's magnetic field, however, is that it seems to reverse its direction every few million years. What causes this to happen is not clear, but a geological analysis proves, with little doubt, that it does occur; and on a geological time scale, quite often. When the reversal of magnetic poles occurs, it does so rather quickly – on the order of 10,000 years – a mere blink of the eye in geological terms.

Finally, the Earth's magnetic field, like its gravitational and electric fields, can and does vary markedly from location to location, and with respect to time. Aircraft pilots who navigate by compass and other directional aids refer to a VFR (visual flight rules) chart. This chart includes diagonal lines that indicate, at any location, by how much the compass heading deviates from actual direction, or in other words, by how much magnetic north varies from geographic, or true, north. A segment of such a chart is shown in Figure 8-48. Pilots must factor in a correction by adding to, or subtracting from, their desired course in order to determine a magnetic heading that will bring them to their destination. These deviations vary with location and time, as the attentive pilot observes.

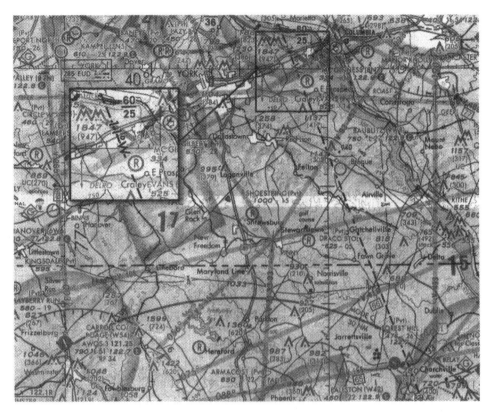

Figure 8–48: A segment of a VFR (visual flight rules) chart used for navigation by pilots. The dashed line with the notation of 11°W indicates that a pilot must add 11° to his or her magnetic heading to ensure the desired heading.

Check your Understanding

1. What is the relationship between a magnetic field and a magnetic force?

2. What are three things that will affect the size of the magnetic force acting on a single positive charge moving through a magnetic field?

3. What is the standard SI unit for magnetic field strength and how is it related to the gauss, G?

4. What is the "right-hand rule" and how is it used to determine the direction of the magnetic force acting upon a positive charge?

5. What type of magnetic field lines would be drawn to indicate a strong, uniform magnetic field?

6. How is the right-hand rule applied when determining the direction of the magnetic force acting upon a wire?

7. Why is there no force exerted on the horizontal segments of a loop placed in a horizontal magnetic field?

8. What is currently considered the likely cause of the Earth's magnetic field?

Summary

Electricity describes the interactions that matter has on other matter, based on the amount of electrical charge it contains. All matter is composed of atoms, which in turn are composed of protons and electrons. Protons and electrons each carry one elementary charge and in matter there are usually as many positive as negative charges, so it appears to be neutral and has no electrical interaction. Only when a charge imbalance exists and the matter acquires a net charge, do electrical interactions occur.

Electrical charges exert a force, F_E, on other charges that varies with the distance between them and the amount of charge present. The equation for this relationship is known as Coulomb's law.

Like charges repel, whereas unlike charges attract. If a concentration of charges exists at opposite ends of a wire or other conductor, then an electrical current will flow as the charge carriers in the wire feel the electrostatic force caused by these unbalanced charges. The amount of current that flows is given by Ohms law, $V = IR$, and is limited by the resistance of the material through which the charge carriers move. Since the motion of charge carriers is restricted by their interaction with the atoms in the material, they dissipate power in the material, given by $P = I V$.

A device that impedes the flow of current and dissipates power is called a resistor. It is possible to arrange individual resistors in various combinations in an electrical circuit. The flow of current through such a circuit is governed by Kirchoff's voltage and current laws. Electronic devices such as computers and stereo amplifiers are constructed using devices made of semiconductor material and devices made of metal (inductors and capacitors) in more complicated electronic circuit arrangements.

Magnetism is a phenomenon characterized by a vector force field, associated with moving electricity, and exhibited by both magnets and electric currents. Magnetic field lines are used to visualize a magnetic field vector, and to align with the orientation of a compass needle placed at that point.

Magnetic flux is the product of some specific surface area and the average magnetic field that is normal to that surface. It is the greatest when the magnetic flux is perpendicular to the area through which it passes. In a similar way, the greatest magnetic force on a conducting wire occurs when it is oriented perpendicular to the field, but no magnetic force occurs on a wire that is parallel to the field.

When a conducting loop of wire is placed in a magnetic field, it is acted upon by a torque. The torque is at its maximum when the loop is exactly parallel to the magnetic field. This torque is the basis on which an electric motor operates. Faraday's law of induction states that a current is induced in a loop as long as a magnetic flux through the loop is changing with respect to time. A voltage is induced as the size of the magnetic field, the area of the loop, the angle ϕ between $\vec{B}$ and the normal to the loop area, or any combination of these parameters, changes over time.

The Earth has a permanent magnetic field of about 1 G or 10^{-4} T. Its source is likely convection currents of ionic material in the Earth's fluid core. Every few million years the Earth's magnetic field reverses. Its gravitational and electric fields vary more rapidly and markedly from location to location. Pilots who navigate by compass must factor in corrections to reach their desired destinations.

Chapter Review

1. An 80 kg person picks up 6×10^{-7} coulombs of electrons walking from the door to the workbench on a dry day. (This is equivalent to charging up to 3,000 Volts!) How many extra electrons did this person pick up? Assuming that the human body is made of only water, what is the ratio of "matched" to "unmatched" electrons in this person's body?

2. An object with 1,000,000 extra electrons gains the same amount of energy being accelerated across a voltage difference of 1 volt as an object with 100 extra electrons that is accelerated over a voltage difference of 10,000 V. Why are their final energies the same?

3. What is the force between a +10 C charge and a –10 C charge at a distance of 10 cm?

4. If it has not run into any other atoms, how fast is a single electron at rest moving after accelerating through a voltage difference of 100 V?

5. Why would an entire string of Christmas lights turn off when just one bulb burns out?

6. A home appliance uses a current of 2.5 A. How many electrons pass through the appliance every second?

7. Silver wires can be thinner than copper wires and have the same resistance because silver is a better conductor than copper. What diameter of silver wire has the same resistance as an equal length of 2-mm diameter copper wire?

8. A space heater dissipates 1,000 W of electrical power. What is the resistance of the 2 m of nichrome heating wire packed inside it, assuming that the average voltage from the wall socket is 120 V?

9. A typical electric dryer uses 35 A of electrical current. What is the voltage difference between the ends of a 5-foot length of electrical cord used operate the dryer, assuming that the wire is 0.2 inches in diameter? How much power is dissipated in the wires (remember that current flows both ways)?

10. A bolt of lightning can carry 40 kA of current over a voltage difference of 10 MV. How many 60-watt light bulbs would be equivalent to this electrical power?

11. You have three resistors, 30 Ohms each. What is the largest equivalent resistance you can make with these three? The smallest?

12. In Figure 8-25, if the resistance of the light bulb is 1.2 Ω, how much power is being dissipated by the bulb? (1 amp × 1 volt = 1 watt)

13. If the incorrectly installed battery in Figure 8-25 were reversed, what would be the power dissipated by the 1.2 Ω light bulb? (1 amp × 1 volt = 1 watt)

14. You have a 1 kΩ resistor. You add a 10 Ω resistor to this resistor in a series combination. By what percentage does your resistance value change (comparing the new equivalent resistance to the original 1 kΩ)?

15. Two 50 Ω resistors are arranged in a series combination and the combination attached to a 9 V battery. How much power is dissipated in each resistor individually?

16. Two 50 Ω resistors are arranged in a parallel combination and the combination attached to a 9 V battery. How much power is dissipated in each resistor individually?

17. Use Kirchoff's laws to determine the difference in voltage between points A and B in the figure below.

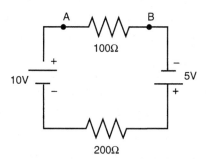

18. Use Kirchoff's laws to determine the current through the 100 Ω resistor in the following figure.

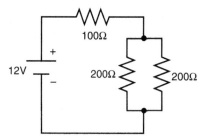

19. A 1,500 W electric fryer, a 1,000 W microwave oven, a 500 W toaster, and a 300 W coffee maker are all connected to the same electrical circuit in a home (120 V RMS service). What are the individual resistances of each of these devices? What is the equivalent resistance of all of them operating at once? How much current will they draw all operating at once? Can they by used together on a circuit fused at 20 A?

20. Stray capacitance (capacitance that appears unintentionally) can be a real problem in fast circuits, because it takes time to move charge onto and off of a capacitor, and therefore to change voltage levels. How might the ribbon cables and circuit boards used in a computer affect the speed at which a computer can operate?

21. In question 1, a person picked up 6×10^{-7} C of charge walking from the door to the workbench on a dry day. From this information, what is the capacitance of the human body?

22. Which has the higher capacitance, two square metal plates 1 cm × 1 cm held 1 mm apart, or two round metal plates 1 m in diameter held 1 cm apart?

23. A computer manufacturer uses a 10-Farad capacitor charged to 5 V to keep the CMOS memory in a computer alive. If the CMOS memory draws 20 μA of current, how long will it be until the voltage on the capacitor drops to 3 V and the CMOS memory becomes unreliable?

24. An electric motor uses 20 A of current while running and then takes 0.002 s to shut off. During this time, it puts a 160 V spike on the power line. What is the inductance of the motor?

25. How much inductance results from wrapping 100 turns of wire around a 1/8 inch diameter nail 2 inches long? Assume that the relative permittivity of the iron in the nail is 200.

26. An electron experiences a magnetic force of 4.65×10^{-15} N when its direction is oriented at an angle of $42.0°$ with respect to a magnetic field of magnitude 3.45×10^{-3} T. Determine the speed of the electron at this time.

27. An electron, a proton, and a neutron each has kinetic energy of 1.1 keV and is projected into a region with uniform magnetic field of 48×10^{-3} T directed out of the page, $\otimes$, as shown below.
    a. Determine the speed of each particle and decide whether the kinetic energy of each particle will change within the region of the magnetic field.
    b. Sketch the approximate path that each particle will follow as it passes through the magnetic field.

28. The rotor described in the DC-Motor Technology Box 8-9 is oriented at an angle of $40°$ with respect to the horizontal. Determine the torque acting upon the rotor, if it can be treated as a rectangular closed loop 12 cm long and 9 cm wide conducting a current of 10 A and operating within a uniform magnetic field of 2.8 T.

29. A powerful electromagnet supplies a field of 1.5 T within a cross-sectional area of 0.3 m^2.
    a. If a coil with 250 turns is placed around the electromagnet, and the electromagnet is powered down causing the field to go to zero in 20 ms, determine the induced *emf*.
    b. If the total resistance across the wire is 15 Ω, calculate the induced current during shutdown.

30. A generator operates on the principle that the changing angle of orientation between magnetic field and windings causes a time-varying magnetic flux through the windings. The change in ϕ with respect to t represents the angular velocity, ω, of the rotor, and the equation takes the form shown from which the maximum induced *emf* can be calculated

$$emf_{max} = NAB\omega$$

The coil of a generator consists of 12 turns of wire, each of area 0.15 m^2, and a total resistance of 10Ω. This coil rotates in a magnetic field of magnitude 0.54 T at a constant frequency of 60 Hz.
    a. Determine the maximum *emf* induced.
    b. Calculate the maximum current induced.

Chapter

9

What is darkness?

Electromagnetic Radiation and Optics

At first the question, "What is light?" seems so simple, yet scientists have struggled for thousands of years to reach an exact answer. The many uses of light and all other forms of electromagnetic radiation are continually expanding. Technological developments such as telescopes and microscopes have also extended the ability to observe, analyze, and use portions of the spectrum that were formerly unknown. All that is known about the universe today, for example, is based on analysis of the electromagnetic radiation collected from distant stars, galaxies, and other radiating bodies using various types of telescopes. Microscopes, on the other hand, operating in the visible, ultraviolet, and X ray portions of the spectrum, have enabled scientists to unlock many of the mysteries of life and disease.

Without visible electromagnetic radiation the sense of sight is lost and other senses such as touch or hearing must be relied upon. When the amount of natural light is inadequate, technology provides artificial light with the flip of a switch. We have become a society dependent on artificial lighting to extend the productive day, provide entertainment, control the flow of traffic, and increase home security. Through all of this, it is interesting to remember that what we call *darkness* is just the absence of visible light.

In this chapter, the abstract nature of all electromagnetic radiation will be discussed. The characteristics and behavior of visible light as it passes through various mediums and reflects from others will be analyzed along with the technology that manipulates it. Line drawings are used to depict light rays, but they are just models intended to show the direction the light is traveling. In reality, a laser beam is the closest thing to a single light ray, but even it has some finite thickness.

Electromagnetic Radiation and Its Spectrum

Objectives

Upon completion of this section, the student will be able to:

■ Explain the nature of electromagnetic radiation.

■ Calculate the amount of energy associated with various frequencies of light.

■ Describe the various regions in the electromagnetic spectrum.

The Nature of Light

Light is defined as electromagnetic radiation that is visible to the human eye. However, what is electromagnetic radiation? In the past, there were two competing theories on the nature of light – the particle theory proposed by Pythagoras (ca. 585–500 BC) and the wave theory of Plato (427–347 BC). Isaac Newton (1642–1727) believed that light was composed of a stream of tiny particles, which he called corpuscles. The **corpuscular theory** was consistent with his observations that light travels in straight lines and forms sharp shadows.

Christian Huygens (1629–1695), a contemporary of Newton, argued that light was composed of waves resulting from molecular vibrations in the radiating medium. His arguments were later strengthened when the British scientist Thomas Young (1773–1829) reported, in 1801, on his now famous double slit experiments, which seemed to prove that light is composed of waves. Additional experimentation followed on the refraction, diffraction, and interference of light that could only be explained by using the wave theory. Consequently, by the latter part of the nineteenth century, only the wave theory of light remained.

The Speed of Light

Measuring the speed of light also proved to be difficult, but creativity and perseverance eventually won out. In 1849, Armand Hippolyte Louis Fizeau (1819–1896) devised a clever apparatus to measure the speed of light, without the use of astronomical observations as had been done in the past. His apparatus used a rotating cogwheel and a fixed mirror 8,633 meters from the wheel. Light was allowed to pass between two teeth of the cogwheel, reflected off a mirror, and return. If the wheel turned fast enough to obscure the reflection, then the reflected beam struck a cog. The time it took the wheel to move the width of one tooth was then equal to the time it took the light to travel twice the distance between the wheel and the mirror. Based on his findings, Fizeau calculated the speed of light to be 3.153×10^8 m/s. Refined techniques have now placed the accepted value for the speed of light at 2.99792458×10^8 m/s, which is typically rounded to 3.00×10^8 m/s.

Electromagnetic Waves and Light

During this same time period studies on the nature of electricity and magnetism were progressing, but no connection had yet been made between **optics** (the study of light) and **electromagnetism** (magnetism developed by a current of electricity). The thread that eventually tied the two fields together was a discovery made in 1845 by Michael

Faraday (1791–1867). He demonstrated that the polarization of light passing through glass could be changed when placed between the poles of a powerful electromagnet.

Based on this observation, James Clerk Maxwell (1831–1879) proceeded to quantify the relationships mathematically. In 1873, Maxwell published a paper concluding that magnetism and electricity are related. When solved, his equation using only electric and magnetic constants predicted an electromagnetic wave that would have the same speed as what had already been determined to be the speed of light. As an extension to his reasoning, he also proposed that light is electromagnetic radiation. It must, therefore, consist of a wave with two components: an electric field and a magnetic field.

There remained, however, an unsolved piece of the light/electromagnetic puzzle. If electromagnetic disturbances are waves, through what medium are they traveling? How can they travel through space, which is by definition mostly empty of matter? A mysterious ether was proposed as the possible explanation, but experimentation failed to prove its existence. In 1905, Albert Einstein (1879–1955) published his special theory of relativity. In it, he stated that electromagnetic disturbances need no medium in which to travel. His bold conclusion, and indeed one of the cornerstones of his theory, was that light always travels through empty space at the velocity, c, regardless of the speed of the light source. Acceptance of this (and experiments have shown it is true) requires us to view light as an independent entity capable of traveling through space on its own.

At this point, you may be wondering how to visualize a light wave. Since no one has ever captured a "piece" of light, the concept of energy fields is useful. It's like trying to capture a "piece" of an energy wave that is traveling down a rope. We can grab the rope, but that is only the medium in which the wave is traveling. Since light can travel without a medium, there may not even be a medium to grab. The best model of an electromagnetic wave is to visualize it as being composed of an electric field and a magnetic field that are vibrating at right angles to each other, as shown in Figure 9-1. It is important to know that you cannot have one of these fields without the other. Experimental results show that a moving or changing electric field creates a changing magnetic field and a moving or changing magnetic field creates a companion electrical field – always.

Understanding that light travels as a wave, it may help to review the wave characteristics presented in Chapter 5. In general, waves travel as disturbances through a medium with a constant speed, specific frequency, f, and wavelength, λ. Using the relationship, $c = f\lambda$, if the speed of a wave in a given medium is known, it is possible to calculate either its frequency or wavelength if the other is known. In air, for example,

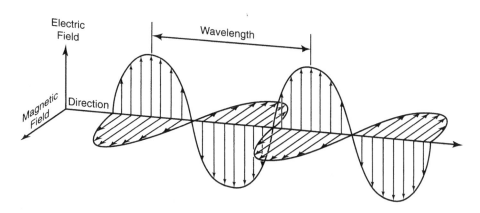

Figure 9–1: An electromagnetic wave is composed of both an electric field and a magnetic field that vibrate at right angles to each other.

the speed of light, c, is typically rounded to 3.00×10^8 m/s; therefore, the equation for it can be written as follows:

$$c = f\lambda$$

$$3.00 \times 10^8 \text{ m/s} = f\lambda$$

In Chapter 5, the hertz, Hz, was defined as the frequency or number of vibrations per second. Therefore, it is possible to calculate the frequency for visible green light that has a wavelength, λ, of 525 nm. For unit agreement, however, it is necessary to first convert the wavelength into its equivalent SI unit.

$$525\,\text{nm} \times \frac{1.00 \times 10^{-9}\text{meters}}{1\,\text{nm}} = 5.25 \times 10^{-7}\text{m}$$

$$c = f\lambda$$

$$f = \frac{c}{\lambda}$$

$$f = \frac{3.00 \times 10^8\,\text{m/s}}{5.25 \times 10^{-7}\,\text{m}}$$

$$f = 5.71 \times 10^{14}\text{ Hz}$$

Once the frequency of a wave is known, it is possible to calculate its energy. In the case of electromagnetic waves, the energy can be determined by multiplying the frequency of the wave by **Planck's constant**, h. Max Planck (1858–1947) derived this constant after he had completed extensive investigations of a phenomenon known as **blackbody radiation**. (A blackbody is an ideal body that completely absorbs all incident radiant energy with no reflection; and that radiates at all frequencies with a spectral energy distribution dependent on its absolute temperature.) In the paper he presented to the German Physical Society in 1900, one of the conclusions drawn was that light energy is absorbed or emitted by matter only in discrete increments.

The energy of an electromagnetic wave is given by the formula, $E = hf$, where $h = 6.626 \times 10^{-34}$ J/Hz or 4.136×10^{-15} eV/Hz. Using the constant, it is possible to calculate the amount of energy associated with red light having a frequency of 3.8×10^{14} Hz.

$$E = hf$$

$$E = \frac{6.626 \times 10^{-34}}{\text{Hz}} \times 3.8 \times 10^{14}\text{Hz}$$

$$E = 2.51 \times 10^{-19}\text{ J}$$

Table 9-1 has been prepared in a similar fashion, showing the energies associated with the various frequencies of visible light.

Color	Wavelength		Frequency	Energy
	Nanometers (nm)	Meters (m)	Hertz (Hz)	Joules (J)
Red	600 – 800	$6.00 – 8.00 \times 10^{-7}$	$5.0 – 3.8 \times 10^{14}$	$3.3 – 2.5 \times 10^{-19}$
Orange	575 – 600	$5.75 – 6.00 \times 10^{-7}$	$5.2 – 5.0 \times 10^{14}$	$3.4 – 3.3 \times 10^{-19}$
Yellow	565 – 575	$5.65 – 5.75 \times 10^{-7}$	$5.3 – 5.2 \times 10^{14}$	$3.5 – 3.4 \times 10^{-19}$
Green	490 – 565	$4.90 – 5.65 \times 10^{-7}$	$6.1 – 5.3 \times 10^{14}$	$4.1 – 3.5 \times 10^{-19}$
Blue	475 – 490	$4.75 – 4.90 \times 10^{-7}$	$6.3 – 6.1 \times 10^{14}$	$4.2 – 4.1 \times 10^{-19}$
Indigo	460 – 475	$4.60 – 4.75 \times 10^{-7}$	$6.5 – 6.3 \times 10^{14}$	$4.3 – 4.2 \times 10^{-19}$
Violet	400 – 460	$4.00 – 4.60 \times 10^{-7}$	$7.5 – 6.5 \times 10^{14}$	$5.0 – 4.3 \times 10^{-19}$

Table 9–1: Energy and frequencies associated with various wavelengths of light.

Technology Box 9–1 ■ Photocells

Photocells that are used in light meters on cameras, night vision goggles, CAT scans, and television cameras, have common roots in the photoelectric effect. The photoelectric effect was first observed in 1886, by Heinrich Rudolf Hertz (1857–1894), a German physicist. He found that when UV light strikes some metal surfaces it causes electrons to be ejected. It was later demonstrated, as shown in the diagram, that if two electrodes – an anode and a light-sensitive metal cathode – are sealed into an evacuated tube and a beam of light shown onto the cathode, an electric current flows. This arrangement is called a phototube and is the forerunner of the modern photocell.

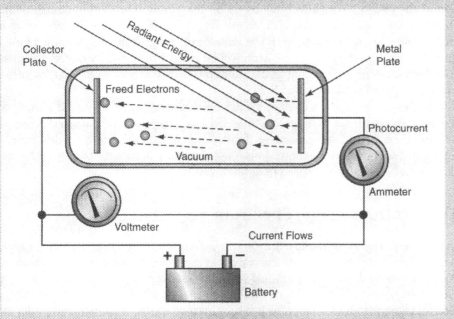

Although the phototube is considered obsolete, the principle survives in the form of a photomultiplier tube, which is used to detect and amplify faint light sources. In this tube, the electrons initially freed from the photosensitive cathode are drawn to the positive electrode as before, only these electrons, in turn, free additional electrons that are drawn to a more positive electrode. After repeating this step several times, a large pulse of current is produced. Besides their use in light meters, photomultipliers are built into such items as television camera tubes, making them sensitive to levels of light far fainter than can be seen by the human eye.

When a battery or other voltage source is connected to the circuit, as shown, a current flows even in the absence of light. When light strikes the cathode, however, the current in the circuit is increased by an amount proportional to the intensity of the light. Thus, this change in current can be used to trigger other electric circuits, such as the ones used to open doors or prevent doors from closing if the light beam is blocked by a person or object.

Modern photoelectric cells are usually constructed from two dissimilar semiconductors. When exposed to light, a voltage is set up across the junction between the two materials. Phototransistors, which are a type of photovoltaic cell, can generate a small current when struck by light that acts like the input current in a conventional transistor. Photovoltaic cells are also used to make solar batteries. Since the current from a photocell can easily be used to operate switches or relays, they are often used as light-actuated counters and burglar alarms. Photocells in such devices are typically called electric eyes.

By 1905, Einstein had extended Planck's idea that the energy carried by electromagnetic radiation is not smoothly spread across the wave, but somehow concentrated at points within it. Einstein proposed that electromagnetic radiation be viewed as consisting of massless bundles of energy, which were later named photons. Although each photon has no mass, it carries a specific amount of energy that is related to its frequency by Plank's constant, $E = hf$. He strengthened the concept by proposing a successful explanation for the **photoelectric effect**, which had been previously observed and noted by Heinrich Hertz (see Technology Box 9-1) in 1887.

Simply put, the photoelectric effect is the emission of electrons from the surface of a metal, when struck by electromagnetic radiation of a sufficiently high frequency. Today technology has put this phenomenon to work in the form of an electric eye that can open supermarket doors, in TV cameras, and in light meters. Einstein's explanation of the photoelectric effect was based on the assumption that an electron either absorbs a whole photon, of a specific frequency, or not. If absorbed, the electron gains an amount of energy equal to hf. If this amount of energy is greater than the work required to free the electron, an electron will be ejected from the surface of the metal. When many atoms and photons of the same type are involved, a steady electric current can result. (See Technology Box 9-1)

The idea that only whole photons can be absorbed, and only one at a time, was revolutionary and again suggested that light is composed of particles. Thus the schizophrenia of the **wave-particle duality** was born. According to this view, all forms of electromagnetic energy appear and disappear in minute bundles (photons) that are transported at the speed of light as waves.

The Electromagnetic Spectrum

In Chapter 6, electromagnetic radiation was introduced as the only form of energy that is not associated with mass and cannot exist in the form of stored energy. Radiant energy can only be detected and studied when it is intercepted by matter and converted into thermal, electric, chemical, or mechanical energy. The energy and wavelength of electromagnetic radiation can thus produce measurements, from which we infer that it is there. In other words, the radiant energy itself cannot be measured, but we can measure the effect of the radiant energy when it interacts with something else in the physical world.

We are generally most aware of only the visible portion of the electromagnetic spectrum because it is responsible for the visual sensations that arise from stimulation of the retina. As shown in Figure 9-2, the wavelength of this portion of the spectrum is quite narrow and ranges from about 390 nm or 3.90×10^{-7} m (purple) to 760 nm or 7.60×10^{-7} m (red).

In 1666, Isaac Newton clearly demonstrated that sunlight is white light composed of all the colors of the rainbow. If you remember the pneumonic device ROY G. BIV, you will have the first letter of all their names: red, orange, yellow, green,

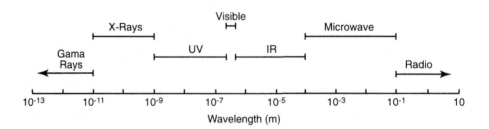

Figure 9–2: The electromagnetic spectrum and its broad subdivisions.

blue, indigo, and violet. In 1800, astronomer William Herschel (1738–1822) made the surprising discovery that the spectrum of the Sun contained more than just the visible colors. Using a prism to split sunlight into a spectrum and three thermometers with blackened bulbs, he used one thermometer to measured the temperature of each color, while placing the other two beyond the spectrum as controls. As he measured the temperatures of the various colors, he first noted that all colors had temperatures higher than the control thermometers. Second, he noted that the temperatures of the colors increased as he moved from the violet to the red end of the spectrum. This prompted Herschel to measure the temperature just beyond the red portion of the spectrum. Here he found a region where no color could be seen, that produced the highest temperature. Herschel had discovered **infrared radiation** (IR), a form of invisible light. After studying Herschel's 1801 discovery of infrared radiation, the German physicist Johann Wilhelm Ritter (1776–1810) observed the effects of solar radiation on silver salts and deduced the existence of radiation just past visible violet light. It was named **ultraviolet radiation** (UV), which means "beyond" violet. Later in the chapter, a more detailed description and some of the technologies based on these regions will be presented.

In summary, it is believed that all electromagnetic radiation can be described as a stream of massless particles, traveling in a wave-like pattern at the speed of light. These bundles of energy, now called photons, differ only in their frequency and, therefore, the amount of energy they carry. In science, classification is both an important and powerful tool. With this in mind, in the following sections each division of the electromagnetic spectrum will be introduced. We will start with the regions of electromagnetic radiation that have the longest wavelengths and lowest energy and move toward those with the shortest wavelengths and the highest energy. It is important to note that this is a continuous spectrum and that the stated wavelengths separating each division are somewhat arbitrary and so vary depending on the source.

Radio Frequency Waves

Theoretically, there is no limit to how long an electromagnetic wave can be. If an electric charge is moved slowly back and forth, it will send out a very long electromagnetic wave, since all electromagnetic waves travel at the great velocity, *c*. There is, however a practical limit to **radio frequency** (**RF**) wavelengths that are useful. As shown in Figure 9-3, the longest RF wavelengths range from thousands of meters down to a fraction of one meter. Power transmission lines emit low frequency, long wavelength RF radiation. Although some remain skeptical, studies attempting to link this source of electromagnetic (**EM**) radiation with a higher incidence of cancer and birth defects have failed to show any correlation.

As shown in Figure 9-3, the Federal Communication Commission, FCC, has identified various frequencies within this portion of the spectrum for specific uses.

The very low frequency (VLF), high (**HF**) to very high frequencies (**VHF**) are used to broadcast radio and television signals. AM radio waves are longer than FM radio waves, for instance. An AM station received at 750 on the dial (750 kHz) uses a

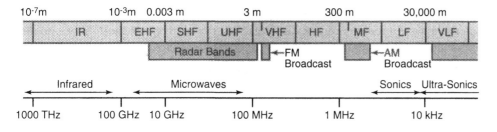

Figure 9–3: Detail of electromagnetic spectrum from radio frequency through infrared.

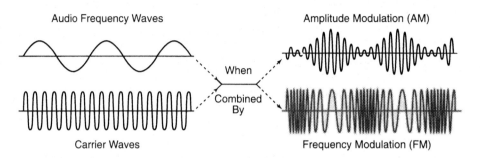

Figure 9–4: Audio frequency and carrier waves (left) can be combined to form an amplitude modulated (AM) (top right) or a frequency modulated (FM) wave (bottom right).

wavelength of about 400 meters and an FM radio station received at 100 on the dial (100 MHz) would have a wavelength of about three meters. These FM waves can be transmitted from an airplane or satellite antenna and received on the ground.

By controlling selective wave properties, it is possible to design electric circuits that will convert audio and video waves into radio frequencies that are transmitted. As shown in Figure 9-4, audio waves can be combined with the **carrier waves** by either **amplitude modulation** (AM) or **frequency modulation** (FM) and transmitted.

The amplitude of the carrier wave in AM transmissions is varied to match changes in the audio wave coming from the radio station. The amplitude of the carrier wave in FM transmissions remains constant, but the frequency of the wave changes to match the audio wave coming from the station. Like any form of communication, if the decoding is simply the reverse of the original encoding, then the character of the original audio signal can be reproduced.

AM broadcasts can be received at great distances because they travel by **ground wave** and **sky wave propagation**, which means they alternately bounce off the ionosphere and the Earth. These waves, therefore, reach beyond the curvature of the Earth. AM signals tend to lose energy by absorption when they contact the Earth. Consequently, they eventually lose power and fade out of range. Another problem with AM signals is sound fidelity. They are affected by static charges and other weather phenomena. The most powerful AM stations operate in the fifty thousand watts range. At this power, a station can be heard up to a thousand miles away under the more favorable conditions of night. The least powerful AM stations, usually used in local settings like college campuses, operate at approximately 250 watts.

Radio broadcasting experienced renewed growth beginning in the 1960s with the licensing of many FM radio stations. Not only do FM waves have constant amplitude, but they also have a better signal-to-noise ratio and are nearly immune to static. FM waves, however, do not travel as far because they penetrate the ionosphere. This means that both antennas must be visible to one another, with no intervening obstructions, because they propagate by **line-of-sight**. All VHF and ultra high frequency (UHF) communications typically use this method of propagation. The advantage of the FM wave is that it makes radio communication in space possible giving astronomers a powerful tool to measure distances from the Earth to the most distant reaches of the universe.

Many astronomical objects are themselves sources of radio frequency (RF) waves. Astronomers can actually "listen" to the stars, planets, and galaxies, much like you and I listen to the radio. All that is needed is a receiver. The trick, however, is to decode the signal received. Obviously, it is much easier to decode a signal for which you know the encryption process. Stars and astronomical objects have proved to be a bit more of a challenge, but incredibly revealing of the large-scale nature of the universe.

Microwaves are the portion of the electromagnetic spectrum in which wavelengths range from about 30 cm to 1 mm (0.30 to 0.001 m), placing them between RF and IR waves, as shown in Figure 9-3. Since microwaves can pass through rain, smoke, and fog, communications devices often use them. For example, beams of microwaves are used to send telephone messages in mountainous areas, where it would be difficult to install and maintain telephone lines. Microwaves also pass through the ionosphere, making them effective for communicating with aircraft and orbiting satellites.

For several decades, microwave ovens have been used to heat and reheat foods and now rival traditional ovens as a standard cooking method. Microwave technology is also being introduced into many other fields. For example, medical uses include a safer method for the removal of cancerous tissue using microwaves rather than surgical techniques. Industry is experimenting with the use of microwaves to decrease air pollutants, disinfect hospital wastes, and clean up contaminated soil. At least one type of burglar alarm uses disturbances in the pattern of a continuous stream of microwaves to detect movement.

Although you may not know how it works, radar is a familiar application of microwaves that was introduced during World War II (1939–1945). You may be familiar with its use at ball games to determine the speed of fastballs or by traffic officers to determine the speed of your car. Radar guns use short bursts of low-frequency microwaves that are reflected from objects. The time between the arrival of one reflected burst and the next allows the speed of the object to be determined using electronic circuitry. Air traffic controllers use reflected microwaves of known frequency to determine the position of airplanes. Meteorology successfully uses **Doppler radar**, which measures the intensity of approaching storms. The more rain droplets present in the clouds, the more microwaves reflected.

One of the newer uses of radar is called **ground penetration radar** or GPR. It is used to map layers of earth without having to disturb the soil. Much like ultrasound, which doesn't disturb the baby in the womb, radar can be sent into the ground and the reflected signal shows layers with different densities. GPR has been correlated with seismic data and found quite reliable. In characterizing areas of contaminated soil, the less invasive a procedure is, the better; so GPR could prove to be a useful tool for this purpose in the near future.

Microwaves

As previously noted, William Herschel discovered that temperatures increase from the blue to the red end of the spectrum. He further found that the highest temperature was just beyond the red region and named it infrared (infra- meaning below red in frequency) or IR radiation. As shown in Figure 9-3, the IR radiation falls between the microwave and visible portions of the electromagnetic spectrum. IR radiation has wavelengths ranging from about 1 mm to 78 μm (1×10^{-3} to 7.8×10^{-7} m).

The most readily detectable IR radiation is in the form of heat. Nearly 60% of the rays from the Sun are in the IR region, but fortunately our atmosphere blocks much of it. A vast number of materials absorb incident IR radiation, which makes them warmer. Any object that has a temperature above absolute zero is also emitting IR radiation, whether it's a glowing ember, a person, or a warm rock. In general, the warmer the object, the greater the amount of IR radiation it will emit. Objects that are not hot enough to give off visible light lose most of their energy in the form of infrared radiation. Glowing charcoal briquettes, for example, give off little visible light, but they emit sufficient infrared radiation to cook hamburgers.

Humans radiate IR radiation at a wavelength of 1.0×10^{-5} m most strongly. Although our eyes cannot see a nearby person "radiating," the nerves in our skin can sense the temperature difference, so we can "feel" their presence. Snakes in the pit viper family (e.g., rattlesnakes, copperheads, and water moccasins) have two heat sensory pits that allow them to use IR imaging to detect warm blooded animals even in the

Infrared

Technology Box 9–2 ■ Microwave Ovens

Among their many uses, microwaves are used to measure the speed of cars and baseballs, transmit telephone and television signals, and treat sore muscles. Their single largest use, however, is in microwave ovens. It is estimated that there are over 80 million in use in the United States, qualifying them as an everyday household appliance.

It all started in 1940, when Sir John Randall and Dr. H. A. Boot perfected a device called a magnetron. Its purpose was to generate microwaves that could be bounced off war machines to detect their presence. The technology was called radar, an acronym for **ra**dio **d**etecting **an**d **r**anging. In 1946, Dr. Percy Spencer, a Raytheon engineer, was performing tests on a magnetron tube when he noted that a candy bar in his pocket had melted. He knew the magnetron produced heat, but he had not felt it. Subsequent testing demonstrated that, indeed, the magnetron could be used for cooking food. Raytheon made the first commercial microwave oven and because of its history named it a Radar Range. They were very large (5.5 ft tall and over 750 lb) and did not sell well. In 1952, Tappan introduced the first home microwave oven, priced at $1,295. Surprisingly, they did sell and the rest is history.

The magnetron, which is at the heart of every microwave oven, generates microwaves by first stepping up 60 Hertz AC voltage to a higher voltage, which is then changed to an even higher DC voltage. In the final step this DC power is converted to microwaves. All home microwave ovens operate in the 915 and 2,450 megahertz frequency range, which is monitored by the Federal Communications Commission (FCC) along with police radar and garage door remote controls.

The microwaves that enter the oven bounce around within its metal interior until they are absorbed by the food. Specifically, the energy is primarily absorbed by fats, sugars, and water, increasing their translational energies, which dissipates as heat in the food. This explains why foods with high water and fat content cook more quickly and why microwaves pass through glass, paper, ceramic, and plastic containers, without heating them.

Microwave cooking tends to be more energy efficient than conventional cooking since the energy heats only the food and not the cooking container and kitchen. In general, microwave cooking does not reduce the nutritional value of foods any more than conventional cooking. In fact, foods cooked in a microwave may retain more of their vitamins and minerals because they are cooked without adding extra water.

dark. There is some evidence that they even have some degree of IR depth perception, due to the use of the two sensory pits.

Infrared detectors were first used during World War II enabling sharpshooters to see their targets in total darkness. Today, infrared technology has many exciting and useful applications. Infrared cameras are used for police and security work as well as in fire fighting. The military relies heavily on the use of infrared radiation for night vision goggles and directing heat-seeking missiles. In the field of infrared astronomy, new and fascinating discoveries are being made about the universe. The remote controls we use to operate our televisions, VCRs, CD players, and garage door openers are a few of the more common uses of infrared technology.

Technology Box 9–3 ■ Lights and LEDs

Thanks to Thomas A. Edison, the incandescent light bulb has been turning night into day, extending our productive and leisure hours, and improving our security for more than 100 years. The typical incandescent light bulb consists of a double spiral of very small diameter (42 microns or 0.0017 inch) tungsten wire, housed in a partially evacuated glass envelope that contains small amounts of nitrogen and argon gases. The tungsten filament is energized by electricity, which heats it to about 2,500°C. From a 60 watt light bulb, about 6 watts is emitted in the form of visible light, while the remaining 90% of the electric power is wasted in the form of invisible infrared light or thermal energy. At this high temperature, tungsten atoms tend to boil off the filament and collect on the inner cooler surface of the glass, causing it to darken with age. It also results in a gradual thinning of the filament and eventual burn out. The typical incandescent light bulb is designed to operate for 750 – 1,000 hours.

Fluorescent lamps are the second popular type of lighting, in part because the individual tubes remain functional 10 to 20 times longer than incandescent bulbs. Within the partially evacuated tube, electrons from the power source are caused to collide with mercury atoms resulting in an emission of ultraviolet light as the electrons change from one energy level to another. The ultraviolet light is then converted to visible light by the layer of white phosphor powders on the inside of the glass envelope. In theory, this whole activity can be performed without creating any thermal energy. Due to some imperfections, however, the tubes do convert some of the electric energy into heat. As a comparison, a 15 watt compact fluorescent tube costs about $1.20/year to operate and yields about 900 lumens of light. A 60 watt incandescent bulb costs about $4.80/year to operate and produces about 855 lumens of light. Simply put, the fluorescent light is a more efficient device for producing visible light.

A new lighting technology is now on the horizon. It is based on the familiar light-emitting diode or LED. Recent discoveries demonstrate that it will soon be practical to use a gallium-based microchip sandwich to produce light. The first chip emits only blue light, but the second chip converts part of the blue light into yellow light. When the complementary blue and yellow colors combine, they are perceived by the human eye as white light. This combination will result in the world's most efficient light source, because it generates only visible light and no thermal energy. Under normal operating conditions, these LEDs are expected to deliver a tremendous cost saving, since they will never burn out or have to be replaced. Since their overall efficiency is about 90%, it is estimated that their use will eventually greatly reduce the nation's total electricity consumption for lighting.

Visible Light

The term light typically refers to that very narrow portion of the electromagnetic spectrum that can be detected by the human eye (see Figure 9-2). The wavelengths of this region range from 780 nm (red) to 390 nm (violet) (7.8×10^{-7} to 3.9×10^{-7} m). In between are the rest of the colors of the rainbow. Unlike paint pigments (red, yellow, and blue), the three **primary colors** of light are red, green, and blue. If these colors are mixed at equal intensities, white light is the result. Although it may be difficult to believe without a demonstration, all other colors of the rainbow are just combinations of these three. A combination of red and green produces yellow, a mixture of red and blue makes magenta, and blue and green produces cyan. This may not come as a surprise since color television and computer monitors use only a combination of red, blue, and green dots to produce all colors. The only color that cannot be made by mixing the primary colors is black. Black is actually the sensation that results when there is either an absence of light or the body absorbs all visible wavelengths of light.

The credit for unlocking the mystery of colored flames belongs to physicist Gustav Robert Kirchoff (1824–1887) and chemist Robert Wilhelm Bunsen (1811–1899). They found that when the light coming from a hot element was passed through a prism, it produced a series of separate, sharply defined lines, not a continuous spectrum. The device they constructed (shown in Figure 4-4) to study this phenomenon is known as a spectroscope. An empty cigar box, prism, and a discarded telescope were all they needed to make their spectroscope. With this instrument they observed and recorded the spectra of every available element and determined that the bright-line spectrum of each element was unique. As related in Special Topics Box 4-3, this method enabled the discovery of helium in the spectrum of the Sun years before it was discovered on Earth. Thus, the birth of spectroscopy gave astronomers a tool that enabled them to determine the elemental composition of stars hundreds of light-years away.

As shown in Table 9-1, each color of the spectrum is associated with a precise amount of energy ($E = hf$). Closer examination of the table reveals that as the wavelength becomes shorter, both the frequency and energy becomes higher. Considering all visible light, therefore, red light has the longest wavelength, lowest frequency, and carries the least amount of energy. In photosynthesis, for example, only very specific wavelengths of red (≈ 670 nm) and blue (≈ 440 nm) light have the exact amount of energy required to boost the electrons in chlorophyll into higher energy levels. Other pigments enhance this light-absorption capacity by capturing, in addition to blue and red wavelengths, yellow and orange wavelengths. None of the photosynthetic pigments, however, absorb the green wavelengths and consequently they are reflected. This is why plants appear green.

Finally, a word about measuring the intensity of visible light. The Latin word for candle is *candela*. As noted in Chapter 2, the candela (cd) is the base SI unit for measuring the intensity of light. After various other attempts, in 1979 the candela was redefined as the luminous intensity of a single-frequency light source operating at a power of $1/683$ watt per **steradian**. This is equivalent to 1.165×10^{-4} W/m^2 or 18.3988 milliwatts over a complete sphere centered at the light source. The light frequency selected, 540 terahertz, THz, corresponds to a wavelength of 555.17 nm (5.5517×10^{-7} m), which is yellow-green light with roughly the same visual brightness as ordinary daylight. It was selected because the human eye is more sensitive to this wavelength than to any other.

Ultraviolet

Ultraviolet light (UV), as shown in Figure 9-5, is that portion of the electromagnetic spectrum that has wavelengths ranging from 390 to 100 nm (3.90×10^{-7} to 1.00×10^{-7} m). Today the UV region has been further subdivided into three smaller regions. **UV-A** is sometimes called near-UV or **black light** (320 – 390 nm). **UV-B** is sometimes referred to as erythemal UV because it is composed of the wavelengths that convert ergosterol in the skin to vitamin D (290 – 320 nm). **UV-C,** which is composed of the shortest UV wavelengths, is the most powerful (160 – 290 nm). It is effective in killing one-cell organisms and, therefore, is sometimes called germicidal UV.

Most electromagnetic radiation falling in the UV range is invisible to the human eye and cannot be detected without use of a scientific instrument. Even on a cloudy day, beach goers need to be aware that UV radiation can penetrate clouds, mist, and fog. Snow skiers should be particularly careful to protect their skin and eyes, since snow reflects up to 80% of the incident UV rays. Staying in the shade does not provide complete protection either, since a considerable amount of UV radiation is scattered by the atmosphere and reflected from nearby surfaces.

Human skin is particularly affected by UV-B radiation and excessive exposure results in a reddening of the skin and sunburn. It has been reported that protecting the

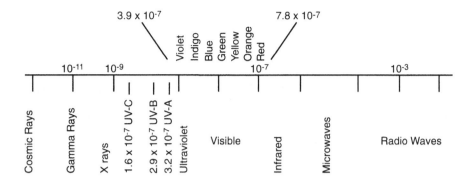

Figure 9–5: Electromagnetic spectrum by wavelength (m).

skin during the first 18 years of life is likely to reduce the risk of cancer by more than 50%. UV radiation carries enough energy to disrupt the structure of DNA molecules. Since DNA is responsible for sending chemical instructions, scrambled DNA can result in the production of cancer cells. More will be said about increasing levels of UV radiation over the Antarctic regions later in the chapter.

X rays

Electromagnetic radiation with wavelengths of approximately 100 – 10 nm (1.00×10^{-7} to 1.00×10^{-8}m) is classified as X ray. In 1895, Wilhelm Conrad Röentgen (1845–1923) discovered these mysterious rays while experimenting with a Crookes tube, an early cathode ray tube (CRT). The "X" assigned to this form of radiation is common practice in mathematics, where unknowns are often identified in this manner.

These rays were soon found to penetrate many solid forms of matter including the skin, muscles, and organs, but not bone. Because of their greater densities, bones, teeth, and the denser portions of our bodies absorb the X rays more efficiently than the softer parts. For this reason, X rays can be used to take pictures of the inner body parts, where the more dense substances appear as shadowy images. Because X rays have very short wavelengths and thus high energies, they are very penetrating and can damage or kill cells. Exposure to a few low-intensity X rays per year is not particularly harmful. However, x-ray technicians must use shielding to protect themselves from the thousands of exposures per year they would otherwise receive.

Industrially, X rays are used to inspect welds in such things as oil pipelines. X-ray images of our Sun yield important information about solar flares that affect most forms of electronic communications. Fortunately, our atmosphere shields us from most extraterrestrial X ray sources, as well as from gamma radiation sources.

Gamma Rays

In the continuos electromagnetic spectrum, gamma rays have the shortest wavelengths we will consider. They range from 10 – 0.001 nm (1.00×10^{-8} m to 1.00×10^{-12} m) and are even more energetic and penetrating than X rays. Gamma rays are produced by radioactive atoms undergoing fission such as during nuclear reactions.

Today medical experts use a variety of gamma ray sources to treat cancer, for diagnostic purposes, and to sterilize equipment and supplies. Industrial personnel use gamma rays to inspect such things as castings and welds. One of the more controversial uses of gamma rays is to kill microorganisms and retard spoilage in the food processing industries. Gamma ray images have enabled astronomers to determine important information on the life and death of stars and other violent processes in the universe.

Undoubtedly, gamma ray technology is in its infancy. As we gain scientific understanding of a given physical process, the potential for its application explodes onto the technological scene. New applications of scientific research are devised every day and electromagnetic radiation is at the center of much of it.

Technology Box 9–4 ■ X-rays

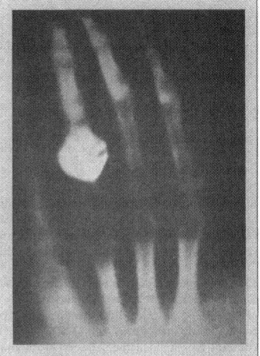

One of the more important noninvasive diagnostic procedures in use throughout the medical and dental professions is the x-ray. Discovered in 1895 by Wilhelm Conrad Röentgen (1845–1923), this short wavelength (1.00×10^{-7} to 1.00×10^{-8} m) high-energy form of electromagnetic radiation was found to pass through a variety of objects. Röentgen's experimentation with these X rays led to their most common modern medical use. He asked Frau Röentgen to place her hand on a photographic plate while he directed the X rays at it for about 15 minutes. The resulting photograph (right) was somewhat startling, since she could see the bones in her hand.

The technology used to produce x-rays today is continuing to provide better images and at the same time limit the amount of radiation exposure to both the patient and the technology professional. Typical exposure times, for example, are now measured in fractions of a second.

Röentgen was awarded the first Nobel Prize for Physics in 1901 for his discovery. He subsequently donated all the prize money to the physics department at Wurzburg, where the interest from the award money continues to provide student financial support.

One of the more important recent developments for the use of X rays is that of the CAT scan, developed in 1972 by Gofrey N. Hounsfield. CAT stands for computerized axial tomography. The equipment rotates 180 degrees around a patient, sending out X ray beams at various locations. The information obtained from these exposures is relayed to a computer, which turns it into a cross sectional picture, which may be viewed on a computer screen.

Photo copyright © SPL/Photo Researchers.

Check Your Understanding

1. Describe the wave by which all electromagnetic radiation is propagated.
2. What information is required to determine the frequency of an electromagnetic wave?
3. What constant is required to determine the energy of an electromagnetic wave?
4. Which type of electromagnetic wave has the longest practical wavelength?
5. What is the difference between AM and FM radio waves?
6. In what form do we most often sense the presence of IR radiation?
7. What are the three primary light colors?
8. What is the most common use for X rays?

9-2 The Behavior of Light Waves

Objectives

Upon completion of this section, you will be able to:

- Describe absorption and reflection of light.
- Explain the difference between real and virtual images.
- Explain the scattering of light.
- Make calculations for wave refraction based on Snell's Law.
- Explain total internal reflection, based on the critical angle.
- Describe diffraction and interference of light waves.
- Explain polarization and dispersion of light waves.

We see objects by the light they either emit or reflect. Light reflecting from white walls makes a room appear brighter and larger than if the walls are a dark color. Reflection of light from the curved surface within a headlight focuses it to a place in the darkness where it is needed to see the road. In general, when light falls on a surface it is partially reflected, while the remainder passes into the material. Within the material, it will be completely absorbed, if the material is **opaque** or only partially absorbed and partially transmitted if it is translucent. For example, depending on the angle of incidence, about 4% of the light rays striking an ordinary **translucent** piece of glass reflect from the front surface. The remainder of the light passes into the glass where some of it is absorbed and about another 4% of the light is reflected upon striking the rear surface. The rest of the light passes through. This explains why you can see your reflection, if the background behind the window is dark.

Absorption

To visualize what happens when electromagnetic radiation encounters matter, imagine a stream of photons (bundles of energy, each with a definite wavelength) encountering a vast array of orderly vibrating atoms. When photons strike an opaque surface, their energy is given up in a process called **light absorption**. Some surfaces absorb light energy and convert it directly into heat, while others may convert it into chemical energy, usually with some heat produced as a result of the energy transformation. This happens, for example, when light hits the film in your camera. You open the shutter for a brief moment and light is allowed to enter. The film reacts chemically to the light photons and thus records the various intensities of light waves and frequencies that strike it. See Technology Box 9-5 for a more complete description of this process.

Reflection

In addition to light absorbers, there are also light sources, reflectors, and transmitters. Stars, light bulbs, flames, and hot glowing objects are examples of light sources. Light reflections do not produce light but light will "bounce" off reflective surfaces. Any object that isn't a perfect absorber (blackbody) is a reflector. If light is reflected from an object with a smooth or even surface, as shown in Figure 9-6 (left), it is called **specular reflection** and is composed of many waves in a single well-defined beam. If the reflecting surface is uneven, as shown in Figure 9-6 (right), each reflected wave will be traveling at a slightly different angle producing a **diffuse reflection**. This means that the directions of the reflected light rays are not perfectly aligned, but are random or disorganized.

Many surfaces produce diffuse reflections, but both types can sometimes be produced by a single surface, under differing conditions. For example, if the surface of a lake is calm, incident light rays are reflected as parallel rays and you can see a clear image of your reflection. If wind is making the surface of the lake rough, either the reflected rays will be scattered in many directions and result in a distorted reflection or you will simply see the surface of the rough water itself.

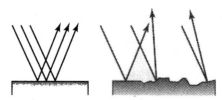

Figure 9–6: Smooth surfaces produce specular reflection (left) while uneven surfaces produce diffuse reflections (right).

Reflective surfaces such as smooth water, glass, and plane mirrors produce virtual images. As shown in Figure 9-7, a **virtual image** is one that appears to be behind the reflective surface as far as the object is in front of the surface. Our brain is tricked because we are accustomed to receiving and interpreting light rays traveling in straight-lines; a **light ray** being defined as the direction the light waves travel. The rays received

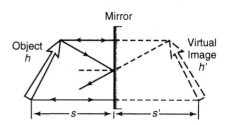

Figure 9–7: Virtual image of an object (arrow) formed by a mirror plane.

by the observer from a virtual image diverge from the virtual arrow and it appears as if it were behind the mirror. Notice that if we try to find the light rays of our image behind the mirror we cannot because they are not actually there. The rays of light do not actually pass through the mirror. This is the essence of a virtual image. The light rays only appear to come from the position behind the mirror. Virtual images cannot be projected on a screen.

The most straightforward discussion of reflected light rays begins with those from a flat mirror. It is understood that we see reflected images because the reflected light enters our eyes. When we analyze exactly how the light is reflected to our eyes, it is clear that the incident light rays reflect at specific angles. The incoming light ray and a line perpendicular to the reflective surface form the **incident angle**, which is labeled as θ_1 in Figure 9-8. The **reflected angle** is then the angle between the perpendicular line and the reflected light ray, which is labeled as θ_2. The angle of incidence and the angle of reflection are always equal. This is known as the **law of reflection**.

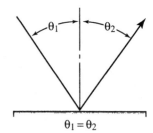

Figure 9–8: The angle of the incident light ray equals the angle of the reflected light ray.

Even in the case of uneven reflecting surfaces, as shown in Figure 9-6 (right), each incident light ray is still reflected at an angle equal to the incident angle. However, due to the irregularities of the surface, the reflected rays result in a diffuse image.

The law of reflection is straightforward if the reflecting surface is flat, but what happens if the reflective surface is curved? Even when the surface is curved the incident and reflected angles are equal with respect to a line perpendicular to the surface of the mirror at the specific point we are examining. Each light ray hitting the surface at that point still follows the law of reflection.

Technology Box 9–5 ■ Photographic Films

The word photography was first used in 1839, derived from two Greek words meaning "light" and "writing." The history leading to modern photography refers often to an ancient device called camera obscura. The term actually applies to any enclosed space that is lit by light passing through only a tiny pinhole. Although the image produced is inverted, it was used extensively by aspiring artists to assist in drawing.

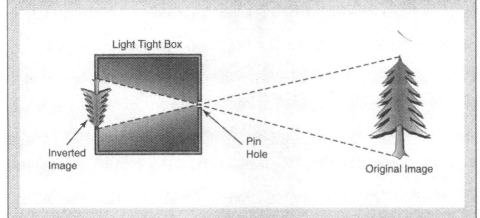

In 1777, Carl Wilhelm Scheele, a Swedish chemist, discovered that when silver chloride absorbs light it turns black. Although unexposed silver chloride dissolves in ammonia water, the exposed silver chloride does not. This suggested that an image formed by exposed silver chloride could be made permanent or "fixed" by washing the unconverted silver salt away.

We now know that when silver halides, such as silver chloride, $AgCl$, or silver iodide, AgI, are exposed to light, the energy of the photons is absorbed by electrons resulting in the silver ions being converted into silver atoms. Chlorine gas, Cl_2, is the other product. The overall equation for the photochemical reaction of silver chloride is: $2\ AgCl + 2\ photons \longrightarrow 2\ Ag + Cl_2$ Although there are billions of ion pairs in a single grain of silver halide, it takes the production of only a few atoms of silver to form to a stable latent image at that site.

Early in the nineteenth century Frenchman Joseph Nicephore Niepce turned his attention to the salts of silver. His big breakthrough came in 1822 when he made a permanent image using camera obscura and silver chloride-coated pewter plate. He then used the vapors from heated iodine crystals to darken the silver and heighten the contrast. This method inspired Louis Daguerre to a similar, but better developing process. In the Daguerreotype process, a polished sheet of silver was attached to a copper plate and made light sensitive by exposing it to iodine vapors, causing silver iodide, AgI to form. Because AgI is more light sensitive than $AgCl$, only 15 to 30 minute exposure times were required to make a suitable impression. The latent image was later developed by treating it with mercury vapor, followed by washes in a salt solution and hot water. The fixed image was permanent and relatively quick, compared to sitting for a portrait to be drawn. This process was so successful that it was purchased by the French government. Daguerre officially unveiled his discovery on August 19, 1839 – the same year an Englishman named William Henry Fox Talbot announced the development of a process he called photography.

When light rays traveling parallel to the principal axis strike a curved surface they experience one of two fates. In general, if it is a **concave mirror**, Figure 9-9 (left), the reflected rays will converge and appear to be originating from a point in front of the mirror. If they strike a **convex mirror**, Figure 9-9 (right), the reflected rays diverge and appear to be originating from a common point behind the mirror.

Concave mirrors have a wide range of applications because of their ability to make light rays converge or **focus**. When converging rays form an image, it is called a **real image**. A real image can be formed on a screen and viewed. Think of an image formed by a convex mirror. If the rays diverge on leaving the mirror, the image cannot be formed on a screen, but can be seen by looking into the mirror. Although it isn't a scientific use, fun houses have amused customers for years with mirrors that have surfaces with varying angles and reflect distorted images. We will have more to say about reflections from curved surface later in this chapter.

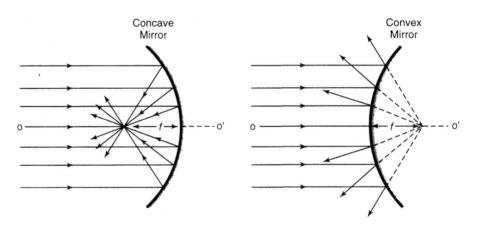

Figure 9–9: Reflections from concave (left) and convex (right) mirrors.

Special Topics Box 9–1 ■ The W.M. Keck Observatory in Mauna Kea, Hawaii

Most modern research telescopes rely on special concave mirror, called parabolic mirrors, to gather sufficient light from distant bodies in the universe to create visible images and photographs. The Keck telescope in Mauna Kea, Hawaii is the largest optical telescope in the world. Its 400-inch, 14-ton, computer-controlled, segmented primary mirror features four times the light-gathering capacity of the 200-inch telescope at Mt. Palomar and can be pointed anywhere in the sky. The Keck telescope is operated by the California Association for Research in Astronomy.

Photo courtesy University of Hawaii.

Scattering

In the previous section, we considered the fate of electromagnetic radiation reflecting from smooth surfaces, such as plane mirrors. In this section, we will explore the fate of electromagnetic radiation striking small objects that are in the same size range as the wavelength of the radiation.

All electromagnetic radiation interacts in some manner with the atoms and molecules it encounters. A special situation arises, however, when the electromagnetic radiation is of the same frequency as an allowable electron transition within an atom or when it matches the vibrational oscillations of a molecule. In these instances, the particle absorbs all of the energy of a photon and is energized from a ground to one of its excited states. Both atoms and molecules in an excited state quickly initiate a re-emission of photons. If a steady supply of energy is provided, the absorption and re-emission of radiation can reach equilibrium.

When electrons that have been excited into a higher energy level return to the ground state they emit a photon. The emitted photons travel in random directions, but their energies are equal to the photon absorbed. When atoms absorb and emit photons, the process is known as **scattering**.

When molecules absorb photons of specific frequencies, the result is in an increase in the vibrational oscillations of the molecule. This added energy is also quickly re-emitted in the form of a photon of the same frequency. Because the molecules are randomly orientated, the photons are scattered in every possible direction; hence the designation **elastic scattering**. Molecules of nitrogen, oxygen, and water in our atmosphere have ground state vibrational frequencies that are most likely to exhibit elastic scattering when the incident radiation is in or near the UV range. The closer the frequency of the incident radiation is to their vibrational frequencies, the more vigorously they will respond. This explains why UV radiation is widely scattered in our atmosphere. It also suggests that light in the blue end of the visible spectrum is more strongly scattered than red.

The lesser amount of red light scattering is demonstrated each day at sunrise and sunset. At these times, the Sun can only be viewed through a great thickness of atmosphere, so it appears as a reddish disk on the horizon. This also explains why, when the Sun is overhead it appears yellow-white. The reason is that very little scattering occurs as it passes through a much thinner layer of atmosphere. The sky appears blue because light travelling laterally scatters blue light downward, while the other wavelengths are transmitted forward. It is interesting to note that if there is no atmosphere, there is no scattering of sunlight and, therefore, the sky appears black as it does on the surface of the moon or in the depths of space.

Refraction

Light travels in straight lines in a homogeneous (uniform) medium. When a light wave passes from one medium to another and enters at an angle, it can be observed to abruptly change its direction if the speed of the wave is different in the second medium. This bending of a light wave as it passes from one medium into another is called **refraction**. When a stick enters water at an angle, it appears to be bent sharply where it enters the water. This is an example of refraction as the light passes from one material (water) to another (air).

As you know, nearly all types of glass, water, oil, plastics, and many other substances are transparent. That doesn't mean that the light passes through them unchanged, however. Not only is the direction of the light wave often changed but so is its velocity. Light waves can pass into a new medium without undergoing a change in direction, but only if they strike perpendicular (normal) to the surface of the new medium.

All photons travel at the speed of light. Yet, as they enter the new medium, the process of absorbing and re-emitting (scattering), affects their net velocity. Therefore, it must be less than c, depending on the medium and the frequency of the radiant energy. If the new medium has a higher **optical density** (passes light more slowly) than the first, then the net speed will be slower.

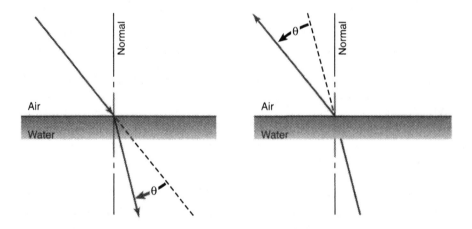

Figure 9–10: A light ray entering an optically denser medium (water, $n = 1.333$) bends angle θ, toward the normal (left) and a light ray entering an optically less dense medium (air, $n = 1.000$) bends angle θ away from the normal (right).

Special Topics Box 9–2 ■ Atmospheric Refraction: Mirages and looming

A commonly observed phenomenon is that light waves bend when they move from one medium into another if they have different refractive indices. An example of this can be seen when a pencil is put into a glass of water and the pencil appears to bend sharply at the point where it enters the water. Air volumes at different temperatures also have different refractive indices. Consequently, the image of an object often appears to waver when viewed through the turbulent, non-homogeneous air rising above a hot object.

On warm days, a layer of hot air often forms just above the pavement on the highway or desert sands, resulting in a mirage. Since light travels faster in the hot expanded air than in the cooler more dense air above it, light rays entering the warm air at an angle will be refracted upward, as shown. One may actually see inverted images of distant objects that are suggestive of the reflections on a pool of water. Mirages are often seen when driving on a hot day. But, when you reach the place that appeared wet, it has moved farther down the road.

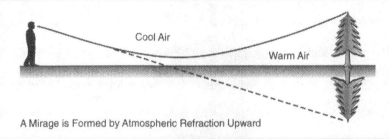

Cool Air

Warm Air

A Mirage is Formed by Atmospheric Refraction Upward

A less common type of mirage is called **looming** and occurs when the atmospheric conditions are reversed. In this instance, the lower layer of air, being next to snow or a cold lake, is cooler than the upper air. Rays of light from a distant object, therefore, are bent downward upon striking the upper warmer layer of air. Consequently, one may see an image of a building or a ship that is actually below the horizon.

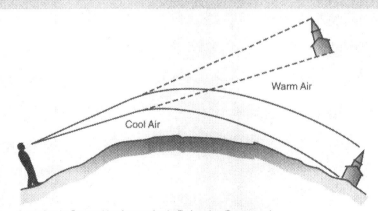

Warm Air

Cool Air

Looming is Caused by Atmospheric Refraction Downward

A ray of light entering the Earth's atmosphere at an angle is always bent toward the normal. We, therefore, can see the Sun when it is slightly below the horizon. This happens at each sunrise and sunset. You can tell that you are looking at a refracted image at these times because it has a flattened shape.

An important property of translucent materials is their **index of refraction**, n. This is the optical property that allows us to predict the angle the light will be refracted in the new medium. The light ray model will again be used to illustrate what happens to light when it enters a medium with a different index of refraction. When it passes from a medium with a lower index of refraction to a medium with a higher index of refraction, it is bent toward the normal. The reverse is also true; light passing from a medium with a higher index of refraction to a medium with a lower index of refraction is bent away from the normal. See Figure 9-10 (right).

To this point, we've seen that light can be reflected at the surface between two media or transmitted through the second medium. In many cases, as shown in Figure 9-11, it does a little of both.

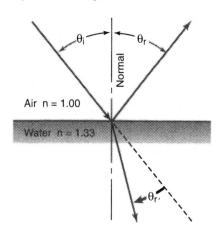

In summary, when considering reflection, the angle of incidence, θ_i, is always equal to the angle of reflection, θ_r, as shown in Figure 9-11. Incident light is refracted when striking a second medium at an angle. The experimental relationship between the angle of incidence and the angle of refraction is known as **Snell's law**. The degree of refraction is dependent upon the ratio of the two indices of refraction of the materials the light is passing through and given by the equation:

Figure 9–11: An incident beam of light striking the flat surface of water at an angle θ_i is refracted at angle θ_r' toward the normal as it passes into the water and reflected at angle θ_r from the surface.

$$n_i \sin\theta_i = n'_r \sin\theta_r$$

For example, if a light ray in water ($n = 1.333$) is incident at an angle of 40° on a piece of glass ($n = 1.517$), what is the angle of refraction for the light ray in the glass?

$$n_i \sin\theta_i = n'_r \sin\theta_r$$

$$\sin\theta_r = \frac{n_i \sin\theta_r}{n'_r}$$

$$\sin\theta_r = \frac{1.333\sin40°}{1.517}$$

$$\sin\theta_r = \frac{1.333(0.64279)}{1.517}$$

$$\sin\theta_r = 0.5648$$

$$\theta_r \cong 34.39°$$

(In the final step, the value of the angle, θ_r, was determined by taking the *inverse* sine of 0.5648.) A mathematical result of Snell's law is a limit to the incident angle that will still be partially refracted. Depending upon the ratio of the indices of refraction, the sine of the angle reaches a point where it is impossible to find a *real* answer to the equation. This value for the angle of incidence is called the critical angle, which will be discussed in the next section.

Total Internal Reflection

As suggested above and shown in Figure 9-12, when the incident light rays strike a glass-air interface at greater and greater angles, an angle is eventually reached at which all the light is reflected and none is transmitted. This is known as the **critical angle**.

The critical angle and resulting total internal reflection is what makes fiber optics work. The integrity of the signal transmitted over a fiber optic cable is dependent upon the total internal reflection to bounce the light waves, and the signal it carries, back and forth along the length of the cable. Total internal reflection is also at work in binocular prisms and other scientific instruments.

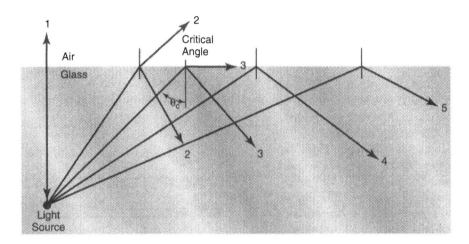

Figure 9–12: Incident ray 1 is perpendicular to the interface, so part of it reflects and the rest passes, without a change in direction. Incident ray 2 is both reflected and refracted. Incident ray 3 strikes the interface at the critical angle, θ_c, so the refracted ray will travel parallel to the interface. Incident rays 4 and 5 experience total internal reflection because they exceed the critical angle.

Technology Box 9–6 ■ Fiber Optics

In the 1790s, the French engineer Claude Chappe developed a successful optical communication system in the form of human operators sending messages from tower to tower using semaphore flags. In the 1840s, it was demonstrated that light could be guided by flowing water. In 1880, Alexander Graham Bell patented a telephone that used light to carry the signals, but his original invention using wire proved to be more practical. By the turn of the nineteenth century, inventors found that bent quartz rods could carry light and patented them as dental illuminators.

During the 1920s, the idea of using bunches of glass fibers to transmit images was granted a patent in the United States. In the mid 1950s, the American physicist Brian O'Brien suggested covering the bare glass fiber with a transparent cladding of lower refractive index to prevent signals from crossing between fibers and other contamination.

Technology Box 9–6 ■ Fiber Optics (Continued)

Today fiber optic cables are composed of microscopic strands of a light-transmitting medium surrounded by several other layers as shown (right). The core carries the light and is composed of a material with a high index of refraction. A thin layer of material with a lower index of refrac-

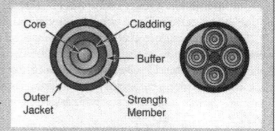

tion, called the cladding, is fused around it. For total internal reflection to occur within the core as shown, the angle of light incident must be greater than the critical angle. The typical core fiber for most light-carrying applications is in the range of 0.001 to 0.003 inches in diameter and has about 85% of its area composed of core material and 15% cladding. Finally, several fiber optic strands are bundled (right) into protective tubes including a strong layer of Kevlar®, the material used in bullet proof-jackets.

Sending information requires a transmitter to first convert the electrical signal into light waves of a specific frequency. Next, the light source launches the optical signal, in the form of data bytes, into the fiber where it travels through the core at the speed of light. At the other end of the cable, a light detector, ad-

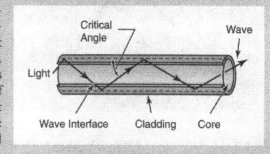

justed to a specific frequency, converts the optical signal back into an electrical signal.

Fiber optics, because of the vast number of light frequencies available, can simultaneously carry thousands of times more information than a copper wire. Current microwave systems, for example, have a maximum data rate of 1.544 megabits per second. Fiber optics systems can operate up to 100 times faster. Other advantages of fiber optics include low transmission error rates, immunity from radio frequency interruptions (RFIs), ease of installation, and they do not carry a current, thereby eliminating the potential for dangerous shocks. Fiber optic cables do require more protection than those containing copper and repairs must be made by a skilled technology professional.

Fiber optics is finding many applications in addition to communications. Arthroscopic and endoscopic surgical procedures, for instance, use fiber optics because they are very small and require only a small incision for the doctor to see into the body.

Source: Jeff Hecht, jeff@jeffhecht.com

References:

City of Light: The Story of Fiber Optics, by Jeff Hecht. Oxford University Press, New York, 1999.

College Physics, Fourth Edition, by Jerry D. Wilson/Anthony J. Buffa. Prentice Hall, Inc., Upper Saddle River, New Jersey 07458, 2000.

Diffraction

From the above discussion, we know that light waves travel in a straight line in a uniform medium, but change directions when there is a change in the medium. Careful observations show, however, that there is also a slight bending or flaring of light waves when they travel either around obstacles or pass through an opening that is of similar dimensions as their wavelength. This bending of waves is called **diffraction**. Not only light waves, but waves of all types undergo diffraction. In fact, this is one of the reasons given when arguing that light must be a wave phenomenon. In Figure 9-13, waves generated on the left side of a barrier are allowed to pass through a narrow opening of approximately the same size as the incident wavelength. The waves that form on the right side of the barrier can be seen to flare in ever widening semicircles.

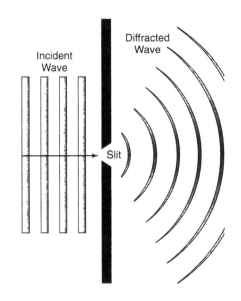

Figure 9–13: Incident waves passing through a slit of approximately the same size as their wavelength generate diffracted waves (right).

Because the wavelength is very small, the size of slit that diffracts visible light best should be equally small. When light passes through such small slits, it casts a fuzzy shadow, indicating that the light did not simply pass through in a straight line, but was diffracted.

Interference

In the introduction to this chapter, it was noted that Thomas Young demonstrated the interference of light, which could only be explained by accepting the wave theory of

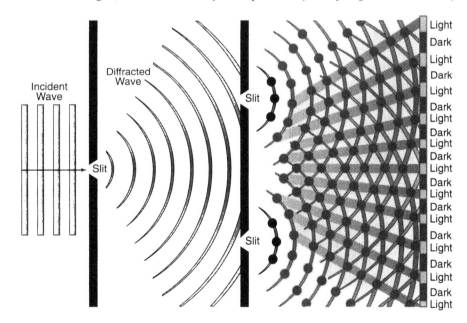

Figure 9–14: Interference pattern formed by light waves from two identical slits.

light. The experiment he conducted is called a double slit experiment because, as shown in Figure 9-14, when the waves from a single source strike two slits, new wavelets are formed. When the two wavelets pass, each continues as though the other is not present. However, at every point in that region, the resultant amplitude of the disturbance is the sum of the disturbance amplitudes created by the individual wavelets; that is, the waves interfere forming an **interference pattern**. If this interference pattern is projected on a screen, it will appear as a series of light and dark bands, called **interference fringes** because they resemble the fringe sometimes used on a rug or clothing. The light bands occur wherever the wavelet amplitudes add constructively, and the dark bands are due to destructive interference.

Another type of interference occurs when light strikes a thin film, such as a soap bubble. At the top surface of the film, some of the light is reflected and some is transmitted. At the bottom surface, light is again reflected and then transmitted back through the film to finally exit. With the light rays bouncing around in this manner, there is opportunity for interaction between all the transmitted and reflected rays. At certain places, the light waves add together, and at others they cancel, creating interference patterns. Soap bubbles, oil on the surface of water, and other thin films capable of reflection can be seen reflecting a glistening rainbow of colors as each color produces its own interference pattern depending on its particular wavelength.

Polarization

Previously it was discussed that electromagnetic radiation is a transverse wave composed of an electric field and a magnetic field that vibrate at right angles to each other. Since the random movement of electrons within the atom is what produces light, the light waves themselves will tend to vibrate in every possible orientation.

In 1669 a curious crystal, known as **Iceland spar** or calcite, was discovered. As shown in Special Topics Box 9-3, this crystal has the ability to produce double images of anything observed through it. The ability of this crystal to separate light was first studied by Huygens. Later others joined in the attempt, but it was Thomas Young who realized that if light were a transverse wave, it would explain the phenomenon.

A **polarizer** is a material with an internal structure such that it will only transmit light waves vibrating in a specific direction. As shown in Figure 9-15, when light passes through the polarizer, the light wave becomes a **linearly polarized wave** because it is composed of waves vibrating in only one plane. It's like threading a jump rope between two slats in the venetian blinds and creating a wave by smoothly jerking the rope side to side.

If the linearly polarized wave next passes through a second polarizer, known as the **analyzer**, that is oriented along the same axis, as shown in Figure 9-15 (left), the light passes and continues unchanged. However, if the analyzer is rotated perpendicular to the plane of the polarizer, then the remaining light waves are absorbed and no light passes, as shown in Figure 9-15 (right).

Sunlight reflecting from the surfaces of roads, water, snow, buildings, etc. is partially polarized in the horizontal direction so our eyes receive lots more horizontally polarized light than vertically polarized light waves. The lenses in polarized sunglasses are oriented to absorb the horizontal and transmit the vertical waves, thus decreasing the unwanted glare of reflected sunlight. Try observing reflected light while slowly rotating your sunglasses 180° or more. Do they seem to detect the polarization of skylight itself?

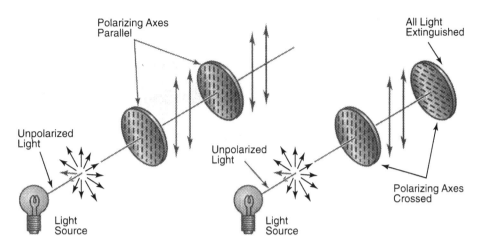

Figure 9–15: Production and extinction of linearly polarized light using a polarizer and analyzer.

Special Topics Box 9–3 ■ Birefringence in Calcite

An interesting phenomenon occurs when we look at an image through a piece of calcite. As shown in the figure, the image appears double. Most materials are isotropic. This means that their internal structure is such that they exhibit properties with the same values, such as the velocity of light transmission, when measured along each axis. If everything in the world were isotropic, there would be symmetry and life would be much simpler.

Some transparent solids are optically anisotropic, meaning that their crystal lattice and properties are not the same along all three axes. In calcite, the crystal is isotropic along two axes, but anisotropic along the third. Crystals having this property are birefringent or double refracting.

A birefringent material has two indices of refraction, one index for the two directions that are isotropic and another along the anisotropic axis. The rays that travel through the crystal, like they would through glass, are called the ordinary rays (o-rays). The other rays are called the extraordinary rays (e-rays). In calcite, the indices of refraction for the o- and e-rays are 1.6584 and 1.4864 respectively. This gives total internal reflection critical angles of 37.08° for the o-rays and 42.28° for the e-rays when in contact with air. When a ray of light enters the calcite crystal, it splits into two rays, one that travels faster than the other. As these two rays exit the crystal, they are bent at two different angles because the angle is affected by the speed of the ray.

When viewing an image through the crystal, you will see it twice. The two light rays in a doubly refracting crystal are linearly polarized in mutually perpendicular directions. This means the two light rays that emerge are vibrating in two planes, perpendicular to one another. How could you verify this fact using sunglasses that have polarized lenses?

Photo copyright © M.Claye/Jacana/Photo Researchers

Dispersion

Refraction and the difference in refractive indices of the different wavelengths of light explain how white light, passing through a prism, is split into a rainbow of colors as shown in Figure 9-16. The violet light experiences a greater change in its direction, i.e. is refracted more, than the red light. **Dispersion** refers to the breaking of white light into the separate colors of the spectrum.

This same phenomenon explains how rainbows are formed. Sunlight passing through water droplets is dispersed into the various colors. See Special Topics Box 9-4.

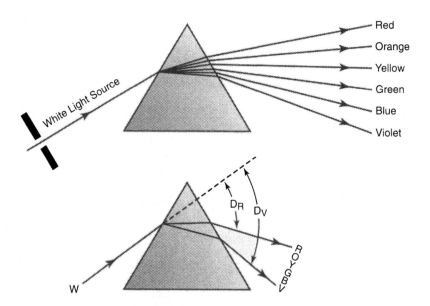

Figure 9–16: The angle that light is refracted when passing from air into glass depends upon its wavelength. Shorter wavelength violet light is refracted the most and longer wavelength red light is refracted the least.

Check your Understanding

1. What is the law of reflection?

2. What is the difference between specular reflection and diffuse reflection?

3. How do real and virtual images differ?

4. What does the property of index of refraction describe?

5. What is Snell's law and its use?

6. How do diffusion, diffraction, and dispersion differ?

7. With respect to total internal reflection, what is the critical angle?

8. Describe how ordinary and polarized light waves differ.

Special Topics Box 9–4 ■ Refraction, Dispersion, and Rainbows

Rainbows are unquestionably the most famous of all optical phenomena occurring in the atmosphere. They can be seen when rain is falling in front of you and the Sun is at your back. As the rays of sunlight enter the raindrop, they are slowed and bent, with the violet wavelength refracting and dispersing more than the red. Although most of the light passes through, some of it strikes the backside of the drop at greater than the critical angle (48° for water). When this happens the light bounces off the back of the drop and is internally reflected back toward our eyes.

Since each wavelength of light bends a bit differently, every color emerges at a slightly different angle: 42° for red and 40° for violet light. This produces a rainbow that is red on the outside and violet on the inside, with all of the other colors in between.

On occasion a larger, fainter secondary rainbow can also be seen above the primary bow. The second bow is the result of sunlight entering the raindrop at an angle that allows it

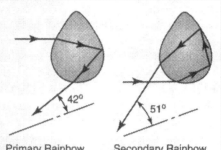

Primary Rainbow   Secondary Rainbow

to make two internal reflections, as shown. Each reflection within the drop weakens the light intensity and results in the dimmer bow. This double reflection also reverses the colors of the second rainbow, with violet now on the outside and red on the inside.

9-3 Infrared and Ultraviolet Radiation

Objectives

Upon completion of this section, you will be able to:

■ Describe various methods of infrared detection and applications.

■ Discuss ultraviolet radiation and its applications.

■ Describe some of the harmful effects of UV radiation.

Electromagnetic energy is pure energy. That is, it is never associated with mass, nor does it ever exist as a stored type of energy. The Sun emits electromagnetic radiation across many wavelengths, but most of the radiant energy leaving the Sun is in the form of either visible light or infrared radiation. In fact, the amount of this energy that strikes the Earth each second is greater than the total energy consumed in the United States each year. It is our good fortune that visible and infrared radiation can travel through space, for this radiation represents the fuel that powers all life.

Unlike visible and IR radiation, much of the shorter wavelength UV radiation is filtered out or scattered in the upper atmosphere. This is especially true at low Sun angles that occur in winter and in the early and late part of each day. UV radiation in the 300 nm range is energetic enough to break carbon-carbon bonds and is largely responsible for many of the damaging effects of sunlight. Although the human eye is not very sensitive to UV radiation, it appears that some insects use it to locate specific flowers. It has been shown that pigeons can recognize patterns when lit only by UV light – an ability they may use to navigate, even on overcast days.

In this section, these spectral regions will be revisited. Included in the discussion will be some of the emerging technologies that are based on these wavelengths.

Infrared Radiation

When you are in the Sun or near a heat source such as a blazing fire or steaming radiator, the sensation of warmth you feel is due to the absorption of infrared radiation. All objects that are warmer than absolute zero emit power in the infrared – even ice cubes. As the temperature of an object increases, the amount of power it gives off rises very rapidly. This is why hot objects radiate so much energy and feel warm from a distance – because the power they emit increases to the fourth power of its absolute temperature. In other words, if the absolute temperature of an object is doubled, the intensity of radiation it emits increases by a factor of 2^4, or 16 times.

Infrared radiation is readily absorbed by most materials, but typically does not penetrate beyond the surface. The photons associated with infrared radiation are converted into thermal energy but increasing the vibrational and translational motion of the molecules results in a temperature increase called heat. Infrared radiation is typically not sufficiently energetic to cause damage to living tissues but has many practical and scientific applications.

Not all objects emit radiation equally, as explained in Chapter 6. If you place your hand very close to someone's face after they have been exercising, you will sense warmth. A slight rise in one person's body temperature can be detected by another because humans are very good emitters of radiation. Our bodies emit radiation at a rate of about eight times our resting metabolic rate or 850 watts. Thus we depend upon absorption of infrared radiation from objects around us to make up the difference. Beyond custom and fashion, clothing also serves the important role of minimizing heat loss from our bodies.

Technology Box 9–7 ■ Infrared Remote Devices

When is the last time you changed the channel on your television by actually pushing a button on the TV? Today, many of us use remote controls – perhaps even several of them – to activate and operate stereos, DVDs, CD players, VCRs, and other electronic devices. To reduce the clutter of several remotes, it is now common to purchase a single, programmable remote control that will operate them all.

A remote control is simply a transmitter of a coded series of pulses of infrared light that is sent to a receiver built into the television or other electronic appliance. The pulses of infrared radiation are generated with special light-emitting diodes (LEDs) operating in the 30-40 kHz frequency range. This range was selected in an attempt to avoid possible interference from other IR sources.

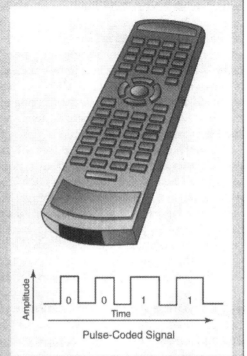

Pulse-Coded Signal

The infrared signal from the remote control to the appliance is transmitted in the form of a binary code that is now similar among all manufacturers. In fact, there are really only three different ways that the manufacturers code the signals. They either vary the length of each pulse, vary the space between pulses, or alter the order between spaces and pulses. A typical series of binary pulses of varying pulse widths is shown.

Infrared Applications

Many common devices, and even a greater number of scientific and technological processes, rely upon the emission, absorption, transmission, and detection of infrared radiation as discussed in Technology Box 9-7. Remote controls emit a signal that is detected by a sensor in a television set; personal electronic organizers communicate with personal computers or other organizers in a similar fashion, by emitting and/or detecting an encoded infrared signal. Since infrared radiation is longer-wave radiation than visible, we cannot see it; and since signals in these devices are transmitted at a very low power, we cannot feel this radiated energy either.

There are really only two fundamentally different types of infrared detectors. One is the **thermal detector**, a device that converts radiated power into a more readily measured parameter as the temperature of the device changes. A thermometer with the bulb painted black (so that it absorbs infrared radiation efficiently) is an example of a thermal detector. As incident radiation increases, the liquid inside the thermometer warms and expands, causing the length of the liquid column to get longer. Thus we have a device that provides a rather crude measurement of radiation. A **bolometer** is a very sensitive thermal detector whose electrical resistance varies with temperature and which is used in the detection and measurement of feeble thermal radiation. It is especially suitable for the study of infrared spectra. A **thermopile** is another type of thermal detector. This device is made of a series of thermocouples, devices that generate a voltage depending upon their temperatures, such that an increase in infrared power causes an increase in thermopile voltage.

The more responsive types of infrared detectors are called **photon detectors** or **quantum detectors**. These generally respond to incident radiation on an electronic level and do not require heating of the detector as with thermal detectors. Thus, they respond much faster than thermal detectors. Photon detectors are manufactured from semiconductors and function on the principal that infrared photons striking the detector will give up their energies to raise an electron to its conduction band where it is free to move along a circuit. Therefore, photon detectors respond to either increasing levels of incident radiation by creating a higher current through, or voltage across the device. This change can subsequently be measured and interpreted to determine the magnitude of the radiation.

A **thermal imager** is a device that uses an array of infrared detectors coupled with special optics and control electronics to produce a thermal image, often called a thermogram or heat picture. A thermogram is composed of many pixels, like a television picture, but the signal associated with each pixel is generated by a single infrared detector. See Figure 6-28 for an example of a thermogram. Although the reprint of the thermogram is in black and white, the original includes colors that depict the different temperature zones. For example, the inside corners of the eyes are considerably warmer than other areas on the face. The extremities, such as the nose, ears, and especially the hair appear cooler than other parts of the face.

The applications of infrared detection and imaging are as widely varied as one might imagine given that many natural and man-made processes involve temperature change. Temperature is a good indicator of whether or not a process is occurring as designed. Industries are using infrared technology in specially designed ovens for drying painted and enameled surfaces, leather, metals, papers, and textiles. In chemistry laboratories, infrared spectroscopy has been used for years to quickly identify unknown organic compounds. The paper, electronics, and steel industries use infrared detection for process control. The power utility industry utilizes infrared imaging to monitor the operating conditions of many different types of equipment such has electric motors and pumps; electric switch gear, motor control centers, and fuses; and valves and steam traps. The power industry also puts the technology to such uses as locating boiler and condenser leaks, monitoring stack emissions, and testing steam tubes within a boiler.

Medical infrared imaging is a useful diagnostic tool. Infrared or heat lamps are used to treat skin diseases and relieve the pain of sore muscles. Infrared imaging is also used to detect heat loss in buildings and in testing electronic systems. Infrared satellites are being used to monitor the weather, to study vegetation patterns, to study geologic features, and to monitor ocean temperatures. Law enforcement agencies use infrared imagers for surveillance. Firefighters with access to infrared vision equipment are able to see through otherwise impenetrable smoke during search-and-rescue operations. The military utilizes land-, air-, and satellite-based infrared systems to characterize battle scenes, and to identify vehicles such as tanks and aircraft. Hand-held infrared-based detectors of chemical weapons agents are also under development.

Finally, infrared detection systems have for many years been designed, constructed, and utilized by the National Aeronautics and Space Administration (NASA) onboard aircraft and the space shuttle to measure gases in the atmosphere. These measurements are critical to providing data to atmospheric scientists needed for validation of environmental models.

The examples provided here make up only a short list of the many applications of infrared detection and imaging in use or under development today. New techniques and improved systems continue to be developed at a remarkable rate. The *International Society for Optical and Instrumentation Engineers* (SPIE) is an excellent resource for those who wish to learn more about commercial and military applications of infrared detection and imaging.

Technology Box 9–8 ■ Using Infrared to Image Carbon Monoxide

Carbon monoxide (CO) is a dangerous gas, a by-product of combustion that is invisible to the human eye, tasteless, odorless, and can cause illness or death if present in sufficiently high concentrations. NASA scientists and engineers developed a technique to determine the amount of carbon monoxide present in the air by taking advantage of the fact that it absorbs infrared radiation only at certain wavelengths.

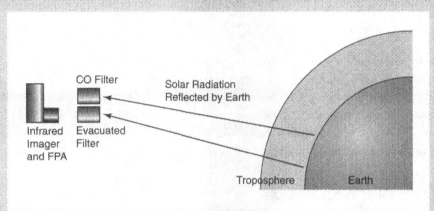

The technique is called Gas Filter Correlation Radiometry, or GFCR. In principle, the technique is rather simple. Infrared radiation from the Sun passes through the atmosphere reaching the Earth where it is reflected back into space. During this process, some of the specific wavelengths of infrared radiation are absorbed by carbon monoxide. An infrared imager situated onboard a spacecraft would measure the reflected signal twice – once after it passes through an evacuated filter and a second time after it passes through a filter containing carbon monoxide gas. In areas with a low concentration of CO, the difference between these two signals is greater, and in locations where concentrations of CO are higher, the difference between these two signals is lower. This inverse relationship exists because greater concentrations of CO in the atmosphere absorb more radiation, leaving less to be absorbed as the radiation passes through the CO filter resulting in a raw signal closer in strength to that corresponding to radiation passing through the evacuated filter.

Ultraviolet Radiation

Like IR, the ultraviolet wavelengths of electromagnetic radiation are invisible to the human eye. As shown in Figure 9-5, the region is composed of wavelengths generally falling between 100 – 400 nm. This places UV radiation between visible violet light and X rays.

UV radiation is more energetic than radio waves, microwaves, and visible light. In fact, the photons of UV radiation have enough energy that they can disrupt certain types of chemical bonds. Consequently, ultraviolet radiation can have damaging effects on both living and nonliving materials, including the ability to kill some microorganisms and degrade polymers.

Earlier in the chapter, it was discussed that the ultraviolet region is subdivided into UV-A, UV-B, and UV-C. Fortunately, most of the UV radiation incident on the Earth is absorbed by ozone in the stratosphere (O_3 + UV-C photons → O_2 + O), 15–30 kilometers above the Earth. There is an abundance of ozone in this region, high-energy radiation with wavelengths shorter than 280 nm is responsible for its production. The absorption of the UV radiation and the formation of ozone happen in two steps: first, the UV radiation breaks the oxygen-oxygen bond in molecular oxygen, O_2, producing two highly reactive single oxygen atoms, O. Second, the single oxygen atoms react with other oxygen molecules producing ozone, O_3. The equations for these two reactions are as follows:

$$O_2 + \text{UV-B photon} \longrightarrow 2\,O$$

$$O + O_2 \longrightarrow O_3$$

Although this process absorbs approximately 90 to 95% of the 200 to 300 nm range of sunlight striking the stratosphere, it is still important to use preventative measures during prolonged exposures to the Sun. (See Special Topics Box 9-5) UV-B and UV-A have been linked to many harmful biological effects, including crop damage, cataracts, photosensitivity reactions, and various types of skin cancer.

Effects of UV Radiation

One important negative effect of UV radiation is on synthetic polymers. A technique was developed at NASA's Lewis Research Center to quantify polymer surface damage that had been induced by exposure to UV radiation. Satellite-based instruments placed in low-Earth-orbit (LEO) were exposed to solar ultraviolet radiation, which caused the polymers to become brittle through bond breaking and crosslinking. This UV damage increased the surface hardness, which in turn affected the durability of the polymer. The technique was used to evaluate changes in the surface hardness of fluorinated ethylene-propylene (FEP) Teflon samples that received varying solar exposures during 3.6 years on the Hubble Space Telescope.

In 1985, a group of British scientists discovered that the levels of ozone over the Antarctic had dropped by between 40 and 50%. This seasonal depletion of ozone that now occurs each spring is referred to as the **ozone hole**. It is believed that chlorofluorocarbons or CFCs are primarily responsible for the decline in ozone in the upper atmosphere. These chemicals were at one time being widely used as propellant gas in aerosol spray cans, refrigerants, and as a blowing agent when making plastic foam. In 1987, the hole reached nearly the size of the United States. This means it was large enough to spread from the pole out over southern Australia, Tasmania, and New Zealand. That year, the ozone layer over the city of Melbourne was 12% less than usual.

Exposure to small amounts of UV radiation help promote good health, but over-exposure can be deadly. Even one-percent reductions in ozone levels can cause devastating effects on humans, animals, and plants. Because Australia is located extremely close to the Antarctic, its hundreds of unique native flora and fauna such as kangaroos, koalas, and platypuses may be in grave danger of becoming extinct.

Australia already has the world's highest incidence of skin cancer. It is estimated that for every 1% increase in biologically active UV rays, there will be an increase of about 3,000 new cases of skin cancer per year. The importance of taking measures to alleviate the conditions that promote further depletion of the ozone layer is clear.

Special Topics Box 9–5 ■ UV Light, Tanning Booths, and Sunscreens

Natural sunlight contains the complete (390 – 100 nm) ultraviolet spectrum that lies between the visible and X ray regions. It is subdivided into UV-A (above 320 nm), UV-B (320 – 290 nm), and UV-C (below 280 nm). UV-A is closest in wavelength to violet light and the least energetic, while UV-C is closest to X rays, the most energetic and, fortunately, mostly blocked by the Earth's atmosphere. All UV radiation possesses enough energy to promote electrons to higher energy levels and, in some cases, ionize the atom. Either event may result in an unwanted chemical reaction or death of the organism.

The effect of UV radiation on humans ranges from vitamin D production and suntans to causing sunburns, premature aging of the skin, and cataracts. Long-term UV exposure increases the chance of cellular DNA damage that can lead to skin cancer or melanomas. In spite of this, young and old alike often believe that they look better, and so feel better about themselves, if they have tanned bodies.

Tanning itself is a UV-driven chemical reaction that occurs within the epidermis, where maturing squamous cells and melanocytes reside. UV-B activates the melanocytes to produce melanin, which protects the underlying dermis and is what gives your skin the tanned appearance. The underlying dermis is responsible for the skin's elasticity, shape, and strength. Elastin and collagen give skin its strength and ability to snap back, and both are extremely vulnerable to, and can be destroyed by, UV-A rays.

A quest for the perfectly tanned body has resulted in commercial tanning salons. In the salon, the customer is exposed to carefully controlled amounts of UV-A and UV-B, which is advertised as perfect balance for developing a deep, dark, healthy-looking tan. Studies show that children and adolescents are harmed more by UV-B rays than adults. It is therefore suggested that you protect your skin by limiting all unnecessary UV exposures; wear protective eye goggles in tanning salons and be aware that certain medications can intensify the harmful effects of UV radiation.

All sunscreens on the market today contain some kind of chemical blocking agent that limits the penetration of UV radiation through the epidermis. The molecules in most sunscreens are composed of alternating single and double carbon bonds and are, therefore, capable of absorbing the high-energy UV photons and re-emitting the energy as heat. No chemical sunscreen blocks 100% of the incident UV radiation. Rather, they are rated by a Sun protection factor (SPF) number. The SPF number is to give you some idea of how long you can stay in the Sun without burning. For example, a SPF 15 sunscreen absorbs about 90% of the incident UV-B radiation. This means that if you normally burn in 10 minutes without sunscreen, you should be protected from burn for 150 minutes. A SPF 30 absorbs about 96.7% and a SPF 40, about 97.5% of incident UV-B radiation. While all sunscreens provide some level of protection against UV-B, no product screens out all UV-A. A final factor that should be considered is the effect of multi-day UV exposures. A significant multi-day exposure to sunlight (e.g., all day Saturday and Sunday) increases the sensitivity of the skin to UV damage on the second day of exposure, even when a sunscreen is used.

Special Materials

The most familiar type of glass is composed primarily of silicon dioxide, SiO_2 and is used to make bottles, windowpanes, lenses, mirrors, and an array of other optical devices. Common glass, however, does not pass all wavelengths of electromagnetic radiation equally. As technology continues to find new applications for IR and UV electromagnetic radiation, there is a need to find other types of glass or glass-like materials to make optical components.

The choice of materials used for optical components depends largely on the range of wavelengths, or the **waveband**, of interest. Common glass, for example, is opaque to the transmission of electromagnetic radiation at wavelengths longer than 2 μm (2×10^{-6} m), so it is not useful for making optical components that operate in the IR waveband. Other types of glass may allow some, but not all, UV wavelengths to pass. In this section, we will take a brief look at some of the materials that are being used for these wavebands. In general, the optical material chosen should be the one that has the fewest objectionable limitations. The following is a list of some of the factors that should be considered:

— a comparison of the operation bandwidth to the spectral range of the material;

— the material with the maximum transmission and minimum reflection, within the required operational range;

— sufficient strength, in the case of instrument windows, to withstand extreme atmospheric pressures;

— sufficient durability, such as scratch resistance or insolubility in water;

— the desired spectral rejection range to block wavelengths other than those that are desired for the application.

While we will not consider detailed information here, a short description is presented for some of the materials that are more commonly used in scientific applications.

Borosilicate crown glass and UV-grade fused silica are the materials most commonly used for standard applications in the visible and ultraviolet wavebands. Common glass lenses block the shortest wavelength UV, transmitting only between 32.0 and 220.0 nm ($3.20 – 22.00 \times 10^{-8}$ m). Fused silica lenses transmit between 19.0 and 250.0 nm ($1.90 – 25.00 \times 10^{-8}$ m). Windows and lenses made from these materials can be coated with an anti-reflection coating, and the coating can be customized to enhance transmission at specific wavelengths. The anti-reflection coating can reduce surface reflections to 1.5 to 2.0% per surface. These materials can be fashioned into lenses of a great many different shapes, diameters, and focal lengths, depending upon the application.

Since the common glass lens described above transmit only the very shortest IR wavelengths, special materials are required for many IR applications. Some of the more common materials include sapphire, calcium fluoride (CaF_2), zinc selenide (ZnSe), and Germanium (Ge). Sapphire and calcium fluoride appear much the same as common window glass; they allow visible light as well as IR wavebands to pass. Zinc selenide, however, transmits only above 550 nm (5.50×10^{-7} m) and germanium appears purplish and is completely opaque to visible light.

The properties of each of the above materials vary considerably. It is common, therefore, for vendors to list the specific characteristics for each window, lens, etc. they sell. The cost of such devices are high compared to common glass; an important consideration when deciding on the material to purchase for a given application.

Check Your Understanding

1. Why is electromagnetic energy referred to as pure energy?

2. At what rate do human bodies emit IR radiation, and by what two ways do we maintain our body temperature?

3. What is measured by a bolometer?

4. What are five considerations that should be taken into account when selecting materials for the optical components for a given IR or UV application?

9-4 Optics

Objectives

Upon completion of this section, you will be able to:

■ Describe the effect on parallel rays of light passing through convex and concave lenses.

■ Draw a ray diagram for thin lenses and explain image type and location.

■ Apply the proper sign convention and solve problems using the thin lens equation.

■ Calculate the linear magnification of an object.

■ Discuss some of the considerations for selecting the appropriate optical material for a given application.

■ Calculate and discuss the image type and distance produced by concave and convex mirrors.

■ Calculate and discuss the image type, distance, and magnification for a two lens optical system.

■ Calculate the angular magnification for a two lens optical system.

■ Explain the reason for spherical and chromatic aberration in lenses.

In this chapter, you have learned that electromagnetic waves, regardless of intensity, wavelength, or energy, can be harnessed to generate effects that serve a multitude of applications. Many modern conveniences, such as lighting, cellular telephones, microwave ovens, and thermal imagers, share at least one common attribute – they rely on the generation, transmission, and/or detection of electromagnetic radiation.

The first microscope (1590) and telescope (1608) were made in Holland. These are only two of many devices such as cameras, projectors, and eyeglasses that use glass lenses to gather and focus light from an object to form a visible image. Lenses and mirrors can be used to create images that appear larger or smaller, nearer or farther, or upright or inverted as compared to the original object.

Lenses

Have you ever noticed how underwater objects, when viewed from above and at an angle to the surface of the water, appear shallower than they actually are? This is caused by refraction, a phenomenon introduced earlier in this chapter. Since our brain processes optical signals assuming they are traveling in a straight line, when refraction occurs we formulate an image that underestimates their true depth, as shown in Figure 9-17.

A lens is simply an optical device that uses refraction to change the apparent location of an object, by altering the path of the light rays originating from

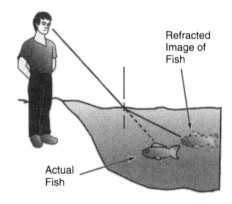

Figure 9–17: Refraction bends light rays causing our eyes to perceive it at a shallower depth.

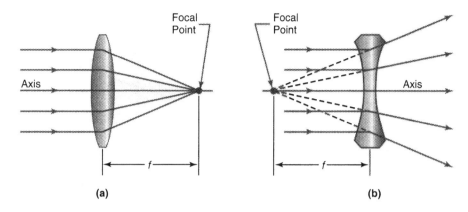

Figure 9–18: Image formation by a converging lens (a) and diverging lens (b).

the object. Figure 9-18a shows a lens through which parallel light rays pass and bends toward the center axis of the lens. It is called a **converging lens** or **convex lens** and the point at which the parallel rays meet is called the **focal point**. The **focal length**, f, is the distance from the center axis of the lens to the object and is always considered a positive number.

Another, less commonly used type of lens, is a **diverging lens**, also known as a **concave lens**. This kind of lens, pictured in Figure 9-18b, causes parallel rays to diverge or move apart. The point from which the rays appear to diverge is called the focal point of the lens and the focal length is again the distance from the center axis of the lens to the focal point.

Image Formation

Figure 9-18 is often called a **ray diagram** because it illustrates the path of the light rays. The horizontal line passing through the center is called the **principal axis** of the lens and the intersection of these rays, after passing through a lens, indicates the effect that the lens has on our perception of the object. The distance at which light rays converge from the lens is related to the size of the image that is produced.

Consider the object placed to the left and beyond the focal length of the converging lens in Figure 9-19a. To keep the drawing uncluttered, only one ray from each point on the image (base and tip) is traced through the lens. In actuality, there are many rays of light leaving each point of the object, and they emanate outward in all directions. Many of these rays, however, are captured and re-directed by the lens. In an ideal converging lens, all rays originating from a single point on the object will converge at a single point on the opposite side of the lens. Furthermore, an ideal converging lens will cause all points of convergence to occur at the same distance from the lens.

As shown in Figure 19a, an inverted real image is formed at a distance, s', to the right of the lens. If a screen were placed at this point a real, inverted image would appear. As previously noted, it is called a real image because the rays actually come together to form the image. The distance, s', is called the image distance and the distance, s, between the object and the lens is the object distance. A sharply focused image will appear only at this point, where the rays converge. If the screen were placed even slightly closer to or farther away from the lens, the image created would appear somewhat blurry or out of focus.

Now consider an object placed closer than the focal length of a converging lens. The rays from the object diverge as they pass through the lens and result in a virtual image behind the object. (Virtual rays and virtual images are usually indicated with dashed lines rather than solid lines and virtual images are always located on the same

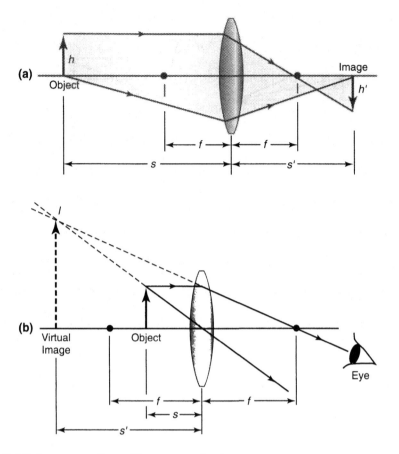

Figure 9–19: Image types formed by a converging lens.

side of the lens as the object.) In the case shown in Figure 9-19b, a screen placed to the right of the lens would serve no useful purpose. The lines representing the diverging rays do converge, however, on the left side of the lens. By looking through the lens as shown in Figure 19b, our eyes would be tricked into believing that an image exists farther from the lens than the object.

You should note that whenever an object is placed at a distance equal to or greater than the focal length of a converging lens, a real image is formed. However, when the object is placed between the lens and its focal point, a virtual image results.

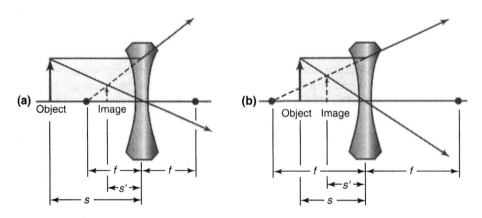

Figure 9–20: Diverging lens form only virtual images of real objects.

As shown in Figure 9-20, diverging lenses also form virtual images. In Figure 9-20a the object has been placed to the left and beyond the focal length of the diverging lens. A virtual image that is smaller than the object is formed at a distance, s', on the left side of the lens. When the object is placed closer than the focal length of the lens as in Figure 9-20b, a virtual image smaller than the object is formed at a distance, s', on the same side of the lens. In conclusion, diverging images produce only virtual images of real objects while converging lens may produce either real or virtual images, depending upon the location of the real object with respect to the focal length of the lens.

Thin-lens Equation

A **thin lens** is defined as one in which its thickness is considered small compared to the distances associated with its optical properties. As shown in the above figures, the actual thickness of each lens is much smaller than either their focal length, radii of curvature, or object and image distances involved.

The ray diagrams shown in Figures 9-18 through 9-21 are very useful in understanding the basic properties of converging and diverging lenses. They are also useful in determining if an image will be real or virtual and in predicting its orientation and approximate location. There is a much more accurate method, however, that can be used to determine their exact position. The image location, s', can be determined exactly using the following equation:

$$\frac{1}{f} = \frac{1}{s} + \frac{1}{s'}$$

in which f and s represent focal length and object distance respectively. We can use this equation to solve for any of the three parameters, given the other two. Many modern optical devices including projectors, telescopes, microscopes, even eyeglasses, use lenses for which the thin-lens equation applies.

The thin-lens equation has its basis in simple geometry. When applying the equation, it is important to consistently use the proper sign convention. When light passes through a thin lens from left to right, then the following convention applies:

— The focal length, f, is positive for converging lenses and negative for diverging lenses.

— The object distance, s, is positive when the object is placed to the left of the lens and negative if it is placed to the right.

— The image distance, s', is positive for real images formed to the right of the lens and negative for virtual images formed to the left.

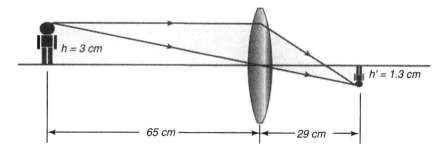

Figure 9–21: Ray diagram shows image produced by toy soldier.

Technology Box 9-9 ■ Nearsightedness and Farsightedness

The conditions called nearsightedness and farsightedness are caused when the eye is unable to focus the rays of light properly. Nearsightedness occurs when the image of a distant object is focused in front of the retina rather than on it. This condition can occur if the cornea and relaxed lens are too curved. When this is the case, objects up close can be seen clearly, but distant objects appear fuzzy and out of focus. In order to correct for this condition, a diverging eyeglass lens is placed in front of the cornea. This causes the rays from an object to diverge slightly as they pass through the lens so that when they pass through the lens of the eye, an image is formed farther back in the eye. The figures below shows image formation in a nearsighted eye, without (a) and with (b) the aid of a diverging eyeglass lens.

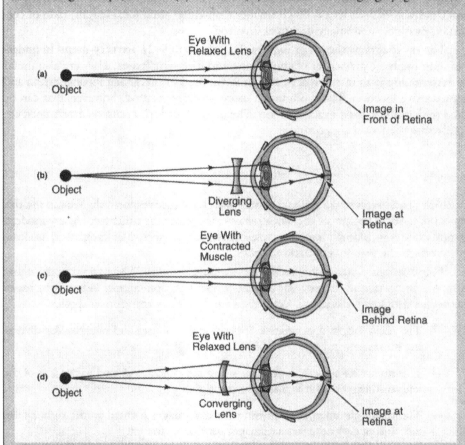

Farsighted people are able to see distant objects clearly, but cannot focus properly on objects that are nearby. A farsighted person must place a book or a newspaper several meters from the eye in order to focus on the print. Otherwise the image produced by the eye is formed behind the retina, and any object held close appears blurred and out of focus. Farsightedness may occur if the lens is unable to curve sufficiently when the muscles of the eye contract.

Eyeglasses incorporating converging lenses are used to correct farsighted vision. The lenses bend light rays toward the axis of the lens, and the rays are converged farther by the lens of the eye. Consequently, the image produced converges on the retina rather than behind it. Image formation is shown for a farsighted person without (c) and with (d) correction.

Note the application of these rules to a calculation involving a toy soldier only 3.0 cm in height, standing on a table top 65 cm from a thin converging lens of 20 cm focal length, as shown in Figure 9-21. What will be the nature of the image formed? First we can apply the thin lens equation to find the image distance and determine whether the image formed is real or virtual by knowing that $f = 20$ cm and $s = 65$ cm.

$$\frac{1}{f} = \frac{1}{s} + \frac{1}{s'}$$

$$\frac{1}{f} - \frac{1}{s} = \frac{1}{s'}$$

$$\frac{1}{20\text{cm}} - \frac{1}{65\text{cm}} = \frac{1}{s'}$$

$$\frac{(65-20)}{1300\text{cm}} = \frac{1}{s'}$$

$$45s' = 1300\text{cm}$$

$$s' \cong 29\text{cm}$$

Note that positive values were used for both the focal length and object distance, according to the sign convention. The problem involves a converging lens and we placed the object to the left of the lens. The image distance is also positive, indicating that a real image would be formed to the right of the lens. We could predicted the real image formation, since the object distance was greater than the focal length.

Magnification

When designing optical devices, we typically must consider more than just where the image will be formed. We also need to consider how much magnification occurs. A slide projector isn't very useful unless it can convert the image on a 35-mm slide into one that measures in meters on a screen. Eyeglasses, in effect, slightly alter the effective magnification of our eyes so that the light rays creating images of objects in our view converge precisely on the retina in the back of our eyes.

Consider the case of an inverted real image, h', produced by a converging lens as shown in Figure 9-22a. The principal ray (Ray 2) and the axis of the lens, form an angle, θ, and the height of the object, h, is the side opposite the angle. The same angle is also formed to the right of the lens by the principal ray and the axis of the lens. The height of the inverted image, h', is the side opposite of this triangle. Since the angle θ is the same for both triangles, the tangent (side opposite divided by side adjacent) function of the two angles must also be equal. This equality can therefore be used to establish a relationship between the object and image distances, s and s', from the lens

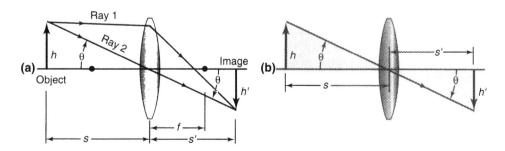

Figure 9–22: A ray diagrams shows the geometrical origin magnification.

and the object and image heights, h and h'. When the tangent of the two angles are applied to the triangles shown in Figure 9-22b, the following results are obtained:

$$\frac{1}{f} = \frac{1}{s} + \frac{1}{s'}$$

$$\tan\theta = \frac{\text{object height, } h}{\text{object distance, } s}$$

$$\tan\theta = \frac{\text{object height, } h'}{\text{object distance, } s'}$$

$$\tan\theta = \tan\theta$$

$$\frac{h}{s} = \frac{h'}{s'}$$

The above equation is especially useful in demonstrating the relationship between distances and heights because the relationship between image and object heights is, after all, what we mean by the term **magnification**.

Using the previous example, we can now calculate how tall the projected image of the toy soldier will be. The answer can be determined by using the above equation and the following calculation:

$$\frac{h}{s} = \frac{h'}{s'}$$

$$h' = \frac{hs'}{s}$$

$$h' = \frac{(3.0\,\text{cm})(29\,\text{cm})}{65\,\text{cm}}$$

$$h' \cong 1.3\,\text{cm}$$

In fact, the image size is reduced in comparison to the object size. The ratio of the image height, h', to the object height, h, is called the **linear magnification**, m, and is defined by the following relationships:

$$\text{linear magnification} = \frac{\text{image height}}{\text{object height}}$$

$$\text{linear magnification} = \frac{\text{image distance}}{\text{object distance}}$$

$$m = \frac{h'}{h} \text{ or } m = \frac{s'}{s}$$

For the example described above, the magnification equals 1.3 cm/3.0 cm, or about 0.43. In other words, the image is a bit less than half the size of the object, the toy soldier. Note that the linear magnification is a dimensionless number, assuming of course that both object and image heights are given in the same units. For example, if $m = 1$, then the object and image are exactly the same size.

If the previously discussed sign convention is properly adopted, then a positive value of m indicates that the image is inverted relative to the object and a negative value of m indicates an upright image. This may seem a little awkward since a positive value of h' indicates an upside-down image, but careful consideration for sign convention and practice will alleviate this.

Mirrors

In the previous section, we considered how lenses can focus and change the direction of light. Now we will turn our consideration to mirrors, a different optical component that acts similarly on light, but does so by reflecting, rather than transmitting and refracting it.

As shown previously in Figure 9-7, the mirror used by most people each day is the plane mirror. It provides an upright, virtual image that is equal in size to the object reflected. It is a virtual image because our visual processing assumes straight-line paths for all light rays and we perceive that the rays converge at a point behind the mirror.

The height of the image produced by a plane mirror is identical to the height of the object, and therefore a plane mirror has a linear magnification $m = h'/h$ equal to one. Consistent with our approach to characterizing lenses, we define magnification of mirrors as the ratio of image to object height.

Plane mirrors are useful in scientific applications only for changing the direction of light rays. A curved mirror, however, has the additional capability to focus the light that strikes it. Curved mirrors can therefore be used to produce images of varying sizes. In other words, a curved mirror can be configured to deliver any degree of linear magnification, within practical limits, depending upon the needs of the application.

Most curved mirrors are spherical in shape. They are constructed by applying a highly reflective coating across the inside or outside of a sphere segment. If the inside of the spherical segment is made reflective, than a concave or **converging mirror** is produced. Alternatively, when the outside of the spherical segment is coated to be reflective, the result is a convex or **diverging mirror**. Examples of each kind of mirror were shown earlier in Figure 9-9.

Both converging and diverging mirrors behave similarly to their converging and diverging lens counterparts. That is, a converging or a diverging lens has a characteristic focal length and produces either a real or a virtual image depending upon the distance between object and mirror. Furthermore, the location of the image can be

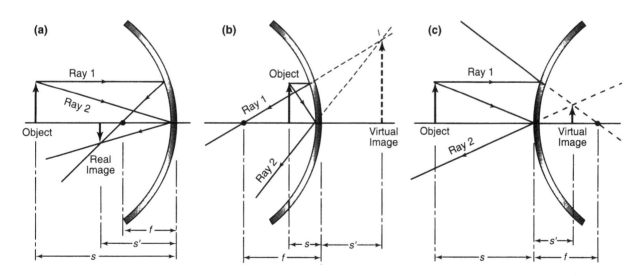

Figure 9–23: (a) An object beyond the focal point of a converging mirror produces an inverted real image. (b) An object between the focal point and a converging mirror produces a virtual, upright image. (c) A diverging mirror always produces an upright virtual image.

determined analytically by applying the mirror equation, which is identical to the thin-lens equation:

$$\frac{1}{f} = \frac{1}{s} + \frac{1}{s'}$$

Consider the three scenarios presented in Figure 9-23. If an object is placed beyond the focal point of a converging mirror, an inverted real image results. However, if the object is brought in closer to the mirror, to a point between the focal point and the mirror, then the image produced is virtual and upright. Finally, if a diverging mirror reflects an object, an upright virtual image is always produced. You can verify each of these scenarios quickly with a shiny polished spoon which, though not exactly spherical, will exhibit the same kind of behavior as described.

Let us consider the example of a dentist examining the cavity in a patient's tooth using a concave mirror of focal length + 1.90 cm. Note that we are using exactly the same sign convention as was applied to the thin-lens equation. If the cavity is 1.40 cm from the surface of the mirror, where is the image located?

By application of the mirror equation, we can determine the distance, s', between the image and the mirror:

$$\frac{1}{f} = \frac{1}{s} + \frac{1}{s'}$$

$$\frac{1}{f} - \frac{1}{s} = \frac{1}{s'}$$

$$\frac{1}{1.90\,\text{cm}} - \frac{1}{1.40\,\text{cm}} = \frac{1}{s'}$$

$$\frac{1}{s'} = \frac{(1.40 - 1.90)\,\text{cm}}{2.66}$$

$$s' = \frac{2.66\,\text{cm}}{-0.50} = -5.32\,\text{cm}$$

Second, it can be determined whether the image is real or virtual. Since the image distance is negative, the image is virtual and appears to be 5.32 cm behind the mirror. Finally, the degree of magnification can be determined by using the equation for magnification. The magnification of the cavity would be given by:

$$m = \frac{s'}{s}$$

$$m = \frac{-5.32\,\text{cm}}{1.40\,\text{cm}}$$

$$m = -3.80$$

The cavity will appear almost four times larger than its actual size. Since the magnification is negative, we expect the image to be in the same orientation as the object viewed.

Composite Optical Systems

Many common instruments involve combinations of lenses and, possibly, other optical components such as mirrors. Two such instruments are microscopes and telescopes.

Both of these instruments produce magnified images that are much larger than would be possible using a single lens.

Since the design and components of many optical systems are quite complex, we will consider only systems involving a combination of two lenses. Therefore, this discussion is intended to serve as an introduction to more advanced study for students who wish to gain a further understanding of optical instruments.

Consider the two-lens system shown in Figure 9-24. The lenses are separated by a distance, d, and have respective focal lengths f_1 and f_2. As light originating at the object, located a distance, s_1, from the lens, passes through the first lens, a real image is produced some distance, s_1', from that lens. The image distance could be calculated using the thin-lens equation and the focal length of the first lens, using the following equation:

$$\frac{1}{f_1} = \frac{1}{s_1} + \frac{1}{s_1'}$$

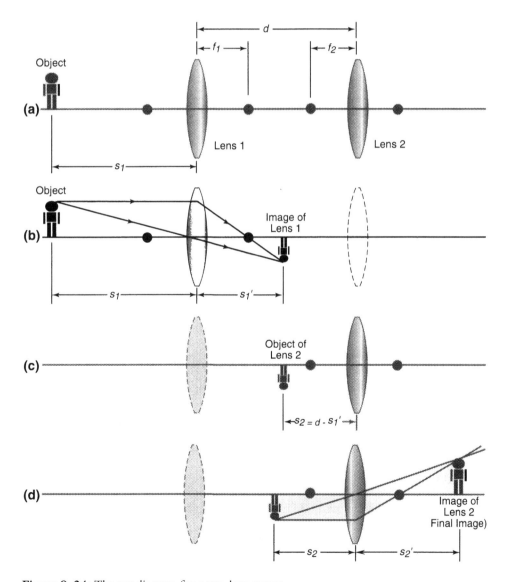

Figure 9–24: The ray diagram for a two-lens system.

The image produced by the first lens then serves as the object for the second lens. As seen in Figure 9-24, the object distance, s_2, equals the difference in the distance between the lenses and the distance of the first image ($s_2 = d - s_1'$) where d is always a positive number. It is important in any optical analysis to track very carefully the sign on each value. In Figure 9-24 the image produced by the first lens is real, and therefore s_1' is positive. This may not always be the case.

The thin-lens equation can again be applied to determine the image distance from the second lens, s_2'. This is the location of the final image relative to the second lens, and provides a means for calculating the magnification. It should be noted that the total magnification of the system is the product of the magnifications of each lens.

With the above relationships clearly in mind, consider an example in which two thin lenses, with focal lengths of 4.8 cm and 7.2 cm respectively, are separated by 25 cm. Where would the image from the first lens have to fall, to form an image on a screen 17 cm behind the second lens? Assign the values 4.8 cm, 7.2 cm, 26 cm, and 17 cm to f_1, f_2, d, and s_2' respectively. We will have to work backwards, in effect, since we know the final image distance from the second lens is 17 cm.

By substitution into the thin-lens equation for the second lens, it can be determined that the object distance from the second lens would be at 12.5 cm to the left of the second lens. This is shown by the following calculations:

$$\frac{1}{f_2} = \frac{1}{s_2} + \frac{1}{s_2'}$$

$$\frac{1}{f_2} - \frac{1}{s_2'} = \frac{1}{s_2}$$

$$\frac{1}{7.2 \text{ cm}} - \frac{1}{17 \text{ cm}} = \frac{1}{s_2}$$

$$\frac{(17 - 7.2) \text{ cm}}{122.4} = \frac{1}{s_2}$$

$$s_2 = \frac{122.4 \text{ cm}}{9.8} = 12.5 \text{cm}$$

Given this value, it is now possible to calculate the distance of the image from the first lens, since it is the difference between the lenses separation and the position of the image, from the second lens.

$$d - s_1' = s_2$$

$$s_1' = d - s_2$$

$$s_1' = 26 \text{ cm} - 13.5 \text{ cm}$$

$$s_1' = 13.5 \text{ cm}$$

In these two steps, we have determined that the image from the first lens would have to fall 13.5 cm beyond the first lens.

Telescopes

A telescope is an optical device that allows us to see very distant objects. A simple telescope consists of two converging lenses separated by a distance that is roughly equal to the sum of their focal lengths. When a distant object is observed (Figure 9-25a), the first lens produces a real image just beyond its focal length as we see in Figure 9-25b. The second lens, or eyepiece, is located so that the image produced by the first lens is just inside the focal point of the second lens. The image produced by the second lens is then a magnified virtual image as shown in Figure 9-25c. The magnification of the telescope equals the product of the magnifications of each lens in the telescope.

Consider a telescope consisting of two lenses that have focal lengths of 20 cm and 5 cm respectively. The distance between the two lenses is 24.5 cm. The telescope is used to produce a virtual image of an athlete 2.0 m tall running at a distance of 55 m away. What is the linear magnification of the telescope and the height of the final image produced? Since the steps to determine the image distances are much the same as used above, they have been summarized in Table 9-2.

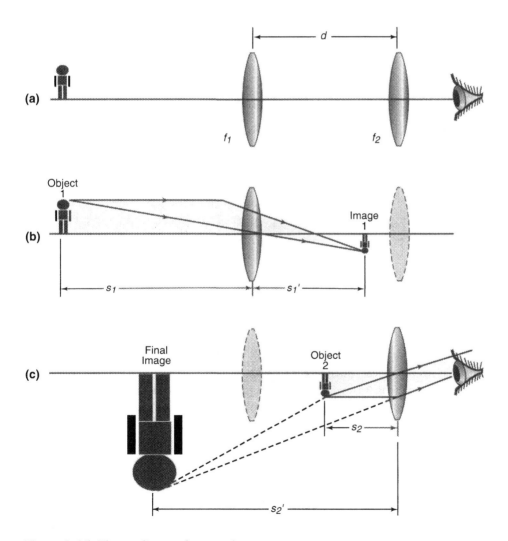

Figure 9–25: The ray diagram for a two-lens system.

First Image Distance (s')	Distance from Lens (s₂)	Second Image Distance (s₂')
$\frac{1}{f_1} = \frac{1}{s_1} + \frac{1}{s_1'}$	$d - s_1' = s_2$	$\frac{1}{f_2} = \frac{1}{s_2} + \frac{1}{s_2'}$
$\frac{1}{20cm} = \frac{1}{5500cm} + \frac{1}{s_1'}$	$24.5\,cm - 20.07\,cm = s_2$	$\frac{1}{5cm} = \frac{1}{4.43cm} + \frac{1}{s_2'}$
	$s_2 = 4.43\,cm$	
$s_1' = \frac{5500cm}{274} = 20.07cm$		$s_2' = \frac{22.14cm}{-0.57} = -38.86cm$

Table 9–2: Calculation of image distances for a telescope.

Recall that the formula for calculating the linear magnification, m, can be either the ratio of the image and object sizes or the ratio of image and object distances. It was also stated that in a two-lens system, the total magnification is the product of the magnification of each lens. Therefore, the total linear magnification of this telescope can be found by completing the following calculation.

$$m = m_1 \cdot m_2$$

$$m = \frac{s_1'}{s_1} \times \frac{s_2'}{s_2}$$

$$m = \frac{20.07\,cm}{5500\,cm} \times \frac{-38.86\,cm}{4.43\,cm}$$

$$m = (0.003649)(-8.772) = -0.032$$

The negative sign reminds us that the final image produced is inverted and virtual. The height of the final image is the product of the object height and total magnification ($h' = m\,h$).

$$h' = m\,h$$

$$h' = \frac{(2m \times 100cm)(-0.032)}{1m}$$

$$h' = -6.4\,cm$$

Although the final image height is only 6.4 cm, which is much smaller than the athlete, it would appear much larger than the athlete when viewed from 55 m. This is because the image is formed very close to the eye. In fact, the image appears only a matter of centimeters away (slightly farther than s_2') from the eye, as can be seen from Figure 9-25c.

Another kind of magnification we often consider is called **angular magnification**, M, a term that accounts not only for the linear magnification of the instrument, but also for the image and object distances. In other words, the angular magnification provides a measure of how large the *image* appears compared to how large the *object* appears.

We define the angular size, θ, of an object as the ratio of its height to its distance, r, from an observer's eye. The angular size of an image as seen through a lens, θ', is defined as the ratio of the height of the final image to the distance between the final image and the last lens. The units of θ and θ' are radians.

$$\theta = \frac{h}{r} \qquad \theta' = \frac{h'}{s'}$$

Given these definitions, the angular magnification is calculated using the equation:

$$M = \frac{\theta'}{\theta}$$

Now reconsider the example of the athlete viewed through a telescope presented above, and calculate the angular magnification.

$$M = \frac{\theta'}{\theta}$$

$$M = \frac{h'}{s'} \times \frac{r}{h}$$

$$M = \frac{-6.4 cm}{-38.8 cm} \times \frac{55 m}{2 m} = 4.53$$

The implication here, of course, is that the image of the runner produced by the telescope appears to be about 4.5 times larger than the runner would appear when viewed by the unassisted eye. This is despite the fact that the linear magnification of the telescope is actually much less than one. In other words, although the image produced by the telescope is much smaller than the actual object, it still appears larger because of its proximity to the eye.

Microscopes

The compound microscope, in its simplest form, is yet another example of a two-lens optical instrument. Unlike the telescope, which permits us to see a very distant, relatively large object in a more detailed fashion, the microscope produces a virtual image of a very small object that is in close proximity. A simple model of the modern optical microscope consists of two lenses, the **objective lens** and the **ocular lens** or eyepiece. Both lenses have very short focal lengths that are typically separated by distances of 10 to 20 cm. Although high-quality microscopes incorporate as many as ten or more optical elements, we can model the principal function of the modern microscope with only two converging lenses, as shown in Figure 9-26.

The function of a microscope is much like that of a telescope. The first lens accepts light from an object and produces a real image. This image serves as the object for the second lens, the ocular lens or eyepiece, and an inverted, virtual final image is produced that is much larger than the object itself. Although the object is very close to the microscope, it is also extremely small, therefore, the total linear magnification produced by a microscope will be much greater than one. This is in contrast to a typical telescope where the total linear magnification is always less than one.

We define the **near point** of the eye as the closest distance of an object upon which the tensed eye can focus, typically a distance of about 25 to 50 cm. Given this information, the angular magnification of the microscope described above can be approximated by the following equation:

$$M \approx \frac{d - f_2}{f_1} \times \frac{\text{near point}}{f_2}$$

Consider the case where a simple, two-lens microscope is to be used to view a red blood cell. The objective and ocular lenses have focal lengths of 0.75 cm and 1.20 cm respectively. The lenses are separated by 20.0 cm. The blood cell is located on a slide a distance 0.85 cm below the objective lens. We wish to determine the angular magnification of the cell, and we assume the near point to be equal to 25 cm.

$$M \approx \frac{(20-1.20)\text{cm}}{0.75\text{cm}} \times \frac{(25\text{cm})}{1.20\text{cm}} \approx 522$$

Application of the above equation indicates that an angular magnification of approximately 522 times would be expected. In other words, the virtual image produced would appear over 500 times larger than the red blood cell as seen with the unassisted eye. Thus, the red blood cell, and many of its details, can be seen clearly through the microscope although they would be invisible to the naked eye.

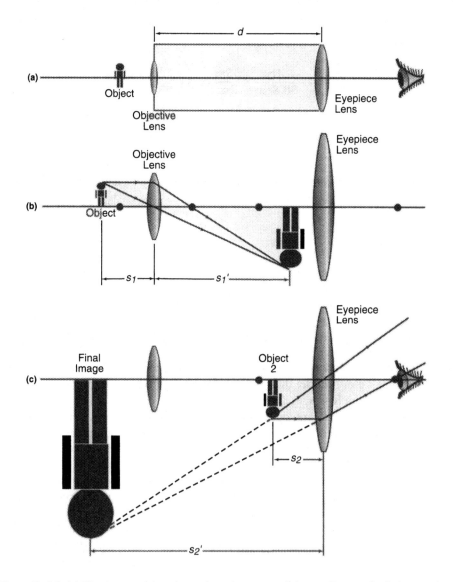

Figure 9–26: (a) Viewing an object through a microscope, (b) a ray diagram for light entering a microscope, and (c) a ray diagram describing the virtual image produced.

Special Topics Box 9–6 ■ What is Astigmatism?

Astigmatism is a defect of the eye caused by a lack of symmetry in the cornea. A normal cornea is uniformly shaped. However, if the cornea is flatter along a vertical axis parallel to its surface than along a horizontal axis parallel to its surface, then light passing through each axis is focused at different distances from the cornea. In other words, the cornea has a focal length that varies with angle around its principal axis.

Astigmatism can be detected by viewing a set of lines such as those in the figure. Viewed with a normal eye, the lines should all appear equally bright. If the lines are viewed with an eye that is affected with an astigmatism, then the lines oriented in one direction may appear fainter and less well focused than those oriented perpendicular to that direction. This condition can be corrected with eyeglass lenses that are less curved in one direction than in another.

Aberration

Aberration, within the context of optical elements, is defined as the failure of a lens or other optical component to produce an exact point-to-point correspondence between an object and its image. Actually, in lenses aberrations come in two forms: spherical aberration and chromatic aberration.

Spherical aberration describes the failure of light rays passing through a lens to focus at a single point. When rays of light move parallel to the principal axis of a spherical lens, they do not all meet at a single point after refraction through the lens. The rays passing through the lens nearer to its edge are refracted more than desired and therefore cross the principal axis a distance closer to the lens than the rays passing through the lens nearer to its middle. Therefore, in a real situation, there is no single focal point. However, there is a single position where the transmitted light rays form the smallest circle. As

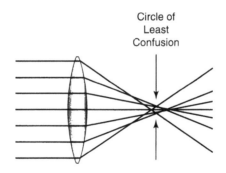

Figure 9–27: Light rays moving parallel to the principal axis of a spherical lens do not cross the axis together, but do produce a circle of least confusion.

shown in Figure 9-27, this is called the **circle of least confusion** where they seem to focus best. Spherical aberration can be minimized by using an aperture (small opening) that permits light to travel only through the center region of a lens. The other choice is to combine several lenses that provide the same net focal length as a single lens, but in which their respective spherical aberrations cancel one another.

Chromatic aberration occurs when the various colors that make up white light refract at different angles. Since blue light refracts more than red light at the surfaces of a converging lens, blue light crosses the principal axis of a lens closer to the lens than does red light. This variation in focal length with respect to color can be corrected by using a compound lens of two different materials as shown in Figure 9-28. Such a lens is called **achromatic** because it can focus all colors of light to the same point.

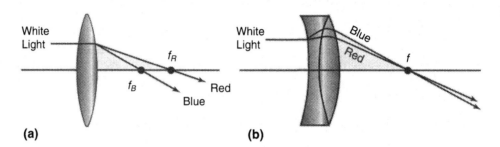

Figure 9–28: (a) Different colors of light focus at different distances from a lens. (b) An achromatic compound lens alleviates this problem.

Lasers

In Chapter 4, we discussed Danish physicist Niels Bohr's hypothesis that electrons can become energized through heating, electrical stimulation, or absorbing light and jump to a higher energy level farther from the nucleus. Later, these "excited" or **metastable** electrons return to a lower energy level by emitting light corresponding to the change in energy levels. The various energy levels within the atom, therefore, determine the permitted wavelengths of light emitted; thus the bright-line spectra for that element.

Some give Einstein the credit for being the father of lasers. In 1921, he was awarded the Nobel Prize for his work on the photoelectric effect and his hypothesis for the existence of photons and stimulated emission. Simply put, he believed that metastable electrons could move to a lower energy level in one of two ways. The first is by the random **spontaneous emission** of a photon. The second is by **stimulated emission**, in which the movement of the metastable electron is triggered by the presence of a photon of the proper frequency.

The first practical application based on stimulated emission was for microwaves and given the name **MASER**, an acronym for *microwave amplification by stimulated emission of radiation*. It was developed in 1954 by Arthur L. Schawlow and Charles H. Townes, who had previously worked on radar bombing systems during World War II. In a paper they published, the possibility of developing stimulated emission devices for the infrared and visible spectrum was predicted.

Not long after (1960), physicist Theodore Harold Maiman developed the first operable *light amplification by stimulated emission of radiation* or **laser**. It was based on optical radiation (about 550 nm) being absorbed by a very low concentration of Cr^{3+} ions in a ruby rod. Following absorption, the electron made a rapid non-thermal transition to a lower metastable level. If the amount of optical radiation pumped is intense enough, a **population inversion** is obtained, meaning that there are more electrons in the metastable state than in the ground state. When an electron of one of these metastable states spontaneously drops back to its ground state, it emits a photon of energy, hf, called the **fluorescent radiation**. As the photon passes by another atom

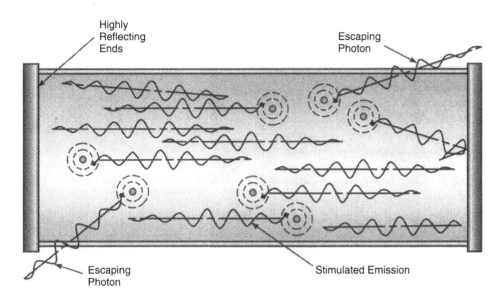

Figure 9–29: A laser cavity with highly reflecting ends shows the stimulated emission of light and the escape of some primary photons through the side-walls.

in the same metastable state, it can immediately stimulate that atom to radiate a photon of the exact same frequency, causing it to return to its ground state as well. These two photons will have the same exact frequency, direction, and polarization. This, again, is the phenomenon of stimulated emission and the two photons, in turn, can stimulate emission of additional photons of the same frequency, direction, and polarization as shown in Figure 9-29. If the conditions are right, a laser is capable of producing a chain reaction resulting in a beam of high-intensity **coherent radiation** or nearly single-wavelength light.

In order to produce a laser beam in a gas, the stimulated emissions are collimated within the vessel such that photons of identical wavelengths are continuously produced, and some are permitted to escape the cavity to contribute to the beam. As shown in Figure 9-29, this is accomplished using highly reflective surfaces at each end of the vessel. There is a buildup of photons within the vessel and only a fraction of those traveling along the axis of the vessel can escape and contribute to the beam. Figure 9-30 shows a helium-neon laser used in the laboratory.

Lasers are used commonly for position and velocity sensing applications and are ideal for aligning optical and mechanical components. In recent years, solid-state lasers have become smaller and are finding applications for such things as supermarket scanners and compact-disc players (see Technology Box 9-10). In fact, scientists have developed a single-atom laser that cannot read a bar code or play music, but it can be used to study how atoms and light interact, thus improving our understanding of the quantum theory.

Figure 9–30: A helium-neon laboratory laser. Photo courtesy Jonathan J. Miles.

Technology Box 9–10 ■ CDs and Lasers

The development of optical digital technology and its application to compact disc (CD) players in 1982 has changed both the way we record music and listen to it. Not only do CDs provide better fidelity and not wear out, but the listener can also jump anywhere on the disk with just a click of a button.

Simply put, the optical digital technology used for CDs is little more than digital information written and retrieved from a CD disk using a laser beam. To produce a CD, the amplitude of the sound wave is digitally sampled thousands of times per second. Digital sampling being similar to finding the individual points along a mathematical curve. Each of these sampling points is then con-

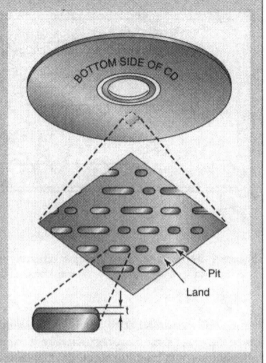

verted into a binary code, consisting of a string of 0's and 1's. This code is then stored as a series of microscopic pits in polycarbonate plastic. The pits are approximately 0.5 microns in width and 0.83 to 3.56 microns in length. As shown in the figure, the remaining unburned area is called land. The pits are imprinted on the disc in a corkscrew pattern, with a distance between tracks of about 1.6 microns.

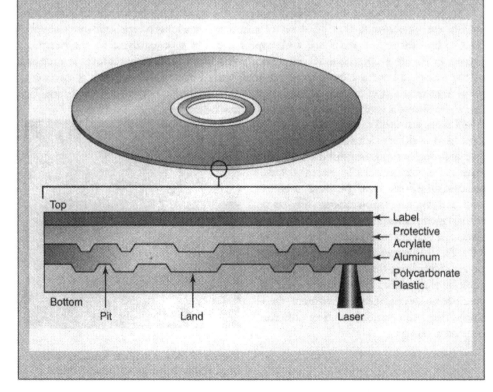

Technology Box 9–10 ■ CDs and Lasers (Continued)

The first step in making multiple copies of a CD is to make a master tape and transfer it to a "glass master." The glass master being a highly polished piece of glass coated with light sensitive material called a photoresist. A laser then burns the master tape pattern into the photoresist and the pattern is later coated with a thin layer of metal. Negative impression molds called "fathers" are made from the glass master using a process known as electroforming. This involves depositing another thin layer of metal over the previous layer. Because the difference between the pits and the land is so tiny, loss of even a few layers of atoms will affect the quality of the copy; therefore, only about 6 "mothers" can be made from a father. The mothers, in turn, are used to make "sons" or "stampers." Again, only about 6 stampers can be made from a mother. Injection molding technology is used to make multiple polycarbonate disks from the stampers. A thin layer of metal is added to the top of the polycarbonate plastic to later reflect the laser light, and a thin layer of acrylate plastic put over that. Finally, the label is added. A single 12 cm CD can store over 783 megabytes of data, which translates to 74 minutes of music.

The music encoded on a CD can be played back by using photodiodes to read the changes in the intensity of laser light reflected from the track. This requires lenses to focus the laser beam, a guidance system to point the laser to the correct place on the rotating disk, and an array of photodiodes. The laser used emits a beam of 780 nm, which is one-fourth the depth of a pit, t. When light falls on a pit region, it must travel down a quarter of a wavelength and reflect back a quarter of a wavelength. This increases its path by one-half wavelength and it, therefore, interferes destructively. Put another way, it is as if no light has struck that region. Light that hits land, however, is reflected back to the photodiode array. In this way, the pits and land are translated back into 0s and 1s, which are converted into an electrical signal, amplified, and sent to the speakers.

Source: "Compact Disc Player" by Ken C. Pohlmann, Professor of Music Engineering Technology, University of Miami.

Check Your Understanding

1. What effects do converging and diverging lenses have on parallel light rays?

2. What is the difference between a real and a virtual image?

3. What is the similarity between the thin lens and mirror equations?

4. What is the difference between linear and angular magnification?

5. How is the thin lens equation applied in a compound optical system?

6. What is the near point of the eye?

7. What are the two types of aberration? How can each be corrected in a lens system?

8. What is stimulated emission?

9. What is meant by the term population inversion?

10. Why is the beam produced by a laser described as coherent radiation?

Summary

Light and all other forms of electromagnetic radiation travel as waves that are composed of electric and magnetic fields that vibrate at right angles to each other. The speed of all electromagnetic waves is typically rounded to 3.00×10^8 m/s. Depending on where they fall in the electromagnetic spectrum, they vary in their wavelengths, frequencies ($c = f\lambda$), and energies ($E = hf$). Sometimes the concept of a massless photon with a definite wavelength traveling at the speed of light is used to describe the smallest bundle of electromagnetic radiation.

The electromagnetic spectrum is typically subdivided into seven regions: radio, microwave, IR, visible, UV, X rays, and gamma rays. As you move from radio to gamma rays, the wavelengths become increasingly shorter and the amount of energy increases correspondingly. The visible portion of the spectrum contains the colors of red, orange, yellow, green, blue, indigo, and violet. Although humans are aware of only the visible portion of the spectrum, through technology, practical uses have been found for all other regions.

When light encounters matter, sometimes its energy is converted directly into chemical energy or thermal energy through a process known as light absorption. Other times the light is reflected from the surface resulting in the formation of either a real or virtual image. If the reflective surface is concave, it can cause the light rays to converge and form a real image. If the surface is convex, the light rays will diverge forming an image that can only be seen by looking into the mirror. The law of reflection states that wherever reflection occurs, the angle of incidence is always equal to the angle of reflection.

When light rays strike objects their frequencies match either an allowable electron transition or molecular vibrational frequency, the particles can absorb all the energy of a photon. They become excited, but quickly re-emitted a photon with the same frequency. Since the particles are in random orientation, the re-emitted photons go in every direction, i.e., scattering occurs. For example, much of the UV entering our atmosphere is scattered in this manner.

Light rays that pass from one medium into another, each with a different index of refraction, experience refraction or bending. This is the reason a stick placed into a pool of water appears to bend at the air-water interface. Light rays always bend toward from the normal when they enter an optically more dense medium, e.g., air into water. The angle of refraction can be calculated by using Snell's law ($n_i \sin \theta_i = n'_{r'} \sin \theta_{r'}$) if the two indices of refraction are known. If the incident angle exceeds the critical angle, then all radiation will be reflected, and none will be transmitted to a second medium. This phenomenon enables fiber optic networks to pass a light signal over long distances.

When light waves pass through an opening that is similar in size to their wavelength, the new wavelets that form on the other side tend to flare in ever widening semicircles and are said to be diffracted. If there are several similar openings, the emerging individual wavelets form an interference pattern. In some areas, the amplitude of the wavelets will reinforce, and in others they will destructively interfere. The resulting light and dark bands are called interference fringes. A similar interference occurs when light strikes a thin film such as a soap bubble or a film of petroleum on water. Each color of the reflecting light ray produces its own interference pattern, and together they result in a rainbow of colors.

Natural sunlight reflecting from various surfaces can become partially polarized horizontally. That is, a part of the random vibrations of the light wave has been eliminated. Sunglasses containing polarizing lenses are designed to allow the vertically polarized light waves to pass, but blocks most of the horizontal waves. In this way, much of the unwanted glare is eliminated.

Prisms and raindrops are two examples of substances that disperse white light into their separate colors. The shorter the wavelength, the greater the degree of refraction occurring, so rainbows, for example, always show red on the top and violet on the inside.

Neither the IR nor UV portions of the electromagnetic spectrum are visible to the human eye. IR radiation is most often perceived as thermal energy and we tend to become aware of UV radiation in the form of sunburns. Many modern day conveniences make use of infrared radiation. Such devices as the remote controls used to operate our TVs, stereos, and garage doors function by either sending or receiving invisible IR radiation.

UV radiation is energetic enough to disrupt chemical bonds. Therefore, it can do great harm to both living organisms and polymers. Most UV radiation is captured in the stratosphere, where it splits ozone molecules into H_2 and O that later recombine to form ozone. The control of certain chemicals believed to be responsible for this depletion of ozone has been enacted nearly world-wide.

Like mirrors, there are also converging and diverging lenses. Converging or convex lenses are those that cause passing parallel light rays to meet at a specific focal point. This type of lens is useful for concentrating a beam of light at a given distance. Diverging or concave lenses cause parallel light rays to move farther apart. Unlike the converging lens that forms a real image, a diverging lens generally forms a virtual image that can only be observed through the lens, and it appears larger and farther from the lens than it actually is. The ratio of the image height, h', to the object height, h, is called the linear magnification, m, and is defined by the ratio of the image height/object height.

Composite optical systems, such as those in microscopes and telescopes, involve a combination of lenses and typically several other components. Both instruments produce magnified images that are much larger than would be possible using a single lens. When considering multi-lens systems, it is important to remember that each lens acts separately, with the image formed by the first lens serves as the object for the second, etc. Therefore, the thin lens equation can be applied as many times as needed to determine the type, size, and position of the final image.

A second kind of magnification is called angular magnification. It accounts not only for the linear magnification, but provides a measure of how large the image appears compared to how large the object appears. Thus, the angular magnification is a ratio of the height of the final image to the distance between the final image and the last lens. When talking about the magnification of a telescope, for example, although the linear magnification is less than one and the image produced is smaller than the actual object, it still appears larger because of its proximity to the eye.

All lenses fail, to varying degrees, to produce an exact point-to-point correspondence between an object and its image. This failure is called aberration. Spherical aberration is the failure of all rays to meet at a single point. The single position that comes closest to the focal point is known as the circle of least confusion. The amount of spherical aberration can be minimized by using a small aperture or a combination of lenses that provides the same net focal length. Chromatic aberration occurs when the various colors making up white light refract at different angles. Compound or achromatic lenses can be used to correct for this deficiency.

Lasers produce nearly single-wavelength beams of light through a process known as stimulated emission. When electrons that have been pumped to a metastable drop back to their ground state, they emit photons. As the photons pass by other atoms in the same metastable state, they can immediately stimulate them to radiate photons of exactly the same frequency. These photons will have the same exact frequency, direction, and polarization. If the conditions are right, a laser is capable of producing a chain reaction resulting in a beam of high-intensity coherent light.

Chapter Review

1. Describe the nature of electromagnetic radiation.

2. If a light wave has a wavelength of 720 nm, what is its speed? What is its frequency?

3. What is blackbody radiation? What influence did it have on the theory of electromagnetic radiation?

4. What is the photon energy associated with light having a wavelength of 593 nm?

5. Describe the photoelectric effect.

6. Name the seven major divisions of the electromagnetic spectrum, and give the range of frequencies they encompass.

7. State the law of reflection for light waves. Use it to determine the angle of reflection for a light wave hitting a reflective surface with an angle of incidence of 24°.

8. Describe scattering and give an example.

9. What is the angle of refraction for a ray of light that travels through a medium with an index of refraction of 1.21, and then enters a medium with an index of refraction of 1.45 at an incident angle of 17°?

10. What is the critical angle between two media having indices of refraction of 1.10 and 1.51? (Hint: Use Snell's Law and let $\sin \theta_2 = 1$.)

11. What is diffraction?

12. What is interference? Why, in terms of wave theory, does it occur?

13. Describe polarized light. Name one use for it other than sunglasses. (You may need to research this in order to answer this question.)

14. What is dispersion? What was its significance in determining the nature of light waves? What part does it play in the formation of rainbows?

15. The distance between a converging lens and the point where incoming parallel light rays bent by the lens converge is called the _____.

16. An object and virtual image of the object formed by a diverging lens appear on the same/opposite (circle one) side(s) of a diverging lens.

17. The ratio of image height to object height and the ratio of image distance to object distance are both called _____.

18. Unlike a plane mirror, a curved mirror can _____ an image by changing the direction of light rays.

19. A simple two-lens telescope produces an image with linear magnification, m, greater/less (circle one) than one and angular magnification, M, greater/less (circle one) than one.

20. Sort the following types of electromagnetic radiation in order by wavelength beginning with the shortest.

Green   Ultraviolet   Microwave   Red   Infrared   Blue

21. Calculate the total power, in watts, emitted by a square blackbody surface, 3 cm along each side, at 350°C.

22. Determine the focal length of a converging lens if the object and image distances, s and s', are 80 and 60 inches respectively. Calculate the linear magnification, m, of this lens as well.

23. A spherical mirror is used to form an image on a screen located 6 m from the object, that is six times larger than the object. Describe the type of mirror used and determine the distance between the mirror and the object.

24. A magnifying glass is used to double the size of an insect being examined, and is held 3.1 cm above the specimen. What is the focal length of the lens?

25. Two converging lenses of focal lengths 8 and 16 cm are located 40.0 cm apart as shown below. The final image is to be located 25 cm to the right of the first lens. How far to the left of the first lens should the object be placed? What is the total linear magnification of the system? Is the final image upright or inverted?

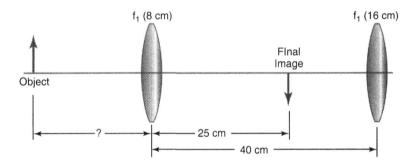

What is O-chem?

Organic Chemistry

There's been a lot of confusion in recent years over the phrase "organically grown." What were these words intended to guarantee the consumer? Since the word organic comes from organism, technically anything that is produced by a plant or an animal could make the claim. This is likely not its intended meaning, however, when it is used to describe food products. Employing the popular connotation rather than the strict definition, most producers use the phrase to inform consumers that the food has been grown without the use of man-made hormones, fertilizers, insecticides, or herbicides. Less scrupulous producers have taken advantage of the strict definition to deceive the public. The US Congress enacted the Organic Foods Production Act of 1990. The U.S. Department of Agriculture is charged with implementing the act and establishing guidelines a producer must follow to legally claim the food was organically produced and handled.

Historically, the terms "organic" and "inorganic" were used to differentiate between those chemical compounds that come from living (organic) sources and those that come from non-living (inorganic) sources. Prior to 1828, all laboratory attempts to make organic compounds without using a living organism failed, leading to the belief that living organisms possessed a special, *vital force,* that had to be present. After Friedrich Wöhler (1800–1882) produced an "organic" chemical called urea without the use of a living organism, however, the **vital force theory** lost credibility and was soon abandoned.

10–1 The Organization of Organic Chemistry

Objectives

Upon completion of this section, you will be able to:

- Explain the difference between inorganic and organic chemistry.

- Identify the element that must be present in a compound for it to be an organic compound.

- Describe and explain the structural feature that separates organic compounds into families.

- Identify two ways to name an organic compound.

- Define isomers and explain what they have in common and how they differ.

Today the distinction between inorganic and **organic chemistry** is based on the presence of the element carbon. In fact, it has been suggested that the term organic chemistry (O-chem) be changed to carbon chemistry. For historical reasons, however, it continues to be called organic chemistry. There are some compounds containing carbon that are not considered organic. These are compounds containing the bicarbonate, HCO_3^-; carbonate, CO_3^{2-}; and cyanide, CN^-, ions; as well as carbon monoxide, CO; and carbon dioxide, CO_2.

One of the other differences between inorganic and organic compounds has to do with the type of bonding present. Most inorganic compounds have ionic bonding, while most carbon-containing compounds have covalent bonds. Early laboratory attempts to make organic compounds failed not because of the absence of a presumed "vital force," but because of a lack of understanding about how to transform covalently bonded molecules. Once this was learned, chemists started adding man-made or **synthetic compounds** to the growing list of organic compounds being discovered from plant and animal sources. Some of the more important synthetic organic compounds include polymers such as plastics and fibers like nylon and polyester. A few statistics will help you understand how the area of organic chemistry has grown. It took until 1960 to identify the first million organic compounds. It took only twelve years more to identify the next million. By 1998, the **International Union of Pure and**

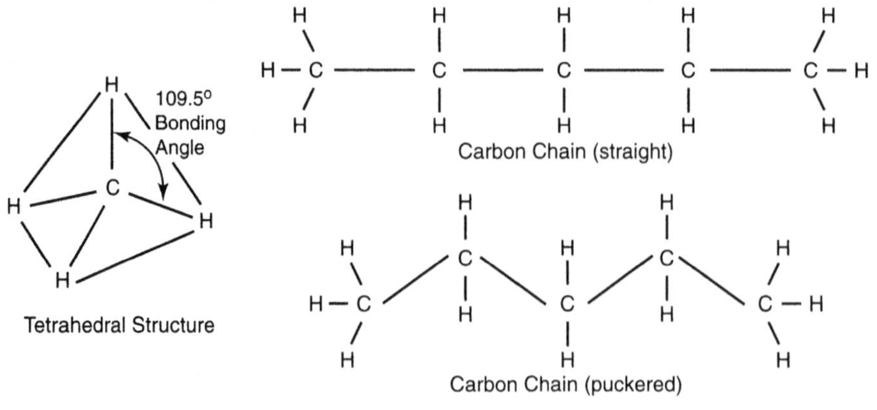

Figure 10–1: Tetrahedral bonding of carbon (left). The rotation of carbon-carbon single bonds allows the chain to take on different shapes (right).

Applied Chemistry (IUPAC) had recorded and devised a naming system for over 18 million organic compounds. Some of these compounds are naturally occurring, but many are synthetic.

Chemically, there is nothing particularly special about the element carbon. It has four electrons in its outer energy level and it typically forms covalent bonds, but so do several other elements in this family. When carbon forms four single covalent bonds, the combined atoms attempt to move as far away from each other as possible. As shown in Figure 10-1, this places carbon in the center of a tetrahedral (four sided) structure with bond angles of 109.5° between each of them. When several carbon atoms bond to each other to form a carbon chain, the 109.5° angles make the carbon chain pucker or zigzag. When the chain is rotated so the puckering is into and out of the page, it appears to be a flat straight line. This is the orientation most often used when writing organic formulas. If the bonds between the carbon are single, they also allow the carbon atoms to twist. The overall shape of a molecule containing several carbon atoms is, therefore, continually changing as it twists about.

Structure of Organic Compounds

How can there be so many different carbon-containing compounds? Although there is no single answer, one explanation lies in the ability of carbon to form long chains that become the backbone of the molecule. The formation of these long chains, however, uses only two of the four electrons that carbon has available for bonding. The other two are available to form single or double covalent bonds. Most often carbon forms single covalent bonds with hydrogen, but it may also bond with a halogen (fluorine, chlorine, bromine, or iodine) as well as with oxygen, sulfur, phosphorous, or nitrogen. It is the different positioning of these bonds, elements, or groups that give each compound its unique properties.

Family Name	Functional Group	Family Name	Functional Group
Alkanes	R – H	Aldehydes	$\overset{\displaystyle O}{\overset{\displaystyle \|}{R-C-H}}$
Alkenes	$\overset{\displaystyle \| \quad \|}{R-C=C-R}$	Ketones	$\overset{\displaystyle O}{\overset{\displaystyle \|}{R-C-R}}$
Alkynes	$R-C\equiv C-R$	Carboxylic Acids	$\overset{\displaystyle O}{\overset{\displaystyle \|}{R-C-OH}}$
Aromatics	⬡	Esters	$\overset{\displaystyle O}{\overset{\displaystyle \|}{R-C-O-R}}$
Alcohols	R – OH	Amines	$\overset{\displaystyle \cdot\cdot}{\underset{\displaystyle \|}{R-N-}}$
Mercaptans	R – SH	Amides	$\overset{\displaystyle O}{\overset{\displaystyle \|}{\underset{\displaystyle \|}{R-C-N-}}}$
Ethers	R – O – R		

Table 10–1: Organic family functional groups.

For just a moment, think about the vast array of animals in the world. Starting at a very early age, someone spent hours teaching you to recognize the difference between a bunny and a kitty. The long ears and fuzzy tail were probably the distinguishing features at first. As your development continued, you could not be fooled by changing the color or length of the ears because you had started recognizing structural differences in their heads, legs, and bodies. Soon you were able to identify pigs, cows, horses, frogs, etc., because you had developed a conceptual understanding of the features that placed each animal into a group. The ability to conceptualize minimizes the amount of information that must be memorized. This simple analogy is intended to help you understand the goal of the approach that will be used for the study of organic chemistry.

The millions of known organic compounds can be subdivided into groups called **organic families**. The compounds in an organic family have similar physical and chemical properties, because they have a common **functional group**, or unique arrangement of atoms or bonding, that gives each family its own set of characteristics. Table 10-1 is a list of the 13 organic families that will be examined in this chapter.

Organic chemistry has its own special shorthand notations. One of the most common is R–. As noted earlier, carbon has the ability to repeatedly bond to itself and form long carbon chains. If all of the bonds are with hydrogen atoms except for one, then this portion of the molecule can be represented by the letter, R –, from the term "radical." A **radical** or **alkyl group** is a fragment of a molecule without a specific length. For example, Figure 10-2 shows the methyl (CH_3–), ethyl (C_2H_5–), propyl (C_3H_7–), butyl (C_4H_9–), and pentyl (C_5H_{11}–) alkyl groups. Each of these segments can be replaced with, R–, in the same way that, n, is used in mathematics. If, n^2, means to square any number, then R–OH means a carbon chain of any length with an oxygen and hydrogen attached.

Alkyl groups that have an –OH group attached to the open position on the carbon chain belong to the alcohol family. If the alcohol functional group were replaced by a –SH group, it would then belong to the mercaptan family. Each time the element or group of elements attached to the R– is changed, it becomes a new family or subfamily.

It is easy to see how changing the functional group can place it in a different family, but what if you want to give it a name? Organic compounds are named by several different methods. **Trivial names** are more like a nickname because they may be based on the source, use, or characteristics of the molecule, not its structure. It's like calling someone Red, because they have red hair. The name isn't very useful unless you are a part of the circle of friends. Several familiar alcohols have trivial names; for example wood alcohol, grain alcohol, and rubbing alcohol. At least you can tell these are all members of the alcohol family, but that is not always the case. These alcohols also have **common names**. Common names are frequently the name associated with the number of carbons in the R– group, followed by the family name. Because wood alcohol has only one carbon in its longest chain, its common name is methyl alcohol.

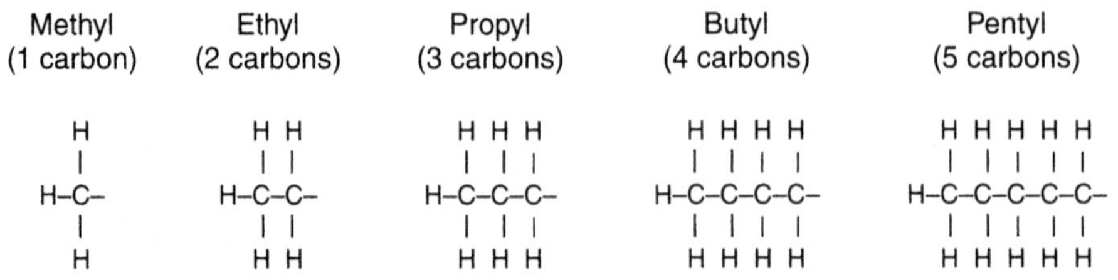

Methyl (1 carbon)	Ethyl (2 carbons)	Propyl (3 carbons)	Butyl (4 carbons)	Pentyl (5 carbons)
H │ H–C– │ H	H H │ │ H–C–C– │ │ H H	H H H │ │ │ H–C–C–C– │ │ │ H H H	H H H H │ │ │ │ H–C–C–C–C– │ │ │ │ H H H H	H H H H H │ │ │ │ │ H–C–C–C–C–C– │ │ │ │ │ H H H H H

Figure 10–2: The methyl, ethyl, propyl, butyl, and pentyl alkyl groups.

In a similar way, grain alcohol – with two carbons – is ethyl alcohol and rubbing alcohol – with three carbons – is isopropyl alcohol.

Even common names soon become troublesome. For example, what does the prefix iso-, in isopropyl alcohol, mean? Whenever a compound has a straight carbon chain and the functional group is at the end, it is considered to be **normal** and n- is sometimes used before the name. Furthermore, if the n- is not used, you are to assume that it is the normal arrangement. If the carbon chain is not straight or the functional group is not attached to its end, then the prefix, iso-, is sometimes used. It tells you that there is something different about the way the molecule is put together. If you examine the following two examples you will discover the difference.

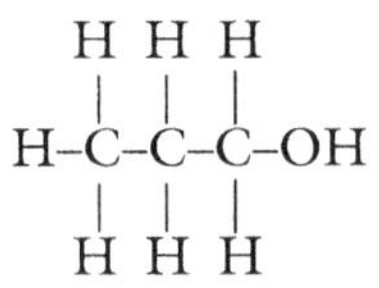

$$\begin{array}{cccc} & H & H & H \\ & | & | & | \\ H- & C- & C- & C-OH \\ & | & | & | \\ & H & H & H \end{array}$$

$$\begin{array}{cccc} & H & H & \quad H \\ & | & | & \quad | \\ H- & C- & C & - C-H \\ & | & | & \quad | \\ & H & OH & \quad H \end{array}$$

n-propyl alcohol 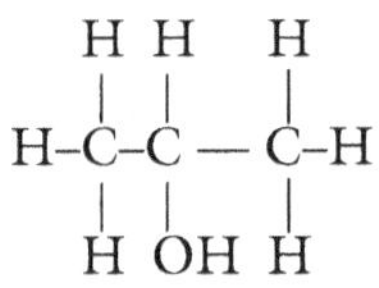   isopropyl alcohol

In the n-propyl alcohol structure, the alcohol functional group (–OH) is attached to the end of the three carbon chain, but in isopropyl alcohol the alcohol functional group is attached to the middle carbon. Both of these molecules have the same **molecular formulas** (C_3H_8O), but they differ in their **structural formulas**. Whenever two compounds have the same molecular formula, but different structural formulas, they are said to be **isomers**; hence the prefix iso-.

This brings us to the second reason for the vast number of organic compounds: isomers. Each time the position of a functional group on the R– group is changed, it becomes another isomer of that substance. Surprisingly, each isomer has its own unique set of chemical and physical properties. Sometimes these differences are slight and at other times, they are great. It is probably obvious, therefore, that writing the molecular formula for a compound is of little usefulness, since it is the structure that determines its properties. Writing full structural formulas, however, is tiresome and takes up a lot of space, so chemists have devised a more compact way of transmitting the essential information. **Condensed structural formulas** group the elements in such a way that it is still possible to understand their placement. Using the above examples, the condensed structural formulas for n-propyl and isopropyl alcohol become:

$CH_3-CH_2-CH_2-OH$

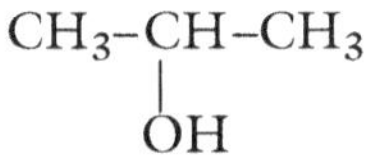

$$\begin{array}{c} CH_3-CH-CH_3 \\ | \\ OH \end{array}$$

As you examine the condensed structural formulas, you will note that the hydrogen atoms surrounding each carbon have been gathered and placed after the carbon symbol. You know that hydrogen can have only one bond and that carbon always has four bonds, so the hydrogen atoms must be attached to the carbon and not to each other. As shown below for the molecular formula C_3H_8O, the use of a straight line, —, to designate the single covalent bond between the atoms is considered optional and may be omitted, but double and triple covalent bonds, if a part of the structure, must be included.

$CH_3\ CH_2\ CH_2\ OH$

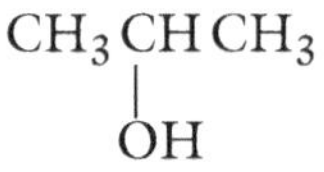

$$\begin{array}{c} CH_3\ CH\ CH_3 \\ | \\ OH \end{array}$$

The isomers of propyl alcohol shown above are an example of **structural isomers**, in which the difference is just in the placement of the functional group. A second kind of isomer, called **functional isomers**, also exists. In functional isomers, the same atoms are again used, but this time the arrangement produces a different functional group placing them in a different organic family. For example, the same molecular formula, C_3H_8O, can be arranged to produce

$$CH_3-O-CH_2CH_3$$

The **general formula** for this isomer is R–O–R, as opposed to R–OH. Referring to the list of general formulas in Table 10-1, you can see that this compound is in the ether family. The common names for these compounds are based on the name of the shortest R– group followed by the name of the longer R– group and finally, the family name: methyl ethyl ether.

Returning to our analogy, the study of organic chemistry is, therefore, similar to our study and understanding of animals. Once you recognize an animal to be a dog, you have certain expectations. Although dogs may vary in size, color, temperament, and abilities, they all bark when excited or scared, they wag their tail when happy, snarl and show their fangs when threatened. They don't have feathers, climb trees, or oink. Similarly, the first thing to recognize in organic chemistry is family membership. General and structural formulas help make it possible to identify the functional group present. Common names often include the family as a part of the name. In the more structured International Union of Pure and Applied Chemistry (IUPAC) naming system, special endings are used to denote family membership. Once the organic family is identified, there are certain physical and chemical properties that are characteristic. By learning the characteristics of each organic family, you will be able to predict the chemical and physical behavior of thousands of different organic compounds.

Check Your Understanding

1. What element must be present for a compound to be classified as organic?

2. What, if any, are the differences between an organic molecule, an organic radical, and an alkyl group?

3. In the common naming system, what do the prefixes methyl-, ethyl-, propyl-, butyl-, and pentyl- designate?

4. Explain how to identify an isomer.

5. Explain the difference between a structural and functional isomer.

10-2 Hydrocarbons

Objectives

Upon completion of this section, you will be able to:

■ Identify the organic families that are hydrocarbons.

■ Describe the general physical properties of the families that are hydrocarbons.

■ Explain the chemical differences between saturated and unsaturated hydrocarbons.

■ Identify and provide common and IUPAC names for simple members of each of the hydrocarbon families.

■ Write and balance chemical equations representing each of the hydrocarbon families.

Natural gas, coal, and petroleum deposits are actually ancient sunlight stored in the form of chemical energy. These "fossil fuels" are all the products of photosynthesis and the organisms that fed on the ancient plant life. After experiencing millions of years of heat and pressure, they contain primarily carbon and hydrogen atoms, and are appropriately called **hydrocarbons**. Today, fossil fuels are the primary source of the energy used for fueling the manufacture of everyday products and conveniences. For purposes of study, this huge array of hydrocarbons has been subdivided as shown in Figure 10-3.

Aliphatic hydrocarbons are one of the two major subdivisions of hydrocarbons and are characterized by straight, branched, or cyclic carbon chains. The aliphatic group is subdivided into three organic families: alkanes, alkenes, and alkynes. The distinctive structural feature between these families is the type of bonding present between two or more of their carbon atoms.

To be a member of the **alkane** family, a molecule must have only single (–C–C–) carbon-to-carbon bonds, with hydrogen atoms attached to all of the other carbon bonds. The carbon chains may be straight, branched, or cyclic (in a circle). To be a member of the **alkene** family, the molecule must have at least one carbon-to-carbon double bond (–C=C–). Again, only hydrogen atoms may be attached to all of the other carbon bonds. Like alkanes, alkenes may also have straight, branched, or cyclic chains. The fewest hydrocarbons are members of the **alkyne** family. To be an alkyne, a molecule must have at least one carbon-to-carbon triple bond (– C≡C –) and only hydrogen atoms attached to all other carbon bonds. Alkynes can have either straight or branched carbon chains, but do not form cyclic arrangements.

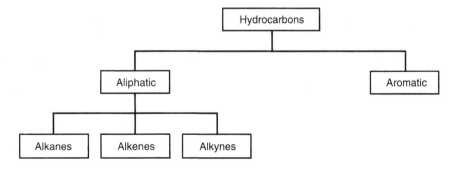

Figure 10–3: Subdivision of hydrocarbons into organic families.

Technology Box 10–1 ■ The Use of IR or NMR to Identify Organic Compounds

As chemists attempt to create unique chemicals and polymers, they need a way to check the actual composition of the chemicals they create. A variety of characteristic properties, such as boiling point, melting point, density, viscosity, refractive index, etc. are used to help in the identification.

One other characteristic that has been found useful is the absorption of different wavelengths of electromagnetic radiation by the chemical bonds present. The two areas of the electromagnetic spectrum that can be absorbed are in the infrared (IR) and the radio frequencies (RF).

Infrared spectrometers, commonly referred to as IRs, can be used to help determine the identity of liquids, solids, gases, and some mixtures. All chemical bonds have some flexibility and a capacity to stretch or bend, if the infrared radiation matches one of its vibrational energies.

An infrared spectrometer, therefore, is a device that shines varying frequencies of infrared light through a sample and measures the amount that is absorbed at each wavelength. From the peaks and valleys on the absorption spectrogram (as shown below), it is possible to determine the types of chemical bonds that are present in the structure. When this information is compared to known standards, it is easy to identify the various components present in the sample.

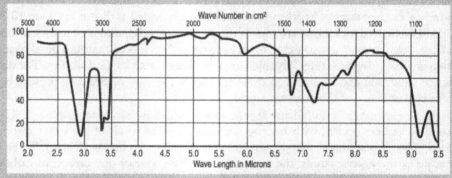

Source: http://www/cbl/leeds.ac.uk/~caypbp/irspec.html.

Similarly, a nuclear magnetic resonance spectrometer, known as NMR, measures the radio waves absorbed by the nuclei of the atoms in a compound. Since chemical bonds hold the structure together, each of the nuclei finds itself in a slightly different environment, which results in their resonating at different energies.

Interpretation of the NMR spectra also identifies the types of chemical bonds present in the compound. As in the case of IR spectroscopy, the resulting NMR spectrogram is compared to a vast computer-based library of known chemicals. Without this "pattern recognition" function performed by computers today, it would be almost impossible to identify complicated chemicals.

The technique called magnetic resonance imaging or MRI, is a powerful medical diagnostic tool that is related to NMR. In MRI, a patient is placed in an intense magnetic field and exposed to low-energy radio waves. In the MRI scan, the nuclei of the atoms are also responding to their chemical surroundings. Rather than trying to interpret the chemical structure of the body, the MRI is used to detect changes in the density of body tissues. Consequently, the presence of tumors and other high-density growths can be detected without exposing the patient to more powerful and potentially harmful X rays.

To be a member of the **aromatic** hydrocarbons family, the molecule must contain one or more benzene rings. The benzene ring is composed of six carbon atoms in a ring, with a special alternating double and single bond arrangement that constantly moves around the ring. The name of this family comes from the strong characteristic aromas of many of its members.

Since the electronegativities of carbon and hydrogen are very nearly equal, they form only nonpolar bonds. The members of all these families are, therefore, nonpolar. This fact dominates the physical properties of all of these families. First, being nonpolar molecules means there is little attraction between neighboring molecules. Since both melting and boiling points are directly related to these attractions, molecules with low attractions, therefore, tend to have low melting and boiling points. The smaller members of each of these families are gases at room temperature, but as the length of their carbon chains increase, so do their boiling and melting points. The majority of these compounds are liquids at room temperature. Only the molecules with long carbon chains are solids at room temperature and even then, unlike ionic solids, they form soft, waxy, greasy solids.

The other noteworthy physical property of all nonpolar molecules is their lack of water solubility. Since water is a polar molecule, it tends to be attracted to only ionic and other polar molecules. All the members of these families are nonpolar; therefore, they are only very slightly soluble in water. This explains the old saying; oil and water don't mix! Nonpolar molecules are, however, soluble in other nonpolar substances. One very common use for some family members is as solvents to remove grease from the surface of metal parts.

Although the physical properties for these families are nearly identical, that is where the similarities stop. The chemical properties of the families are quite different. The alkanes are distinguished by having no functional group and being the most chemically inert (non-reactive) of all organic families. The presence of either a double or triple carbon-to-carbon bond functional group makes the alkenes and alkynes chemically reactive families. Both the double and triple bonds can be broken and form bonds with other molecules. The unique arrangement of the six carbon atoms and alternating double-single nature of the electrons moving around the ring, give the aromatics chemical properties midway between that of the alkanes and the alkenes. More will be said about these properties later.

Alkanes

The alkanes are the least complicated of all organic families, because they do not have a functional group. Members of this family are represented by the general formulas R–H and C_nH_{2n+2}. They are also sometimes called the **paraffins** or **saturated hydrocarbons**. Paraffins (L. *parum* – little; *affinis* – affinity), describe the family's lack of chemical reactivity. **Paraffin wax**, used to make such things as waxed paper and candles, is a mixture of alkanes with chains in excess of twenty carbons. The name, saturated hydrocarbons, refers to their inability to add more hydrogen atoms.

Alkanes are nonpolar molecules. They have low melting and boiling points, are only slightly soluble in water, are less dense than water, and have characteristic gasoline-type (hydrocarbon) odors. The first ten members are listed in Table 10-2. These names should be memorized as soon as possible, since their bolded prefixes are used throughout organic chemistry to indicate the number of carbon atoms present.

General Description

Alkane Name	Molecular Formula	Condensed Structural Formula
Methane	CH_4	CH_4
Ethane	C_2H_6	$CH_3\,CH_3$
Propane	C_3H_8	$CH_3\,CH_2\,CH_3$
Butane	C_4H_{10}	$CH_3\,CH_2\,CH_2\,CH_3$
Pentane	C_5H_{12}	$CH_3\,CH_2\,CH_2\,CH_2\,CH_3$
Hexane	C_6H_{14}	$CH_3\,CH_2\,CH_2\,CH_2\,CH_2\,CH_3$
Heptane	C_7H_{16}	$CH_3\,CH_2\,CH_2\,CH_2\,CH_2\,CH_2\,CH_3$
Octane	C_8H_{18}	$CH_3\,CH_2\,CH_2\,CH_2\,CH_2\,CH_2\,CH_2\,CH_3$
Nonane	C_9H_{20}	$CH_3\,CH_2\,CH_2\,CH_2\,CH_2\,CH_2\,CH_2\,CH_2\,CH_3$
Decane	$C_{10}H_{22}$	$CH_3\,CH_2\,CH_2\,CH_2\,CH_2\,CH_2\,CH_2\,CH_2\,CH_2\,CH_3$

Table 10–2: Straight-chain alkanes and their names.

The smaller alkanes are primarily used as fuels. For example, the first member of the family is known as natural gas and may be piped to your home for heating and cooking. It is also known as sewer gas, swamp gas, and when found in mines, firedamp, but its chemical name is methane. Propane (C_3H_8) and butane (C_4H_{10}) are sold as bottled gas for homes and RVs. Pentane (C_5H_{12}), hexane (C_6H_{14}), heptane (C_7H_{16}), octane (C_8H_{18}), nonane (C_9H_{20}), and decane ($C_{10}H_{22}$) are major components of gasoline.

As the length of the carbon chain increases, it is used for jet airplane fuel (JP5) or kerosene ($C_{12}H_{26}$), diesel oils ($C_{16}H_{34}$–$C_{18}H_{38}$), lubricating oils, and greases. The very largest members of this family are the tar-like components of roofing and asphalt paving materials

Nomenclature

Since this family is the least complicated organic family, we will use it as a model to explain the various systems of naming. As previously noted, trivial names are only valuable to those who regularly use and are familiar with the substance. This category also includes trade names, which may be proprietary products that have been named by their manufacturer. Unless you are familiar with the product, these names will be of little help in determining its structure or family association.

Common or Trivial System

The use of n- (for normal) before the name of a hydrocarbon indicates an unbranched carbon chain. If a compound has the arrangement $(CH_3)_2CH-$ on one end of a chain with no other branching, the prefix iso- is added to the name of the compound.

$$CH_3\,CH_2\,CH_2\,CH_2\,CH_3$$
n-pentane

$$\overset{\displaystyle CH_3}{\underset{}{\overset{|}{CH_3\,CH\,CH_2\,CH_3}}}$$
isopentane

Alkanes with one hydrogen atom removed are called alkyl groups or **side chains**. As shown in Table 10-3, the -ane ending of the alkane is changed to -yl.

				$\overset{\textstyle |}{}$
CH$_3$–	CH$_3$CH$_2$–	CH$_3$CH$_2$CH$_2$–	CH$_3$CH$_2$CH$_2$CH$_2$–	CH$_3$CH$_2$CHCH$_3$
methyl	ethyl	propyl	butyl	sec. butyl

CH$_3$–CH– $\quad$	 CH$_3$	CH$_3$–CHCH$_2$– $\quad$	 CH$_3$	CH$_3$–$\overset{\textstyle |}{\underset{\textstyle |}{\text{C}}}$–CH$_3$ CH$_3$
isopropyl	isobutyl	t-butyl

Table 10–3: Common alkyl group names.

IUPAC Rules for Alkanes

Common names are typically used for the first few members of each family. However, due to the existence of isomers, they soon fail to provide enough naming alternatives. It was recognition of this problem that lead the first international group of chemists to meet in Geneva, Switzerland, in 1892, to develop a systematic method for naming organic compounds. The committee name was changed from its original, International Union of Chemists (IUC), to the International Union of Pure and Applied Chemists (IUPAC) in 1919.

The rules adopted, which have been repeatedly modified, now provide us with an orderly naming system known as the Geneva or IUPAC System. The name of any compound, and most certainly that of a complex structure, is unmistakable when the IUPAC rules are followed. The first few rules are listed below, followed by examples. For a more detailed listing of IUPAC rules, consult a reference or organic chemistry text.

IUPAC Rules for Naming Simple Alkanes

1. Determine the longest continuous carbon chain in the compound. The name of this chain is the last part of the complete name.

2. Number the carbon atoms in the chain from one end to the other, starting at the end closest to the first substituent (side) group. If that group is equidistant from each end, then look for a second or third substituent group until the smallest set of numbers is found.

3. The position of each substituent is designated by the number of the carbon atom to which it is attached. Hyphens are used to separate the number from the name of the substituent.

4. If identical groups appear more than once, the number of the carbon to which each is bonded is given each time. If identical groups appear on the same carbon, the number is repeated. Numbers are separated from each other by commas. The number of identical groups is indicated by prefixes of di-, tri-, tetra-, penta-, etc.

5. The last substituent named becomes one word with the parent (longest carbon chain) hydrocarbon.

6. Substituent groups are to be placed in alphabetical order (ethyl before methyl, etc.), before prefixes are added.

Attempt to follow the above rules when considering the following IUPAC names.

$$\underset{1}{CH_3}\underset{2}{CH_2}\underset{3}{CH}\underset{4}{CH_2}\underset{5}{CH_3}$$
$$|$$
$$CH_3$$

3-methylpentane

$$\underset{5}{CH_3}\underset{4}{CH_2}\underset{3}{CH}\underset{2}{\overset{CH_3}{\underset{|}{CH}}}\underset{1}{CH_3}$$
$$|$$
$$CH_3$$

2,3-dimethylpentane

$$\underset{4}{CH_3}\underset{3}{CH_2}\underset{}{CH}\underset{}{CH_2}\underset{2}{\overset{CH_3}{\underset{|}{CH}}}\underset{1}{CH_3}$$
$$|$$
$$\underset{5}{CH_3}\underset{6}{CH_3}$$

4-ethyl-2-methylhexane

2,3,4-trimethylhexane   1,2-dimethylcyclohexane   3,3-diethyl-2,6-dimethylheptane

In the example named 3,3-diethyl-2,6-dimethylheptane, it may first appear that it could also be named 5-ethyl-5-isopropyl-2-methylheptane. However, IUPAC rules state that if two or more carbon chains are competing for longest, use the one that gives *more smaller* side chains. The preferred name has four side chains and the second (incorrect) name has only three.

Halohydrocarbons

As we will learn later, members of the alkane family will react with halogens (chlorine, bromine, etc.) to form substituted hydrocarbons. A halohydrocarbon is one in which one or more halogen atoms have replaced hydrogen atoms on the parent hydrocarbon. When these substitutions take place, the halogens must also be identified by name and position along the carbon chain. In the IUPAC System, the name of each of the halogens is shortened to the following prefixes: fluoro-, chloro-, bromo-, and iodo-. All of the other IUPAC rules apply as equally to the halogens as they do to the alkyl groups.

In Table 10-4, several of the well-known halogenated hydrocarbons are listed, along with their common and IUPAC names and use.

Formula	Common Name	IUPAC Name	Original Uses
$CHCl_3$	Chloroform	1,1,1-trichloromethane	Anesthetic
CHI_3	Iodoform	1,1,1-triodomethane	External antiseptic
CCl_4	Carbon tetrachloride	1,1,1,1-tetrachloromethane	Early anesthetic, Solvent
CH_2Cl_2	Methylene chloride	1,1-dichloromethane	Solvent
CCl_2F_2	Freon-12	1,1-dichloro-1,1-difluoromethnane	Refrigerant, Propellant
$CH_3 - CCl_3$	Trichloroethane (TCA)	1,1,1-trichloroethane	Solvent
$CH_2F - CF_3$	Freon-134a	1,1,1,2-tetrafluoroethane	Refrigerant
$CH_3 - CH_2Cl$	Ethyl chloride	1-chloroethane	Local anesthetic
$CF_3 - CHBrCl$	Halothane (Fluothane)	2-bromo-2-chloro-1,1,1-trifluoroethane	General anesthetic

Table 10–4: Common halogenated hydrocarbons.

Cycloalkanes

As previously noted, it is possible for alkanes to exist as closed rings, called **cyclo-alkanes**. The shortest carbon chain that can form a ring has only three carbons and is called cyclopropane. Several of the smaller cycloalkanes tend to be highly reactive (flammable) substances. Carbon's 109.5° bond angle must be greatly distorted for this ring to form, so the molecule will willingly react to open the ring and reduce the strain. Nevertheless, cyclopropane has found use as an anesthetic and a starting material for the production of other organic compounds.

Molecules that contain a closed ring no longer conform to the general formula (C_nH_{2n+2}), since two hydrogen atoms must be removed to close the ring. This gives them the same ratio of carbon and hydrogen atoms (C_nH_{2n}) as alkenes.

Within the IUPAC System, the prefix cyclo- is added to the name to indicate that it forms a closed ring. If the longest carbon chain is in a ring, then the cyclo- prefix is added just before the parent name. The numbering of the ring starts at a substituent group and numbers in the direction that will give the lowest set of numbers. If the longest chain is a straight chain, then the cyclo- prefix is added to the side chain name. Under no circumstances should the numbering cross between a carbon chain and a carbon ring. Examine each of the compounds listed below.

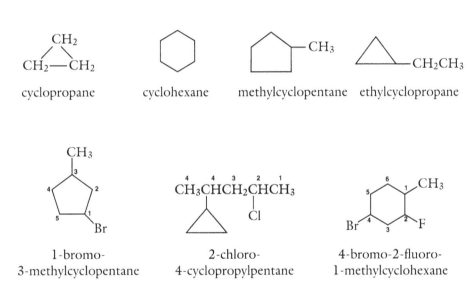

Chemical Reactions

Members of the alkane family are the least reactive of all organic families. They do not react with concentrated sulfuric acid, H_2SO_4, strong bases, such as sodium or potassium hydroxide, NaOH or KOH, or sodium metal, Na. They do burn, or combust, in the presence of oxygen and under chemically severe conditions will undergo substitution reactions with members of the halogen family. As the name suggests, a **substitution reaction** is one in which a bond must be broken and an atom removed, before a new bond can be formed to add a different atom.

Combustion with oxygen is the reaction that explains the usefulness of all hydrocarbons as fuels. If sufficient oxygen is present, all of the carbons are converted to carbon dioxide and all hydrogen atoms combine with oxygen to form water. When insufficient oxygen is present (limiting reagent), it may produce varying amounts of additional products, including carbon, C, often seen as a black cloud in the exhaust of diesel engines, and carbon monoxide, CO. Here, we will consider only the products of complete combustion.

Combustion: $R{-}H + O_2 \rightarrow CO_2 + H_2O + Energy$

Example: $CH_3\,CH_2\,CH_3 + 5O_2 \longrightarrow 3CO_2 + 4H_2O + 531\ kcal/mole\ of\ C_3H_8$

Example: $2\ \underset{}{\bigcirc}\!\!\overset{CH_3}{} + 21\ O_2 \rightarrow 14\ CO_2 + 14\ H_2O$

The balancing of the last equation requires doubling to eliminate the use of $10\text{-}\frac{1}{2}$ moles of oxygen. This doubles all of the other coefficients giving 2, 21, 14, and 14 respectively. Although this is an exothermic reaction, the energy term has not been included.

As noted above, substitution reactions are similar to what happens in many sporting events: one player must leave the playing arena before a new player enters. In the case of halogenation reactions, the conditions that are required to break the existing bond involve the use of high-energy ultraviolet (*uv*) light and temperatures between 250 and 400°C. In the terminology typically used, a **halogenation reaction** is one in which one or more halogens are attached to the parent molecule. To be more specific, the term **chloronation** or **bromonation,** which are the two most common, specify which halogen is used. The following example is for the bromonation of ethane.

Halogenation: $R - H + X_2 \xrightarrow[\Delta]{uv} R - X + H - X$

Example: $CH_3 - CH_3 + Br_2 \xrightarrow[\Delta]{uv} CH_3 - CH_2Br + H{-}Br$

The chemistry of the cyclic aliphatic alkanes is essentially the same as the chemistry of straight-chain alkanes. As we examine the chemical reactions of other organic families, it is important to remember the general lack of chemical reactivity of this family. Therefore, when other more reactive organic functional groups are present and react, the alkane-like portion of their carbon chains will likely remain unchanged.

Alkenes

General Description

The second family of aliphatic hydrocarbons is the alkenes. They are found, along with alkanes, in oil and coal deposits. Alkenes can be easily identified by having at least one carbon-to-carbon double bond in their carbon chain. The general formulas for alkenes are C_nH_{2n} and $R{-}CH{=}CH{-}R$. This family is also known as the **olefins** (L. *oleum* – oil; *ficare* – to make) and they are one of the **unsaturated hydrocarbons**. Again, the terms saturated and unsaturated are a reference to the ability to add more hydrogen atoms to the molecule. Since alkenes have a double bond as their functional group, they have the ability to add another molecule of hydrogen.

Alkenes are also nonpolar molecules. This means that their physical properties are the same as those of the alkanes: low melting and boiling points and less dense than water with very low solubility in water. Chemically, however, the alkenes are very different from the alkanes. The double bond functional group is the site of most chemical reactions, which tend to be of the addition type.

Alkenes are also one of the components of gasoline, but their primary uses are as **feedstocks**, or starting materials, for making a vast array of other organic compounds.

Nomenclature

When smaller $(C_2{-}C_5)$ alkenes are given common names, the -ane ending of the alkane is changed to -ylene. Pentylene and amylene are both used for the 5-carbon member. Table 10-5 lists common and IUPAC names for the first few members.

As you examine the structural formulas and corresponding names, several things become apparent. First, there is now another type of isomer. These are positional isomers, meaning that it is the position of the double bond that is the only difference. Careful examination of the α-butylene and β-butylene structures reveal that the double bond has been moved from after the first carbon to after the second carbon. In the common system, rather than numbering the longest carbon chain, Greek letters are used to identify the carbons: alpha, α; beta, β; gamma, γ; delta, Δ, etc. The second difference is the ending used for the longest chain. The common naming system uses -ylene to denote that the molecule is an alkene. The common names are generally used for the first four members of this family, but IUPAC names are used for all others.

The IUPAC names listed in Table 10-5 have the same features as the common names. However, they use numbers to locate the position of the double bond and they have a consistent -ene ending. Below, the IUPAC rules for naming alkenes are summarized.

Structural Formula	Common Name	IUPAC Name
$CH_2 = CH_2$	ethylene	ethene
$CH_2 = CH-CH_3$	propylene	propene
$CH_2 = CH-CH_2-CH_3$	α-butylene	1-butene
$CH_3-CH = CH-CH_3$	β-butylene	2-butene
$CH_3-C = CH_2$ | CH_3	isobutylene	2-methyl-1-propene or just methylpropene
$CH_3-CH_2-CH_2-CH=CH_2$	α-pentylene or α-amylene	1-pentene

Table 10–5: Examples of common and IUPAC names of alkenes.

IUPAC Rules Changes for Naming Alkenes

1. Find the longest continuous carbon chain that includes the carbon-to-carbon double bond.

2. Number the longest chain, starting at the end closest to the double bond. If the double bond is in the middle of the chain, then start from the end closest to a substituent.

3. List again, by number, the substituent groups in alphabetical order. Just before the longest chain name, give the number of the carbon that the double bond follows.

4. Change the -ane ending of the corresponding alkane to -ene.

5. When naming cycloalkenes, the two carbon atoms holding the double bond must be numbered as one and two. You can number either clockwise or counterclockwise around the ring. Again, the goal is to use the smallest set of numbers.

Using the rules listed above, examine the following structures and their IUPAC names.

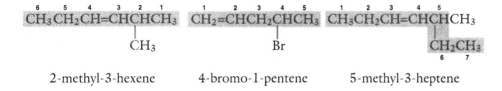

2-methyl-3-hexene   4-bromo-1-pentene   5-methyl-3-heptene

CH₂CH₃ structure: 3-ethylcyclopentene (cyclopentene ring with ethyl group, carbons numbered 1-5)

$$\overset{5}{C}H_3$$
$$|$$
$$\overset{4}{C}H_3$$
$$|$$
$$CH_3 - \overset{}{C} = CH - CH_3$$
$${}_{3}{}_{2}{}_{1}$$

3-methyl-2-pentene

$$CH_3CH_2CH_2\overset{\overset{Cl}{|}}{\underset{3}{C}} \overset{4}{C}HCH_2CH_2CH_3$$

4-chloro-3-propyl-2-heptene

3-ethylcyclo-
pentene

It should be noted that when two or more double bonds are found in the longest carbon chain, the number of each is given and an appropriate prefix di-, tri-, etc. added prior to the -ene ending.

Examples:

$$\overset{4}{C}H_2 = \overset{3}{C}H - \overset{2}{C}H = \overset{1}{C}H_2$$

1,3-butadiene

$$\overset{1}{C}H_2 = \overset{2}{C}H \underset{\underset{CH_3}{|}}{\overset{3}{C}} = \overset{4}{C}H\overset{5}{C}H_2\overset{6}{C}H_2\overset{7}{C}H_3$$

3-methyl-1,3-heptadiene

1,3-cyclopentadiene (cyclopentadiene ring, carbons numbered 1-5)

Chemical Reactions

With the exception of combustion, the double bond is the site of all other chemical reactions for this family. The general reaction for the combustion of an alkene is:

Combustion: $R - CH = CH - R \quad + \quad O_2 \rightarrow CO_2 \quad + \quad H_2O + Energy$

Example: $CH_3 - CH = CH - CH_3 + 6\,O_2 \rightarrow 4\,CO_2 + 4\,H_2O$

Because of the nature of the reacting substance, the addition reactions for this family are subdivided into two groups: symmetrical and unsymmetrical additions. As implied by the names, **symmetrical additions** are those in which the substance being added is symmetrical or has two equal halves, e.g., Br–Br or H–H. **Unsymmetrical additions** are those in which the substance being added does not have two equal parts, e.g., H–Cl or H–OH.

Symmetrical Additions

Earlier it was noted that unsaturated means that there is room on the molecule to add more hydrogen. The reaction in which hydrogen is added to a carbon-to-carbon double bond is called **hydrogenation**. To improve the rate of this reaction a catalyst is used. A **catalyst** is a substance that changes the rate of a reaction without being used up. To speed up he hydrogenation reaction, the metals platinum, Pt, or nickel, Ni, are typically used.

Hydrogenation: $R - CH = CH - R \; + \; H_2 \overset{Pt}{\longrightarrow} R - H$

            alkene (unsaturated)   alkane (saturated)

Example: $CH_3 - CH = CH_2 \quad + \quad H_2 \overset{Pt}{\longrightarrow} CH_3 - CH_2 - CH_3$

Hydrogenation plays an important role in food industries. Most vegetable oils are highly **polyunsaturated**, meaning that they contain many double bonds. Although these polyunsaturated oils are easier to digest, they also tend to react with moist air and become rancid. To prolong shelf life, polyunsaturated vegetable oils are often partially hydrogenated, turning them into less digestible semi-solids at room temperature. These products are sold as margarines and vegetable shortenings.

Halogenation reactions break the double bond and add a halogen atom to each of the carbons. When writing general formulas for halogenation, X_2, denotes the use of any halogen.

Halogenation: $R - CH = CH - R + X_2 \longrightarrow R - CHX - CHX - R$

Example: $CH_2 = CH - CH_2{-}CH_3 + Br_2 \longrightarrow CH_2 - CH - CH_2{-}CH_3$
$$\qquad\qquad\qquad\qquad\qquad\qquad\qquad\qquad\quad \underset{\displaystyle Br}{|} \ \ \underset{\displaystyle Br}{|}$$

Unsymmetrical Additions

Adding a symmetrical reagent poses no problem, because each of the carbons participating in the double bond gets one-half of the adding reagent. When an unsymmetrical reagent adds, however, there are two possible products. The hydration of an alkene below will be used to illustrate the choices.

Hydration

When an alkene is mixed with water in the presence of an acid catalyst, the double bond is easily broken, resulting in the formation of an alcohol. **Hydration reactions** occur during digestion and offer a partial explanation for why polyunsaturated oils are a dietary preference.

Hydration: $R{-}CH = CH_2 + H_2O \xrightarrow{\ H^+\ } RCH_2{-}CH_2{-}OH \quad$ or $\quad RCH{-}CH_3$
$$\qquad\qquad\qquad\qquad\qquad\qquad\qquad\qquad\qquad\qquad\qquad\qquad\qquad\quad \underset{\displaystyle OH}{|}$$

As shown above, if the hydrogen portion of the water adds to carbon that has only one hydrogen attached, the –OH portion of the molecule must add to the carbon at the end of the molecule. On the other hand, if the elements of water are reversed, then the product is an alcohol with the –OH group attached to the second carbon from the end. You may ask, which is correct or does it result in a 50:50 mixture of isomers?

The Russian professor, Markovnikov, studied these reactions extensively and found that although you do get some of both isomers, there is clearly a preference. His observations are now incorporated into a statement known as **Markovnikov's rule**. This rule states that whenever the reagent, H–Y, adding to an alkene releases a hydrogen ion the entering hydrogen always attaches to the carbon atom that already has *more* hydrogen attached. Simply stated: them that have – get!

The following model, therefore, works for predicting the favored product for the unsymmetrical additions we will study: $CH_3{-}CH{=}CH_2 + H{-}Y \longrightarrow CH_3{-}CHY{-}CH_3$, where Y can be any atom or group of atoms. Finally, when both carbon atoms involved in the double bond have the same number of hydrogens attached, Markovnikov's rule is not needed. The application of Markovnikov's rule to the hydration of propene yields the following mixture

Example: $2\,CH_3{-}CH{=}CH_2 + 2\,H_2O \xrightarrow{\ H^+\ } CH_3CH_2{-}CH_2{-}OH + CH_3{-}CH{-}CH_3$
$$\qquad\qquad\qquad\qquad\qquad\qquad\qquad\qquad\qquad\qquad\qquad\qquad\qquad\qquad\qquad\qquad\quad \underset{\displaystyle OH}{|}$$

$\qquad\qquad\qquad$ propene $\qquad\qquad\qquad$ n-propyl alcohol $\quad$ isopropyl alcohol

The two isomers possible are n-propyl alcohol and isopropyl alcohol. However, application of Markovnikov's rule indicates that isopropyl alcohol is the *favored* product.

Hydrohalogenation

As suggested by the name, **hydrohalogenation** is the addition of hydrogen and a halogen, e.g., H–Cl or H–Br to an alkene. Again, Markovnikov's rule must be employed to determine the favored isomer.

Hydrohalogenation: $R–CH=CH_2 + H–X \longrightarrow R–CHX–CH_3$ (favored product)

Industrially, **halocarbons** (halogenated hydrocarbons) have been produced for many years and in great quantities. They have found uses as solvents, refrigerants, propellants, fire retardants, and pesticides, to name but a few. Several of these compounds have become major environmental concerns, due to their environmental persistence, toxicity, and tendency to accumulate in the fatty tissue of organisms and their ability to damage the ozone layer.

Sulfonation

The carbon-to-carbon double bond in alkenes is also broken by the action of concentrated sulfuric acid, H_2SO_4 during **sulfonation** reactions. Since alkanes do not react with sulfuric acid, this provides an easy way to distinguish between them. The general reaction for the reaction is:

Sulfonation: $R–CH=CH_2 + H–O–SO_3H \longrightarrow R–\underset{\underset{OSO_3H}{|}}{CH}–CH_3$ (favored product)

Note that the bond forms between the carbon and one of the oxygen atoms. The products of this addition may be named as alkyl hydrogen sulfates, e.g., isopropyl hydrogen sulfate.

Polymerization

One of the more important reactions of this family is **polymerization**. To understand polymerization, let us think for a moment about how trains are formed. Trains are composed of an engine (an impurity or catalyst), many cars (**monomers**), and, in the past, a caboose (an impurity or catalyst). The railroad workers hook many cars behind an engine and, finally, add a caboose. How polymerization starts is mostly a mystery, but chemists do know what conditions tend to favor its occurrence. It starts when a catalyst or impurity provides a pair of electrons to one of the monomers, forming a bond. This combination is, in turn, attracted to and forms a bond with a neighboring monomer molecule, and so on. In an instant, 500 to 1,000 molecules become connected. Eventually, a pair of electrons is given to a catalyst or impurity and, just as strangely as this started, the polymerization process stops. The giant molecules that are the products of these reactions are called **polymers.** They are inconsistent and may vary greatly in the lengths of their carbon chains.

One of the simplest polymers is called polyethylene, because its monomer is ethylene (ethene) gas. Polyethylene is a chemically inert plastic that is used to make many products. For example, it would probably be more accurate to refer to the packaging of milk or orange juice as a polyethylene container rather than a plastic container. Examples of several familiar polymers are shown below. Note the only differences between each of them are the atoms or groups that are attached to the carbon backbone.

$$\text{Polyethylene: n } (CH_2=CH_2) \xrightarrow{\text{heat/pressure/catalyst}} \text{catalyst–}(–CH_2–CH_2–)_n\text{ –catalyst}$$

$$\text{ethylene} \qquad\qquad\qquad \text{polyethylene}$$

where n = 00 to 1,000.

It should be noted that as the electrons of one monomer molecule are used to connect to the growing chain, the double bond becomes a single bond. Although they are formed from and named as alkenes, it is apparent that the resulting polymers are actually alkanes. This is why the chemical properties of polyethylene are that of an alkane.

Polyvinyl chloride: $n\,(CH_2 = CH) \xrightarrow{heat/pressure/catalyst} catalyst-(-CH_3\ CH-)_n\ -catalyst$
$\qquad\qquad\qquad\qquad\quad |\qquad\qquad\qquad\qquad\qquad\qquad |$
$\qquad\qquad\qquad\qquad\ Cl \qquad\qquad\qquad\qquad\qquad\qquad Cl$

$\qquad\qquad\qquad\quad$ vinyl chloride $\qquad\qquad$ polyvinyl chloride (PVC)

Polypropylene: $\quad n\,(CH_2 = CH) \xrightarrow{heat/pressure/catalyst} catalyst-(-CH_2\ CH-)_n\ -catalyst$
$\qquad\qquad\qquad\qquad\qquad |\qquad\qquad\qquad\qquad\qquad\qquad |$
$\qquad\qquad\qquad\qquad\ CH_3 \qquad\qquad\qquad\qquad\qquad\quad CH_3$

$\qquad\qquad\qquad\qquad$ propylene $\qquad\qquad\qquad$ polypropylene

Teflon: $\qquad\quad n\,(CF_2 = CF_2) \xrightarrow{heat/pressure/catalyst} catalyst-(-CF_2-\ CF_2-)_n\ -catalyst$
$\qquad\qquad\quad$ tetrafluoroethylene $\qquad\quad$ polytetrafluoroethylene (PTFE)
$\qquad\qquad\qquad\qquad\qquad\qquad\qquad\qquad$ or Teflon®

The major point to understand from these examples is that each time the substituent atoms or groups attached to the carbon backbone are different, the polymers exhibit a unique set of physical and chemical properties. All that remains, therefore, is to find applications where these properties fit a unique need.

Cycloalkenes

The physical and chemical properties of **cycloalkenes** are similar to the rest of the family.

General Properties

4-methylcyclopentene $\quad$ 1-ethyl-1,4-cyclohexadiene $\quad$ isopropylcyclopropane
$\qquad\qquad\qquad\qquad\qquad\qquad\qquad\qquad\qquad\qquad\qquad$ 2-cyclopropylpropane

Alkadienes and Alkatrienes

The physical and chemical properties of molecules containing two or more double bonds are similar to the rest of the family. **Dienes** and **polyenes** occur in many places in nature, i.e., turpentine, natural rubber, and the yellow coloring matter in carrots.

General Properties

$CH_2=CH-CH=CH-CH=CH_2 \qquad\qquad CH_2=C-CH=CH_2$
$\qquad\qquad\qquad\qquad\qquad\qquad\qquad\qquad\qquad\qquad\qquad |$
$\qquad\qquad\qquad\qquad\qquad\qquad\qquad\qquad\qquad\qquad CH_3$

$\qquad$ 1, 3, 5-hexatrine $\qquad\qquad\qquad\qquad$ 2-methyl-1, 3-butadiene
$\qquad\qquad\qquad\qquad\qquad\qquad\qquad\qquad\qquad$ (isoprene)

Alkynes

General Properties

The least common aliphatic hydrocarbons are the alkynes. They can be easily identified because they have at least one carbon-to-carbon triple bond ($R-C\equiv C-R$) in their structure and they conform to the general formula, C_nH_{2n-2}. The other name for this family is the **acetylenes**, after its most familiar member – acetylene.

The alkynes are very similar to the alkenes in most ways. Because they are nonpolar compounds, they have low melting and boiling points. They also have very low water solubility. Chemically, they resemble the alkenes, but since they have a triple bond, they can add a second molecule of hydrogen, halogen, water, etc.

The first member of the family, acetylene, can be made easily using a man-made substance called calcium carbide, CaC_2, and water (see Technology Box 10-3). This method was used for years to generate acetylene for oxyacetylene welding equipment and carbide lamps. Acetylene gas is highly explosive and usually contains an impurity that gives it a disagreeable garlic-like odor. Today, most oxyacetylene welding torches are fueled by bottled acetylene. However, due to its explosive nature and shock sensitivity, acetone is also put into the cylinders to improve handling safety. The members of this family are used as starting feedstocks for a variety of other manufactured organic compounds.

Technology Box 10–3 ■ Calcium Carbide and Acetylene Production

The English chemist Edmund Davy is credited with the 1836 discovery of acetylene gas. Marcelin Berthelot, a French chemist, successfully prepared acetylene by placing carbon and hydrogen in an electric arc in 1862. However, it was the discovery of a way to make calcium carbide, CaC_2, in 1892 by the American chemist, T. L. Wilson, which was the turning point for the manufacture and use of acetylene. Wilson found that he could make calcium carbide by heating a mixture of limestone and carbon to a temperature of 2,800°C, in an electric furnace, according to the following equation:

$$2 \, CaCO_3 + 5 \, C \longrightarrow 2 \, CaC_2 + 3 \, CO_2$$

When water is brought into contact with the calcium carbide, the following spontaneous reaction takes place, producing calcium hydroxide and acetylene gas:

$$CaC_2 + 2 \, H_2O \longrightarrow Ca(OH)_2 + CH{\equiv}CH$$

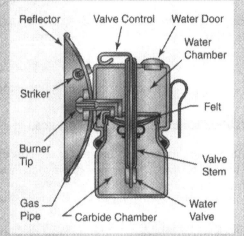

Since acetylene is an unstable compound, it is dangerous to store and handle. The fact that water and calcium carbide produce acetylene on demand, however, soon led to many uses. When acetylene burns in the air, it produces a large amount of light in the form of a bright yellow flame. The carbide lantern became the technology developed for lighting on horse-drawn carriages, the first automobiles, and for miners.

As shown in the figure, a carbide lantern is composed of two compartments, an upper compartment for storing water and a lower compartment where the dry calcium carbide is placed. When light is desired, a valve is opened to allow water to drip on the calcium carbide. The gas produced in the lower compartment is carried by a small tube to the gas jet that is in the area backed by the reflector. The acetylene gas is then ignited producing the light.

When the light is no longer needed, the water is turned off, causing the reaction stop and the lamp to go out. The ability to start and stop the production of acetylene can be repeated, until the water or calcium carbide is used up. At that point, the lower compartment can be cleaned of the residual calcium hydroxide and both compartments recharged.

Although the original uses for the carbide lantern have been largely replaced by safe electric lighting, spelunkers (people who make a hobby of studying and exploring underground caves) continue to keep the technology alive.

The second major use of calcium carbide-generated acetylene is in the welding and cutting of steel. Although the design principles are much the same, acetylene generators used for welding and cutting purposes must be much larger (30 – 50 gallons), since the volume of acetylene required is much larger. By using a specially designed torch, the acetylene and oxygen are mixed prior to burning. Oxyacetylene torches can produce a 3,315.5°C flame, which is used mostly for cutting steel.

Technology Box 10–3 ■ Calcium Carbide and Acetylene Production (Continued)

Due to the dangers associated with these large acetylene generators, the welding industry today supplies acetylene gas in pressurized cylinders. Because compressed acetylene gas will break down and explode, acetone is used in the cylinders to stabilize it. Acetone absorbs acetylene and releases it when the cylinder is opened and the pressure reduced. This greatly increases the margin of safety for handling and using cylinders of acetylene.

Source: http://wasg.iinet.net.au/lamppics.htm

Nomenclature

Common Names

Although, there are only a few, the small alkynes may be named as derivatives of the simplest member of the family – acetylene.

$$CH \equiv CH \qquad\qquad CH \equiv C-CH_3 \qquad\qquad CH \equiv C-CH_2\,CH_3$$

     acetylene   methylacetylene (allylene)   ethylacetylene

IUPAC Rule Changes for Alkynes

Alkynes are named like alkenes, except the ending -ene of the parent hydrocarbon is changed to -yne. The longest chain containing the triple bond is found and numbered from the closest end.

$$\overset{4}{C}H_3\overset{3}{C}H_2\overset{2}{C} \equiv \overset{1}{C}H$$

$$\overset{1}{C}H_3-\overset{2}{C}H_2\overset{3}{C} \equiv \overset{4}{C}-\overset{5}{C}H-CH_3$$
$$\underset{\underset{CH_3\ \ CH_3}{\diagdown\ \diagup}}{\overset{6}{C}H_2}$$

$$\overset{1}{C}H_3-\overset{2}{C} \equiv \overset{3}{C}-\overset{4}{C}H-\overset{5}{C}H_3$$
$$\underset{CH_3}{|}$$

    1-butyne   5,6-dimethyl-3-heptyne   4-methyl-2-pentyne

Chemical Reactions

Members of this family are quite reactive, undergoing many of the same reactions as alkenes. Due to the triple bond, however, they can add two moles of hydrogen, two moles of halogen, etc.

Hydrogenation: $R-C \equiv CH + 2\,H_2 \xrightarrow{\ \ Pt\ \ } R-H$

Halogenation: $R-C \equiv CH + 2\,X_2 \longrightarrow R-\underset{\underset{X}{|}}{\overset{\overset{X}{|}}{C}}-\underset{\underset{X}{|}}{\overset{\overset{X}{|}}{C}}H$

Technology Box 10–4 ■ Bakelite: The world's first fully synthetic plastic

Just after the turn of the twentieth century, the Belgian born chemist Dr. Leo Baekeland, was attempting to find a replacement for shellac. At that time, shellac was obtained by heating and filtering the resinous secretions of a southern Asian insect. The resulting material, called varnish (a mixture of shellac and alcohol), was widely used for preserving wood products. Baekeland, however, knew that shellac was also a good electrical insulator and he envisioned a day when its demand by the emerging electrical industry would outstrip the natural supply. His solution to the problem was to find a way to make shellac synthetically.

Initial experiments involving the heating of phenol (carbolic acid) and formaldehyde (using an acid catalyst) produced a shellac-like liquid that could be used to preserve wood much like varnish. Further heating, however, turned the liquid into gummy goo. After nearly three years of failures, he stumbled upon a process that produced a substance (see figure) that had some rather remarkable properties. He gave it the name "Bakelite." The key to his success was the use of a heavy iron vessel that acted like a pressure cooker. With it, he was able to control the formaldehyde-phenol reaction, while under pressure.

After heating, the product was a hard, translucent, moldable substance – the world's first plastic. In 1909, Baekeland unveiled Bakelite at the meeting of the New York chapter of the American Chemical Society.

For the next ten years, Bakelite was used mostly for making a wide array of electrically non-conducting products, such as light bulb sockets and electrical outlets. Eventually, it was discovered that it could also be dyed and soon it was being used to make jewelry, household, and novelty items. During the 1930s and 1940s, Bakelite was a common item in the family home. It might be in the form of handles on the flatware, a napkin holder, salt and pepper shakers, or a serving tray. Over the years most of these items have become collector's items, with original prices of less than $10 soaring to over a $1,000.

Aromatics

The last organic family of the hydrocarbons group is the aromatics. They are also known as the **arenes, coal tar compounds**, and the **benzene ring compounds**. All members of this family contain a unique structure called the benzene ring. When ben-

General Description

zene was first discovered and analyzed by Michael Faraday in 1825, it was assumed to have multiple bonding because it was composed of six carbon atoms and only six hydrogen atoms. When its chemical reactivity was tested, however, it behaved more like an alkane than an alkene.

This puzzle was not solved until the mid-nineteenth century when Friedrich August von Kekulé proposed a solution. While dreaming, he imagined a snake taking its own tail into its mouth and, thereby, forming a ring. When he awoke and attempted to draw the structure, placing the six carbon atoms into a ring, he found he still needed to include three double bonds. Equally puzzling, he found that he could put the double bonds in two different locations. Finally, it was proposed that the six carbon atoms in the benzene are being held with unique alternating double and single bonds that are constantly trading places around the ring. In truth, the bonds are neither single nor double, but somewhere in between – a hybrid. Today, the benzene ring is represented by either one of its two contributing **resonance structures** or by its **resonance hybrid**.

Aromatic hydrocarbons are found in the black, viscous, pitch-like material called coal tar. It is a by-product of the process used to convert bituminous coal into **coke**. Some of these valuable chemicals are now obtained in large quantities from petroleum refining.

Though unsaturated, aromatic hydrocarbons usually react in a manner similar to the saturated hydrocarbons. Bromine will not add to benzene under the same conditions that give a rapid reaction with alkenes. Members of the family undergo combustion and substitution reactions very easily.

Benzene and many of its **derivatives** (related compounds) are used as solvents and starting materials for making other chemicals. For example, trinitrotoluene (TNT) is less shock-sensitive than nitroglycerine and is widely used as an explosive. The use of phenol as a germicide was introduced in 1867 by the pioneer of antisepsis, Dr. Joseph Lister. Although its use was very successful in stopping the spread of disease, even at low concentrations, it is irritating to the skin. Phenol is also used in the production of drugs, dyes, and the early man-made polymer, Bakelite.

Nomenclature

Common Names

Disubstituted benzene compounds use the prefixes ortho-, meta-, and para-, (abbreviated o-, m-, and p-, respectively) to locate the positions of the substituents relative to one another. Many members of this family have common names of long standing. Over time, the IUPAC System has adopted these names and uses them as parent hydrocarbon names.

o-dichloro-benzene   m-dichloro-benzene   p-dichloro-benzene   o-xylene   p-nitrophenol

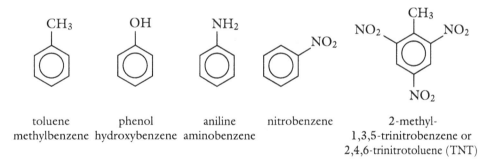

toluene   phenol   aniline   benzoic acid   benzaldehyde

(These common names are also accepted IUPAC names.)

IUPAC Rule Changes for Aromatics:

1. If the benzene ring is the longest continuous chain, the numbering should start at the constituent group that will produce the lowest set of numbers.

2. If the benzene ring is not the longest continuous chain in the molecule, it may be named as a substituent group. When named as a substituent group, its name is phenyl-.

3. The parent name may be either benzene or one of the IUPAC accepted common names; e.g., toluene, phenol. If one of the accepted names is used, then the numbering of the ring must start with the functional group of that substance.

4. The prefixes ortho-, meta-, and para- are not a part of the IUPA system; therefore, they are used only in the common system to locate relative positions.

toluene   phenol   aniline   nitrobenzene   2-methyl-
methylbenzene   hydroxybenzene   aminobenzene   1,3,5-trinitrobenzene or
                                          2,4,6-trinitrotoluene (TNT)

2-hydroxy-1,3,5-trinitrobenzene   1,2-dimethylbenzene   1,3,5-trimethylbenzene
or 2,4,6-trinitrophenol   or 2-methyltoluene   or 3,5-dimethyltoluene
picric acid (common name)   o-xylene (common name)   mesitylene (common name)

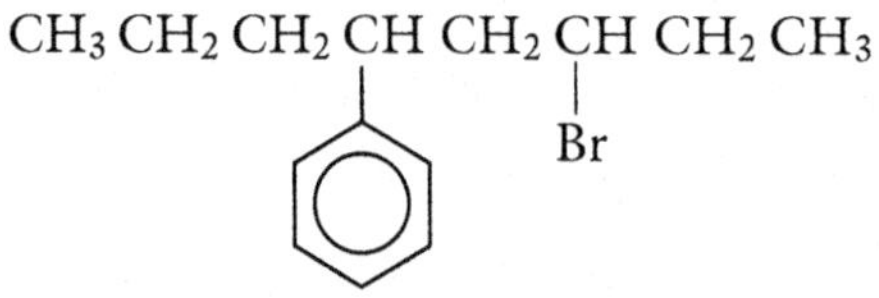

$$CH_3\ CH_2\ CH_2\ CH\ CH_2\ CH\ CH_2\ CH_3$$

Br

3-bromo-5-phenyloctane

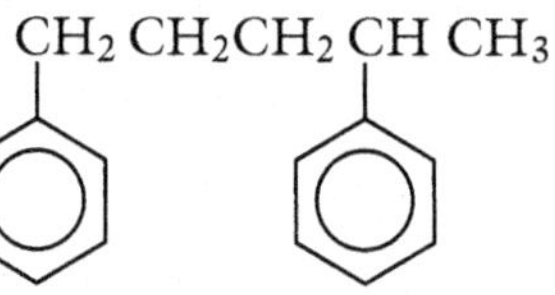

$$CH_2\ CH_2CH_2\ CH\ CH_3$$

1,4-diphenylpentane

1,4 diphenylbenzene

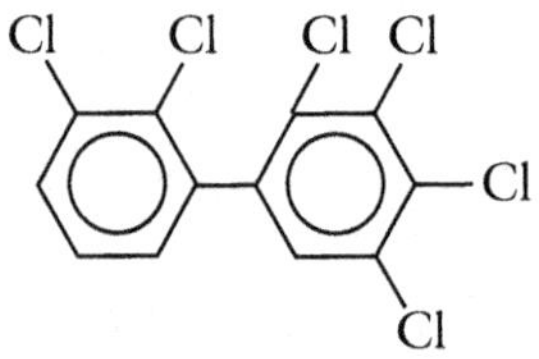

One of the 209 possible
polychlorinated biphenyl (PCB) isomers

Chemical
Reactions

Combustion Reaction

Combustion: $2\ \bigcirc\ +\ 15\ O_2\ \rightarrow\ 12\ CO_2\ +\ 6\ H_2O$

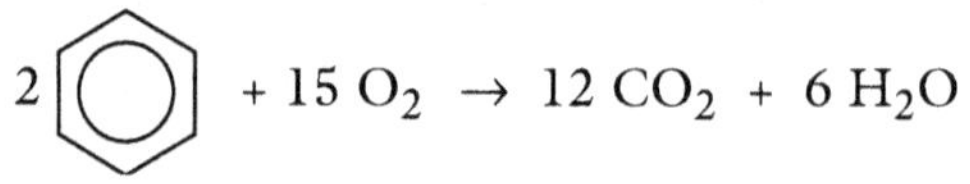

Addition-type Reactions

Hydrogenation: $\bigcirc\ +\ 3H_2\ \xrightarrow{\ Pt\ }\ \bigcirc$

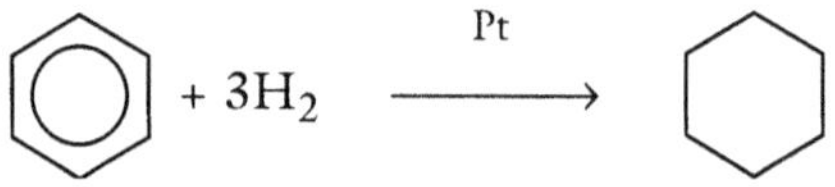

Bromonation or
Chloronation: $\bigcirc\ +\ 3Cl_2\ \xrightarrow{\ uv\ }$

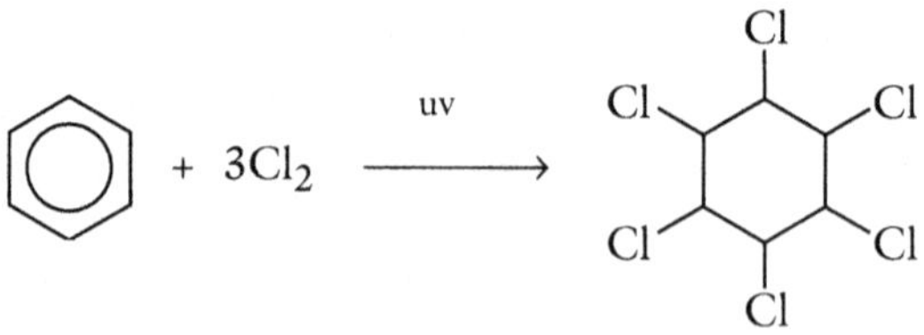

Substitution Reactions

Halogenation (Chloronation)

In the presence of iron or iron(III) chloride, benzene undergoes a typical alkane-like
substitution reaction with the halogens, chlorine, and bromine.

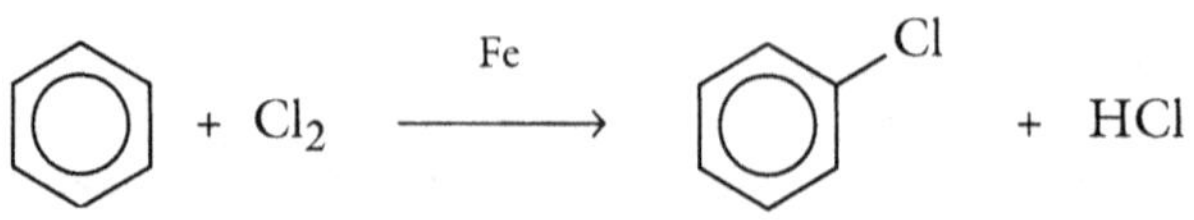

$$\bigcirc\ +\ Cl_2\ \xrightarrow{\ Fe\ }\ \bigcirc\!-Cl\ +\ HCl$$

Nitration (HNO_3)

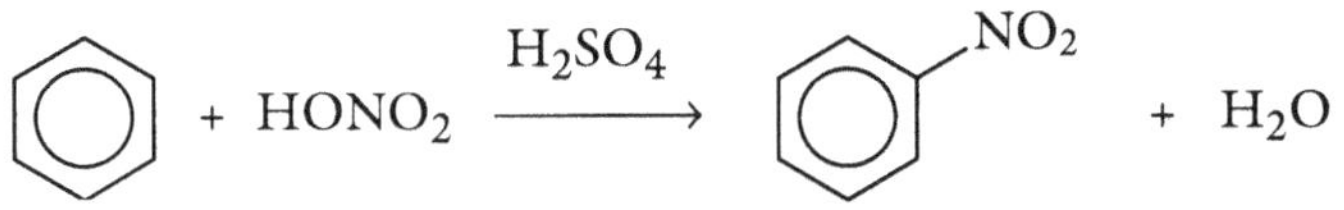

Sulfonation (H_2SO_4)

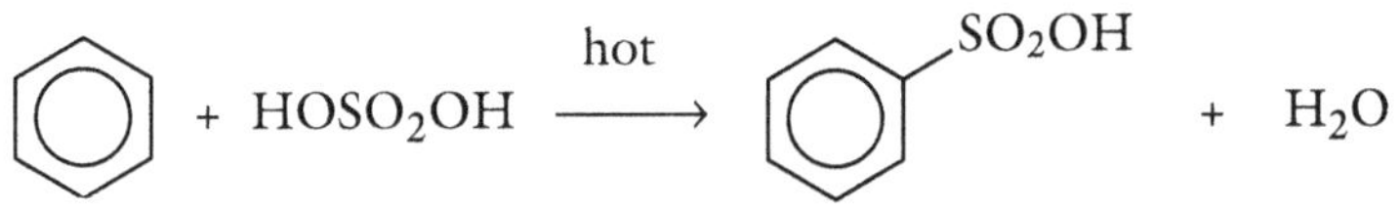

benzenesulfonic acid

Check Your Understanding

1. Describe the water solubility of the organic families that are hydrocarbons.

2. What is the first step in naming any organic molecule?

3. What is the name of an alkyl group with the formula, CH_3CH_2-?

4. When carbon atoms form a ring, what prefix is used before the last name?

5. Which of the four organic families (alkanes, alkenes, alkynes, and aromatics) is least chemically reactive and why?

6. Which of the four organic families (alkanes, alkenes, alkynes, and aromatics) has the most similar chemical behaviors and why?

7. How does a substitution reaction differ from an addition reaction?

8. When is it necessary to use Markovnikov's rule?

9. What is a polymer?

10. When and with which family are the prefixes, ortho-, meta-, and para- used?

Objectives

Upon completion of this section, you will be able to:

■ Identify members of the alcohol, mercaptan, and ether families by either general formulas, names, or from structural formulas.

■ Describe the physical properties of an alcohol, mercaptan, and an ether.

■ Name small members of each of the alcohol, mercaptan, and ether families with a common and IUPAC name.

■ Write and balance chemical reactions typical for the alcohol, mercaptan, and ether families.

Alcohols

General Description

The general formulas for **alcohols** are R–OH and $C_nH_{2n+1}OH$. As such, they may be viewed either as alkyl derivatives of water or as hydroxyl derivatives of hydrocarbons. In either case, the –OH functional group does not ionize and they do not produce basic solutions.

Alcohols are polar molecules. In chemistry, "likes dissolve likes" is a rule-of-thumb. As shown in Figure 10-4, the presence of the –OH group, as in water, makes hydrogen bonding possible. The four smallest alcohols are water-soluble liquids and have pleasant odors. Five to ten carbon alcohols are liquids at room temperatures, but as the carbon chain increases, the water solubility rapidly diminishes. Those with twelve or more carbon atoms are water insoluble solids at room temperature. Dodecyl (C_{12}) alcohol, for example, is used in dripless candles.

Alcohols also have much higher melting and boiling points than the corresponding alkanes. The carbon atom to which the –OH group is bonded is called the **alcoholic carbon**. This carbon can form three other bonds, resulting in three possible situations. Because each of these situations has an effect on the alcohol's chemical behavior, they are divided into three subclasses: primary, 1°, secondary, 2°, and tertiary, 3°. When only one carbon is bonded to the alcoholic carbon, it is a primary alcohol. If the alcoholic carbon has two carbon atoms bonded to it, it is a secondary alcohol; and when it has three other carbon atoms bonded to it, it is a tertiary alcohol. Examine the following examples and note that: 1 carbon bond = 1°; 2 carbon bonds = 2°; 3 carbon bonds = 3°.

Alcohols undergo a wide range of reactions involving the –OH group. Many of these are **elimination reactions**. This name is used because they remove the elements

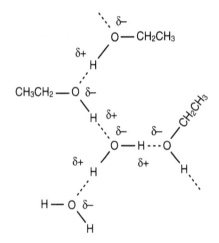

Figure 10–4: Hydrogen bonding: water and ethyl alcohol.

of water (H and –OH) between two molecules. The organic fragments join forming a new compound, and water is always the other product. The subclass of the alcohol involved, however, may have an effect on how the process occurs.

$$
\begin{array}{ccc}
\begin{array}{c} H \\ | \\ R - C - OH \\ | \\ H \end{array} &
\begin{array}{c} R \\ | \\ R - C - OH \\ | \\ H \end{array} &
\begin{array}{c} R \\ | \\ R - C - OH \\ | \\ R \end{array} \\[4pt]
\text{primary } (1°) & \text{secondary } (2°) & \text{tertiary } (3°)
\end{array}
$$

As will be seen in the reactions section, alcohols are industrially important compounds for a variety of reasons. First, they are excellent organic solvents. Second, they are used in large quantities as antifreeze and antiboil mixtures. Finally, they are the major starting points for the manufacture of many other organic compounds, including polyesters.

Nomenclature

Common Names

The common names for alcohols are generally composed of their alkyl group name followed by the family name, *alcohol*. Trivial names that have been in use for a long time are also found. For example, the first member, wood alcohol, was named after its source. Grain alcohol is the product of all fermentation processes. Alcohols with three carbon atoms are called propyl alcohol, including rubbing or isopropyl alcohol. All butyl alcohols have four carbons and five-carbon alcohols are called either pentyl or amyl alcohols.

$$
\begin{array}{ccc}
CH_3-OH & CH_3CH_2-OH & CH_3CH_2CH_2-OH \\
\text{methyl alcohol} & \text{ethyl alcohol} & \text{n-propyl alcohol}
\end{array}
$$

$$
\begin{array}{cc}
\begin{array}{c} OH \\ | \\ CH_3- CH-CH_3 \end{array} & CH_3CH_2CH_2CH_2CH_2-OH \\[4pt]
\text{isopropyl alcohol} & \text{n-amyl or n-pentyl alcohol}
\end{array}
$$

IUPAC Rule Changes for Alcohols

1. Determine the longest continuous carbon chain to which the –OH group is attached.

2. Number the longest chain giving the lowest possible number to the position of the hydroxyl group.

3. The positions of each substituent are indicated by number. The number of the carbon bonded to the hydroxyl group is given just before the last name.

4. The -ane family ending is replaced with -ol.

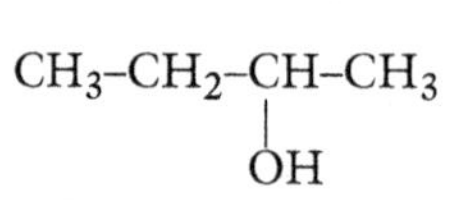

$$CH_3-CH_2-\underset{\underset{OH}{|}}{CH}-CH_3$$

2-butanol

$$CH_3-\underset{\underset{CH_3}{|}}{\overset{\overset{OH}{|}}{C}}-\overset{\overset{Br}{|}}{CH}-CH_3$$

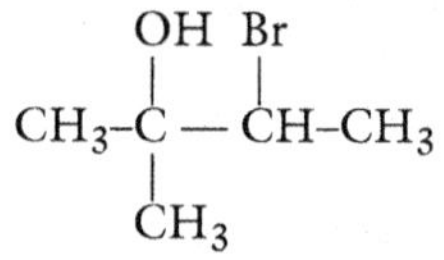

3-bromo-
2-methyl-2-butanol

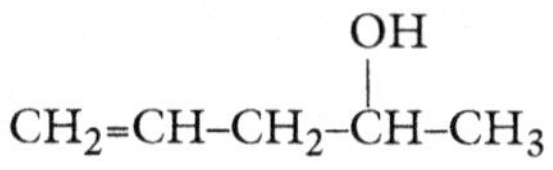

$$CH_2=CH-CH_2-\underset{\underset{OH}{|}}{CH}-CH_3$$

4-pentene-2-ol (alcohol name)
4-hydroxy-1-pentene (alkene name)

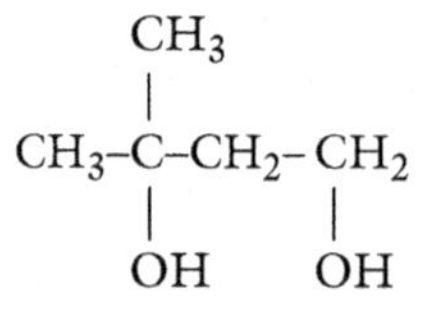

$$CH_3-\underset{\underset{OH}{|}}{\overset{\overset{CH_3}{|}}{C}}-CH_2-\underset{\underset{OH}{|}}{CH_2}$$

3-methyl-1,3-butanediol

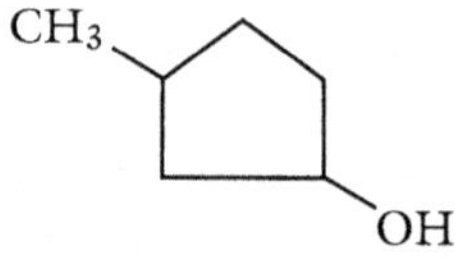

3-methylcyclopentanol

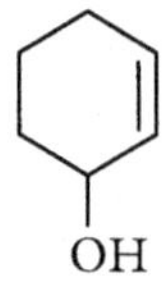

2-cyclohexenol (alcohol name)
3-hydroxycyclohexene (alkene name)

Chemical Reactions

Combustion

In recent years, much attention has been given to finding cleaner burning fuels for cars. One **oxygenated fuel** that achieved favorable results was **gasohol,** a mixture of 80–90% hydrocarbons and 10–20% methanol or ethanol. As the name implies, oxygenated fuels are those that have one or more oxygen atoms incorporated into the fuel molecule and, therefore, provide for more complete combustion.

Combustion: $R-OH + O_2 \rightarrow CO_2 + H_2O$

Example: $CH_3CH_2-OH + 3\ O_2 \rightarrow 2\ CO_2 + 3\ H_2O$

Oxidation

The symbol (O) indicates that a chemical is used to provide single oxygen atoms during a chemical reaction. The dichromate, $Cr_2O_7^{2-}$, and permanganate, MnO_4^- ions are strong oxidizing agents and will convert 1° and 2° alcohols into carbonyl compounds ($>C=O$). When solutions containing the orange dichromate ion are used, they are converted to the blue-green Cr^{3+} ion if a 1° or 2° alcohol is present. When a spectrophotometer is used to analyze the amount of color change, it can be used as a quantitative measurement for blood alcohol level. The technology is called a **breathalyzer test**.

Primary Alcohols + (O) $\rightarrow$ Aldehydes + Water

$$1°\ R-OH\ +(O)\quad \rightarrow R-\overset{\overset{O}{\|}}{C}-H\quad +\ H_2O$$

Secondary Alcohols + (O) $\rightarrow$ Ketones + Water

$$2°\ R-OH\ +(O)\quad \rightarrow R-\overset{\overset{O}{\|}}{C}-R\quad +\ H_2O$$

Tertiary Alcohols + (O) $\rightarrow$ No Rx, because no hydrogen atom is attached to the alcoholic carbon.

Esterification

Esterification is a natural process that is responsible for the odors associated with many fruits. Although, esters, such as wintergreen and aspirin, are prepared and used in the food and drug industries, the largest use of the reaction today is in the production of polyester resins. For example, most wash and wear clothing now contains a significant percentage of polyester fibers. Polyethylene terephthalate (PET) is a polyester plastic that is rapidly replacing the use of glass bottles. Other polyesters serve as the glue that holds together fiberglass fabric to form strong materials that are replacing the use of metals for such things as underground fuel storage tanks.

$$\text{Esterification:} \quad R\text{–OH} + HO\text{–}\overset{\displaystyle O}{\overset{\|}{C}}\text{–}R' \rightarrow R\text{–O–}\overset{\displaystyle O}{\overset{\|}{C}}\text{–}R' + H_2O$$
$$\text{Alcohol} + \text{Carboxylic Acid} \rightarrow \quad \text{Ester} \quad + \text{ Water}$$

Note the ′ after the "R" in the equation above indicates that the length of the carbon chain on one side of the equation doesn't have to be the same as the length of the carbon chain on the other side.

Ether Synthesis

It can be said that the discovery of diethyl ether and its anesthetic properties moved surgery from the barber's chair to the doctor's office. Diethyl ether can be made by eliminating the water between two ethyl alcohol molecules. The use of H_2SO_4, to aid in the removal of the water and a temperature of 140°C are the favored conditions.

$$\text{Ether Synthesis:} \quad R\text{–OH} + HO\text{–}R' \xrightarrow[\text{heat}]{H_2SO_4} R\text{–O–}R' + H_2O$$

Dehydration

Alcohols can also undergo **dehydration reactions** producing an alkene and water. Depending on the subclass, the order of ease of dehydration is 3° > 2° > 1°. Primary alcohols like ethanol are, therefore, the most difficult to dehydrate, requiring 170 – 180°C and 100% H_2SO_4.

$$\text{Dehydration:} \quad R\text{–}\overset{\displaystyle H}{\underset{\displaystyle H}{\overset{|}{\underset{|}{C}}}}\text{–}\overset{\displaystyle H}{\underset{\displaystyle OH}{\overset{|}{\underset{|}{C}}}}\text{–}R \xrightarrow[\text{heat}]{H_2SO_4} R\text{–}\overset{\displaystyle H}{\overset{|}{C}} = \overset{\displaystyle H}{\overset{|}{C}}\text{–}R + H_2O$$

Glycols

Alcohols containing two –OH groups are known as **glycols** or **diols**. The addition of a second polar –OH group increases the melting and boiling points and makes them more water-soluble. A mixture of ethylene and propylene glycol are sold as automobile radiator antifreeze. Glycols are used in the manufacture of polyesters. Glycols can undergo two esterification reactions – one at each hydroxyl group. This ability is essential for a polymerization reaction to continue.

General Description

Common names for the glycols are based on the alkene from which they can be made, i.e., ethylene glycol made from ethylene. In the IUPAC System, the positions of the

Nomenclature

–OH groups are designated by numerical prefixes and the ending, -diol, added to the parent hydrocarbon name. Several examples follow, using both the common and/or IUPAC systems.

$$CH_2–CH_2$$
$$\quad|\quad\quad|$$
$$OH\quad OH$$

ethylene glycol
1,2-ethanediol

$$OH\quad OH$$
$$\quad|\quad\quad|$$
$$CH_3–CH–CH_2$$

propylene glycol
1,2-propanediol

$$OH\quad\quad\quad OH$$
$$\quad|\quad\quad\quad\quad|$$
$$CH_3–CH–CH_2–CH_2$$

butylene glycol
1,3-butanediol

$$CH_3–CH–CH–CH_2–CH_2–CH_3$$
$$\quad\quad\quad|\quad\quad|$$
$$\quad\quad OH\quad OH$$

2,3-hexanediol

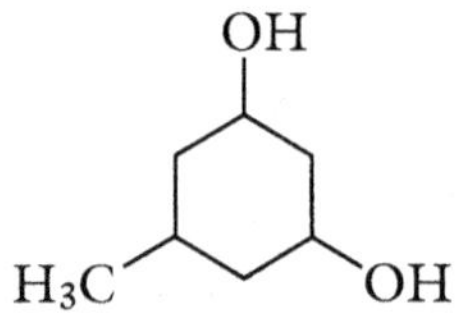

5-methyl-1,3-cyclohexanediol

Chemical Reactions

Ethylene and propylene glycols have a sweet taste. Given the opportunity, pets will drink antifreeze and the results are usually fatal. When glycols are oxidized, they produce the corresponding dicarboxylic acid. When this happens internally, it interrupts the Krebs cycle, ceasing all energy production in the body and resulting in death.

Oxidation: glycol + (O) → dicarboxylic acid + water

Example: $HO–CH_2CH_2–OH + 4\,(O) \longrightarrow$
$$\quad\quad\quad\quad\quad\quad O\ \ O$$
$$\quad\quad\quad\quad\quad\quad || \ \ ||$$
$$\quad\quad\quad\quad HO–C–C–OH + 2\,H_2O$$

ethylene glycol   oxalic acid

Glycerol

General Description

Glycerol, or **glycerin**, is a by-product of the soap-making process. It is a viscous, sweet-tasting liquid. Because of the three –OH groups present, it has high melting and boiling points and is soluble in all proportions in water. Glycerol is an excellent **humectant** (moisture-retaining agent). It is used on tobacco products and in the cosmetic industries to help products retain moisture. Digestion of simple fats and oils, called triglycerides, produce glycerin.

Nomenclature

Glycerol and glycerin are both common names. The IUPAC System names it as an alcohol and adds a -triol ending to the name of the longest carbon chain.

$$CH_2–OH$$
$$\quad|$$
$$CH–OH$$
$$\quad|$$
$$CH_2–OH$$

glycerol or glycerin
1,2,3-propanetriol

$$CH_2–O–NO_2$$
$$\quad|$$
$$CH–O–NO_2$$
$$\quad|$$
$$CH_2–O–NO_2$$

nitroglycerin 1,2,3-trinitropropane
(vasodilator & dynamite)

Mercaptans

The **mercaptans** or **thioalcohols** are moderately polar molecules. They are less polar than the corresponding alcohols, but more polar than the alkanes. As such, their melting points, boiling points, and water solubility fall between the alcohols and alkanes. The general formula for mercaptans, R–SH, makes them appear to be alkyl derivatives of H–SH (H_2S, rotten egg gas). In fact, several of the smaller members have equally disagreeable odors. The odor of skunk, onions, garlic, and the substance added to natural gas to give it an odor are all members of this family.

The –SH group, also known as the **sulfhydryl group**, will undergo a number of chemical reactions. Under conditions of very mild oxidation, two mercaptan molecules will unite forming a **disulfide bond**. These bonds are particularly important in maintaining the shapes of enzymes and other large protein molecules. Hair is particularly rich in disulfide bonds. Many people periodically spend large sums of money to have these bonds chemically broken and reformed, a process referred to as a permanent.

In addition, sulfhydryl groups are present on certain enzymes and will react with heavy metal ions, like Pb^{2+}, Hg^{2+}, Ag^+, Zn^{2+}, and Cu^{2+}. This renders the enzyme inactive, making these metals toxic to most living organisms.

General Description

Common System

Nomenclature

In the common naming system, mercaptans are named by using the alkyl group name and adding the family name, *mercaptan*.

$$CH_3-SH \qquad CH_3-CH_2-SH \qquad CH_3-CH-CH_3$$
$$\qquad\qquad\qquad\qquad\qquad\qquad\qquad | $$
$$\qquad\qquad\qquad\qquad\qquad\qquad\qquad SH$$

methyl mercaptan   ethyl mercaptan   isopropyl mercaptan

IUPAC Rule Changes for Mercaptans

The IUPAC rules are the same as for naming the alcohols, except the -ol ending of the alcohol is replaced by a -thiol ending for mercaptans.

cyclopentanethiol   3-ethylcyclopentanethiol   3-chloro-1-pentanethiol

Salt Formation: (Pb^{2+}, Hg^{2+}, Ag^+, Zn^{2+}, Cu^{2+})

Chemical Reactions

2 enzymes – SH + Pb^{2+} ⟶ enzyme – Pb – enzyme + 2 H^+
  (active)   (biologically inactive)

                      (very mild conditions)
Oxidation: R–SH + HS–R′ ⟶ R–S–S–R′ + H_2O

                      (very mild conditions)
Reduction: R–S–S–R′ (H) ⟶ R – SH + R′– SH

Ethers

General Description

Ethers are nonpolar molecules. With general formulas of R–O–R and C_nH_{2n+1}–O–C_nH_{2n+1}, ethers are functional isomers of the corresponding alcohols. They may be considered as either dialkyl derivatives of water or alkyl derivatives of alcohols. The two alkyl groups attached to the **ether linkage**, –O–, make them more like alkanes than alcohols. They have boiling points that are lower than the corresponding alcohol; very slight water solubility; and are readily soluble in organic solvents. They do not exhibit hydrogen bonding. Ethers have anesthetic properties, since they are nonpolar, fat-soluble substances that depress the central nervous system. Diethyl ether is an excellent anesthetic, but is irritating to the air passages and frequently results in nausea.

Ethers are the second most inert organic family, so their reactions are few. They do not react with sodium metal, Na, strong acids like H_2SO_4, strong bases like NaOH or KOH, or mild oxidizing agents. When mixed with air, ethers burn with an explosive force and they will form dangerous **ether peroxides**, R–O–O–R.

Nomenclature

Common Names

In the common naming system, both groups bridged by the ether linkage are named, followed by the family name, *ether*. Sometimes the prefix di- is used, if both alkyl groups are the same (symmetrical ethers), but this is not necessary.

$$CH_3\text{–O–}CH_3 \qquad C_2H_5\text{–O–}C_2H_5$$

methyl ether
or dimethyl ether

ethyl ether
or diethyl ether

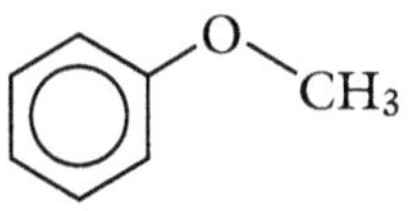

methyl phenyl ether 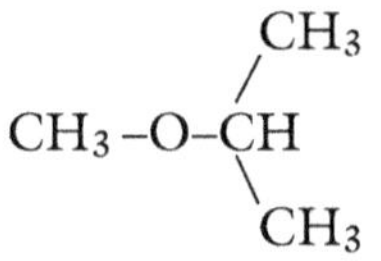   methyl isopropyl ether   vinyl ether or divinyl ether

IUPAC Rule Changes for Ethers

The larger alkyl group bonded to the ether linkage is named like a substituted hydrocarbon. The smaller alkyl group and oxygen is called an **alkoxy substituent**. The IUPAC system, therefore, treats ethers as alkoxyalkanes. The alkoxy side chains have names like methoxy, ethoxy, propoxy, etc., which are alphabetized and located by number, just like any other substituent group.

$$CH_3\text{–O–}CH_2\text{–}CH_2\text{–}CH_3 \qquad CH_3\text{–}CH_2\text{–O–}CH_3$$

1-methoxypropane

methoxyethane

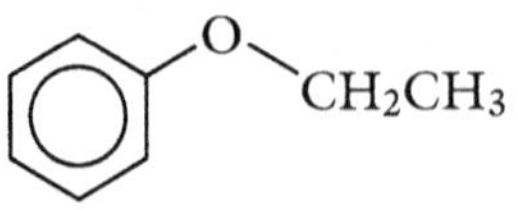

ethoxybenzene

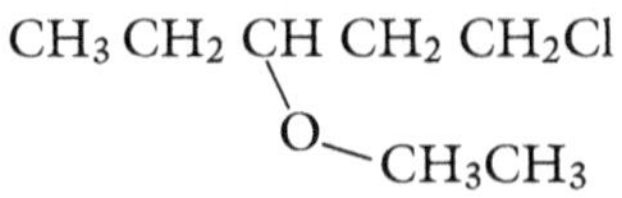

1-chloro-3-ethoxypentane

Technology Box 10–5 ■ The Future of MTBE

The Clean Air Act Amendments (CAAA) of 1990 established a program that allowed for the use of reformulated gasoline to reduce the pollutants in auto exhaust. The pollutants of concern were the small volatile organic compounds (VOCs) that produce ozone and carbon monoxide. Further reduction in the use of known cancer-causing chemicals such as benzene and 1,3-butadiene was also desired.

Gasoline traditionally consisted of dozens of different hydrocarbons, including the many isomers of hexane, for example. Under ideal conditions and the presence of an excess of oxygen, the combustion of hexane produces only water, carbon dioxide, and energy, according to the equation:

$$2\ C_6H_{14} + 19\ O_2 \rightarrow 14\ H_2O + 12\ CO_2 + energy.$$

If there is an insufficient amount of oxygen present, however, the combustion products are water and a mixture of CO_2, CO, and carbon, which is sometimes seen in the exhaust gases. As the amount of oxygen available for combustion decreases, so too, does the amount of energy produced. In addition, research indicates that if there is a deficiency of oxygen, other carbon-containing compounds are also produced. Although there is no agreement on the exact mechanisms of the combustion process, it is agreed that where oxygenated fuels are used, there is an accompanying decrease in air pollutants

The CAAA stated that reformulated fuel must contain at least 2.0% oxygen by weight. The two most popular oxygenates are ethyl alcohol and methyl tertiary butyl ether, **MTBE**. By 1999, the Environmental Protection Agency, EPA, estimated that over 85% of all reformulated gasoline contained MTBE. At the same time, it had become apparent that MTBE pollution is threatening the nation's drinking water supply. Although the amount of MTBE entering the environment each day appears to be small, it is coming from a variety of sources, including spills, watercraft use, and leaks in storage and gasoline dispensing equipment.

Unfortunately, MTBE brings its own health-related problems, including toxicity and complaints of nausea, dizziness, and eye irritation. It also has enough water solubility that when spilled or leaked, it moves with the flow of water into the environment.

On March 24, 2000, EPA published an advance notice to begin rule making to eliminate or limit the use of MTBE in gasoline and to treat it as another environmental hazard. It is known that oxygenates improve the performance of fuels and reduce exhaust emissions, but the challenge remains to find an alternative to the continued use of MTBE.

Chemical Reactions

Combustion

Earlier it was noted that oxygenated fuels are in wide use today. Another additive in use is methyl tertiary butyl ether or MTBE, which has been shown to reduce smog-forming emissions by as much as 15 percent. This all seemed very promising until it was discovered that MTBE is finding its way into groundwater supplies. MTBE migrates somewhat more rapidly than other gasoline components and is more costly to remove. Biodegradation, which will work for most petroleum clean-ups, will not work for MTBE. In California, where it was the preferred oxygenate additive, MTBE is being phased out and its use will be banned by 2002. Nationally, the EPA has taken action to limit the use of MTBE.

Combustion: $R–O–R' + O_2 \longrightarrow CO_2 + H_2O$

Example: $CH_3CH_2–O–CH_2CH_3 + 6\,O_2 \rightarrow 4CO_2 + 5\,H_2O$

Peroxide Formation

Ethers readily form explosive peroxides when exposed to the air. These compounds are extremely unstable, requiring only slight contact such as the unscrewing of the bottle cap to detonate them. Manufacturers take steps to improve the margin of safety by supplying them in metal cans or brown bottles, adding an oxidation inhibitor, and recommending that they remain refrigerated and in the dark when not in use. Even then, opened containers of ether should not be kept for more than a few months.

Peroxide Formation: $R–O–R' + (O) \longrightarrow R–O–O–R'$
(extremely unstable)

Check Your Understanding

1. What is the –OH group effect on the physical properties of alcohols?

2. What structural feature distinguishes 1°, 2°, and 3° alcohols? Name one type of reaction where the alcohol subclass makes a difference in the product(s) produced.

3. Which chemical reaction produces either natural fruit odors or synthetic fibers?

4. Poisoning is the result of a reaction between a heavy metal and which functional group that is present on many enzymes?

5. What is the difference between the IUPAC endings for alcohols and mercaptans?

6. What is the second most inert organic family?

10-4 Aldehydes and Ketones

Objectives

Upon completion of this section, you will be able to

■ Identify members of the aldehyde and ketone families by either general formulas, names, or from structural formulas.

■ Describe the physical properties of aldehydes and ketones.

■ Name small members of each of the aldehyde and ketone families with a common and IUPAC name.

■ Write and balance chemical reactions typical for the aldehyde and ketone families.

Aldehydes

The **aldehydes** are the first of several organic families containing the **carbonyl group**, (see formula below). The symbols δ^- and δ^+, are again included to show the polarity of this group. Because of the higher electronegativity of oxygen, the electrons are drawn away from the carbon atom, resulting in a partial positive charge on the carbon atom and a partial negative on the oxygen atom.

$$
\begin{array}{cc}
\underset{\text{carbonyl group}}{\overset{\displaystyle O\delta^-}{\underset{\displaystyle}{\overset{\displaystyle \|}{-C-\delta^+}}}} & \underset{\text{general formula for aldehydes}}{\overset{\displaystyle O}{\overset{\displaystyle \|}{R-C-H}}}
\end{array}
$$

As shown by their general formulas (above), aldehydes have only one hydrogen atom connected to the carbonyl carbon and a carbon chain of varying lengths on the other side. Aldehydes are moderately polar, with moderate melting and boiling points like the mercaptans. Only the first four members are soluble in water. Smaller members, like formaldehyde, have unpleasant odors, but some of the larger members are used for artificial flavorings. For example, citral (oil of lemon) is an unsaturated aldehyde (3,7-dimethyl-2,6-octadienal) found in the oils of citrus fruits. Benzaldehyde is alternately used as either artificial cherry or almond flavoring.

The aldehyde functional group is quite sensitive to oxidation and cannot, therefore, be stored in contact with the air for long periods. This functional group is also the site where addition reactions occur. Some of the sugars are **aldohexoses**, which have six carbons and an aldehyde functional group. Glucose is a nutritionally important example that is also known as dextrose or blood sugar.

Common Names

Nomenclature

The common names for aldehydes were derived from the carboxylic acid they produced upon oxidation. When applied to the aldehydes, the "-ic acid" ending is changed to an "aldehyde."

$$
\underset{\text{formaldehyde}}{\overset{\displaystyle O}{\overset{\|}{H-C-H}}} \qquad
\underset{\text{acetaldehyde}}{\overset{\displaystyle O}{\overset{\|}{CH_3-C-H}}} \qquad
\underset{\text{propionaldehyde}}{\overset{\displaystyle O}{\overset{\|}{CH_3-CH_2-C-H}}} \qquad
\underset{\text{n-butyraldehyde}}{\overset{\displaystyle O}{\overset{\|}{CH_3(CH_2)_2-C-H}}} \qquad
\underset{\text{benzaldehyde}}{\overset{\displaystyle O}{\overset{\|}{\bigcirc\!\!\!\!\!\!\hexagon-C-H}}}
$$

IUPAC Rule Changes for Aldehydes

Aldehydes, in the IUPAC system, are named by dropping the -e from the name of the hydrocarbon and adding -al. If the carbon chain needs to be numbered to identify side chains, you must always start with the carbonyl carbon.

$$\underset{\text{ethanal}}{\text{CH}_3\text{--C--H}} \qquad \underset{\text{3-chlorobutanal}}{\text{CH}_3\ \overset{\text{Cl}}{\underset{|}{\text{CH}}}\ \text{CH}_2\text{--C--H}} \qquad \underset{\text{4-bromo-2-pentenal}}{\text{CH}_3\text{--CH Br--CH=CH--C--H}}$$

(with $\overset{\text{O}}{\underset{||}{}}$ carbonyls above each)

Chemical Reactions

Methods of Preparation

Oxidation of 1° Alcohols:

$$1°\ \text{R--OH} + (O) \rightarrow \text{R}-\overset{\text{O}}{\underset{||}{\text{C}}}-\text{H} + \text{H}_2\text{O}$$

Example:

$$\underset{\text{1-propanol}}{\text{CH}_3\ \text{CH}_2\ \text{CH}_2\ \text{OH}} + (O) \longrightarrow \underset{\text{propanal}}{\text{CH}_3\ \text{CH}_2\ -\overset{\text{O}}{\underset{||}{\text{C}}}-\text{H}} + \text{H}_2\text{O}$$

This method for preparing aldehydes does not give a good yield, but can be used as a confirmatory test for primary alcohols (1° ROH).

Oxidation

The aldehyde functional group is extremely sensitive to oxidation reactions. Exposure to air is enough to cause the oxidation of many aldehydes. Because glucose has an aldehyde functional group, having a way to identify its presence in urine specimens is important. Several different methods have been developed and named. Regardless of the chemicals used, it is the ease with which aldehydes can be oxidized that make the tests work.

Oxidation:

$$\text{R--}\overset{\text{O}}{\underset{||}{\text{C}}}\text{--H} + (O) \rightarrow \text{R--}\overset{\text{O}}{\underset{||}{\text{C}}}\text{--OH}$$

Identification tests for the aldehyde functional group:

Silver Mirror or **Tollen's Test:**

$$\text{R--}\overset{\text{O}}{\underset{||}{\text{C}}}\text{--H} + (O) \xrightarrow[\text{heat}]{\text{AgNO}_3\ \text{NH}_3\ \text{H}_2\text{O}} \text{R--}\overset{\text{O}}{\underset{||}{\text{C}}}\text{--OH} + \text{Ag} \downarrow$$

Benedict's and **Fehling's Test:**

$$\text{R--}\overset{\text{O}}{\underset{||}{\text{C}}}\text{--H} + (O) \xrightarrow[\text{sodium citrate or tartrate}]{\text{Cu}^{2+}} \text{R--}\overset{\text{O}}{\underset{||}{\text{C}}}\text{--OH} + \text{Cu}_2\text{O} \downarrow$$
$$\text{(brick red)}$$

Hemiacetal/Acetal Formation

Aldehydes and alcohol molecules will undergo a series of reactions, in which an alcohol and aldehyde first add together, producing a **hemiacetal**. This can then be followed by an elimination reaction between the hemiacetal and a second molecule of alcohol forming an **acetal**. These reactions are of particular interest, since the hemiacetal reaction is how glucose forms a six-member ring, and the second reaction is how two or more glucose molecules join to form maltose or starch molecules.

$$
\text{Hemiacetal Formation:} \quad
\begin{matrix} O \\ \| \\ R{-}C{-}H \end{matrix}
\; + \;
\begin{matrix} H{-}O \\ | \\ R' \end{matrix}
\; \rightleftharpoons \;
\begin{matrix} OH \\ | \\ R{-}CH{-}O{-}R' \end{matrix}
$$

The double arrows used in this equation indicate that this reaction reverses easily.

$$
\text{Acetal Formation:} \quad
\begin{matrix} OH \\ | \\ R{-}CH{-}O{-}R' \end{matrix}
\; + \;
\begin{matrix} H{-}O \\ | \\ R'' \end{matrix}
\; \underset{}{\overset{H^+}{\rightleftharpoons}} \;
\begin{matrix} O{-}R'' \\ | \\ R{-}CH{-}O{-}R' \end{matrix}
\; + \; H_2O
$$

This reaction will also reverse, but water and acid must be present. The reverse reaction is the first step in the digestion of many complex carbohydrates, such as starch.

Ketones

The **ketones** are also a moderately polar family. Ketones have general formulas and physical properties that are similar to the aldehydes (see below). They have melting and boiling points that are lower than the alcohols of similar size and only the first few members of the family have some water solubility. In contrast to the smaller aldehydes, ketones are fragrant and used in perfumery and as flavoring agents.

General Description

$$
\begin{matrix} O \\ \| \\ R - C - R' \end{matrix}
$$

general formula for ketones

Smaller ketones, such as acetone and methyl ethyl ketone (MEK) are important solvents. For example, acetone is used as a solvent for fingernail polish, super glue, and model airplane cement. Ketones are much more stable toward oxidation than aldehydes; therefore, they fail to give positive tests with Tollen's, Fehling's, and Benedict's reagents. Sugars that are α-**hydroxyketones**, such as fructose, however, do give a positive test and are therefore called reducing sugars.

Common Names

Nomenclature

The common names for ketones are formed by naming the shorter and then the longer of the two alkyl groups on each side of the carbonyl carbon. The family name, *ketone*, is then added at the end.

$$
\begin{matrix} O \\ \| \\ CH_3{-}C{-}CH_3 \end{matrix}
\qquad\qquad
\begin{matrix} O \\ \| \\ CH_3\,CH_2{-}C{-}CH_3 \end{matrix}
$$

dimethyl ketone (acetone)   methyl ethyl ketone (MEK)   diphenyl ketone

IUPAC Rule Changes for Ketones

In the IUPAC system, ketones are named by dropping the -e from the parent hydrocarbon name, and adding -one. The longest carbon chain is numbered, giving the lowest possible number to the carbonyl carbon atom and that number precedes the last name.

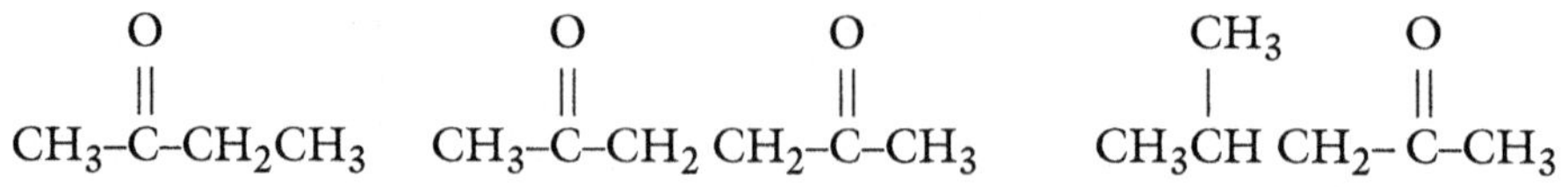

$$\underset{\text{2-butanone}}{\overset{\overset{\displaystyle O}{\underset{\displaystyle \|}{}}}{CH_3-C-CH_2CH_3}} \qquad \underset{\text{2, 5-hexanedione}}{\overset{\overset{\displaystyle O}{\underset{\displaystyle \|}{}}\qquad\overset{\displaystyle O}{\underset{\displaystyle \|}{}}}{CH_3-C-CH_2\ CH_2-C-CH_3}} \qquad \underset{\text{4-methyl-2-pentanone}}{\overset{\overset{\displaystyle CH_3}{\underset{\displaystyle |}{}}\quad\overset{\displaystyle O}{\underset{\displaystyle \|}{}}}{CH_3CH\ CH_2-\ C-CH_3}}$$

Chemical Reactions

Preparation

Unlike aldehydes, oxidation of a secondary alcohol is a good method of preparing ketones and gives good yields. It can also serve as a confirmatory test for secondary alcohols (2° ROH).

$$\text{Oxidation of 2° ROH:}\quad \underset{\text{2-propanol}}{\overset{\overset{\displaystyle OH}{\underset{\displaystyle |}{}}}{CH_3-CH-CH_3}} + (O) \longrightarrow \underset{\text{2-propanone}}{\overset{\overset{\displaystyle O}{\underset{\displaystyle \|}{}}}{CH_3-C-CH_3}} + H_2O$$

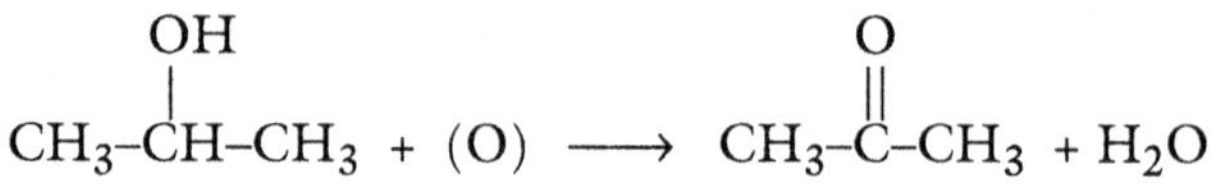

Ketones will also undergo a number of addition-type chemical reactions. This method is used by the **ketohexose**, fructose, to form a five-member ring structure. Acetone is a product of fat metabolism and during diabetes mellitus large quantities of acetone may be passed in the urine and breath. Ketones can also be reduced, producing the corresponding secondary alcohol.

Check Your Understanding

1. What is the carbonyl group and what effect does it have on the polar character of organic molecules?

2. What are the common names of aldehydes based on?

3. What property of the aldehyde functional group makes it possible to test for glucose in urine samples?

4. In which properties, physical or chemical, are aldehydes and ketones more alike?

5. What are the IUPAC naming system endings for aldehydes and ketones?

10-5 Carboxylic Acids and Esters

Objectives

Upon completion of this section, you will be able to:

- Identify members of the carboxylic acid and ester families by either general formulas, names, or from structural formulas.

- Describe the physical properties of carboxylic acids and esters.

- Name small members of each of the carboxylic acid and ester families with a common and IUPAC name.

- Write and balance chemical reactions typical for the carboxylic acid and ester families.

Carboxylic Acids

Carboxylic acids contain a reactive structure, whose name is derived from the *carbonyl* and *hydroxyl* groups it contains. The **carboxyl group** structure, $-COOH$, is also written as

$$
\begin{array}{c}
O\ \delta^- \\
\|\quad\ \ \delta+ \\
-C-O-H
\end{array}
$$

The highly electronegative carbonyl oxygen atom ($\delta-$) decreases the electron density on the carbonyl carbon. This, in turn, shifts the electrons of the hydroxyl group oxygen toward the carbonyl carbon and away from the end of the hydrogen atom. The weakened bond between the oxygen and hydrogen ($\delta+$) atom aid in the release of a proton or hydrogen ion. **Resonance** (electron shifts) between the two oxygen atoms is also a contributing factor to its ability to donate a hydrogen ion. Using the Arrhenius definition, carboxyl groups are acids, because they release hydrogen ions in water.

General Description

The carboxylic acids are polar molecules. As was true in the alcohol family, carboxylic acids can participate in hydrogen bonding, not only with water, but also with each other. This has an effect on their melting and boiling points, as well as their water solubility. The smaller family members (C_1 to C_3), are sour smelling, water-soluble liquids. The two-carbon acid, acetic, is responsible for the odor and taste of vinegar. The **monocarboxylic acids** (containing only one carboxyl group) show a gradation of properties with increasing length of the carbon chain.

For example, C_1 to C_{10} members are liquids at 25°C. Longer chain acids are odorless white solids of waxy consistency. C_1 to C_3 have sharp odors and are water-soluble. C_4 is only slightly soluble in water and has the odor of rancid butter. C_6, C_8, and C_{10} are reminiscent of the odor of goats. Monocarboxylic acids with more than 10 carbons are known as **fatty acids**. They may also contain one or more carbon-to-carbon double bonds, in which case they are called **unsaturated fatty acids**.

Carboxylic acids donate a H^+ (a proton) to water and bases. The typical chemical reactions for this family involve the loss of the hydrogen ion from the end of the carboxyl group, forming a **carboxylate ion**, $R-COO^-$, and a hydrogen ion, H^+. Carboxylic acids are weak acids, with only a small percent of the molecules undergoing ionization. This is the reason we can tolerate the sour taste of vinegar or lemon juice on our foods.

Nomenclature

Common Names

Many carboxylic acids have common names of long standing and are associated with their discovery. A few examples would include:

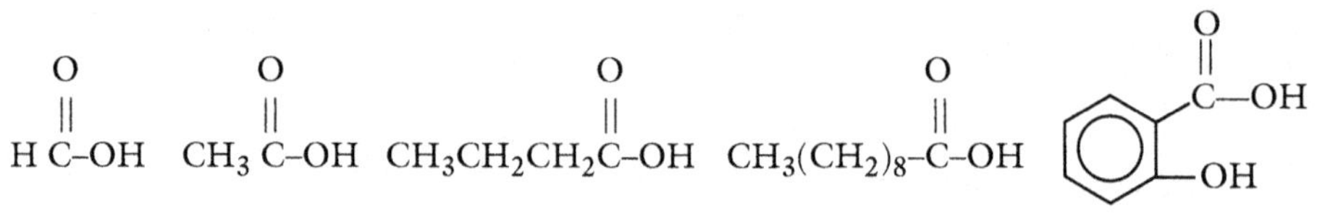

formic acid   acetic acid   butyric acid   capric acid   salicylic acid
(ants)   (sour, acrid)   (rancid butter)   (goat)   (willow tree)

IUPAC Rule Changes for Carboxylic Acids

Carboxylic acids are named by selecting the longest carbon chain, starting at the end with a carboxyl group. The carbon in the carbonyl carbon is automatically designated number one carbon. Pick the name of the alkane corresponding to the longest carbon chain, drop the -e, and add the -oic ending, followed by the word acid. All substituents on the acid chain are located by the use of numbers.

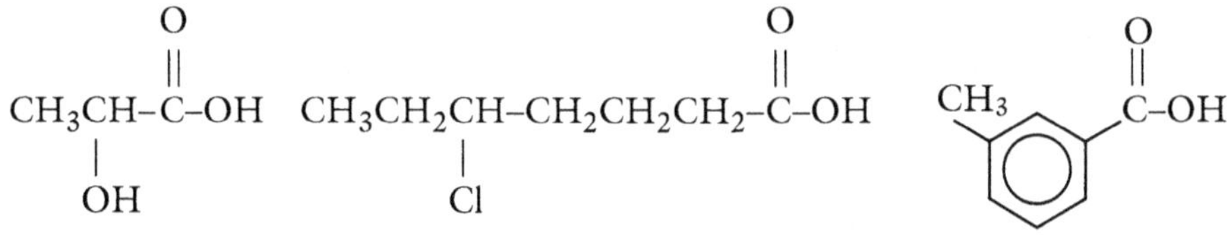

2-hydroxypropanoic   5-chloroheptanoic acid   3-methylbenzoic acid
acid (lactic acid)

Chemical Reactions

Neutralization or Saponification

The reaction of an organic acid with a base produces a salt. Either neutralization or saponification can be used to name this reaction, based on the length of the carboxylic acid's carbon chain. If the chain has more than 10 carbons, the product is not only a salt, but a soap. These reactions are, therefore, frequently called **saponification reactions**.

The conversion of the carboxyl group to its salt or soap makes the end of the molecule ionic and, therefore, water soluble. Although soap making was once a thriving industry, most uses for soaps have now been replaced with **surfactants**, called **detergents**. There are many different types of detergents in use, but they all resemble soap in having a long alkane-like hydrocarbon tail and some type of ionic head. Detergents have the advantage of not forming insoluble soap scums when used in hard water.

Neutralization/Saponification: $R-\overset{\overset{\displaystyle O}{\|}}{C}-OH + HO-Na^+ \longrightarrow R-\overset{\overset{\displaystyle O}{\|}}{C}-O^- Na^+ + HOH$
(a salt or soap)

Strong acids will convert salts of carboxylic acids back to the carboxylic acid and a salt.

Acid Preparation from Salt: $R-\overset{\overset{\displaystyle O}{\|}}{C}-O^-Na^+ + H^+Cl^- \longrightarrow R-\overset{\overset{\displaystyle O}{\|}}{C}-OH + Na^+Cl^-$

Esterification

One of the most widely used nonprescription drugs is aspirin. It is an ester made by the reaction between an alcohol functional group of salicylic acid and acetic acid, in the presence of an acid catalyst. The reaction is given below.

$$\text{Esterification:} \quad R\text{--}CH_2\text{--}OH + HO\overset{\displaystyle O}{\overset{\|}{\text{--}C}}\text{--}R' \longrightarrow R\text{--}CH_2\text{--}O\overset{\displaystyle O}{\overset{\|}{\text{--}C}}\text{--}R' + HOH$$

Example:

salicylic acid	acetic acid	acetylsalicylic acid	water
(alcohol)	(acid)	(aspirin ester)	

Amide Formation

Carboxylic acids react with ammonia, or 1° and 2° amines, to produce amides (see Table 10-1). This reaction is important for a variety of reasons, including the synthesis of protein in living organisms and the manufacture of polyamides, called nylons.

$$\text{Amide Formation:} \quad R\overset{\displaystyle O}{\overset{\|}{\text{--}C}}\text{--}OH + H\text{--}\overset{\displaystyle}{\underset{\displaystyle H}{N}}\text{--}H \longrightarrow R\overset{\displaystyle O}{\overset{\|}{\text{--}C}}\text{--}\overset{\displaystyle}{\underset{\displaystyle H}{N}}\text{--}H + HOH$$

Dicarboxylic Acids

Many organic molecules contain more than one carboxyl group (see below), along the carbon chain. The **dicarboxylic acids** are also polar molecules. The second carboxyl group increases their polar nature and they have higher melting and boiling points and greater water solubility than the corresponding monocarboxylic acids.

General Description

$$HO\overset{\displaystyle O}{\overset{\|}{\text{--}C}}\text{--}R\overset{\displaystyle O}{\overset{\|}{\text{--}C}}\text{--}OH$$

general formula for dicarboxylic acids

Technology Box 10–6 ■ How useful is PET?

Where can PET be found around the home? The acronym, PET, stands for poly-ethylene terephthalate, known also as plastic, polyester, Dacron®, and Mylar®. The polyester, PET, is made by the polymerization of terephthalic acid and ethyl-ene glycol, according to the following equation:

terephthalic acid   ethylene glycol   polyethylene terephthalate

    The resulting polyester, PET, is versatile and used in a dizzying array of con-sumer products. One of these is as a miracle fiber sold under the trade name Da-cron®. Most ski clothing consists of an outer layer of nylon, followed by a filler layer of very fine polyester fibers, which create a dead-air space and traps body heat. Since both of these fabrics can withstand the rigors of repeated washing and dry-ing, they not only keep you warm, but they look good and are easy to keep clean.

    Dacron® has also found medical applications. A tightly woven fabric made from Dacron® can be used to replace a diseased or damaged aorta. Hernias in the abdominal wall can be surgically repaired using a Dacron® patch to strengthen the area. When blended with cotton or other natural fiber, Dacron® provides us with wash-n-wear garments. Once a pleat or crease has been set in Dacron® with a hot iron, the garment returns to its original shape, after each laundry cycle. When a garment's label identifies that it has a polyester content, it does not nec-essarily mean that it contains Dacron®. It may be one of many polyesters that have found wide application in the fabrics industry.

    PET has other uses. When it is melted and blown into molds, it becomes re-cyclable soft drink, peanut butter, salad dressing, shampoo, dishwashing liquid, and water bottles. In this form, it is identified as either PET or PETE plastic. In what seems magical to some, much of it is now being recycled and later trans-formed into items such as sleeping bag filler, doormats, carpeting, and tennis ball containers. We know, however, that Dacron® and PET are the same: it's just a matter of either drawing into threads or blowing it into a mold.

    In 1952, DuPont, the maker of PET, introduced it in the form of a polyester film called Mylar®. It was a flexible, exceptionally strong, durable film that was in-ert to water, unaffected by oil, grease, and most aromatic compounds. Even at temperatures ranging from −100° to +300°C, it remained clear and flexible. Since Mylar® contained no plasticizers, it does not become brittle or yellow with age.

    In addition, Mylar® can be laminated, metallized, cut or punched, and has excellent electrical insulating properties. It is used in applications such as the membrane touch switches that are used on clothes dryers, microwave ovens, food mixers, industrial instrumentation, and hand-held electronic games.

    When metallized, it is used to make energy-saving window shades, car win-dow tinting, and balloons. Due to its superior strength, even when very thin, My-lar® is used as the backing material for applications such as carbon ribbons used on typewriters, printers, and audio and videotapes. The clarity of Mylar® makes it an excellent choice for packaging food products, toys, hardware, paper products, and soft goods. Its heat resistance makes it ideal for oven roasting bags and boil-in-bag dinners. You will definitely find PET to be useful around the house.

Source: http://www.dupont.com

Common Names

The first member of the dicarboxylic acids is oxalic acid (HOOC–COOH), which is found in many leafy green vegetables (spinach and rhubarb) and is the best known of these acids. If consumed, oxalic acid is highly corrosive to mucous membranes and poisonous. Fortunately, cooking destroys most of it. It is also used as a bleaching agent and rust remover.

Succinic acid, $HOOC–(CH_2)_2–COOH$, is a metabolic intermediate in the citric acid cycle. (Citric acid is a tricarboxylic acid.) There is no system involved in the common names for the dicarboxylic acids. They have evolved over time and, if needed, must be memorized. The following are some examples:

$$HO-\overset{O}{\overset{\|}{C}}-CH_2\overset{O}{\overset{\|}{C}}-OH \qquad HO-\overset{O}{\overset{\|}{C}}-(CH_2)_3-\overset{O}{\overset{\|}{C}}-OH \qquad HO-\overset{O}{\overset{\|}{C}}-(CH_2)_4-\overset{O}{\overset{\|}{C}}-OH$$

malonic acid   glutaric acid   adipic acid

IUPAC Rule Changes for Dicarboxylic Acids

Dicarboxylic acids are named by locating the longest carbon chain, starting and ending with a carboxyl group. The carbon chain is numbered from one end to the other, keeping the numbers of substituent groups as small as possible. An ending of -dioic is added to the alkane name and the word *acid*.

$$HO-\overset{O}{\overset{\|}{C}}-\underset{\underset{CH_3}{|}}{CH}\ \overset{O}{\overset{\|}{C}}-OH \qquad HO-\overset{O}{\overset{\|}{C}}-(CH_2)_6-\overset{O}{\overset{\|}{C}}-OH \qquad HO-\overset{O}{\overset{\|}{C}}-\hexagon-\overset{O}{\overset{\|}{C}}-OH$$

2-methyl-1,3-propanedioic acid   octanedioic acid   1,4-benzenedicarboxylic acid (terephthalic acid)

Esters

Esters are moderately polar molecules. As shown by their general formulas (see below), they resemble carboxylic acids, but the addition of another carbon chain reduces their polarity and ability to undergo hydrogen bonding. The ester functional group, however, imparts enough polarity to the molecules to make the smallest ($<C_5$) molecules water soluble. They also have melting and boiling points that are similar to the corresponding aldehydes and ketones.

$$R - \overset{O}{\overset{\|}{C}} - O - R' \text{ or } R - COO-R'$$

general formula for esters

Many esters have pleasant fragrances. The odors of esters are often described as fruit-like and, indeed, they are responsible for the odors of many fruits. Esters are also used in making perfumes. Waxes are composed of large esters and are used for making shoe and car polishes.

Esters can be hydrolyzed in the presence of acid catalysts or enzymes to produce an acid and alcohol. Esters also undergo saponification in the presence of aqueous NaOH or KOH, yielding the free alcohol and the sodium or potassium salt of the carboxylic acid, which may also be a soap.

Nomenclature

Common Names

Simple esters are named like salts. The alkyl group from the alcohol is named first, followed by the name of the acid with the -ic ending changed to -ate.

$$
\underset{\text{methyl acetate}}{CH_3-O-\overset{\displaystyle O}{\overset{\displaystyle \|}{C}}-CH_3}
\qquad
\underset{\text{ethyl acetate}}{CH_3-\overset{\displaystyle O}{\overset{\displaystyle \|}{C}}-O-CH_2CH_3}
\qquad
\underset{\text{ethyl butyrate (pineapple odor)}}{CH_3\,CH_2-O-\overset{\displaystyle O}{\overset{\displaystyle \|}{C}}-CH_2\,CH_2\,CH_3}
$$

methyl salicylate (oil of wintergreen)

IUPAC Rule Changes for Esters

The alkyl group from the alcohol is named first, followed by the IUPAC name of the acid with only the -ic ending changed to -ate.

$$
\underset{\text{ethyl ethanoate}}{CH_3CH_2-O-\overset{\displaystyle O}{\overset{\displaystyle \|}{C}}-CH_3}
\qquad
\underset{\text{pentyl ethanoate (banana)}}{CH_3-\overset{\displaystyle O}{\overset{\displaystyle \|}{C}}-O-CH_2CH_3}
\qquad
\underset{\text{methyl-3-methylpentanoate}}{CH_3CH_2\,\underset{\underset{\displaystyle CH_3}{|}}{CH}\,CH_2-\overset{\displaystyle O}{\overset{\displaystyle \|}{C}}-O-CH_3}
$$

Chemical Reactions

Acid Hydrolysis

Two of the more important reactions of esters involve **hydrolysis** (water splitting) of the ester linkage. In one case, the reaction is catalyzed by an acid and in the other by a base. Acid hydrolysis is essentially the reverse of esterification. After a bottle of aspirin has been repeatedly opened in the humid bathroom air, it is common to note a sour, vinegar-like, aroma. This is due to the acid hydrolysis of the ester linkage in aspirin, resulting in the formation of some acetic acid. In the digestion of triglycerides, and catalyzed by enzymes, this type of reaction also occurs.

When hydrolysis occurs in a basic solution, the ester linkage is again broken. However, the carboxylic acid formed reacts with the base, resulting in the salt of the acid being produced. Because polyester fabrics contain many ester linkages, they can be seriously damaged when either strong acids or bases are spilled on them.

$$
\text{Acid hydrolysis:}\quad R-\overset{\displaystyle O}{\overset{\displaystyle \|}{C}}-O-R' \;+\; H_2O \;\underset{\text{or enzyme}}{\overset{H^+}{\longrightarrow}}\; R-\overset{\displaystyle O}{\overset{\displaystyle \|}{C}}-OH \;+\; HO-R'
$$

Basic Hydrolysis:

$$R-\overset{\overset{\displaystyle O}{\|}}{C}-O-R' + NaOH \xrightarrow[heat]{H_2O} R-\overset{\overset{\displaystyle O}{\|}}{C}-O^- Na^+ + HO-R'$$

Inorganic Esters

Inorganic esters may also be prepared by the reaction of an alcohol with an inorganic acid. Nitroglycerine is such an ester; the product of the reaction between glycerin (1,2,3-propanetriol) and nitric acid (HNO_3). The structural formula for nitroglycerin is shown in the section on glycerin. Certain polyphosphate esters, such as adenosine triphosphate, ATP, are the energy-rich molecules that provide much of the energy for doing muscular work.

Check Your Understanding

1. Why is a molecule considered an acid, if it contains a carboxyl group?

2. What is the typical odor of smaller acids and of smaller esters?

3. What are the similarities and differences between a soap and a detergent?

4. What does the term hydrolysis mean?

5. Why can polyester fabrics be damaged by both strong acids and bases?

10-6 Amines and Amides

Objectives

Upon completion of this section, you will be able to:

■ Identify members of the amine and amide families by either general formulas, names or from structural formulas.

■ Describe the physical properties of amines and amides.

■ Name small members of each of the amine and amide families with a common and IUPAC name.

■ Write and balance chemical reactions typical for the amine and amide families.

Amines

General Description

The general formulas for the **amines** are: $R-\overset{..}{N}H_2$; $R_2-\overset{..}{N}H$; and $R_3-\overset{..}{N}$. Like the alcohols, amines are subdivided by the number of alkyl groups attached into primary, 1°, secondary, 2°, and tertiary, 3°, amines. Amines are considered alkyl or **aryl** derivatives of ammonia, NH_3.

Amines are polar molecules. As shown above, all subclasses of amines retain an unshared pair of electrons (..) and can, therefore, accept a hydrogen bond at a minimum. Those amines that have one or more hydrogens attached to the nitrogen can also start hydrogen bonds with other polar compounds or water.

The smaller members of the amine family are water-soluble gases with an unpleasant ammonia- to fish-like odor. Those containing from 3 to 11 carbons are liquids, while larger members are solids. Although the odors of smaller amines are unpleasant, their salts (called alkylammonium salts) are odorless.

The amines are **organic bases**, or **Lewis bases**, because only three of the five electrons in the outer shell of the nitrogen atom are used in bonding. Structures deficient by an electron pair (**Lewis acids**) may combine with amines, share the electron pair, and produce salts. They are also considered bases by the Arrhenius definition because, like ammonia, they cause the production of hydroxide ions when dissolved in water ($NH_3 + H_2O \rightarrow NH_4^+ + OH^-$).

Some amines occur naturally in the decomposition of nitrogen-containing products such as protein, and are responsible for the smell of decaying flesh. Some of the most valuable drugs obtained from plant extracts are characterized by the presence of one or more basic nitrogen atoms in their structure. These compounds are generally classed as alkaloids, including morphine and other important pain relievers.

Nomenclature

Common Names

In the common system, amines are identified by naming each of the alkyl group(s) attached to the nitrogen atom, starting with the smallest and working up. The family name, *amine*, is then added to the end of the last alkyl name.

$$CH_3\,CH_2-NH_2$$

ethylamine

$$\overset{\displaystyle CH_3}{\overset{\displaystyle |}{CH_3-N-CH_3}}$$

trimethylamine

$$\overset{\displaystyle CH_3}{\overset{\displaystyle |}{CH_3\,CH_2-N-CH_3}}$$

dimethylethylamine

IUPAC Rule Changes for Amines

In the IUPAC System, amines are regarded as amino-, alkylamino- or dialkylaminoalkanes. The longest carbon chain attached to the nitrogen atom is the parent hydrocarbon. If no alkyl groups are attached to the other nitrogen positions, the $-NH_2$ is named as *amino*.

aminobenzene	methylaminoethane	3-dimethylaminopentane

NH_2 (on benzene ring)

$CH_3 - N - CH_2CH_3$
$\quad\quad |$
$\quad\quad H$

$CH_3\ CH_2\ CH\ CH_2\ CH_3$
$\quad\quad\quad\quad |$
$\quad\quad CH_3 - N - CH_3$

Salt Formation

Chemical Reactions

Like ammonia, amines have basic properties. They can use the unpaired electrons to combine with the hydrogen ion and produce a water-soluble, odorless salt. Amino acids are the building blocks of all protein. During the digestion of protein, each amino acid molecule that leaves the stomach takes along, in ionic form, one molecule of hydrochloric acid, HCl. This explains, in part, why drinking milk may temporarily alleviate the symptoms of an ulcer.

Salt Formation:
$$R-\overset{..}{N}H_2\ +\ H^+Cl^-\ \longrightarrow\ R-\overset{\overset{\displaystyle H^+}{|}}{N}H_2\ Cl^-$$

an amine $\quad\quad$ an ammonium chloride salt

Amide Formation

Ammonia, NH_3, primary, 1°, and secondary, 2°, amines will react with carboxylic acids to form amides. Tertiary, 3°, amines, fail to do so because they lack a hydrogen atom attached to the nitrogen. Whether the amino and carboxyl groups are on the same or different molecules makes no difference in how the reaction occurs.

Amide Formation:
$$R-\underset{\underset{\displaystyle H}{|}}{N}-H\ +\ HO-\overset{\overset{\displaystyle O}{||}}{C}-R'\ \longrightarrow\ R-\underset{\underset{\displaystyle H}{|}}{N}-\overset{\overset{\displaystyle O}{||}}{C}-R'\ +\ H_2O$$

Amides

General Description

As the **amide** general formulas show, the nitrogen may be unsubstituted, monosubstituted, or disubstituted with carbon chains.

$$R-\overset{\overset{\displaystyle O}{||}}{C}-NH_2 \quad\quad R-\overset{\overset{\displaystyle O}{||}}{C}-NHR \quad\quad R-\overset{\overset{\displaystyle O}{||}}{C}-NR_2$$

Amides are polar molecules. Unsubstituted and monosubstituted amides have at least one hydrogen attached to the nitrogen and can start hydrogen bonds. Disubstituted amides can accept hydrogen bonds, but they cannot start new ones. Due to their ability to hydrogen bond, amides have high melting and boiling points and are somewhat water soluble. Methanamide, for example, is a water soluble liquid, but all others are odorless crystalline solids at 25°C.

Technology Box 10–7 ■ Bulletproof vest, anyone?

Nylon is one of the most impressive and versatile of all man-made materials. Its origin dates back to the 1930s when a group of researchers at DuPont, led by Dr. Wallace H. Carothers (1896–1937), made the first nylon, which was marketed as nylon 66. It got its name because it was made by the polymerization of a 6-carbon dicarboxylic acid (adipic acid) and a 6-carbon diamine (hexamethylenediamine). Nylon 66 is a man-made polyamide.

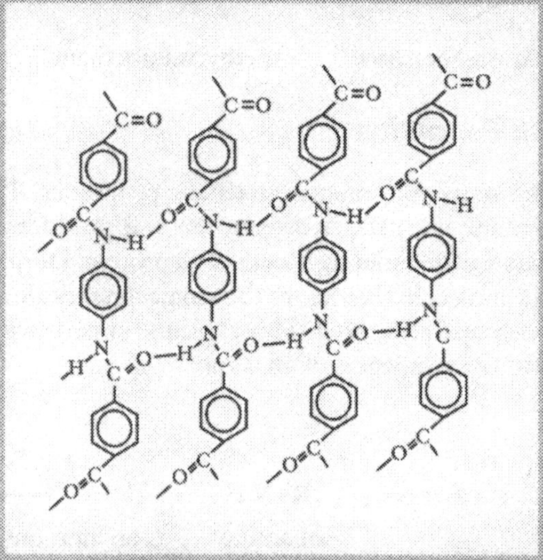

During World War II, the United States developed extensive nylon production capabilities. It was found that nylon could be melted and extruded through spinnerets to form thin, but extremely strong threads that resembled silk. Nearly all of the early production of nylon was used for making parachutes. Later, production turned to making nylon stockings, underwear, and other sheer garments, virtually eliminating the need for silk. By the early 1950s, nylon fibers had been found to produce a much stronger tire than could be produced using rayon fibers. It was also discovered that nylon could be cast into blocks and machined like a metal. Many nylon gears are used in household appliances to reduce the annoying noise caused by metal gears.

Some of the more recent high-tech nylons include applications for the production of car parts, bulletproof vests, and flame resistant suits, helmets, and gloves. For example, the trade name of the nylon used in bulletproof vests and a wide variety of other industrial applications is KEVLAR®. It was introduced in the early 1970s, and considered one of the most important man-made organic fibers ever developed. As shown in the diagram, it is the para-aramide (p-aromatic amide), poly-paraphenyleneterephthalamide. Its chains are highly oriented with strong cross-linking between the chains, which result in its unique combination of properties. It has very high tensile strength, but low weight, high chemical resistance, and toughness. It is electrically non-conducting and flame resistant. All of these properties make it an excellent choice for use in bulletproof vests.

Another closely related product used in flameproof suits and gloves has the trade name NOMEX®. It is another form of nylon and similar to KEVLAR®. Fibers of NOMEX® consist of long rigid molecular chains produced of polymetaphenylenediamine, which do not flow or melt upon heating. NOMEX® is both chemically and thermally very stable and will not decompose or char until temperatures in excess of 350°C are reached.

Source: http://www.dupont.com/

The interaction of an amine and a carboxylic acid is the essential feature of protein synthesis. Proteins are huge molecules whose backbones involve amide linkages.

This amide linkage, $-\overset{\overset{\displaystyle O}{\|}}{C}-N-$, is also known as the **peptide bond.** It is the reactive site on the molecule. It should be noted, however, that the carbonyl group attached to the nitrogen of the amine group destroys the basic character of the amine.

Common Names

Nomenclature

The common names for simple amides are produced by dropping the -ic and acid of the corresponding acid and adding the family name, *amide*. For amides that have one or more hydrogen atoms substituted by alkyl groups, the prefix N- is added before the name of the alkyl group and then the -ic acid is dropped and -amide ending added.

$$\overset{\overset{\displaystyle O}{\|}}{H-C}-NH_2 \qquad \overset{\overset{\displaystyle O}{\|}}{CH_3-C}\underset{\underset{\displaystyle H}{|}}{-N}-CH_3 \qquad \overset{\overset{\displaystyle O}{\|}}{H-C}-N-(CH_3)_2$$

formamide   N-methylacetamide   N,N-dimethylformamide

IUPAC Rule Changes for Amides

The IUPAC names for the amides are developed using the same method as for the common names, except for IUPAC acid names drop the -oic acid and adds -amide.

$$\overset{\overset{\displaystyle O}{\|}}{CH_3C}-NH_2 \qquad \bigcirc\!\!\!\!\!\!-\overset{\overset{\displaystyle O}{\|}}{C}\overset{\overset{\displaystyle H}{|}}{-N}-CH_3 \qquad \overset{\overset{\displaystyle O}{\|}}{CH_3C}-N-(CH_3CH_2)_2$$

ethanamide   N-methylbenzamide   N,N-diethylethanamide

Acid Hydrolysis

Chemical Reactions

For all practical purposes, the only significant reaction that occurs to proteins during digestion is acid hydrolysis. The reaction is essentially the reverse of amid formation, except the salt of the amine is produced. Basic hydrolysis also breaks the amide bond. However, in a basic solution, the salt of the carboxylic acid results.

Acid Hydrolysis: $R-\overset{\overset{\displaystyle O}{\|}}{C}\underset{\underset{\displaystyle R}{|}}{-NH} + H_3O^+ + Cl^- \longrightarrow R-\overset{\overset{\displaystyle O}{\|}}{C}-OH + \underset{\underset{\displaystyle R}{|}}{NH_3^+} Cl^-$

Basic Hydrolysis: $R-\overset{\overset{\displaystyle O}{\|}}{C}-NH_2 + Na^+ + OH^- \longrightarrow R-\overset{\overset{\displaystyle O}{\|}}{C}-O^- Na^+ + NH_3$

Silk and wool are two natural fibers that are protein and, therefore, contain amide bonds. In the early 1930s it was discovered that the use of a dicarboxylic acid and a **diamine** resulted in a polymerization reaction that produced a man-made substitute for silk. This polymer was named nylon-66, because it was made from the six carbon dicarboxylic acid (adipic acid) and the six carbon diamine (hexamethylenediamine). See the Technology Box 10-7 to learn more about nylon.

Check Your Understanding

1. What is the difference between a primary, secondary, and tertiary amine?

2. Why are amines considered organic bases?

3. What is the IUPAC method for naming amines?

4. What is a peptide bond? What is its other name?

5. Which reaction best explains protein digestion?

Summary

The study of organic chemistry includes both naturally occurring and man-made carbon compounds. Over the years, millions of compounds have been separated from living materials or synthesized in the laboratory. For purposes of simplification, organic compounds have been subdivided into groups known as organic families. All organic families, except the alkanes, have a unique structural feature, called a functional group, which gives them their characteristic physical and chemical properties. The melting points, boiling points, and water solubility of the different families are closely related to the polar nature of their functional group. In all instances, as the length of the carbon chain increases the melting points and boiling points increase and their solubility in water decreases.

Three methods are in use for naming organic compounds. Trivial names are used when the chemicals are very familiar. Common names typically identify each of the alkyl or aryl groups and have the family name as their last name. The International Union of Pure and Applied Chemistry (IUPAC) System now provides a systematic naming method that is used worldwide. The system generally identifies the longest continuous carbon chain within a molecule; its name, along with the appropriate family ending, constitutes the last name. All substituents attached to the longest carbon chain are located by numbers and listed, in alphabetical order, before the last name.

The chemical reactions of the organic families are many and varied. The important point to remember is that if you know how one family member reacts, it is reasonable to expect that all other family members react in a similar way.

Chapter Review

1. Describe the element that is present in all organic compounds and identify its characteristics.

2. Write a paragraph or two that describes, in general, how the IUPAC naming system works.

3. Explain the difference between a trivial name and a common name.

4. Describe three different types of isomers. Use examples and give each of the isomers appropriate IUPAC names.

5. What would be the correct IUPAC name for each of the following structures?

   a. CH_3–CH–CH_3
      |
      Cl

   b. CH_2Cl–CH_2–CH_2Cl

   c. CH_3–CH_2Br

   d. CH_3–CH–CH–CH_3
         | |
         CH_3 CH_3

6. Draw correct condensed structural formulas for each of the following:

   a. 2,2-dichloropentane
   b. 3-ethyl-2-methyloctane
   c. 1,2-dimethylcyclohexane
   d. 3-cyclopropylpentane

7. Complete and balance the following reaction for the combustion of pentane.

   $$CH_3CH_2CH_2CH_2CH_3 + 5O_2 \longrightarrow$$

8. What would be the correct IUPAC name for each of the following structures?

   a. CH_2 = CH–CH_2Br

   b. CH_2 = C–CH_2Cl
            |
            CH_3

   c. CH_2=CHBr

   d. CH_3 – C = C–CH_3
            | |
            CH_3 CH_3

9. Draw correct condensed structural formulas for each of the following:

   a. 2,3-dimethyl-2-pentene
   b. 3-bromo-4-methyl-2-octene
   c. 1,2-dichlorocyclohexene
   d. 3-iodo-1,3-pentadiene

10. Complete and balance the following equation:

   $$CH_3\text{–}CH\text{=}CH\text{–}CH_2\text{–}CH_3 \ + \ H_2 \xrightarrow{Pt}$$

11. Through use of an example, explain how Markovnikov's rule determines the product of addition reactions.

12. Write a brief description of the polymerization process. Be sure to include the terms monomers and polymers.

13. Explain the accepted theory for the arrangement of electrons within the benzene ring and its effect on the chemical properties of aromatic compounds.

14. What would be the correct IUPAC name for each of the following structures?

   a. $CH_3CH_2CH_2$–OH

   b. CH_3CHCH_2–OH
         |
         CH_2–CH_3

c. $CH_2Br-CH_2-CH_2-OH$

d. $CH_3CHCH_2-CH-CH_3$
       | |
      CH_3 OH

15. What are the differences in the way a 1°, 2°, and 3° alcohol reacts with a strong oxidizing agent? Use chemical formulas to explain your answer.

16. During an esterification reaction, an alcohol and a carboxylic acid react to form an ester and water. Write the general formula for this reaction and give a specific example.

17. What is the difference between the alcohol and mercaptan functional groups? Explain what factor is most responsible for the major differences in their physical properties.

18. What are the common names for each of the following ethers?

    a. $CH_3CH_2-O-CH_3$

    b. $CH_3-CH-O-CH-CH_3$
            | |
          CH_3 CH_3

    c. $CH_3-O-CH_2CH_2CH_3$

    d. $CH_3-CH-O-CH_2CH_3$
            |
          CH_3

19. The chemical properties of the ethers are most like those of which other organic family? Explain your answer and give examples.

20. Write the general structural formula for each organic family that contains the carbonyl group.

21. Explain why the ease with which the aldehyde group can be oxidized has medical importance. Write general reactions and name several examples of this reaction.

22. Using the Arrhenius definition, explain why the presence of the carboxyl group in a molecule makes it behave as an acid.

23. Explain the difference between a neutralization and saponification reaction.

24. Explain the similarities and differences between polyesters and polyamides. Give examples of both kinds of polymers.

25. Do some research and determine why the amide linkage is also known as a peptide bond. What is its role in protein synthesis and digestion?

Does exposure to nuclear radiation make you radioactive?

Nuclear Radiation, Reactions, and Energy

Nuclear radiation is a broad term encompassing several different types. Each type reacts differently with atoms and is potentially harmful to living organisms. The everyday use of radioactive substances has been one of the most controversial developments of modern science. Nuclear power, once advertised as "...too cheap to meter," was supposed to be the solution to society's craving for energy. We now know that nuclear power plants are as expensive to operate as any other type and even more expensive to shut down. While not the miracle solution it was once thought to be, nuclear power continues to provide a reliable source of electrical energy for millions of people around the world.

Many myths surround nuclear radiation. One widespread myth is that exposure to any kind of radiation will make you radioactive, and people even joke that you will "glow in the dark." Misconceptions such as this are one of the issues holding back at least one promising use of radiation: killing bacteria and other microorganisms that cause food to spoil. As long as customers associate radiation with danger, supermarkets will remain reluctant to offer irradiated products.

Nuclear radiation can be dangerous in the wrong circumstances, but used properly it can provide significant benefits. The power of the atomic bomb, for example, poses the potential for tremendous loss of life, but the use of other radioactive substances for medical purposes has saved many lives. Accurate information and knowledge are the keys to making sound judgements as a society about the uses of nuclear radiation and energy.

11-1   Energy in the Nucleus

Objectives

Upon completion of this section, you will be able to:

■ Identify the components of the nucleus.

■ Calculate the equivalence between mass and energy.

■ Define the term "isotope" and explain the different types of isotopes.

■ Write the symbolic notation for an isotope.

You might wonder if there is a difference between the terms "atomic energy" and "nuclear energy." The answer is no. The term "nuclear energy" is, however, not only more contemporary, but it is also more correct. Scientists started making great strides in the understanding of radioactivity soon after the turn of the twentieth century, but it was the use of the atomic bomb near the end of World War II that brought it to the public's attention. As afraid as people were of the atomic bomb, they were also fascinated by atomic energy, and interested in the use of radioactive materials for peaceful purposes. On August 1, 1946, not quite a year after the end of World War II, President Harry S. Truman signed the Atomic Energy Act, transferring the control of atomic energy from military to civilian hands and establishing the Atomic Energy Commission (AEC) to foster and control the peace time development of atomic science and technology. The word "atomic" even made its way into the popular culture.

Scientists have known for a long time that the nucleus of the atom is the source of atomic energy, but the popular press has only recently picked up the term. Now, whether we are referring to nuclear weapons or nuclear medicine, the pro- or anti-nuke lobbies, we are using the correct term – nuclear – to describe this source of energy.

Mass and Energy

The atom is composed of protons, neutrons, and electrons in a precise arrangement, and the vast majority of the mass of an atom (approximately 99.95%) is concentrated in the nucleus. Because protons and neutrons reside in the nucleus, they are known as **nucleons**.

Most of the volume of the atom is occupied by electrons. As shown in Figure 11-1, if the atom were the size of a large beach ball, the nucleus would be smaller than a single grain of sand. During ordinary chemical reactions, atoms interact with each

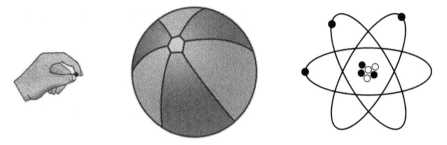

Figure 11–1: The nucleus contains 99.95% of the mass of an atom. The usual artist's rendition of an atom (right) is not to scale – if it were, the nucleus would be much smaller.

other by sharing or exchanging electrons. The phenomenon of radioactivity, however, originates in the nucleus of the atom. The electrons surrounding the nucleus take no part, although electrons do sometimes appear and disappear inside the nucleus – a topic for later discussion.

Energy is the key to understanding nuclear reactions. The nuclear power industry likes to point out – quite correctly – that the nuclear energy available in one pound of uranium is the same amount of energy made available by burning many train carloads of coal. When we describe a nuclear reaction, we will use reaction equations similar to those used for chemical reactions. Like any other equation, the left and right sides – energy, mass, charge, etc. – must balance. There is one difference between chemical and nuclear reactions: in nuclear reactions, mass is freely exchanged for energy according to Einstein's famous equation:

$$E = mc^2$$

where E is energy, m is mass, and c is the speed of light in a vacuum. Therefore, in a nuclear reaction equation, mass plus the equivalent amount of energy on one side must balance mass plus the equivalent amount of energy on the other side. Since the speed of light is a very large quantity, the constant, c^2, is immense. A nuclear power plant can power a sizable city for a year by converting just 40 grams (the equivalent of one thimble) of uranium into 1 billion kilowatt-hours of electrical energy.

Knowing that the nucleus of each element contains protons and that like charges repel, it would seem logical that the enormous electric force repelling the individual protons would cause the nucleus to quickly break apart. In nature, however, stable nuclei are the rule.

To this point, we have dealt with only the two most often encountered forces in the universe – gravitational and electromagnetic. The effects of gravitational force are obvious. The effects of electromagnetic force are responsible for not only the electrical appliances we take for granted, but also for the chemical reactions that keeps us alive. There are, however, two more forces in the universe – the strong nuclear force and the weak nuclear force. These two forces are important only between the particles within the nucleus of an atom where distances

Figure 11-2: Converting one thimbleful of uranium fuel to energy yields enough energy to supply a sizable city with electrical power for a year.

are extremely small. The Bohr radius (5.3×10^{-11} m), which is the typical radius of an atom, is so small that 1,000,000 atoms could line up across the width of the period at the end of this sentence. The distances between particles in the nucleus of an atom are over 5,000 times smaller than this!

The **strong nuclear force** is an attractive force that holds the collection of protons and neutrons together. If the separation between a proton and its neighbors is increased by a factor of 10, however, the strong nuclear force would be dramatically reduced and the repulsive force of the positive charges would overwhelm it.

Suppose we could assemble a nucleus from a supply of protons and neutrons. As the last proton being added gets nearer and nearer the rest of the positive charges, the repulsive force becomes greater and greater. It is similar to rolling a ball up a hill that is getting progressively steeper. When the particles get close enough, the repulsive force is suddenly overcome and the proton falls into place with the rest of the protons

Figure 11–3: Protons and neutrons in the nucleus are in a stable arrangement because the strong nuclear force attracts them. Removing a proton or neutron from the nucleus would be like removing a ball from the bottom of a crater – it would take work to accomplish.

and neutrons. The strong nuclear force has overcome the electrostatic force. It would be as if the ball rolled into a crater at the top of the hill. It took work to get it there, but now it is in a stable arrangement and it will take work to get it out of the crater (see Figure 11-3).

Bringing the last neutron into the nucleus would be a little different. Because it is an uncharged particle, there would be no repulsive force, but just like the proton, as it gets very close it, too, will feel an attractive force – the strong nuclear force – and fall into place.

Although atoms, neutrons, protons, and electrons are all extremely small, scientists have been able to determine their masses very precisely. For example, the mass of a proton, m_p, is 1.6726×10^{-27} kg, while the mass of the neutron, m_n, is slightly more at 1.6749×10^{-27} kg. By comparison, the mass of the electron, m_e, is 9.1094×10^{-31} kg, which is about 2,000 times less massive than either the proton or neutron.

The most common isotope of carbon is carbon-12. It is composed of 6 protons and 6 neutrons in its nucleus – a total of 12 nucleons – plus 6 electrons in their various energy levels. The mass of one atom of carbon-12 has been found to be 19.9265×10^{-27} kg. However, if we total the individual masses of the subatomic particles, we get a slightly different answer:

6 protons	$= 6 \times 1.6726 \times 10^{-27}$ kg $=$	10.036	$\times 10^{-27}$ kg
6 neutrons	$= 6 \times 1.6749 \times 10^{-27}$ kg $=$	10.049	$\times 10^{-27}$ kg
6 electrons	$= 6 \times 9.1094 \times 10^{-31}$ kg $=$	0.0055	$\times 10^{-27}$ kg
Total	$=$	20.0905	$\times 10^{-27}$ kg

For some reason, the mass of the atom is less than the sum of all of its parts. Mass is missing! This is not a matter of poor measurements – the measurements are both precise and accurate.

To understand what has happened, recall that Einstein told us mass and energy are equivalent. Therefore, we can conclude that negative mass is the equivalent of negative energy! That's what appears to have happened in this case. It's like negative mass has been added – or equivalently, that the atom has gained some negative energy. We're used to thinking of energy as a positive number, but that is not always the case. Kinetic energy must be positive, but potential energy can be negative when compared to that of a free particle. If there is any proof necessary that negative energy is more than just a bit of mathematical bookkeeping, here it is.

The nucleons that make up the carbon-12 atom are under the influence of the strong nuclear force. Because that force is attractive, it would require work to disassemble the nucleus. It is similar to moving balls (Figure 11-3) out of the pit – that, too, requires work. The balls in the pit, like the nucleons, have negative energy. If this were not so, every nucleus would be unstable. However, we know that even the unstable nuclei of radioactive elements, may remain assembled for a long time before radioactive decay occurs.

In addition to illustrating the concept of negative energy, this example demonstrates that the kilogram is not a convenient unit for measuring the mass of atoms and their components. Another unit, the atomic mass unit, u ($1 \text{ u} = 1.66053873 \times 10^{-27}$ kg), is commonly used. On this scale the proton, neutron, and electron masses are, respectively:

$$m_p = 1.0072635 \text{ u}$$

$$m_n = 1.0086486 \text{ u}$$

$$m_e = 0.00054858 \text{ u}$$

By definition, the mass of the carbon-12 atom is 12 u exactly. When we add up the masses of the components of this nucleus again, using atomic mass units, it becomes:

6 protons	$= 6 \times 1.0072635 \text{ u}$	$=$	6.043581 u
6 neutrons	$= 6 \times 1.0086486 \text{ u}$	$=$	6.051892 u
6 electrons	$= 6 \times 0.0005486 \text{ u}$	$=$	0.003291 u
Total		$=$	12.098764 u

This calculation shows that 0.09876 u of mass is missing from carbon-12. The missing mass is equivalent to a negative contribution to the potential energy. To get the mass back would require providing an equivalent amount of positive energy to break up the carbon nucleus. That amount of energy would be given by:

$$E = mc^2$$

$$E = (0.09876 \text{ u}) (1.66054 \times 10^{-27} \text{ kg/u})(2.998 \times 10^8 \text{ m/s})^2$$

$$E = 1.474 \times 10^{-11} \text{ J}$$

From a human perspective, that is a very, very small amount of energy – equaling approximately 1/100,000,000 the kinetic energy of a paper clip after falling one centimeter. From the perspective of the nucleus, however, this is a huge amount of energy. Unless something, such as a fast-moving neutron, comes along and delivers this amount of energy, the carbon-12 nucleus will remain bound together. For this reason, this is often referred to as the **binding energy**, E_B, of the nucleus.

The electron volt, eV, which is equal to $1.602176462 \times 10^{-19}$ J (updated in 1998) is a unit of work or energy that is accepted for use with SI units. Since the energies involved in nuclear reactions are large, the mega-electron volt (1,000,000 eV) or MeV, is more practical, although it is still a small unit relative to the joule, since 1 MeV $= 1.6 \times 10^{-13}$ J. When this unit is used to determine the binding energy for the carbon-12 nucleus, the results are:

$$E_B = \frac{1.474 \times 10^{-11} \text{J}}{(1.602 \times 10^{-13} \text{J})/\text{MeV}} \cong 92.00 \text{ MeV}$$

To calculate the amount of energy (measured in MeV) that is equivalent to one atomic mass unit (1 u), it would be: 1 u $\Leftrightarrow$ 931.49 MeV. This will be handy later in analyzing the reactions responsible for nuclear energy.

Isotopes

We already know that to maintain charge balance, all atoms have the same number of electrons as protons. The number of protons determines the atomic number, A, and the name of the element. In other words, $A = 2$ is helium, $A = 6$ is carbon, and so forth. Atoms of the same element that differ in the number of neutrons are called isotopes. Since neutrons do not have a charge, altering their number does not change the number of electrons required to maintain charge balance. Therefore, having the same number of electrons means that the element's isotopes all have the same chemical properties.

Hydrogen is the simplest element in nature. The most common isotope of hydrogen is unique in that its atoms do not contain a neutron. About one percent of the naturally occurring hydrogen atoms also have a neutron in their nucleus. This isotope of hydrogen, called **deuterium**, is sometimes referred to as **heavy hydrogen**. Although the term heavy hydrogen is no longer in common use, adding a neutron to a hydrogen nucleus doubles the mass. Since both isotopes still have only one valence electron, they still bond with oxygen in the same way.

Water molecules made with deuterium atoms in place of the more common hydrogen atoms, then, are known as heavy water. It has even been given its own chemical symbol, D_2O, where the D symbol stands for deuterium. It is obviously a heavier molecule than H_2O by the mass of two neutrons. As we will see later, the deuterium atoms in heavy water can undergo fusion – the same type of reaction that is the energy source for the Sun. The possible use of heavy water in making atomic weapons gave it great strategic value during World War II.

The number of nucleons determines the atomic mass, Z. The only major difference between the nuclei of so-called light (hydrogen, helium, carbon, etc.) and heavy (uranium, plutonium, etc.) elements are the number and arrangement of these nucleons.

In chemistry, it is customary to identify an element by its symbol. When nuclear reactions and nuclear energy are involved, additional information about the composition of the nucleus is often needed. This information can be included, as superscripts and subscripts, to the left of the symbol. For example, most carbon atoms have 6 protons and 6 neutrons. This information can be included by placing its atomic number, A, at the lower left-hand corner and its atomic mass, Z, at the upper left-hand corner of the symbol. The results are $^{12}_{6}C$.

Atomic Composition		
Number of nucleons (Z)	:	12
Number of protons (A)	:	6
Number of neutrons (Z – A)	:	6
Symbol, including nucleons	:	$^{12}_{6}C$

It is true that this representation contains redundant information. If the atomic number is six, then it is always carbon. This form, however, reminds us that carbon has 6 protons, allows us to calculate the number of neutrons, and may save us from having to look up this information. The atomic mass information may also be conveyed by using a carbon-12 or C-12 designation. Although helpful, this form may still result in someone having to look up the atomic number.

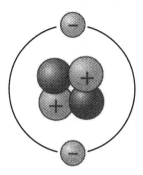 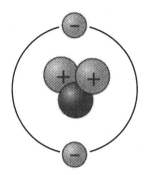

Figure 11–4: The most common (99.9999%) naturally occurring isotope of helium, He-4 (left), has 2 protons and 2 neutrons. The other stable isotope of helium is He-3, with two protons and only one neutron. Because they have identical electron arrangements, they have identical chemical character.

Since an element's identity does not depend on the number of neutrons, it might seem that changing the number of neutrons would have very little effect. Nothing could be farther from the truth. Neutrons are affected by the strong nuclear force, but not the electrostatic force. Therefore, having neutrons in the nucleus actually helps bind it together. Unfortunately, this simple picture is only true up to a point, because protons and neutrons also fit together in the nucleus in more subtle and complex patterns. If you add too many neutrons to an atom's nucleus, you eventually reach the point of diminishing returns and the nucleus actually becomes less stable. This is due to another short-range force called the weak nuclear force.

Recently, the **weak nuclear force** has been shown to be another manifestation of the electromagnetic force. For every atom, there are a few – one, two, or more – arrangements of protons and neutrons that are stable. These are generally arrangements in which the ratio of protons and neutrons are nearly equal (1:1). For the more massive elements, this balance tends to shift in favor of a greater proportion of neutrons. As previously noted, the various isotopes of an element tend to exhibit the same chemical properties, but their nuclear properties can vary drastically. Figure 4-9 shows all the stable isotopes for the known elements; many others are unstable and referred to as radioactive, although some of their nuclei stay together for decades or longer. Others may last only minutes or a fraction of a second. The radioactive isotope of carbon, known as carbon-14, is used in carbon dating (see Technology Boxes 4-4 and 11-2). On average, this isotope lasts for thousands of years before undergoing radioactive decay. Fortunately, all common elements have stable isotopes. It is the relatively rare, radioactive isotopes that are the basis for nuclear science.

Check Your Understanding

1. What two components of the nucleus are also known as nucleons?

2. Which has more mass outside the nucleus – a proton, neutron, or an electron?

3. What is binding energy?

4. How do deuterium and hydrogen atoms differ?

5. What is heavy water?

6. What does the atomic number tell us about the composition of an atom?

7. The subtraction of the atomic number from the atomic mass is equal to what?

8. When writing the nuclear symbol for any isotope, where is the atomic mass placed?

11-2   **Types of Radiation**

Objectives

Upon completion of this section, you will be able to:

■ Identify and describe the differences between the major types of nuclear radiation.

■ Describe radioactive decay events using nuclear formulas.

■ Describe how radiation interacts with matter.

■ Discuss the relationship between half-life and radioactivity.

Various types of ray guns have been a staple of science fiction for years. You might be surprised to learn that one type of ray gun is about to become a reality. The U. S. Air Force is having a fleet of aircraft built that can shoot down missiles with a ray of light called a laser beam. The prototype for this aircraft has been around for 20 years, but only recently has laser technology advanced to the point of making it practical.

Although no one has yet made a gamma ray gun, gamma rays are quite real. The word ray shares its root with both radiation and radial. It describes something that is emitted by a source and moves away from that source in a straight line – if the medium through which it moves does not change. Visible light from a bulb travels in a straight line radially outward in all directions. In this section, we will use the term radiation to describe certain specific products from nuclear reactions and nuclear decays. In many cases, these could more accurately be called types of ionizing radiation, since they carry enough energy to strip the electrons away from the nuclei of any matter they happen to encounter. This accounts for the damaging effects they have on both inert and living matter.

When scientists first discovered the phenomenon of invisible types of radiation, they found that certain materials – e.g. pitchblende, which is a uranium ore – are radiation sources. Although invisible, their effects can be seen on photographic film even if it is sealed in an envelope. The radiation can be stopped by some materials, particularly if they are dense. This was one of the first properties scientists noted. They also found that some sources give out more penetrating radiation than others.

Eventually, scientists classified three types of radiation – alpha, beta, and gamma – and named them by using the first three letters of the Greek alphabet. Because these share the property of traveling in straight lines, they were initially called alpha rays, beta rays, and gamma rays. Today, the terms alpha radiation and beta radiation have been substituted because we know they are made up of the same atomic particles from which atoms are built. It is still proper to speak of gamma rays, however, because they are electromagnetic radiation.

Alpha Radiation

Of the various types of nuclear radiation, **alpha radiation** is the most massive and the least penetrating. A single sheet of paper or the top layer of dead skin cells can stop most alpha radiation. Because it is easy to stop, it is also easy to protect against it. Alpha radiation sources can be dangerous, however, but usually only when inhaled or ingested. The particles comprising alpha radiation are composed of two protons and two neutrons, which make them identical to the helium nucleus. Alpha particles have an atomic mass of 4 and an atomic number of 2. Since alpha particles contain two positively charged particles but no electrons, each alpha particle has a positive charge 2 times that of the elementary charge, e.

The symbol, α, is sometimes used to represent the alpha particle, but it is also represented by the symbol for helium, ^4_2He, complete with the designations for atomic number and mass. Alpha radiation is composed of two protons and two neutrons, but no electrons, just like a helium nucleus.

Alpha Radiation	
Symbol :	^4_2He
Atomic mass :	4
Atomic number :	2
Charge :	+2
Penetration :	low

Any radioactive element that emits an alpha particle is losing two if its protons and two of its neutrons. In doing so, it is transmuted into another element! Alchemists tried to accomplish such transmutations using a chemical reaction (see Technology Boxes 4-5 and 11-1), but since chemical reactions deal only with the electron arrangements of an atom, their efforts failed. To actually transmute an element, you must change the number of protons in its nucleus. The following is the equation for a polonium-210 nucleus emitting an alpha particle:

$$^{210}_{84}\text{Po} \rightarrow \,^{206}_{82}\text{Pb}\star + \,^4_2\text{He}$$

In this example, the polonium atom turns into an atom of lead as it emits the alpha particle. As noted before, both the atomic masses ($210 = 206 + 4$) and atomic numbers ($84 = 82 + 2$) remain balanced. What is not obvious from the equation is the amount of energy the alpha particle takes with it. The binding energy of the polonium atom is not equal to that of the lead atom and the alpha particle. There is extra energy on the right-hand side of the equation that temporarily stays with the Pb nucleus – denoted by the asterisk – but most of it goes with the alpha particle.

The amount of energy an alpha particle takes away from a reaction varies according to the isotope involved, but it is the same for all atoms of that isotope. Typically, this energy is on the order of 1 MeV in the form of kinetic energy. In other words, the alpha particle exits the nucleus moving very fast. In fact, it starts at a significant fraction of the speed of light.

Because of its charge, an alpha particle interacts with other charged particles in the atoms of any material it encounters. The electrons in these materials are attracted to the positively charged alpha particles and, in some cases, they are ripped from their parent nuclei. This process results in a charge imbalance or ionization of the material and creates positively charged ions. Alpha particles, therefore, are one kind of **ionizing radiation.**

Tearing apart an atom (Figure 11-5) takes energy, so the alpha particle is slowed down. After doing this repeatedly, it eventually loses all of its energy and stops in the material. Unfortunately, the ionization process scrambles the chemistry of the atoms that it affects. If these atoms are part of living tissue, that tissue can be destroyed or

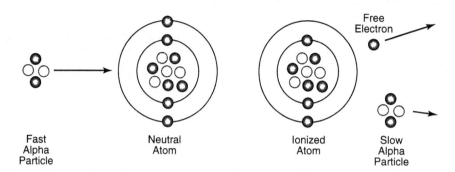

Free
Electron

Fast
Alpha
Particle

Neutral
Atom

Ionized
Atom

Slow
Alpha
Particle

Figure 11–5: Alpha particles lose energy when they encounter, ionize, or remove electrons from other atoms.

changed into a malignant (cancerous) form. For this reason, any of the radiation types that cause ionization are potentially dangerous to our health.

Beta Radiation

Beta radiation is generally more penetrating than alpha radiation, but less so than gamma radiation. It takes several sheets of aluminum foil to stop beta radiation. Because it is also a type of ionizing radiation, beta radiation can potentially be dangerous to our health. For example, **beta burn** is a type of radiation damage that is similar in many ways to a burn from a hot object. Beta particles are actually nothing more than electrons, fast moving electrons originating from the nucleus.

The symbol, β^-, is used for the beta particle, which has an atomic mass of 0. An atomic number is not usually assigned to a beta particle because it is not a nucleon. Effectively the number would be –1 because, like any other electron, each beta particle has a negative electrical charge of 1 times the elementary charge, e. Some of its important properties follow.

Beta Radiation	
Symbol :	β^-
Atomic mass :	0
Atomic number :	–1
Charge :	negative
Penetration :	moderate

It would be perfectly logical, at this point, to wonder how an electron could be ejected from a nucleus. After all, electrons remain far away from the nucleus in energy levels. However, electrons, protons, and neutrons are related in other ways.

Imagine what would happen if an electron and proton could be brought together. The negative charge of the electron would cancel the positive charge of the proton, producing an electrically neutral particle. If you carefully add the masses of the proton and electron, you will find that the new particle would be slightly less massive than an actual neutron. Since an actual neutron has slightly more mass, it also has slightly more potential energy. However, there's the matter of binding energy, and Einstein told us we can swap mass for energy and vice versa.

The reverse of the above process occurs with beta radiation. A nucleus that emits a beta particle finds itself with one less neutron and one more proton. This is another example of transmutation. An equation that describes a typical beta radiation event is the following:

$$^{14}_{6}C \rightarrow ^{14}_{7}N + \beta^{-}$$

In this reaction, radioactive carbon-14 decays into nitrogen-14 and emits a beta particle. Notice that the atomic mass does not change, but the atomic number does. This means that the carbon atom is transmuted into a nitrogen atom. This extra positive charge is balanced by the negative charge of the beta particle. This is why we can think of the beta particle as having a negative atomic number.

As with the alpha particle, the emitted beta particle acquires most of the energy released by this reaction, so it leaves the nucleus at a high speed. Because it is less massive than an alpha particle, the beta particle travels much faster than an alpha particle of equal kinetic energy. Beta particles also lose their energy when they encounter and ionize other matter.

Not all radiation that is classified as beta radiation is negatively charged. There are also positively charged electrons called **positrons**, β^{+}. Other than charge, the positron has the same characteristics as an electron. A positron is a type of **antimatter**. (Antimatter is real and not just a figment of a science-fiction writer's imagination.) When an antiparticle encounters its corresponding particle, the two destroy each other by converting both of their masses into energy. Because we live in a world composed mostly of matter, antimatter does not exist very long before it bumps into and destroys matter – it does, nevertheless, briefly exist.

Technology Box 11–1 ■ Can lead be turned into gold?

During the Middle Ages, the forerunners of modern chemists and scientists flourished; they were known as alchemists. What they were trying to accomplish is similar to what modern materials scientists want – to improve naturally occurring substances. One improvement they were trying to accomplish was to change base (worthless) metals into gold. If they could do this, they would become rich beyond their wildest dreams. The alchemists tried all sorts of processes and in their approach stumbled onto information that would later form the foundation for modern chemistry. Unfortunately, the search that was uppermost in their minds was doomed to failure.

The problem with the approach the alchemists took was trying to change lead into gold using only chemical processes. During a chemical process, the various elements that make up the compounds are rearranged, but the elements themselves remain unchanged. To change lead to gold would mean changing the nucleus of a lead atom into the nucleus of a gold atom. The nuclear symbols for the stable isotopes of lead is $^{206}_{82}Pb$, and gold is $^{197}_{79}Au$.

To turn lead into gold, then, the lead nucleus would have to lose three protons and six neutrons. In fact, modern scientists have been able to accomplish this by using a large complicated machine called an accelerator or atom smasher.

The accelerator operates by firing elementary particles, at very high speeds, into a lead target. The reaction between the high-speed particles and the lead nuclei causes some of the nuclei to break apart and transmute into gold nuclei. Unfortunately, these machines are so expensive to run that making gold from lead is a losing proposition! For the money you would spend to transmute lead into a little gold, you can buy a lot of gold on the open market.

Positron emission is similar to emission of an electron, except that when a nucleus emits a positron it ends up with one more neutron and one fewer proton. The following is such a reaction:

$$^{24}_{12}\text{Mg} \rightarrow ^{24}_{11}\text{Na} + \beta^+$$

At this point, we need to make a confession – neither of the last two equations was completely balanced. During **beta decay** (both electron and positron), there is another particle that is emitted. It is a peculiar particle that took a long time to discover.

After observing many beta decays and totaling the mass, energy, and momentum of the particles they could detect, scientists realized that the energy and momentum did not balance. There was something missing from the right-hand side of the equation. To achieve balance, scientists proposed the existence of another particle. If the particle could be found, it would account for these unbalanced quantities.

It soon became clear why it was hard to detect this particle – it has no charge and its mass is too small to measure. This particle is called the **neutrino,** which means little neutral one. Neutrinos are the least reactive particles known to man. Most neutrinos, even ones carrying large amounts of energy, can pass through the entire Earth without undergoing a single reaction. The fact that they were ever detected is a tribute to scientific persistence.

Gamma Radiation

Gamma radiation is different in character than either alpha or beta radiation. Whereas alpha and beta radiation are composed of matter, gamma radiation is high-energy electromagnetic radiation. These are truly rays, a form of "light" we cannot physically see. It is true that we can think of it as being composed of tiny energy packets, called photons, but photons are not particles in the classical sense.

It is theoretically possible to catch an alpha particle and a couple of beta particles and assemble them into a helium nuclei. It is not even theoretically possible to do anything of the sort with gamma rays because they are only energy; they have no material substance so they cannot be stopped or collected. They are always moving at the speed of light! Because they have no atomic mass, no atomic number, and no electric charge, just the symbol γ is used to represent them. What makes gamma rays the most penetrating and potentially one of the most dangerous forms of radiation is that they possess a lot of electromagnetic energy in a very compact packet.

Gamma Radiation	
Symbol :	γ
Atomic mass :	0
Atomic number :	0
Charge :	none
Penetration :	very high

Nuclei typically emit gamma rays in conjunction with other types of radiation. They help carry the energy away from a reaction. For example, in the alpha decay reaction that produced lead-206 above, the resulting lead nucleus was left in an **excited** state – that is, with excess energy. Very quickly, the lead nucleus releases that energy in the form of a gamma rays, according to the following equation:

$$^{206}_{82}\text{Pb}^{\star} \rightarrow\ ^{206}_{82}\text{Pb} + \gamma$$

The time that the lead nucleus remains in the excited state is very short, so from the perspective of an observer, the beta and gamma radiation appear to be emitted from the polonium nucleus simultaneously. The gamma rays that are emitted from this or any other radioactive decay, are unique to the particular nuclei involved in the sense that the energy and the frequency of the gamma radiation emitted is characteristic of each decay process. The gamma rays, therefore, serve as fingerprints that can be used to identify the isotope that is involved. In other words, a radioactive source in a sealed container could be identified by simply looking at the spectrum of gamma rays it emits.

As you learned in Chapter 9, light (electromagnetic radiation) is a wave composed of an electric field wave and a magnetic field wave. Like any other wave, it has a wavelength, λ, and a frequency, f. The wavelength and frequency of the radiation are related, so if you know one you can calculate the other by using the equation, $\lambda f = c$. In this equation, c is the speed of light. What is perceived as different colors of light are, in fact, electromagnetic waves of different wavelengths and frequencies.

Stationary charges are the source of electric fields and moving charges create magnetic fields as presented in Chapter 9. Another way to create an electric field is by changing a magnetic field. Similarly, a magnetic field can be created by changing an electric field. In fact, changing electric and magnetic fields create each other. This phenomenon can exist in space without any charges nearby and represents stored energy in electromagnetic waves. Because these fields are sometimes localized to a tiny region of space, we can think of them as packets of energy. The exchange of this energy with the atoms of material always occurs in discrete amounts, and these packets are termed photons. It is possible to have many packets in the same place carrying a large amount of energy, but there is a limit to how small an amount of energy can be packaged. A **quantum** or photon is the name given this smallest packet of electromagnetic energy. This amount of energy, E, depends on the frequency, f, of the light. The relationship between the energy in one photon and the frequency of the light is, $E = h f$, where h is Planck's constant and equal to 6.6261×10^{-34} joule second or 4.141×10^{-15} eV second. On a human scale an electron volt, eV, or a million electron volts, MeV, is a tiny amount of energy, but on the scale of atoms, 1 MeV is a huge amount of energy. Gamma rays consist of photons of light that come in energy packets ranging from 0.10 to a few MeVs.

The size of the energy packet explains why gamma rays are so penetrating. The energies associated with chemical bonds are on the order of a few electron volts, as are the energies of visible light photons. Although there are transparent materials through which visible light can pass with very little interaction, when visible light meets most materials it is either absorbed or reflected. The energies of the bonds in most materials and the energies of visible light are a good match. It is very easy for a photon to give all of its energy to a single electron in a single atom, exciting the atom and generally warming the material.

Gamma rays, on the other hand, possess much more energy than they could possibly give to a single electron; but they can give a little of their energy to one electron, a little to another, and so forth, many times over. When an atom absorbs visible light photons there is not enough energy transferred to remove the electron from around its nucleus; whereas gamma rays can easily ionize many atoms. If these ionization events happen within the human body, the chemistry of the body is altered on a very basic level. The body can repair a certain amount of damage, but when the chemistry of too many cells becomes scrambled, the damage can be overwhelming or even fatal.

This is why high-level radiation sources are surrounded by thick layers of dense shielding materials when transported or stored. The idea is to ensure that the energy of the photons from the gamma rays is expended in the atoms of the shielding material before it reaches a human receptor.

Neutron Radiation

Neutron radiation doesn't make the alpha, beta, gamma list of radiation types originally discovered by scientists. The neutrons emitted from radioactive materials do, however, constitute an important type of radiation. The atomic symbol for the neutron is $_0^1 n$. The neutron, having no charge, has no atomic number, but does have an atomic mass of 1, as shown below.

Neutron Radiation	
Symbol :	$_0^1 n$
Atomic mass :	1
Atomic number :	0
Charge :	none
Penetration :	high

The electric field emanating from the charge on alpha and beta particles means that they interact with other charged particles at long range. Neutrons, on the other hand, exhibit no such long-range interactions, making them potentially very damaging.

The interaction of a neutron with other particles can be likened to a collision between larger objects. A collision between an electron and a neutron, for example, has about the same effect on the motion of the neutron as a ping-pong ball would have on the motion of a bowling ball. When a neutron reaches the nucleus of an atom, its most significant interaction occurs. The chance of an encounter between a nucleus and a neutron is not great because nuclei are small, so a neutron must pass through a considerable thickness of material before a collision is likely to occur. This means that neutron radiation is also highly penetrating.

Because neutrons interact most significantly with the nuclei of other atoms, this type of radiation can be especially damaging in two ways. First, the popular myth that any material exposed to radiation becomes radioactive itself is not true in most cases, although it can be in the case of neutron radiation. The closer the mass of a fast moving neutron comes to a nucleus, the more of the neutron's energy is transferred. Since most nuclei are heaver than a neutron, the neutron typically loses only part of its energy during each collision. After several collisions, however, the neutron has lost enough kinetic energy that it can now be captured by a nucleus. In other words, after the neutron has been sufficiently slowed, the strong nuclear force (attractive) causes it to "join" the other nucleons. Unfortunately, this usually upsets the delicate balance and, in turn, causes it to become an unstable nucleus (radioactive). This process is called **neutron activation** and is a major consideration when repairing or decommissioning nuclear power plants. A nuclear power plant can't simply be shut down and forgotten, even if the nuclear fuel has been removed. Some of the other materials in the plant will likely remain radioactive (hot) for years due to neutron activation.

Special Topics Box 11–1 ■ How is half-life determined?

There are many principles of physics and chemistry that can be demonstrated with household materials – most nuclear radiation phenomena are an exception. There is one demonstration, however, that illustrates the phenomenon of half-life in an entertaining way.

Count the number of M&M candies from a large bag and put them in a bowl. Record this number along with the time when you spill the bowl of candies onto a table. Pick up only those candies that have the M&M logo showing and put them back into the bag. Count and record the candies remaining on the table as you put them back into the bowl. Also, record how long it took you to complete this first trial.

Spill the remaining candies in the bowl and again remove only the ones with the logo showing. Count the number of candies remaining. Repeat the procedure until most of the candies are gone.

The candies that fall with the logo showing represent radioactive nuclei that have undergone a nuclear decay. On average, half of the candies should fall with the logo up each time. The time it takes for you to spill, sort, and count them the first time will be used as one half-life. If you plot the number of candies you started with and then the number remaining after each half-life, you should find the plot is similar to Figure 11-7.

Note that exactly half of the candies will not fall with the logo up each time. Sometimes there will be a few more, sometimes a few less. This is to be expected, especially since we are not working with a very large sample. Remember that radioactive decay is a random process. On average, you would expect to lose half of the nuclei you start with to radioactive decay after one half-life, but these statistics average out only after many trials or by using a very large sample size.

You will also find that there are some candies that never get picked up. This is typical. It may mean that some candies were missed during the logo printing process. Similarly, if you were monitoring the activity of a source, you would also expect to pick up some background radiation from other radioactive materials. This means that the rate of radioactive decay that you observe would never truly go to zero. When measuring the activity of a radioactive source, you would subtract this background radiation from your measurements.

The second way neutron radiation can be damaging is that it can cause exposed materials to weaken or become brittle. Every time a neutron collides with a nucleus, the atom recoils. This causes it to move a little out of position relative to its neighboring atoms, altering the structural integrity of the material on an atomic level. Designers of nuclear power plants must carefully choose the materials they use, especially those that will be near the core of the reactor. Even after years of exposure, these materials must remain strong so the plant can continue to operate safely.

Half-life

Alpha, beta, gamma, and neutron radiation all result from nuclear events. Predicting when these events will occur, however, is problematic. Suppose two atoms of the same radioactive isotope could be placed side by side for observation. The two nuclei are identical in every way as far as modern science can detect, and yet there is no way to

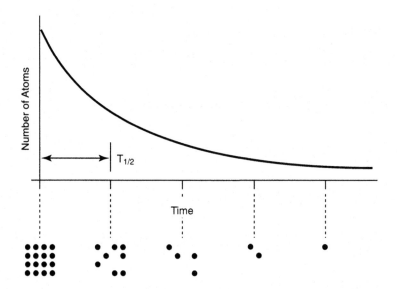

Figure 11–6: One-half of the original sample disintegrates at the end of each half-life.

predict which of the two will undergo radioactive decay first. One may decay a milli-second after it is selected and the other not for a thousand or more years. If this bothers you, you are in good company. Radioactive decay is a quantum mechanical process and its randomness bothered Einstein, too. "God does not play dice with the universe," he once said. Nevertheless, as far as we can tell today, when an atom will decay is com-pletely random. In large samples it is easy to measure the number of disintegrations quite precisely, so probability becomes a powerful tool for accurately predicting the rate of radioactive decay.

It is a fact of nature that some radio-active isotopes tend to undergo nuclear decay in a shorter length of time than others. If a sample is observed long enough, theoretically one-half of the ra-dioactive atoms will decay. This time may be short or long, but for any given isotope it is well-defined and called its half-life, $T_{\frac{1}{2}}$. If observation continues to the end of the second half-life, one-half of the remaining atoms will have de-cayed. This leaves only one-quarter of the original sample. If observation con-tinues through yet another half-life, one-half of the remaining atoms will have decayed, leaving only one-eighth of the original sample, and so forth (see Figure 11-6). When the number of at-

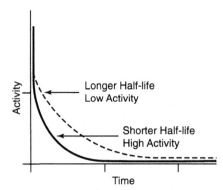

Figure 11–7: Given the same sample sizes of two different materials, the one with the shorter half-life will have the higher activity. Since more atoms of this isotope will decay in a given time period.

oms remaining is plotted as a function of time, it results in an exponential decay curve. Mathematically it can also be shown that the half-life does not depend on how many atoms are remaining. Half-life depends only on the kind of atom.

For each atom, there is a probability that its nucleus will undergo a radioactive de-cay during a given time interval. The **activity**, R, of the source is, therefore, propor-tional to the number of atoms present multiplied by this probability. In other words, if there are twice as many atoms, there will be twice as much activity. Obviously, the activity of the source decreases with time. This is because the number of unstable nu-

Technology Box 11-2 ■ What is carbon-14 dating?

Radioactive carbon dating can be used to determine the age of materials that were once living. Photo copyright © Francois Gohier.

It is impossible to totally avoid all radioactive materials. For example, carbon-14 is radioactive and constitutes a part of all living objects. It has a half-life of 5,730 years and is present to approximately one part per trillion in the atmosphere. This is not enough to cause a radiation hazard, but it is enough to be measured. Living organisms continually exchange carbon atoms with their environment, primarily in the form of $^{14}_{6}CO_2$.

When an organism dies, the exchange process stops. The carbon-14 atoms continue to decay, but they are not replaced. Therefore, the ratio of carbon-14 to carbon-12 isotopes starts to fall. After one half-life (approximately 5 millennia), the ratio will be one part per two trillion. Because the decay is exponential, the age of the object can be determined by accurately measuring the amount of carbon-14 remaining.

clei remaining has decreased. A plot of these variables also produces the same type of exponential decay curve.

Activity is measured by the number of radioactive decays per second. In other words, it is the rate that atoms are changing or transmuting into other isotopes. It tells us that radioactive isotopes with a short half-life will be highly radioactive at first because many of their atoms decay in a short time period. (Figure 11-7) Conversely, isotopes with a long half-life are low-activity sources. If you want higher activity from this isotope, then you will have to start with a larger sample.

Check Your Understanding

1. Which type of nuclear radiation has the most massive, least penetrating particles?

2. What is the name of the process that results in one element being changed into another element?

3. What is the difference between beta particles and positrons?

4. What are the characteristics of the neutrino?

5. What type of radiation is composed of high-energy photons?

6. What is neutron activation?

7. What does the half-life of a radioactive isotope describe?

11-3 Nuclear Reactions

Objectives

Upon completion of this section, you will be able to:

- Describe the difference between fission and fusion reactions.

- Read and understand nuclear equations.

- Calculate the amount of energy released in a nuclear reaction given its starting and ending constituents.

Just as balanced equations can be written for chemical reactions, they can also be written for the nuclear reactions that produce the different types of radiation. The major difference is that in a nuclear equation the atomic numbers must balance and so, too, must the atomic masses. In general, there are two types of nuclear reactions: fission and fusion.

Fission reactions typically split a nucleus into two smaller nuclei and release a few neutrons, $_{0}^{1}n$. The process usually begins with the capture of a free neutron. The fission of a uranium-235 nucleus into two smaller nuclei is described in the formula:

$$_{0}^{1}n + _{92}^{235}U \rightarrow _{40}^{95}Zr + _{52}^{139}Te + _{0}^{1}n + _{0}^{1}n$$

Totaling the atomic masses $(1 + 235 = 95 + 139 + 1 + 1)$ and the atomic numbers $(92 = 40 + 52)$ on both sides of this equation, demonstrates that they balance. In other words, the total number of protons and neutrons do not change.

Fusion reactions work the opposite way. In a fusion reaction, two nuclei join to become one nucleus. For example, when two deuterium nuclei fuse, they produce one helium-3 nucleus and a free neutron. The equation for the reaction is:

$$_{1}^{2}D + _{1}^{2}D \rightarrow _{2}^{3}He + _{0}^{1}n$$

Again, the atomic masses $(2 + 2 = 3 + 1)$ and atomic numbers $(1 + 1 = 2)$ balance. As with fission reactions, fusion reactions can also release free neutrons. Both types of nuclear reactions can release large amounts of energy.

Energy from Fission Reactions

Fission is the reaction that drives the world's nuclear power plants. Heavy nuclei, such as uranium or plutonium, are usually involved. Since these heavy nuclei have many protons and neutrons, they can split in many different ways. For example, in addition to the previous example and many other possible combinations, uranium-235 can fission in the following ways:

$$_{0}^{1}n + _{92}^{235}U \rightarrow _{38}^{90}Sr + _{54}^{143}Xe + _{0}^{1}n + _{0}^{1}n + _{0}^{1}n$$

$$_{0}^{1}n + _{92}^{235}U \rightarrow _{36}^{92}Kr + _{56}^{141}Ba + _{0}^{1}n + _{0}^{1}n + _{0}^{1}n$$

$$_{0}^{1}n + _{92}^{235}U \rightarrow _{39}^{99}Y + _{53}^{135}I + _{0}^{1}n + _{0}^{1}n$$

Typically, a nucleus is split approximately in half. The products of each reaction are different and are themselves usually radioactive.

Special Topics Box 11–2 ■ The Chain Reaction

Long before the nuclear age, there was a legend that describes the numbers involved in a chain reaction. According to the legend, the game of chess was devised for the ruler of ancient India. Delighted at the game, the ruler offered its inventor anything he could name. In return, he replied, "Oh, sire, I do not wish for much – simply place one grain of rice on the first square of the chessboard, two on the second, four on the third, and so forth, doubling the number of grains for every new square. This will be my reward." The ruler thought this a small price to pay for the game, until he ordered the payment to be carried out. He quickly found that it would take all of the rice his country could produce – and more – to fulfill the bargain. So, as rulers were prone to do, he solved the problem in a straightforward way, by beheading the inventor.

Whether this legend is true or not, it does indicate how quickly a chain reaction can yield a very large number of neutrons. Try this demonstration. Take a measuring cup full of rice. Place the grains in piles: 1, 2, 4, 8, 16, 32, and so forth. How long does it take to become more convenient to measure the rice in spoonfuls? How many generations does it take until the measuring cup of rice is used up?

If this were a nuclear fission reaction, how many generations would it take to have Avogadro's number of neutrons free in the material? If it takes, on average, 1×10^{-8} seconds for a nucleus to capture a neutron, fission, and release two neutrons, how long would it take to get Avogadro's number (6.02×10^{23}) of free neutrons?

Notice, also, that each fission reaction starts with the addition of a free neutron and produces two or three free neutrons, in addition to the smaller nuclei. This is the reason nuclear reactions maintain what is known as a **chain reaction**. Each of these neutrons can initiate new fission reactions. Simple mathematics shows that if the first fission reaction releases 2 neutrons, they, in turn, can cause the release of 4 neutrons, then 8, then 16, then 32, etc. As Figure 11-8 shows, this series will quickly bloom into a very large number of free neutrons. On average, the fission of each uranium-235 produces 2.5 neutrons. After n generations, the number of possible fission events, N,

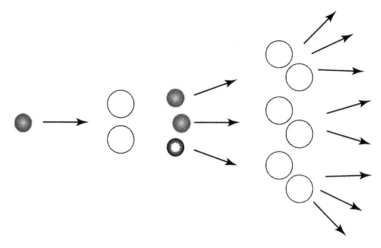

Figure 11–8: When uranium-235 undergoes fission, it releases either 2 or 3 uncombined neutrons. If each of these neutrons also initiate fission in another uranium atom, a chain reaction occurs.

would be given by $N = (2.5)^n$, indicating that the growth in the number of free neutrons is exponential. This calculation assumes that all of the neutrons released initiate more fission reactions. In fact, some will exit the material before interacting with another nucleus. Even with these losses, such an uncontrolled chain reaction has the potential to release tremendous amounts of energy. This kind of uncontrolled chain reaction is what happens when an atomic bomb is detonated. For peaceful uses, a neutron absorbing material, in the form of **control rods**, is placed in with the nuclear fuel to keep the chain reaction under control.

Nuclear fission is used for electrical power generation because it releases energy. The process, however, is totally different from that of a chemical reaction. Returning to the equation for the fission of uranium-235, researchers have measured the masses of each of the components with great precision. Let's compare what goes into the reaction with what comes out. The equation and masses represented by the equation are:

$$\,_{0}^{1}n + \,_{92}^{235}U \rightarrow \,_{40}^{95}Zr + \,_{52}^{139}Te + \,_{0}^{1}n + \,_{0}^{1}n$$

$\,_{92}^{235}U$	=	235.0403231 u	$\,_{40}^{95}Zr$ =	94.9080427 u
$\,_{0}^{1}n$	=	1.0086486 u	$\,_{52}^{139}Te$ =	138.9347300 u
			$\,_{0}^{1}n$ =	1.0086486 u
			$\,_{0}^{1}n$ =	1.0086486 u
Totals	=	236.0489717 u	=	235.8600699 u

In this case 0.1889018 u is missing from the products, but thanks to Einstein we know that the mass doesn't just disappear – it turns into energy. The missing mass in the equation indicates that the products have even more negative energy; they lost that much of their mass/energy by releasing it as useful energy. The energy equivalent of the missing mass can be calculated using the following relationship:

$$(0.1889018 \text{ u}) (931.49 \text{ MeV/u}) \Leftrightarrow 175.96 \text{ MeV}$$

Therefore, the typical fission reaction for uranium-235 should produce approximately 176 MeV of energy. However, even more energy results because most of the fission products themselves are not stable nuclei. The decay of these nuclei eventually ends when each decays into a stable nucleus. This series of steps is called a **decay chain**. Literally hundreds of possible decay chains start with the fission of one uranium-235 nucleus. The fission of one uranium-235 nucleus and its decay chain products releases an average of about 200 MeV of energy.

The energy from these fission reactions heat the fuel in the reactor. If the fuel is not cooled, it can quickly reach its melting point – the so-called meltdown phenomenon for a nuclear reactor. Getting the heat energy out of the reactor core is, therefore, not only a matter of preventing a meltdown; it is also the mechanism for harnessing the nuclear energy itself.

In a coal-fired electrical power plant, the energy released from burning coal is used to heat water into steam, which is then used to turn a turbine connected to a generator. As it cools and condenses back into water, the steam loses energy, which has become electrical energy due to the action of the turbine. A nuclear power plant works in exactly the same way, except that instead of using the energy from burning coal as the source of energy to heat water into steam, it uses the energy from fissioning nuclei (see Figure 11-9). Rather than cooling the reactor core with the same water

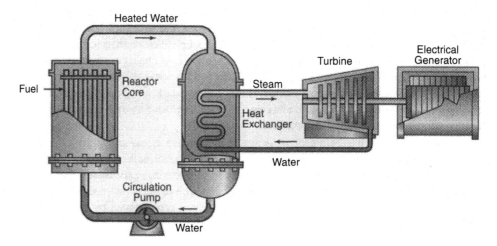

Figure 11–9: The energy released by fission is used to boil water into steam and turn a turbine generator.

that turns the steam turbines, in a nuclear power plant a closed-cycle heat exchanger is typically used to isolate the two systems. As with any energy transfer, some energy is lost in the process. This lost energy appears as waste heat in various parts of the system, which is why nuclear power plants typically have large cooling towers associated with them.

Energy from Fusion Reactions

Fusion reactions release energy for the same reason that fission reactions do: the mass of what is left after the reaction is less than the mass that was present at the beginning of the reaction. Take, for example, the fusion of the two deuterium nuclei that we previously used. The equation and the masses represented on both sides of the equation are as follows:

$$\,^{2}_{1}D + \,^{2}_{1}D \rightarrow \,^{3}_{2}He + \,^{1}_{0}n$$

$^{2}_{1}D$	=	2.014108 u	$^{3}_{2}He$	=	3.0160293 u
$^{2}_{1}D$	=	2.014108 u	$^{1}_{0}n$	=	1.0086486 u
Totals	=	4.028216 u		=	4.0246779 u

As in the fission reaction, there is some mass missing from the products – in this case 0.0035381 u. The energy equivalent of the missing mass is the energy released by this reaction.

$$(0.0035381 \text{ u}) (931.49 \text{ MeV/u}) \Leftrightarrow 3.30 \text{ MeV}$$

Unlike fission reactions that happen spontaneously when enough fissionable material is assembled, fusion reactions happen only under certain circumstances. Simply putting deuterium gas in a bottle will not start a fusion reaction. When atoms get close together, their electrons repel each other. Therefore, the first step in achieving a fusion reaction is ionization. Although there are equal numbers of positive and negative charges remaining in the container, the mixture is now more of a particle soup than a conventional gas.

Figure 11–10: In magnetic confinement fusion, strong magnetic fields keep the charged particles of the plasma away from the walls. In this picture of the National Spherical Torus Reactor, several magnetic field coils are clearly visible. Photo courtesy Princeton Plasma Physics Laboratory.

One way to ionize a gas is by passing an electrical current through it. This heats the gas and gives its atoms enough energy to ionize into plasma (see Chapter 4). The heating not only forms a plasma, but the nuclei also become more energetic (move faster). This is important because slow moving, like charged nuclei simply push each other away. Consequently, they never get close enough for the attractive, strong nuclear force to exert its effect. If the gas is hot enough, however, the approaching nuclei overcome the repulsive forces and collide. When this happens, there is a chance they may fuse and release energy.

The temperature required for deuterium to have a reasonable chance of undergoing a fusion reaction is extremely high – over 100,000,000°C. This is far above the melting point of any known material from which a container could be fashioned. The real problem, however, is not keeping the plasma from melting its container, it is keeping the container from cooling its contents, thereby quenching the reaction. The solution was to build a container without any physical walls! Such a container can be formed by using strong magnetic fields (see Figure 11-10). Because moving, charged particles curve in the presence of magnetic fields, such fields can turn them back from the physical walls of a container and keep them confined. There are many possible configurations of so-called **magnetic bottles** and, although they eventually leak, they can keep their components confined long enough for some of the nuclei to fuse.

Merely keeping the plasma away from the walls, however, does not keep it hot enough to cause the fusion reaction to continue. Like all hot objects, the plasma continuously radiates heat away. This heat could be replaced by passing an electric current or microwaves through it. This means using some of the energy produced by the fusion reaction to keep the plasma hot enough to continue reacting.

The D-D reaction is not the only possible fusion reaction; there are many others. The one shown in Figure 11-11, which has kept scientists busy for decades, is the fusion reaction between deuterium and another hydrogen isotope called tritium. **Tritium** ($_1^3T$) is the isotope of hydrogen that has two neutrons and, therefore, three nucleons (hence the "tri" in the name). Tritium atoms have a mass of 3.0160493 u. The equation for the deuterium-tritium or D-T fusion reaction would be as follows:

$$_1^2D + _1^3T \rightarrow _2^4He + _0^1n$$

The mass that is turned to energy by this reaction is equivalent to 17.6 MeV of energy. The energy is shared between the newly formed helium nucleus and the free neutron. The neutron gets the larger share (14.1 MeV) and the remaining (3.5 MeV) energy goes to the helium nucleus. In a magnetic confinement system, these particles have two very different destinies. The positively charged helium nucleus remains trapped in the magnetic bottle. Eventually, it will collide with other particles in the plasma and give up its energy, which helps keep the plasma hot. The uncharged neutron is unaffected by the magnetic bottle and can escape, carrying its energy with it. If the neutron is stopped in another material, known as a **blanket**, that material will get hot and the energy can be recovered by running a cooling fluid through the blanket. In a steam generator, the heated fluid heats water to produce steam that drives a turbine generator, similar to the scheme used in fission plants (see Figure 11-9).

Such fusion energy schemes have been on the drawing boards for decades. Although simple in principle, the technical challenges of keeping a plasma together within a magnetic field, heating it enough to start the fusion reaction, and keeping the reaction going using only internal heat have occupied hundreds of scientists for years. In the laboratory, fusion reactions have demonstrated that they can produce more energy than they consume (see Figure 11-12). These reactions, however, only lasted for a short time. There are technical hurdles that must be solved before a practical fusion power plant can be built and operate reliably for years.

For a time, fusion energy was promoted as clean nuclear power. Fusion reactors, it was claimed, would not generate the quantities of high-level radioactive waste that

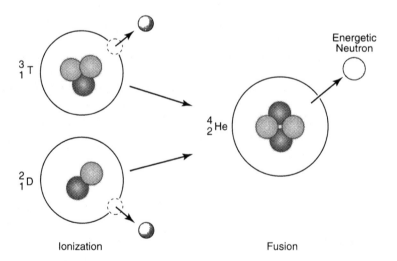

Figure 11–11: If an ionized gas is hot enough, the nuclei can reach very high speeds and collide hard enough to fuse into one.

Figure 11–12: A deuterium and tritium (D-T) fusion reactor – the Tokamak Fusion Test Reactor (TFTR) at the Princeton Plasma Physics Laboratory – successfully demonstrated a fusion reaction that produced more energy than was required to run it. Photo courtesy Princeton Plasma Physics Laboratory.

fission reactors do. Tritium, however, is itself radioactive and, because of its small size, tends to infest much of the equipment used to handle it. More importantly, the neutrons released by the fusion reaction are radiation hazards. They can also make the materials surrounding the structure radioactive at a low, but still hazardous level, through neutron activation.

As with most technologies, there is no such thing as a free lunch. Nevertheless, the prospect of "burning" seawater for energy makes fusion energy a worthwhile research pursuit. Meanwhile, another hydrogen fusion reactor – the Sun – continues to supply our energy needs both directly and indirectly and is expected to do so for the next few billion years.

Check Your Understanding

1. What are the two types of nuclear reactions and what are their basic differences?

2. What is the difference between a chain reaction and a decay chain?

3. What function is served by control rods?

4. What is one way to produce plasma?

5. How does a magnetic bottle confine plasma?

6. What is the purpose served by a blanket in a fusion reactor?

11-4 Radiation Protection

Objectives

Upon completion of this section, you will be able to:

■ Identify the three maxims of radiation safety.

■ Discuss the effects of time, distance, and shielding on radiation exposure.

We are continually exposed to natural radiation sources. Usually, these sources are at such low levels that our bodies can repair the damage with no apparent ill effects. There are a few instances, however, where natural sources can pose a hazard. Perhaps one of the most common exposures to a natural source of radiation is from the accumulation of radioactive radon gas in our homes. Well-insulated buildings intended to conserve energy do not "breathe." If there is a source of radon, it tends to accumulate and may reach dangerous levels. **Radon**, an alpha particle emitter, is a particularly dangerous gas because breathing brings it deep into the lungs, where there is no natural shielding. Fortunately, radon is easy to detect and, in most cases, a modest investment in a ventilation system will solve the accumulation problem.

The only time that radiation usually poses a hazard is near a radioactive source. Such sources can be handled safely and effectively if the three radiation safety maxims – time, distance, and shielding – are kept in mind. The goals are to minimize exposure time, maximize the distance from the source, and maximize shielding from the source as much as is practical.

Time

If you were told that there was a radioactive source nearby, your first question should be, "Just how radioactive is it?" As previously noted, the quantity that measures the radioactivity of a source is called activity, R. The activity is simply the number of radioactive decay events happening per second. Each of these events will cause one or more types of radiation to be emitted – alpha particles, beta particles, gamma rays, or, perhaps, several neutrons.

The SI unit for activity is the **becquerel**, Bq, which is equal to one decay per second. A source emitting only a single decay per second is not very radioactive. In fact, it would be hard to distinguish it from the natural background radiation of radioactive elements in the soil, cosmic radiation, and so forth. If you wished to purchase a radioactive source from a commercial supplier, you would probably order a source by the **curie**, Ci. The relationship between the becquerel, and the curie are the following:

$$1 \text{ Ci} = \frac{3.7 \times 10^{10} \text{ decays}}{s} = 3.7 \times 10^{10} \text{ Bq}$$

Fortunately, unless you have a license from the Nuclear Regulatory Commission, a supplier would not sell a 1 Ci source! If you do have a license, you will need the proper facilities to safely store such a radioactive (hot) source. Most teaching laboratories use sources that have an activity of less than 1 microcurie (1 μCi) to as low as 1/1,000 of a microcurie. A 1 μCi radioactive source still undergoes 37,000 radioactive decays per second, which is usually plenty for demonstration purposes.

If you place a radiation detector (see Technology Box 11-3) next to a 1 μCi source, however, it will not count 37,000 radioactive events in a second. This is because radiation is emitted from a point source uniformly in all directions – a situation known as **isotropic**. Therefore, only some of the radiation will strike the detector, but how much depends on the detector's cross-sectional area, A, and the distance, r, that it is from the source (see Figure 11-13).

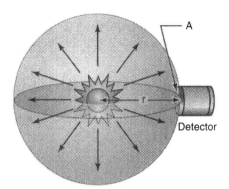

Isotropic Radioactive Source

Figure 11–13: Isotropic radiation sources emit radiation in all directions. A detector placed near the source will have a geometric efficiency that is less than one and, therefore, will detect only a part of the radiation coming from the source.

Theoretically, the radiation from a source would fall anywhere on the surface of an imaginary sphere of radius, r. The probability that it hits the detector depends on what fraction of the surface of the imaginary sphere is within the range of the detector. The geometry efficiency, $\xi_{geometry}$, is based merely on the geometry of the situation. The following equation is used for calculating the geometry efficiency:

$$\xi_{geometry} = \frac{A}{4\pi r^2}$$

Detection efficiencies, ξ_t, can range from a few percent to as low as a fraction of a percent. Another consideration is the efficiency of the detector. Detection of radiation can be difficult. Every type of detector is different and usually responds differently to different radiation types. Detector efficiencies, $\xi_{detector}$, are typically on the order of 10% to 30%. The overall detection efficiency, therefore, is the product of these two effects:

$$\xi_t = \xi_{geometry}\,\xi_{detector}$$

The efficiency, therefore, must be taken into account when determining the activity, R, of a source. If N events are detected over time, t, coming from a particular source, the activity of the source can be calculated using the formula

$$R = \frac{N}{\xi_t t}$$

Because the overall efficiency, ξ_t, can be very low, the activity of the source may be high even if the number of events actually detected is modest.

The activity, R, is actually a rate. The total number of radioactive products the detector sees is given by the equation, $N = R\xi_t\, t$, so that if a source with activity, R, is placed near it, the total amount of radiation that the detector receives depends on its exposure time. The same holds true with a human being. One way to minimize the amount of radiation is to minimize exposure time.

Technicians who take x-rays leave the room or go behind a shield when performing the x-ray. The minimal risk to the person receiving an infrequent x-ray is far outweighed by the benefits of proper medical and dental care. But the technician who administers dozens of x-rays every day, five days a week, throughout the year must take care to minimize their exposure times.

Technology Box 11–3 ■ How can you determine your level of exposure to radiation?

There are many types of radiation detectors. They can be subdivided into two broad categories – those that send a signal to an indicator or recorder immediately and those that are examined after a period of exposure.

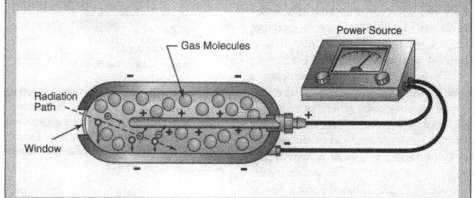

A Geiger-Mueller tube detects radiation when the gas inside is ionized.

Detectors that send a signal to an indicator or recorder are most often used for research purposes, as well as real-time monitors. There are many variations, but most operate on the same principal: radiation ionizes atoms. If a voltage is applied across the material, generating an electric field, the electrons will feel a force that sweeps them to an electrode where they are collected and counted by electronic circuits.

The information that these detectors yield can be as simple as whether or not radiation was detected. The clicking of a Geiger counter is an audible indicator of a radiation event. It is also possible to have a computer count these radiation events and save that data to a file. More sophisticated electronics attached to sensors can gather more information such as how much energy the radiation contained

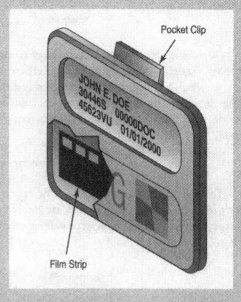

A film badge darkens when exposed to radiation. The amount of darkening is proportional to the exposure. Film badges, collected and developed once a month, are commonly used to monitor long-term exposure of workers.

There are other variations on this type of real-time detector. Another scheme uses transparent materials that emit pulses of light when exposed to radiation. These materials are called **scintillators**. The light is measured by very sensitive photodetectors. A small flash of light indicates that radiation is present, while the amount of light indicates how much energy the radiation left in the material.

Technology Box 11-3 ■ How can you determine your level of exposure to radiation? (Continued)

The second category of radiation detectors may be as simple as a piece of photographic film sealed in a light-tight container. Such film badges are often used to monitor the long-time exposure of a worker who works with radioactive material or works in an area where there are radioactive sources. Badges are typically exchanged monthly for new ones. The more radiation the badge has received, the darker the developed film. An optical instrument is used to measure the level of darkening and technicians compare that to a standard to determine the amount of exposure the worker has received. This, of course, presumes that the worker was wearing the badge during all potential periods of exposure!

A second device that measures accumulated exposure is the quartz fiber dosimeter, which looks something like a thick fountain pen. They are charged periodically with an electrical charge. Exposure to radiation causes the charge to leak off. The devices have a tiny meter built in, which the user can read by holding the device up to the light and noting the position of a needle on a scale. The less charge remaining on the dosimeter, the greater the exposure to radiation. Such badges are typically issued to workers or guests who need to go into a potential radiation environment for a short period of time. Although these devices do not sound a warning if they encounter a single radioactive particle, the user can check the total radiation exposure up to that time.

A popular joke is that wearing a radiation badge protects against radiation. Unfortunately, this is not true. Keeping a continuous watch on one's exposure, however, is a key element in assuring that radiation sources are used safely.

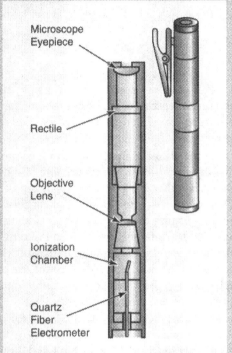

Microscope Eyepiece

Rectile

Objective Lens

Ionization Chamber

Quartz Fiber Electrometer

When a charged quartz fiber dosimeter is exposed to radiation, ionization decreases the charge in proportion to the exposure.

The activity measures only how many radioactive decays the source undergoes every second. One radioactive isotope may emit alpha particles, another beta particles, and still others may be a source of neutrons. If they each undergo the same number of decays per second, they each have the same activity, R. In determining how much radiation is too much, it is not just the number of radioactive decays that determines the **dose**, but the energy that these radiation products bring with them.

The unit of radiation dose is the **rad**. The rad is equivalent to 0.01 joules per kilogram of absorbing tissue. The more active the radiation source, the longer the exposure time, or the more energetic the radiation, the greater the radiation dose.

Returning to the radioactive source discussed earlier, if an 80-kg person is exposed for one minute to the 1 Ci source that is emitting 1 MeV gamma rays, what is the dose in rads? Assume that all of the gamma ray energy is absorbed by the person's body and that the exposure is roughly uniform. The dose would be determined by the following calculation:

$$\text{dose} = \frac{\text{activity} \times \text{energy/type} \times \text{time}}{\text{mass}}$$

$$\text{dose} = \frac{(3.7 \times 10^{10}\,\text{gamma ray})(1 \times 10^{6}\,\text{eV/gamma ray})(1.6 \times 10^{-19}\,\text{J/eV})(60\,\text{s})}{80\ \text{kg}}$$

$$= 0.0044\ \text{J/kg}$$

$$= 0.0044\,\frac{\text{J}}{\text{kg}} \times \frac{1\ \text{rad}}{0.01\ \text{J/kg}}$$

$$= 0.44\ \text{rad}$$

This 0.44 rad dose is approximately 10% of the maximum allowable yearly dose recommended by the U.S. government for occupationally exposed workers. A 1 μCi gamma source would have to be handled for 1,000,000 minutes to deliver the same dose.

In the above example, we chose gamma radiation, one of the forms of radiation least dangerous to living tissue. The unit of the rad is an acceptable unit of dose for matter, whether it is a pipe in a nuclear power plant or the plumber servicing the pipe. When it comes to the effect on living tissue, however, some types of radiation are more damaging than others. When dealing with living tissue, there is another unit of dose that is commonly used called the **rem**. The rem and rad are related, but the rem takes into account the amount of damage that various types of radiation can do to living tissue. This is accomplished by the use of a factor called the **relative biological effectiveness** (RBE), which may differ for each type of radiation. The conversion between rad and rem, therefore is 1 rem = 1 rad × RBE.

As shown in Table 11-1, RBE = 1 for gamma rays, so that the person in the previous example received both a 1 rad and a 1 rem dose. If it had been a neutron source, the RBE could have been as high as 10 rem, since neutrons of the same energy are that much more damaging to living tissue than gamma rays.

Radiation	RBE
X rays	1
Gamma rays	1
Beta particles	1
Alpha particles (into the body)	10 – 20
Neutrons: For immediate radiation injury For cataracts, leukemia, and genetic changes	 1 4 to 10

Table 11–1: Table of relative biological effectiveness.

As noted earlier, one way of reducing radiation dosage is by reducing exposure time. This is impossible, however, when the radiation source is inside your body. Although it is not very likely that you would ever accidentally ingest a high-level radiation source, it remains one of the concerns of nuclear power plant safety. Take, for example, an accidental release of radioactive strontium-90. If this strontium-90 contaminates the grass eaten by a cow, it enters the food chain. Since the chemistry of strontium is like that of calcium, it will be incorporated into the milk. If a child drinks this contaminated milk, it will become a part of their bone tissue. Strontium-90 has a half-life of nearly 30 years. Once it becomes a part of the bone tissue, it will give the person years of elevated radiation exposure, which leads to an increased risk of cancer. For this reason, the isotopes that are selected and ingested for medical applications (see Technology Box 11-4) are those with short half-lives (hours or days) so they will quickly decay after serving their medical purpose.

Technology Box 11–4 ■ How is hyperthyroidism detected and treated?

In the medical field, two important uses of radioactive isotopes involve the diagnosis and treatment of disease. The fact that certain organs and glands in the body tend to collect specific elements makes this possible. The thyroid gland, for example, tends to collect trace amounts of iodine from our diet and incorporate it into the thyroid hormone, thyroxin. In some people, the thyroid gland is over-active, leading to a condition called hyperthyroidism. Both the diagnosis and treatment of this condition involves the use of radioactive iodine-131.

The patient is first given an oral dose containing iodine-131, which joins the iodine already present in the thyroid. Twenty-four hours later, the amount of radioactive iodine uptake by the thyroid is determined by placing a detection tube above the thyroid gland and measuring the amount of radiation. Since the amount of radiation present is directly proportional to the activity of the thyroid, it is easy to determine if the thyroid activity has been high, low, or normal. A patient with hyperthyroidism will have a higher than normal level of radioactive iodine, whereas a patient with a hypothyroidism will record low values.

If an above-normal level confirms hyperthyroidism, the patient is then given a therapeutic dosage of iodine-131, which will result in an even higher radiation count. This radiation will destroy more of the thyroxin producing cells, thereby reducing production and bringing the hyperthyroidism under control.

Because the iodine-131 is concentrated mostly in the thyroid gland, the rest of the body receives little radiation. The nuclei of the iodine-131 atoms are quite unstable and have a half-life of only 8 days. Therefore, only a few days after treatment the radiation exposure dies away.

In fact, most radioactive isotopes used in medical applications have a short half-life, so the radioactive exposure that a patient receives is of relatively short duration.

One automobile manufacturer proudly advertised that a supplier of radioactive isotopes to local hospitals had chosen its vehicles for the company's fleet. The commercial continued by stating that the company needed reliable cars, because if any of them broke down while transporting the isotopes, the product could disappear before their eyes!

Distance

Another way to limit your exposure is to stay away from the radiation source. This sounds simple, but how far is far enough? Fortunately, a little basic geometry again provides the answer. As in the case of the radiation detector, radiation emitted from a **point source** is isotropic (emitted uniformly in all directions). The amount of radiation striking a detector is the same regardless of its position on the surface of the imaginary sphere that surrounds the source. It is only the distance, r, that the detector is from the source that makes a difference.

One way to quantify this exposure is the radiation **flux, Φ**. Flux is the units of radiation particles (including photons) per unit time per unit area. Since the activity, R, of the source is measured in decay events per unit time, the flux is just the activity of the source divided by the area of the imaginary sphere. The following equation can be used to calculate flux:

$$\Phi = \frac{R}{4\pi r^2}$$

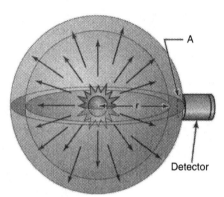

Isotropic Radioactive Source

Figure 11–14: The amount of radiation that a detector of a fixed size receives decreases as that detector gets farther away from the source.

This equation assumes that each decay event emits one particle of radiation. As shown in Figure 11-14, the number of decay events remains the same no matter where the detector, or person, is located. The same number of events per unit time is simply spread out over a larger and larger imaginary sphere as distance from the source increases. Eventually, r can become so large that the flux of particles due to the source becomes negligible.

This relationship, where the r^2 factor appears in the denominator, is so common that it has its own name, the **one over r-squared law**. It applies not only in the context of nuclear radiation, but to the intensity of light from a light bulb and the sound energy from a source like a jackhammer.

Shielding

As mentioned previously, when administering a medical x-ray, the technician either leaves the room or goes behind a lead-lined wall or a thick yellowish piece of lead-impregnated glass. Lead is often the substance used for radiation shielding. But, why is it effective? Are there other materials that work? The answers to these questions lie in the statistics behind the shielding process.

Let's first consider neutron radiation. Neutrons feel no forces from either protons or electrons. They must collide with a particle to interact with it. This type of collision is known as a **hard sphere collision**. It is similar to a ball on a billiard table – either the cue ball strikes the object ball or not. The cue ball passing by the object ball a millimeter away has no more effect than missing it by a meter.

A neutron can interact with either electrons or the nuclei of an atom. However, its interaction with an electron is not very eventful because of the large mass difference. Even if the neutron hits a nucleus, it will be slowed down only a little. To slow down the free neutrons within a nuclear reactor requires the use of a moderator so that the neutron's energy may be absorbed; a **moderator** being a substance with a low atomic mass that is capable of reducing the speed of neutrons by absorption. For this reason, graphite and hydrogen-rich materials like water and heavy water are materials of choice, rather than materials like lead.

As previously noted, the nucleus occupies only a tiny portion of the atom's volume. If a neutron were fired at an imaginary sheet of material exactly one atom thick, almost certainly the neutron would pass through the material without striking a nucleus. As shown in Figure 11-15, the total cross-section of all nuclei amounts to only a small fraction of any material's total area.

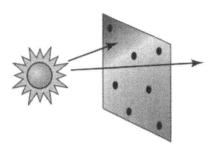

Figure 11–15: The cross-sectional area of the nucleus is what shield us from radiation. Each atom in the material blocks radiation over one small area.

If the imaginary sheet were made 100 atomic layers thick, the chance of the neutron's hitting a nucleus would increase by a factor of approximately 100. If additional layers of atoms continued to be added, a point where the sheet was half covered with nuclei would eventually be reached. When the material becomes this thick, the neutron would have a 50-50 chance of making it through. This thickness of material is known as the **half-thickness,** or $X_{\frac{1}{2}}$. The chance of a neutron passing through this thickness of material without striking a nucleus is 50 percent.

Even with a larger neutron flux, on average only half of them will make it through this thickness of material, and the other half will not. If we place another half-thickness of material behind the first, however, only half of the remaining neutrons will make it through this material. This is shown graphically in Figure 11-16.

If the graphs in Figure 11-6 and 11-16 look similar, it is no coincidence. Mathematically, it can be shown that the factor that determines the half thickness, $X_{\frac{1}{2}}$, is how large and densely packed the nuclei are for a given material. Since the nuclei are composed of protons and neutrons, this means that the half thickness depends on the density of the nucleons.

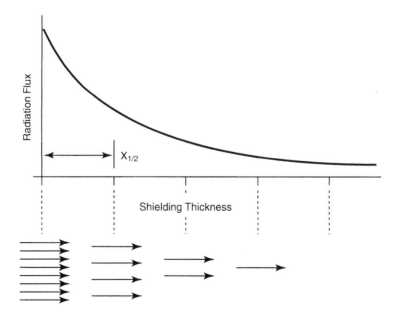

Figure 11–16: The half-thickness, $X_{\frac{1}{2}}$, of a material is defined as the thickness, on average, that half of the radioactive particles make it through.

Fortunately, the density of a material is a well-known quantity. Lead is a very common choice for a shielding material because it is dense. Therefore, a relatively thin sheet of lead packs a large number of neutrons and protons. In other words, the half-thickness of lead is small. Other, less dense materials can also shield against neutrons – it just requires a much thicker piece. More dense materials are preferred for radiation shielding because sufficient shielding can be achieved with a thinner sheet.

The neutron was chosen as an example because it is easy to see that the neutron must make a direct collision with the nucleus to be stopped. The size of the nucleus is just its cross-sectional area. The symbol that nuclear scientists have adopted for this cross-section is σ (sigma), and like any other cross-section, it has the units of area. Assuming the nucleus is roughly a sphere, its cross-sectional area is given by $\sigma = \pi r^2$. If the radius, r, of the nucleus is approximately 10^{-14} m, this means that $\sigma \approx 10^{-28}$ m^2. A very small area, indeed. Scientists have adopted another, more convenient unit to measure the nuclear cross-section. It is called the **barn**, where 1 barn $= 10^{-28}$ m^2. Incidentally, the unit name barn comes from the old line about not being able to hit the broad side of a barn – a sure indication that nuclear scientists also have a sense of humor!

Other types of radiation interact with the atoms in the shielding materials in different ways. Alpha and beta particles, for example, have long-range interactions with the charged particles in the atoms of the shielding materials. Gamma rays interact in another way. Nevertheless, all of these types of radiation are somewhat blocked when they encounter a thickness of the material.

Regardless of the type of interaction with the shielding material, some specific thickness blocks one-half the radiation. Therefore, the half-thickness is a valid quantity to assign to any shielding material. The half-thickness of a given material will vary with both the type and energy of the radiation. For example, the half-thickness of aluminum for 1 MeV beta radiation may be less than a millimeter, while the half-thickness of the same aluminum for 1 MeV gamma rays may be several centimeters.

How Much Protection is Enough?

We have just discussed three ways of protecting a person from the potentially harmful effects of radiation – time, distance, and shielding. A fair question might be, "How much protection is enough?" In principle, more shielding should always be better. On the other hand, there are radiation sources all around us – in the soil, water, air, and from cosmic radiation. This so-called **background radiation** gives the average person living in the United States a dose of approximately 0.3 rem per year. The amount varies. For example, the residents of Denver receive an additional dose because the mountains contain relatively high concentrations of radioactive minerals and because there is less atmosphere above them to shield them from cosmic rays. In practice, if exposed to a particularly active artificial radiation source, once enough shielding is in place to bring the radiation from this source below the background level, additional shielding will provide little additional benefit.

The Nuclear Regulatory Commission has established very conservative (low) guidelines for allowable radiation exposure. Exposure to radiation entails a risk, but so does driving a car to the grocery store. On the other hand, the risk of not doing so – starvation – far outweighs the likelihood of being killed in an automobile accident. There are ways to reduce our risks while driving, just as there are ways to reduce risk when dealing with radioactive sources. The key is to follow sensible guidelines based on the principles of time, distance, and shielding to reduce the risk of handling these materials to acceptable levels.

Check Your Understanding

1. What is the quantity that measures the radioactivity of a source called?

2. What is the SI unit for measuring activity?

3. What is the relationship between the curie and the becquerel?

4. What is the relationship between the rem and the rad?

5. What does radiation flux describe?

6. What factor does the term relative biological effectiveness (RBE) take into account?

7. What is the half-thickness of a material and what does it describe?

8. What is background radiation and some of its sources?

Summary

Chemical reactions involve the electrons around the nucleus, whereas nuclear reactions involve the strong and weak nuclear forces that act on the nucleons (protons and neutrons). According to Einstein's famous equation, $E = mc^2$, mass and energy are equivalent. When the mass of each nucleon and electron within an atom are totaled and compared to the mass of the atom, some mass is missing. The missing mass is equivalent to a negative contribution to the total potential energy, where one atomic mass unit (1 u), is equivalent to 931.49 MeV of energy.

Scientists originally identified three types of radiation: alpha, beta, and gamma rays. Alpha particles are the most massive and least penetrating. When a nucleus emits an alpha particle, it becomes another element; a process known as transmutation. Beta particles are less massive than alpha radiation, but more penetrating. When a beta particle is emitted from the nucleus, a new proton results, again transmuting the element. Positively charged electrons called positrons can also be emitted from the nucleus. Positrons are a type of antimatter; when they encounter a beta particle, their combined masses are converted into an equivalent amount of energy. All forms of beta decay are accompanied by yet another particle, the neutrino. The neutrino was difficult to detect because it has no charge and practically no mass. During beta decay reactions it helps conserve both energy and momentum. Gamma radiation is composed of tiny high-energy packets called photons. Because they are not composed of matter in the classical sense, they are the most penetrating.

Another important type of radiation is the neutron. Neutrons are massive compared to electrons, so their important reactions occur only when they strike the nucleus. Such contact may cause the nucleus to become unstable resulting in fission, or it may convert an atom into a radioactive isotope. Called neutron activation, this process is a major concern when decommissioning nuclear power plants.

All naturally occurring elements have radioactive isotopes. The half-life of a radioactive isotope is the amount of time required for one-half of the atoms in the original sample to undergo nuclear decay. These isotopes can undergo two types of nuclear reactions: fission and fusion. In fission reactions, the nucleus typically splits into two smaller nuclei of approximately equal sizes and releases several free neutrons. If not absorbed by a neutron absorbing material, these free neutrons can cause an uncontrolled chain reaction to occur. Nuclear power plants, therefore, use control rods to prevent meltdowns.

Fusion reactions, on the other hand, join two nuclei into one larger nucleus and may release several free neutrons. Because plasma is required, fusion reactions are much more difficult to start, maintain, and contain. The extremely high temperatures necessary for these reactions (over 100,000,000°C) require magnetic bottles to keep the components confined.

There are three ways to protect against radioactive exposure: time, distance, and shielding. The dose of radiation may be reported in either rads or rems. The rem takes into account the relative biological effectiveness (RBE) of the radiation type (1 rem = 1 rad × RBE).

Radiation flux is the activity of the source divided by area. The amount of exposure, therefore, can be reduced by moving farther from the source. The half-thickness of a shielding material is that thickness through which only 50% of the radiation has a chance of passing. This is considered an acceptable measure of the protection for any shielding material, regardless of the type of interaction it has with the radiation. The half-thickness of a material varies with both in the type and energy of the radiation.

Reducing radiation exposure to an artificial source below that of the background radiation is considered excessive. People living in different areas of the world are routinely exposed to varying amounts of background radiation, depending on the composition of the soil, water, and air in the region.

Chapter Review

1. What two elementary particles are collectively referred to as nucleons?

2. Complete the sentence: Isotopes have the same number of _____ but a different numbers of _____ in their nuclei.

3. The atomic symbol for the most common isotope of nitrogen is $^{14}_{7}N$. What is the atomic mass and atomic number of this isotope?

4. Compute the following energies in joules, then convert your results into mega-electron-Volts (MeV). Recall that the formula for kinetic energy is $K = \frac{1}{2}mv^2$. You will also find the conversion from miles to kilometers, 1 mile = 1.609 kilometers, useful.
   a. The kinetic energy of a 1 gram insect flying at 0.5 meters/second.
   b. The kinetic energy of a 0.145 kg, 90 mph fastball.

5. Verify the statement that converting 40 g of matter to energy is the equivalent of 1 billion kilowatt hours. Show your work. (Hint: c = 3 × 108 m/s.

6. Giving your answer in atomic mass units, how much mass is missing from the following atoms? Give this answer in MeV, what is the binding energy associated with the nucleus of each of the following atoms?
   a. Nitrogen-14 (7 protons, 7 neutrons, 7 electrons) mass = 14.0030740 u
   b. Oxygen-16 (8 protons, 8 neutrons, 8 electrons) mass = 15.9949146 u
   c. Aluminum-27 (13 protons, 14 neutrons, 13 electrons) mass = 26.9815384 u
   d. Silicon-28 (14 protons, 14 neutrons, 14 electrons) mass = 27.9769265 u

7. For each of the following isotope names give the nuclear symbol, the atomic number, the atomic mass, the number of protons, the number of neutrons, and the number of electrons in the atom.
   a. Sodium-22
   b. Cobalt-60
   c. Strontium-90
   d. Uranium-235

8. Given the following information, determine the binding energy associated with the deuterium, $_1^2D$, and tritium, $_1^3T$ nuclei. Give your answer in MeV.

   Mass of the hydrogen atom: 1.0078250 u

   Mass of the deuterium atom: 2.0141018 u

   Mass of the tritium atom: 3.0160493 u

9. Explain the difference between fission and fusion reactions.

10. How can you tell that the following equation is correct – that is, that only one neutron is released in the D-T fusion reaction: $_1^2D + _1^3T \rightarrow _2^4He + _0^1n$?

11. If, after all of the nuclear reactions are taken into consideration, each fission of uranium-235 yields 200 MeV of energy, what fraction of the mass of the original atom will have been lost? How much uranium, in kilograms, would you need to provide 1 billion kilowatt-hours of energy?

12. Assuming that, on average, a fission event in uranium-235 yields 2.5 neutrons, how many "generations" of fission events would you need to release Avogadro's number of neutrons? How much energy, in kilowatt-hours, would have been released by this number of fissions, if, on average, one fission event yields 200 MeV of energy?

13. There are two stable isotopes of lithium, $_3^6Li$ and $_3^7Li$ with a mass of 6.0151223 u and 7.0160040 u, respectively. The following fusion reactions between lithium and neutrons are important in the production of tritium. What is the energy yield from each? (The mass of the $_2^4He$ nucleus is 4.0026032 u). Give your answer in MeV.

    a. $_0^1n + _3^6Li \rightarrow _1^3T + _2^4He$

    b. $_0^1n + _3^7Li \rightarrow _1^3T + _2^4He + _0^1n$

14. Pound for pound (or, kilogram for kilogram), which reaction gives more available energy, fission of uranium-235 or fusion of deuterium and helium? Assume that the former yields 200 MeV of recoverable energy per reaction, while only the 14.1 MeV neutron energy is recoverable from the latter.

    a. Which type or types of radiation have no mass?

    b. Which type or types of radiation have no charge?

    c. Which type of radiation is the most massive?

    d. Which type of radiation is the least penetrating?

15. The mass of an alpha particle is less than the mass of a helium atom by the mass of an electron. The mass of an alpha particle is 4.002055 u. How fast, in meters per second, is an alpha particle with 1 MeV of kinetic energy moving? You may assume that $K = \frac{1}{2}mv^2$.

16. What is the wavelength of a 1 MeV and a 5 MeV gamma ray?

17. Name the three maxims of radiation safety.

18. How much of the original quantity of a radioactive isotope is left after two half-lives?

19. If the half-thickness of a particular material for a particular type of radiation is 1 mm, how thick a slab of this material is required to reduce the radiation flux to 1/8 of its original value?

20. A radiation detector has a sensor with a circular cross section 10 cm in diameter. It is placed 1 meter from a 1 μCi source. This detector detects 10 radiation events every second. What is the detector efficiency $\xi_{detector}$?

21. How many events per second would the detector in question 20 detect on average if it were placed 0.5 meters from the source? How many would it detect 10 cm from the source?

22. What is the flux of radiation when the detector in question 20 is 1.0 m from the source? When it is 0.5 meters away? When it is 10 cm away?

23. Sodium-22 has a half-life of 2.61 years. If you purchased one of these sources with an activity of 1 μCi today, how many decays per second would the source be emitting 1 year from now? What would it emit 3 years from now?

24. Strontium-90 has a half-life of 29.1 years. If an area were contaminated with this radioactive isotope today, how long would one have to wait until the radiation levels have dropped to 1/10 of their current value?

25. You measure 1,000 radioactive decay events in 1 second from a cesium-137 source. You place 0.1 mm of aluminum between the source and the detector and measure 800 decay events per second. What type of radiation is this likely to be? What is the half-thickness of aluminum for this type of radiation? What count rate would you expect if you used 0.5 mm of aluminum as shielding?

The Quantum Mechanical Model of the Atom

Bohr's simple planetary model of the atom continues to be a highly useful tool for visualizing both an atom's structure and bonding capabilities. Shortly after its introduction in 1913, however, several failings of the model became apparent and the search for a better model resumed.

In 1924, Louis de Broglie, a young French physicist, proposed a hypothesis that became the basis for a whole new model. His hypothesis, which was later demonstrated, proposed that it was possible for an electron to simultaneously exhibit the characteristics of a particle and a wave. For reasons that go beyond the level of this discussion, the idea that an electron could appear to be in different regions around the nucleus simultaneously (as derived from the Schrödinger wave equations) allowed scientists to conceive a new model. The result is called the quantum mechanical model.

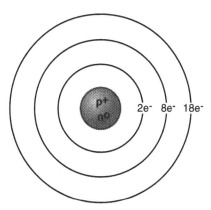

Bohr's planetary model

Using this model, Schrödinger found that he could describe each of an atom's electrons using just three quantum number designations: n, l, and m_l. It was later found that electrons appear to spin on their own axis, just like the Earth spins on its own axis as it orbits the sun. This necessitated the addition of a fourth quantum number, m_s, called the spin quantum number, which was arbitrarily assigned the opposing values of $+\frac{1}{2}$ and $-\frac{1}{2}$.

In summary, each electron within an atom needs four quantum numbers to completely describe its energy, location, and behavior. The following table summarizes the quantum numbers indicating both their function and permitted numerical values.

Quantum Symbol	Quantum Number	Function	Numerical Values
n	Principal	Distance of electron from nucleus	$n = 1, 2, 3 \ldots$ (whole numbers)
l	Secondary	Shape of electron's orbit	$l = 0, 1, 2, \ldots (n-1)$
m_l	Magnetic	Orientation of orbit in space	$m_l = -1 \ldots 0 \ldots +1$
m_s	Spin	Spin on the electron	$m_s = +\frac{1}{2}$ or $-\frac{1}{2}$

Several other guiding principles were developed to assist in determining the energy of electrons and their placement in multi-electron atoms.

- **Aufbau principle** – an entering electron always goes to the lowest (n + l) energy position available.

- **Pauli exclusion principle** – no two electrons within the same atom can have the exact same set of four quantum numbers.

- **Hund's rule** – when equal energy levels are available, each level must be half-filled before pairing can occur.

To describe the one electron in the hydrogen atom, the first consideration is for the Aufbau principle, mandating that the electrons always fill the lowest energy position available. That would be when n = 1. Once this decision is made, both the l and m_l values are limited to zeros (l = 0 and m_l = 0) and the remaining $m_s = +\frac{1}{2}$ or $-\frac{1}{2}$ arbitrarily chosen.

Similarly, the two electrons in a helium atom would have the same first three quantum numbers. Limited by the Pauli exclusion principle, however, the electrons would select opposite spin quantum numbers.

Hydrogen		Helium		
n	= 1	n	= 1	1
l	= 0	l	= 0	0
m_l	= 0	m_l	= 0	0
m_s	= $+\frac{1}{2}$	m_s	= $+\frac{1}{2}$	$-\frac{1}{2}$

With the addition of the second electron in helium, the first energy level (n = 1) is filled with a spin pair of electrons. As elements with additional electrons are considered, it becomes necessary to introduce the second energy level (n = 2) and employ various other filling principles to specify their exact locations.

Lithium				Beryllium				
n	= 1	1	2	n	= 1	1	2	2
l	= 0	0	0	l	= 0	0	0	0
m_l	= 0	0	0	m_l	= 0	0	0	0
m_s	= $+\frac{1}{2}$	$-\frac{1}{2}$	$+\frac{1}{2}$	m_s	= $+\frac{1}{2}$	$-\frac{1}{2}$	$+\frac{1}{2}$	$-\frac{1}{2}$

With completion of the element boron, the n = 2, l = 0 area is now completed with a second spin pair of electrons. Elements beyond boron with additional electrons must introduce the new area, n = 2, l = 1. This new area (l = 1) is said to be a triply degenerate orbital (broken into three equal areas). Hund's rule specifies that each of these areas must receive one electron before spin pairing can start.

Boron						Carbon							Nitrogen							
n	=	1	1	2	2	2	n	=	1	1	2	2	2	2	n	=	1	1	2	2	2	2	2
l	=	0	0	0	0	1	l	=	0	0	0	0	1	1	l	=	0	0	0	0	1	1	1
m_l	=	0	0	0	0	-1	m_l	=	0	0	0	0	-1	0	m_l	=	0	0	0	0	-1	0	$+1$
m_s	=	$+\frac{1}{2}$	$-\frac{1}{2}$	$+\frac{1}{2}$	$-\frac{1}{2}$	$+\frac{1}{2}$	m_s	=	$+\frac{1}{2}$	$-\frac{1}{2}$	$+\frac{1}{2}$	$-\frac{1}{2}$	$+\frac{1}{2}$	$+\frac{1}{2}$	m_s	=	$+\frac{1}{2}$	$-\frac{1}{2}$	$+\frac{1}{2}$	$-\frac{1}{2}$	$+\frac{1}{2}$	$+\frac{1}{2}$	$+\frac{1}{2}$

Notice that each n = 2, l = 1 orbital is now one-half filled. As the remaining three elements in this Period are completed each of the unpaired electrons will become spin paired.

Oxygen									Fluorine										Neon										
n	=	1	1	2	2	2	2	2	n	=	1	1	2	2	2	2	2	2	n	=	1	1	2	2	2	2	2	2
l	=	0	0	0	0	1	1	1	l	=	0	0	0	0	1	1	1	1	l	=	0	0	0	0	1	1	1	1
m_l	=	0	0	0	0	-1	-1	0	$+1$	m_l	=	0	0	0	0	-1	-1	0	0	$+1$	m_l	=	0	0	0	0	-1	-1	0	0	$+1$	$+1$
m_s	=	$+\frac{1}{2}$	$-\frac{1}{2}$	$+\frac{1}{2}$	$-\frac{1}{2}$	$+\frac{1}{2}$	$-\frac{1}{2}$	$+\frac{1}{2}$	$+\frac{1}{2}$	m_s	=	$+\frac{1}{2}$	$-\frac{1}{2}$	$+\frac{1}{2}$	$-\frac{1}{2}$	$+\frac{1}{2}$	$+\frac{1}{2}$	$-\frac{1}{2}$	$+\frac{1}{2}$	$-\frac{1}{2}$	m_s	=	$+\frac{1}{2}$	$-\frac{1}{2}$	$+\frac{1}{2}$	$-\frac{1}{2}$	$+\frac{1}{2}$	$-\frac{1}{2}$	$+\frac{1}{2}$	$-\frac{1}{2}$	$+\frac{1}{2}$	$-\frac{1}{2}$

With the final pairing of the n = 2, l = 1 electrons in argon, the second energy level is now full with a total of eight (8) electrons. Any element beyond argon, therefore, would require starting the n = 3 energy level and repeating the filling pattern of the second energy level.

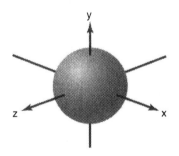

Chemists have found that by carefully examining an element's electronic configuration much can be determined about its ability to form chemical bonds. Writing the four quantum numbers for each of the element's electrons, however, tends to be a tedious and time-consuming task even for the chemist.

Fortunately, there is a shorthand way to convey the same information. It is based on the use of the n = 1, 2, . . . quantum numbers to indicate the distance from the nucleus. The l = 0, 1, etc. numbers are replaced with a corresponding s (spherical), p (perpendicular), d, and f to describe the shapes of the various electron orbitals. The ml = –1 ... 0 ... +1 quantum numbers are used to tell the orientation of the orbitals along the axis. Since l = 0 is a spherical orbital, it does not have any orientation; therefore, ml always equals 0 when l = 0. But when l = 1, then ml can equal either –1, 0, or +1, indicating that this orbital is triply degenerate and will have orbitals along the x, y, and z axes. In a similar manner, the d and f orbitals have increasingly complex orbitals and orientations; but again, this goes beyond the scope of this discussion.

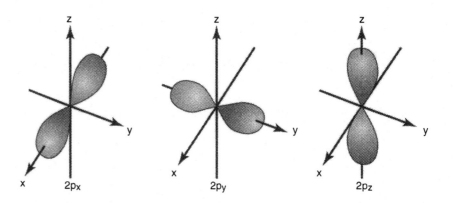

Employing *s p d f* Notations

The electron configuration for each of the above elements, can now be written in a shorthand notational form. They are:

H	$1s^1$			C	$1s^2$	$2s^2$	$2p_x^1$	$2p_y^1$
He	$1s^2$			N	$1s^2$	$2s^2$	$2p_x^1$	$2p_y^1$	$2p_z^1$
Li	$1s^2$	$2s^1$		O	$1s^2$	$2s^2$	$2p_x^2$	$2p_y^1$	$2p_z^1$
Be	$1s^2$	$2s^2$		F	$1s^2$	$2s^2$	$2p_x^2$	$2p_y^2$	$2p_z^1$
B	$1s^2$	$2s^2$	$2p_x^1$	Ne	$1s^2$	$2s^2$	$2p_x^2$	$2p_y^2$	$2p_z^2$

The electronic configuration for all other elements can be written by simply applying the following mnemonic device:

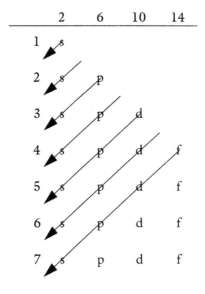

Although, the use of $s\,p\,d\,f$ notation is more complicated than the simple energy levels of the Bohr model, it does allow chemists to explain variable oxidation numbers and predict the shape of covalently bonded compounds. Today, chemists use the Bohr model when it explains the behavior; when it fails they switch to the quantum mechanical model to provide the extra detail needed.

Appendix

B

Inorganic Nomenclature

Background

Chemical compounds were often named for their appearance, occurrence, or their properties in the past. This method lead to such names as Paris green, oil of vitriol, blue vitriol, butter of tin, salt peter, baking soda, plaster of Paris, ammonia, salt, lye, and water, to name a few. The problem with this method of naming chemical compounds is that it relies solely on rote memorization. Although most of these colorful names have since been replaced with more chemically descriptive names, we continue to use a few of them.

In the latter part of the nineteenth century, chemists from across the world started meeting periodically to discuss the method for naming the growing number of chemical organic compounds. That group, which later (1919) became known as the International Union of Pure and Applied Chemists (IUPAC), is now recognized as the world-wide authority for naming both inorganic and organic compounds. Their division of inorganic chemistry is responsible for establishing and maintaining the rules for naming all inorganic compounds. The diagram and rules that follow summarize their efforts toward the naming of simple inorganic compounds.

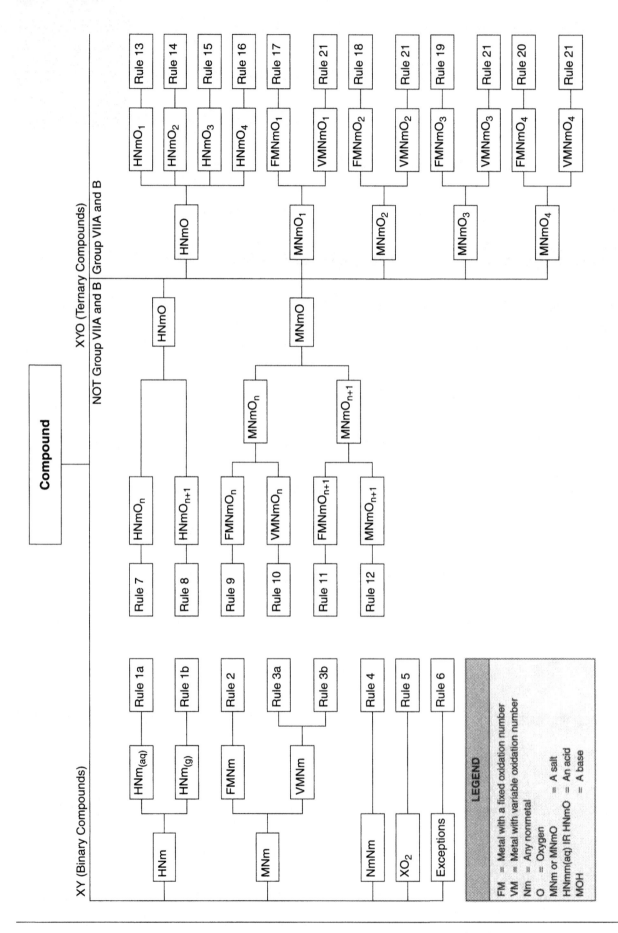

I. Binary compounds (**XY**) Compounds containing two different elements.

A. Binary hydrogen compounds (HNm)

Binary hydrogen compounds dissolve in water to form acids. The method used to name the compounds when they are a gas (g), differs from the method used to name the corresponding acid resulting from dissolving the gas in water (aq). Because these compounds are generally used in their aqueous (water) form, the absence of (g) generally suggests that they are in their acid form.

Rule 1a   *Binary acid names start with the prefix hydro- followed by the stem of the nonmetal's name, plus an -ic acid ending.*

Rule 1b   *Binary hydrogen compounds in their gaseous form are named with the word "hydrogen" followed by the stem of the nonmetal's name, plus an -ide ending.*

Examples:

$HCl_{(aq)}$	hydrochloric acid	$HCl_{(g)}$	hydrogen chloride
$H_2S_{(aq)}$	hydrosulfuric acid	$H_2S_{(g)}$	hydrogen sulfide
$HCN_{(aq)}$	hydrocyanic acid	$HCN_{(g)}$	hydrogen cyanide

B. Binary salts composed of a metal and a nonmetal (MNm)

1. Metal has constant oxidation number and a nonmetal type

Rule 2   *Name the metal, followed by the stem of the nonmetal's name, plus an -ide ending.*

Examples:

NaBr	sodium bromide	K_2S	potassium sulfide
$CaCl_2$	calcium chloride	CaI_2	calcium iodide
AgF	silver fluoride	K_2O	potassium oxide

2. Metal has a variable oxidation number and a nonmetal type

Rule 3a   *Use the English name of the metal, followed by its oxidation number in parenthesis. (Express oxidation numbers in Roman numerals.) This is followed by the stem of the nonmetal's name, plus an -ide ending.*

Examples:

$FeCl_3$	iron(III) chloride	Cu_2O	copper(I) oxide
$SnBr_4$	tin(IV) bromide	HgI_2	mercury(II) iodide
FeO	iron(II) oxide	PbS	lead(II) sulfide

Rule 3b   *An older method of expressing the oxidation state of metals with variable oxidation numbers uses the stem of its Latin name, with an ending of -ous for the lower oxidation state, and an ending of -ic for the higher oxidation state. This is followed by the stem of the nonmetal's name, plus an -ide ending.*

Examples:

$FeCl_3$	ferric chloride	Cu_2O	cuprous oxide
$SnBr_4$	stannic bromide	HgI_2	mercuric iodide
FeO	ferrous oxide	PbS	plumbous sulfide

C. Binary compounds composed of two nonmetals   (NmNm)

When two nonmetals combine, they may do so in several ratios. To determine the number of atoms of each kind present, Greek prefixes are used.

mono- = one	penta- = five	nona- = nine
di- = two	hexa- = six	deci- = ten
tri- = three	hepta- = seven	hemi- = one-half
tetra- = four	octa- = eight	

Rule 4   *A Greek number prefix precedes the name of the first nonmetal. (Generally, the prefix is omitted if it is mono-.) Then the Greek number prefix precedes the stem of the second nonmetal name, plus the -ide ending.*

Examples:

CO	carbon monoxide	SO_2	sulfur dioxide
NO_2	nitrogen dioxide	P_2O_5	diphosphorous pentoxide
$AsCl_3$	arsenic trichloride	N_2O_4	dinitrogen tetroxide

D. Peroxides (XO_2)

Hydrogen and several of the Group IA and IIA metals combine with oxygen to form compounds containing the peroxide ion, O_2^{2-}. Peroxides can be distinguished from oxides by determining the apparent oxidation number of the positive ion. If the positive ion's oxidation number appears to be higher than the element's Group number, then it should be named as a peroxide.

Rule 5   *Once it has been determined that the oxidation number of the positive ion is higher than the element's Group number, give the positive ion's name, followed by the negative ion's name, peroxide.*

Examples:

H_2O_2	hydrogen peroxide	Na_2O_2	sodium peroxide
BaO_2	barium peroxide	Li_2O_2	lithium peroxide

E. Binary compound rule exceptions

For historical reasons, several polyatomic ions were given names that make naming compounds containing them violations of the general rules for naming binary compounds. They are the hydroxides, OH^-, cyanides, CN^-, and ammonium, NH_4^+, compounds.

Rule 6   *If compounds contain the OH^-, CN^-, or the NH_4^+ ions, they are considered as exceptions to the binary compound rules, and have -ide endings.*

Examples:

KOH	potassium hydroxide	NH_4OH	ammonium hydroxide
NH_4Cl	ammonium chloride	$Ca(OH)_2$	calcium hydroxide
$NaCN$	sodium cyanide	$Al(CN)_3$	aluminum cyanide

II. Ternary compounds (XYO) Compounds containing three different elements.

A. Ternary acids (HNmO)

    1. Ternary acids with less oxygen ($HNmO_n$) type

        Rule 7   Use the stem of the nonmetal's name, plus -ous acid ending.

        Examples:

HNO_2	nitrous acid	H_3PO_3	phosphorous acid
H_2SO_3	sulfurous acid	H_3AsO_3	arsenous acid

    2. Ternary acids with more oxygen ($HNmO_{n+1}$) type

        Rule 8   Use the stem of the nonmetal's name, plus -ic acid ending.

        Examples:

HNO_3	nitric acid	H_3PO_4	phosphoric acid
H_2SO_4	sulfuric acid	H_3AsO_4	arsenic acid
H_2CO_3	carbonic acid	H_3BO_3	boric acid

B. Ternary salts composed of a metal and nonmetal + oxygen ($MNmO_n$)

    1. Metal with fixed oxidation number and polyatomic ion with less oxygen type

        Rule 9   Give the name of the metal, followed by the stem of the nonmetal's name, plus an -ite ending.

        Examples:

$LiNO_2$	lithium nitrite	$CaSO_3$	calcium sulfite
Na_3PO_3	sodium phosphite	$BaAsO_3$	barium arsenite
KNO_2	potassium nitrite	$Al_2(SO_3)_3$	aluminum sulfite

    2. Metal with variable oxidation number and polyatomic ion with less oxygen type

        Rule 10 Use the same rules for naming the metal ion in the salt as you did when naming binary salts with variable oxidation numbers. [See Rule 3a or Rule 3b] Once again, the stem of the nonmetal name, plus an -ite ending.

        Examples:

$FeSO_3$	iron(II) sulfite	or	ferrous sulfite
$CuNO_2$	copper(I) nitrite	or	cuprous nitrite
$Cu(NO_2)_2$	copper(II) nitrite	or	cupric nitrite

C. Ternary salts composed of a metal and nonmetal + more oxygen ($MNmO_{n+1}$)

    1. Metal with fixed oxidation number and polyatomic ion with more oxygen type

        Rule 11 Give the name of the metal, followed by the stem of the nonmetal's name, plus an -ate ending.

        Examples:

$LiNO_3$	lithium nitrate	$CaSO_4$	calcium sulfate
Na_3PO_4	sodium phosphate	$BaAsO_4$	barium arsenate

2. Metal with variable oxidation number and polyatomic ion with more oxygen type

Rule 12 Use the same rules for naming the metal ion in the salt as you did when naming binary salts with variable oxidation numbers. [See Rule 3a or Rule 3b] Once again, the stem of the nonmetal's name, plus an -ate ending.

Examples:

$FeSO_4$	iron(II) sulfate	or	ferrous sulfate
$CuNO_3$	copper(I) nitrate	or	cuprous nitrate
$Cu(NO_3)_2$	copper(II) nitrate	or	cupric nitrate

III. Compounds of Group VII A and B (XYO)

The elements in Groups VII A and B on the periodic table have the ability to combine with oxygen in four different ratios. Therefore, the previously discussed -ous and -ic acid endings are each given an additional meaning by using a hypo- prefix. A hypo- prefix is added for the one with less oxygen than the -ous ending. A per- prefix is added for the one with more oxygen than the -ic ending.

A. Acids of Group VII A and B elements (HYO_1, O_2, O_3, or O_4)

1. Acids with Group VII A or B element and only one oxygen (HYO_1) type

Rule 13 Give the prefix hypo- before the stem of the Group VII element's name, plus an -ous acid ending.

2. Acids with Group VII A or B element and two oxygen (HYO_2) type

Rule 14 Give the stem of the Group VII element's name, plus an -ous acid ending.

3. Acids with Group VII A or B element and three oxygen (HYO_3) type

Rule 15 Give the stem of the Group VII Element's name, plus an -ic acid ending.

4. Acids with Group VII A or B element and four oxygen (HYO_4) type

Rule 16 Give the prefix per- before the stem of the Group VII element's name, plus an -ic acid ending.

Examples:

$HClO$	hypochlorous acid	$HBrO$	hypobromous acid
$HClO_2$	chlorous acid	$HBrO_2$	bromous acid
$HClO_3$	chloric acid	$HBrO_3$	bromic acid
$HClO_4$	perchloric acid	$HBrO_4$	perbromic acid

B. Ternary salts of Group VII A and B elements $(MYO_1, O_2, O_3,$ or $O_4)$

1. Metal with fixed oxidation number and Group VII A or B element and only one oxygen (MYO_1) type

   Rule 17 Give the name of the metal followed by the prefix hypo- before the stem of the Group VII element's name, plus an -ite ending.

2. Metal with fixed oxidation number and Group VII A or B element plus two oxygen (MYO_2) type

   Rule 18 Give the name of the metal followed by the stem of the Group VII element's name, plus an -ite ending.

3. Metal with fixed oxidation number and group VII A or B element plus three

   Oxygen (MYO_3) type

   Rule 19 Give the name of the metal followed by the stem of the Group VII Element's name, plus an -ate ending.

4. Metal with fixed oxidation number and Group VII A or B element plus four

   Oxygen (MYO_4) type

   Rule 20 Give the name of the metal followed by the prefix per- before the stem of the Group VII A or B element's name, followed by an -ate ending.

   Examples:

$NaClO$	sodium hypochlorite	$KMnO$	potassium hypomanganite
$NaClO_2$	sodium chlorite	$KMnO_2$	potassium manganite
$NaClO_3$	sodium chlorate	$KMnO_3$	potassium manganate
$NaClO_4$	sodium perchlorate	$KMnO_4$	potassium permanganate

5. Metals with variable oxidation numbers and Group VII A or B elements

   Rule 21 When naming compounds that have metals with variable oxidation numbers, follow Rule 3a or Rule 3b for determining the oxidation number of the metal, and then either Rule 17, Rule 18, Rule 19 or Rule 20 to determine the remaining portion of the name.

   Examples:

$Fe(BrO)_2$	iron(II) hypobromite	or	ferrous hypobromite
$Fe(BrO_2)_3$	iron(III) bromite	or	ferric bromite
$CuBrO_3$	copper(I) bromate	or	cuprous bromate
$Cu(BrO_4)_2$	copper(II) perbromate	or	cupric perbromate

Appendix

C

Derived Units

Examples of SI Derived Units		
Derived Quantity	**SI Derived Unit**	
	Name	**Symbol**
Area	square meter	m^2
Volume	cubic meter	m^3
Speed, velocity	meter per second	m/s
Acceleration	meter per second squared	m/s^2
Wave number	reciprocal meter	m^{-1}
Mass density	kilogram per cubic meter	kg/m^3
Specific volume	cubic meter per kilogram	m^3/kg
Current density	ampere per square meter	A/m^2
Magnetic field strength	ampere per meter	A/m
Amount of substance concentration	mole per cubic meter	mol/m^3
Luminance	candela per square meter	cd/m^2
Mass fraction	kilogram per kilogram, which may be represented by the number 1	kg/kg = 1

SI Derived Units with Special Names and Symbols				
Derived Quantity	**SI Derived Units**			
	Name	**Symbol**	**Expression in other SI Units**	**Expression in SI Base Units**
Plane angle	radian [a]	rad	–	$m \cdot m^{-1} = 1$ [b]
Solid angle	steradian [a]	sr [c]	–	$m^2 \cdot m^{-2} = 1$ [b]
Frequency	hertz	Hz	–	s^{-1}
Force	newton	N	–	$m \cdot kg \cdot s^{-2}$
Pressure, stress	pascal	Pa	N/m^2	$m^{-1} \cdot kg \cdot s^{-2}$
Energy, work, quantity of heat	joule	J	$N \cdot m$	$m^2 \cdot kg \cdot s^{-2}$
Power, radiant flux	watt	W	J/s	$m^2 \cdot kg \cdot s^{-3}$
Electric charge, quantity of electricity	coulomb	C	–	$s \cdot A$
Electric potential difference, electromotive force	volt	V	W/A	$m^2 \cdot kg \cdot s^{-3} \cdot A^{-1}$
Capacitance	farad	F	C/V	$m^{-2} \cdot kg^{-1} \cdot s^4 \cdot A^2$
Electric resistance	ohm	W	V/A	$m^2 \cdot kg \cdot s^{-3} \cdot A^{-2}$
Electric conductance	siemens	S	A/V	$m^{-2} \cdot kg^{-1} \cdot s^3 \cdot A^2$
Magnetic flux	weber	Wb	$V \cdot s$	$m^2 \cdot kg \cdot s^{-2} \cdot A^{-1}$
Magnetic flux density	tesla	T	Wb/m^2	$kg \cdot s^{-2} \cdot A^{-1}$
Inductance	henry	H	Wb/A	$m^2 \cdot kg \cdot s^{-2} \cdot A^{-2}$
Celsius temperature	degree Celsius	°C	–	K
Luminous flux	lumen	lm	$cd \cdot sr$ [c]	$m^2 \cdot m^{-2} \cdot cd = cd$
Illuminance	lux	lx	lm/m^2	$m^2 \cdot m^{-4} \cdot cd = m^{-2} \cdot cd$
Activity (of a radionuclide)	becquerel	Bq	–	s^{-1}
Absorbed dose, specific energy (imparted), kerma	gray	Gy	J/kg	$m^2 \cdot s^{-2}$
Dose equivalent [d]	sievert	Sv	J/kg	$m^2 \cdot s^{-2}$

(a) The radian and steradian may be used advantageously in expressions for derived units to distinguish between quantities of a different nature but of the same dimension.

(b) In practice, the symbols rad and sr are used where appropriate, but the derived unit "1" is generally omitted.

(c) In photometry, the unit name steradian and the unit symbol sr are usually retained in expressions for derived units.

(d) Other quantities expressed in sieverts are ambient dose equivalent, directional dose equivalent, personal dose equivalent, and organ equivalent dose.

This information taken from the Foundation of Modern Science and Technology from the Physics Laboratory of NIST: http://physics.nist.gov/cuu/Units/index.html.

Glossary

α–**Hydroxyaldehydes:** Aldehydes that have a hydroxyl group attached to the first carbon (α-carbon) after the carbonyl carbon.

Aberration: The failure of a lens to produce an exact point-to-point correspondence between an object and its image.

Absolute pressure: A pressure measured relative to a vacuum or zero pressure.

Absolute zero: Theoretically, the lowest possible temperature matter can have.

Absorption spectra: The reverse of a bright-line spectrum: when white light passes through a substance dark lines appear in the spectrum at those frequencies at which the element can absorb radiant energy.

AC: Short for "alternating current." An electrical circuit in which the voltage supplied is continually changing in a sinusoidal relationship with time.

Acceleration: The rate of change of velocity (speed and/or direction) with respect to time.

Accuracy: How much a measurement differs from the true value of the quantity it is attempting to quantify.

Acetal: An organic molecule characterized by having two ether linkages attached to the same carbon, $R–CH(–O–R)_2$. An acetal linkage is used to connect two molecules of glucose forming maltose.

Acetylenes: Another name for the alkynes based on the most common member, acetylene.

Achromatic: A lens that eliminates chromatic aberration by combining different glasses in such a way that one lens cancels the dispersion of the other.

Acid: Arrhenius definition: a substance capable of producing hydrogen ions, H+, in water.

Action-reaction pairs: Pair of forces that are equal, opposite in direction, and never act on the same body.

Activation energy: The amount of energy required for a chemical reaction to occur. Catalysts change the activation energy, but do not change the difference in energy between reactant and product.

Active metal: Those metals found above hydrogen in the electromotive series.

Alcoholic carbon: The carbon to which the hydroxyl group (–OH) is attached.

Alcohols: The organic family characterized by the formula R–OH.

Aldehydes: The organic family characterized by the formula R–CHO.

Aldohexoses: A six carbon sugar with an aldehyde functional group.

Aliphatic hydrocarbons: A major group of hydrocarbons characterized by having either a straight or branched carbon chain. The alkanes, alkenes, and alkynes are aliphatic hydrocarbons.

Alkaline: Synonymous with the word "base."

Alkane: The organic family characterized by the formula, R–H. Alkanes are the only organic family that does not have a functional group.

Alkene: The organic family characterized by the formula R–CH=CH–R, and also known as the olefins.

Alkoxy substituent: An alkyl group that is attached to a carbon chain by an ether linkage, –O–R. Molecules containing an alkoxy group are known as ethers.

Alkyl group: An alkane group with one hydrogen missing. The –ane ending of the alkane name is dropped and replaced by an –yl ending; e.g., eth*ane* to eth*yl*.

Alkyne: An organic family characterized by the formula R–C≡C–R, and also known as the acetylenes.

Alloys: A solid or liquid mixture of two or metals, or of one or more metals with certain nonmetallic elements, such as in carbon steel.

Alpha particles: A subatomic particle containing two protons and two neutrons but no electrons. This is the same composition as the nucleus of a helium-4 atom.

AM: See amplitude modulation.

Amide: An odorless organic family characterized by the formula R–CO–N<.

Amide bond: Also known as the peptide bond; the bond between the carbonyl carbon and the nitrogen, e.g., –CO—N<.

Amines: The organic family characterized by the formula R–N– and known as the organic bases. The smaller members of this family have a fishy odor.

Amorphous: A solid having no real or apparent crystalline form, shapeless.

Ampere: The base unit of electrical current defined as the quantity of charge that passes through a length of wire per unit time.

Amplitude: The extent that a wave rises above or falls below the medium's resting or mean position.

Amplitude modulation: In AM radio, the amplitude of the carrier signal is varied to match changes in the audio wave from the station.

Analog: A mechanism having any number of possible values within a given range. An amplifier is a type of analog circuit.

Analog scale: A scale in which a measurement is made by comparing the position of an indicator to its location on a printed scale.

Analyzer: A second polarizer that can be rotated to determine the degree of polarization. When the axis of the analyzer is perpendicular to the polarized wave, no light passes, but when its axis is the same, all polarized light passes.

Angular magnification: A type of magnification that provides a measure of how large the image appears compared to how large the object appears.

Angular momentum: A vector quantity that is the product of the angular velocity of the body or system and its moment of inertia with respect to the rotation axis, and that is directed along the rotation axis.

Angular motion: The circular motion of an object around an axis.

Angular velocity: A measure of the rate of rotation around an axis usually expressed in radians or revolutions per second or per minute.

Anhydrous salt: A normally hydrated salt that has lost its water of hydration, for example $BaCl_2$ vs. $BaCl_2 \cdot 2H_2O$

Anions: A negatively charged ion, such as Cl^-, that is attracted to the positively charged anode.

Anode: The positive electrode of an electrochemical cell. Negatively charged anions are attracted to the anode.

Antilog: The number corresponding to a logarithm; more properly, antilogarithm.

Aqueous: Something made from, with, or in water.

Arenes: Another name for the organic family known as the aromatics, coal tar compounds, and benzene ring compounds.

Aromatic: The organic family characterized by having at least one benzene ring in their structure.

Aryl: A side chain containing the benzene ring structure, such as a phenyl group.

Astronomical unit: The average distance between the sun and earth. One AU equals 93,000,000 miles or 149,000,000 kilometers.

Atom: The smallest part of matter to retain the physical and chemical characteristics of an element. Atoms are composed of protons, neutrons, and electrons.

Atomic mass unit (amu or u): An arbitrary unit of mass based on the proton and equal to $1.6605402 \times 10^{-24}$ g.

Atomic number: The number of protons in the nucleus of an atom. Elements are arranged on the periodic table based on increasing atomic numbers.

Average velocity: One-half the sum of the starting and ending velocities.

Avogadro's number: The number of atoms in a g-mole of an element. Avogadro's number is equal to 6.02×10^{23}.

Balanced forces: Equal and opposite forces acting on the same body.

Base: Arrhenius definition: a substance capable of producing hydroxide ions (OH^-) in water.

Batteries: Electrochemical devices that generate electric current by using an oxidation-reduction reaction to convert chemical energy to electrical energy.

Benedict's test: A reagent similar to Fehling's, it is a single solution used to test for the presence of reducing sugars, especially for the detection of diabetes.

Benzene ring compounds: Another name for the organic family known as the aromatics, arenes, and coal tar compounds.

Best-fit line: A data trend line that is determined by linear regression.

Binary acid: An acid composed of only two elements; HCl, for example.

Binary logic: Arguments that can be answered by either no or yes and have values of either 0 or 1 in the base-two numbering system.

Binding energy: The energy associated with assembling the nucleus of an atom.

Black light: See UV-A.

Blackbody: A body with an emissivity of one that absorbs all incident radiation and emits the maximum theoretical power, given its temperature.

Blackbody radiation: The radiation emitted by a perfect absorber of all incident radiation.

Blanket: The neutron-absorbing material surrounding a fusion reaction and used to recover the heat produced.

Boiling point: The temperature at which a pure substance makes a transition from the liquid to the gaseous state.

Bolometer: A sensitive thermal detector whose electrical resistance varies with the temperature.

Boltzman's constant: The average kinetic energy of the particles in a gas; equal to 1.38×10^{-23} joules/K.

Boundary: The real or imaginary surface that separates a system from its surroundings.

Boyle's law: States that the absolute pressure and volume of a gas are inversely proportional ($PV = k$) at constant temperature.

Breathalyzer test: A brand of breath analyzer used to determine the blood alcohol level of a suspected drunk driver. It is based on the ability of the orange dichromate ion to be reduced to the blue-green chromium(III) ion, in the presence of a primary alcohol.

Bright-line spectra: A unique set of colored lines emitted by an element as electrons move from a higher energy back to the ground state.

Bromonation: The addition or substitution of a hydrogen for a bromine atom on an organic molecule.

Buffer: A salt solution that prevents pH changes from occurring when small amounts of either acid or base are added.

Candela: The unit of luminous intensity in the SI system.

Capacitance: A quantity that measures how effectively a capacitor can store charge.

Capacitor: An electrical device whose function is to store charge. The amount of charge a capacitor can store for a given voltage is indicated by its capacitance.

Carbonyl compounds: Organic compounds containing the carbonyl group ($>C=O$) as a part of their structural formula.

Carbonyl group: A moderately polar organic functional group, characterized by $>C=O$, where the oxygen end has a partially negative charge and the carbon end has a partially positive charge.

Carboxyl group: A polar organic functional group characterized by –COOH, in which the hydrogen bond has been weakened and can, therefore, be removed by the action of water; an acid group.

Carboxylate ion: The name given to the negative ion resulting from the ionization of the carboxyl group, –COO–.

Carboxylic acids: The organic family characterized by the formula R–COOH. Molecules containing more than one carboxyl group are known as dicarboxylic acids, tricarboxylic acids, etc.

Carrier waves: The high-frequency electromagnetic wave to which audio signals from a broadcast station are added.

Catalyst: A substance that changes the rate of a chemical reaction without being used up.

Cathode: The negatively charged electrode and source of cathode rays in a cathode ray tube (CRT). The electrode at which reduction (gain of electrons) occurs.

Cathode rays: A stream of electrons emanating from a cathode in an evacuated tube.

Cations: A positively charged ion such as Na^+, which is attracted to a negatively charged cathode.

Causal: A series of measurements in which one variable is dependent on or related to a second measurement.

Celsius: The temperature scale in which absolute zero is $-273.13°$, the freezing point of water is $0°$, and the boiling point of water is $100°$.

Centrifugal force: The fictitious force that an object moving in a circular path exerts on the body constraining the object, and that acts outwardly away from the center of rotation.

Centripetal force: The force that keeps an object moving in a circular path, and that is directed inward toward the center of rotation.

CGS system: A self-consistent set of units in which the centimeter, gram, and second are the base units for length, mass, and time, respectively.

Chain reaction: A self-perpetuating event in which the neutrons released by the disintegration of one radioactive nucleus trigger similar disintegration of several other nuclei, and so forth.

Charge: A measure of the property that is the source of electrical forces in matter. The unit of charge in the SI system is the coulomb.

Charge carrier: An elementary particle in an atom (proton or electron) that has the property of having one unit of electrical charge.

Charles' law: States that the volume of a gas is directly proportional to its absolute temperature $(V/T = k)$ if the pressure is constant.

Chemical changes: Changes that result in the production of one or more new substances.

Chemical equation: A type of mathematical expression that quantitatively represents a chemical reaction by means of chemical symbols, and which lists the reactants and products. Equations must be balanced to obey the law of conservation of atoms.

Chemical formula: A shorthand representation that shows both the number and kind of atoms present in one molecule.

Chemical properties: A set of properties that describes how a substance reacts with another substance.

Chemistry: The branch of physical science that is the study of the composition of matter and the changes it can undergo.

Chlorofluorocarbons: Substituted hydrocarbons, in which fluorine or chlorine atoms have replaced hydrogen; e.g., Freon, CCl_2F_2.

Chloronation: The addition or substitution of a hydrogen for a chlorine atom on an organic molecule.

Chromatic aberration: The property of a lens to refract list from a single point but but of different wavelengths to different foci.

Circle of least confusion: The point at which the rays passing through a lens seem to focus best.

Circular motion: Also called angular motion; the motion of an object travelling around an axis.

Closed circuit: An electrical circuit with at least one closed path around which current can flow.

Closed system: A system in which the mass remains constant; mass flow across the boundary of the system is prohibited.

Coal tar compounds: Another name for the organic family known as the aromatics, benzene ring compounds, or the arenes.

Coefficient of friction: A numerical value between zero and one related to the frictional forces between the stationary and moving surfaces of two objects.

Coefficients: The numbers used in front of reactants or products to balance a chemical equation.

Coherent radiation: A beam of high intensity radiation of nearly single-wavelength produced by a laser.

Coke: The carbon residue remaining from the destructive distillation of bituminous coal, petroleum, or coal tar pitch.

Colligative properties: Those properties such as vapor pressure, viscosity, melting and boiling points, that are affected by the number of moles of solute per kilogram of solution and not by the chemical nature of the solute.

Colloidal dispersions: Homogeneous solutions in which the solute particles are of colloidal size and result in cloudy or milky appearance. Solute particles will not pass through a dialyzing membrane.

Combination reaction: A type of chemical reaction in which elements or two smaller molecules are combined, forming a more complex substance.

Common names: A system of naming used for simple organic compounds that typically uses the name of each of the alkyl or aryl groups, followed by the family name; e.g., methyl ethyl ether or ethyl alcohol.

Commutator: A series of bars connected to the armature coils of a generator or motor. Its function is to cause a unidirectional current output in a generator and the reversal of the current into the coils of a motor.

Compound: A pure substance that is composed of two or more elements in a particular ratio; e.g., NaCl.

Concave lens: See diverging lens.

Concave mirror: See converging mirror.

Concentrated solution: A relative term indicating that the solution contains more solute per volume or mass than a dilute solution.

Concentration: The amount of solute in a given volume or mass of mixture or solution.

Condensation point: The temperature at which a pure substance changes from a gaseous state to a liquid state, and the same temperature as the boiling point.

Condensed structural formulas: A style of writing organic structural formulas in which all the atoms or groups of atoms attached to a single carbon are written directly behind that carbon, followed by the next carbon and its attached atoms or groups; e.g., $CHCl_2-CH_3$.

Conduction: The transport of thermal energy within a body or between two bodies in direct contact.

Conductor: A material in which charge carries readily; metals are good conductors.

Conservation of angular momentum: The law stating that the angular momentum of a rotating object remains unchanged unless it is acted upon by an external, unbalanced torque.

Control rod: A rod composed of a neutron-absorbing material that is placed in the fissionable material to reduce the number of free neutrons, and therefore control the rate of the chain reaction.

Control volume: See open system.

Convection: The transfer of heat by the relative motion of a fluid over a surface at a different temperature.

Converging lens: Also known as a convex lens; it causes the parallel light rays passing through it to be bent toward the center of the lens.

Converging mirror: Also known as a concave mirror; like a converging lens, it causes the reflected light rays to meet at a focal point.

Conversion factor: A factor that relates a number in one unit to its equivalent in another unit; e.g., 1 inch = 2.54 cm.

Convex lens: See converging lens.

Convex mirror: See diverging mirror.

Corpuscular theory: The theory advanced by Isaac Newton that light is composed of a stream of tiny particles.

Corrosion: The degradation of a metal or alloy through an oxidation-reduction reaction in which the oxidizing and reducing agents are in direct contact.

Coulomb: The unit of electric charge in the SI system equal to the quantity of electricity transferred by a current of one ampere in one second. It takes 6.25×10^{18} elementary charges to make one coulomb.

Coulomb's law: States that the force of attraction or repulsion acting along a straight line between two electric charges is directly proportional to the product of the charges and inversely proportional to the square of the distance between them.

Covalent bond: A bond resulting from the sharing of one or more pairs of electrons between two nuclei.

Critical angle: The angle beyond which all incident light striking a surface is reflected and none is transmitted.

Cryogenic: Of or relating to very low temperatures.

Crystal: The normal form of a substance in its solid state. Crystals have characteristic shapes and cleavage planes due to the arrangement of their small particles, which form a particular pattern, called a lattice.

Crystal structure: The geometric arrangement of the crystals of a substance.

Curie: A unit of radioactive activity equal to 3.7×10^{10} decays/s.

Current: The phenomenon of charges in motion, such as the flow through electrical wiring. The base electrical unit is the ampere.

Cycloalkanes: Members of the alkane family that have their longest carbon chain in the form of a closed ring. Their general formula is C_nH_{2n}, and cyclo– is added just before their last name.

Cycloalkenes: Members of the alkene family that have their longest carbon chain in the form of a closed ring. Their general formula is C_nH_{2n-2} and cyclo– is added just before their last name.

DC: Short for "direct current," an electrical circuit in which the voltage supplied by a battery or power supply is constant over time.

Decay chain: Sequence of radioactive decays that a radioactive isotope passes through to reach stability.

Decomposition reaction: Type of chemical reaction driven by heat or electrolysis, in which a more complex substance is broken down into its elements or simpler substances.

Dehydration: The process of removing of water from a substance.

Dehydration reactions: Chemical reactions in which water or the elements of water (H and –OH) are removed from between two reacting molecules.

Density: A measure of the mass of an object divided by its volume.

Derivatives: Compounds produced from a parent compound; e.g., CH_3Br is a derivative of CH_4.

Derived units: Units built from the seven fundamental SI units to measure other quantities; e.g., newton.

Detergents: Substances that reduce the surface tension of water. The most common synthetic detergents (syndets) are linear alkyl sulfonates, in which the alkyl group contains more than ten carbons.

Deuterium: An isotope of hydrogen composed of one proton, one neutron, and one electron.

Diamine: Members of the amine family that contain two amino groups, $-NH_2$.

Diatomic: A molecule composed to two atoms; e.g., Br_2.

Dicarboxylic acids: Members of the carboxylic acid family that contain two carboxyl groups, –COOH.

Dielectric constant: A constant used in Coulomb's law, magnitude of which is dependent on the substance used between the charged plates in a capacitor.

Dienes: Members of the alkene family that contain two carbon-to-carbon double bonds.

Diffraction: Bending of light waves after passing the edge of an object or passing through a narrow slit.

Diffuse reflection: Radiant energy reflected from an uneven surface leading to reflected light rays traveling in many different directions.

Diffusion: The act of one gas passing through another.

Dilute solution: A term of relative concentration signifying there is less solute per volume or mass than is in a concentrated solution.

Dimensional analysis: The use of units to determine if a correct series of mathematical operations have been used.

Diode: A semiconductor device in which current can only flow one way, made by juxtaposing a layer of n-type material with one of p-type.

Diols: Members of the alcohol family that contain two hydroxyl groups, –OH.

Dipole: The separation of electrical charges, as within a chemical bond or molecule.

Diprotic: An acid capable of ionizing and releasing two hydrogen ions per molecule; e.g., H_2SO_4.

Dispersion: The separation of white light into a rainbow of colors, based on the differences in the refraction of various wavelengths of light.

Displacement: A vector quantity indicating both distance and direction from a starting point.

Dissociation: The process by which a polar solvent such as water dissolves an ionic solid resulting in a solution that contains solvated ions and is an electrolyte.

Disulfide bond: An organic molecule containing an –S–S– linkage; they play an important role in maintaining the shape of protein molecules. Hair and fingernails are rich in disulfide bonds.

Diverging lens: Also known as a concave lens, this type of lens causes parallel light rays to move farther apart.

Diverging mirror: Also known as a convex mirror; like a diverging lens, it causes reflected light rays to spread farther apart.

Dopant: An impurity added to a semiconductor to change its electrical properties.

Doppler radar: A radar system used to measure the intensity of approaching storms based on the reflection of microwaves by water droplets.

Dose: The amount of radiation absorbed by a person or object and measured in units of energy received per unit mass.

Double replacement: A type of chemical reaction in which the reacting substances trade their ions, these reactions tend to go to completion when a gas, insoluble salt, or slightly ionized molecule such as water, is formed.

Ductile: A property of a substance that makes it capable of being drawn into wire or hammered into a thin sheet. A characteristic typical of metals.

e/m: Charge (e) to mass (m) ratio. The modern value for the e/m ratio of an electron is 1.759×10^{11} C/kg.

Elastic scattering: The scattering of photons by randomly oriented molecules.

Electrical current: The movement of electrical charges from one location to another.

Electrically neutral: A substance having no net charge.

Electrode: Either of two substances having different electromotive activity, which enable an electric current to flow in the presence of an electrolyte.

Electrolysis: The process of decomposing water and other inorganic compounds by means of an electric current.

Electrolyte: A solution containing ions, making it capable of conducting electricity.

Electromagnet: A magnet containing a core of magnetic material; e.g., iron surrounded by a coil of wire through which an electric current is passed to magnetize the core.

Electromagnetism: The generation of a magnetic field by a current of electricity.

Electromotive force: The potential difference or voltage derived from an electrical source per unit quantity of electricity passing through the source.

Electromotive series: Also known as the activity series, it is an arrangement of metals based on their abilities as reducing agents. Those metals found above hydrogen are called active metals.

Electron-dot formula: Also known as Lewis structures, these formulas are a representation of the valence electrons in a molecule or ion that show the location of the electrons and chemical bonds.

Electronegativity number: An arbitrary set of numbers assigned by Dr. Pauling to describe an element's attraction for electrons. Fluorine was assigned the highest (4.0) and the inert gases the lowest (0) electronegativities.

Electrons: One of the three basic building blocks of atoms, characterized by a very small mass (9.109×10^{-28}g) and a charge of minus one.

Electron-volt: The energy that one electron gains or loses when it experiences a voltage change of 1 V.

Electrostatic force: The force that one charge exerts on another at a distance.

Electrostatic potential: Also known as voltage, this property of space indicates an accumulation of excess charge. Positive charges are attracted to regions of lower potential and away from regions of higher potential. The opposite is true of negative charges.

Element: The fundamental part of matter that cannot be further subdivided by ordinary chemical reactions. Atoms are the smallest part of an element.

Elementary charge: The amount of charge present on one charge carrier (proton or electron) in an atom, equal to 1.6×10^{-19} coulombs. An amount of charge cannot be less that one elementary charge.

Elimination reactions: Type of chemical reaction in which the elements of water (H and –OH) are removed from a molecule or from between two molecules.

EM: See electromagnetism.

Emf: See electromotive force.

Emissivity: The property of a radiating surface represented by a dimensionless value falling between 0 and 1.

Endothermic: A chemical reaction dependent upon a constant energy source in order to continue; e.g. photosynthesis.

Energy: The capacity for doing work.

Energy levels: The concentric orbits around a nucleus that are occupied by electrons in the planetary model of the atom.

Energy of motion: See mechanical energy.

Energy shells: See energy levels.

Engineering: The application of the principles of physical science and mathematics to the properties of matter and the sources of energy in order to design useful things.

English system: The system of units that includes such units as the inch and the foot as measures of length and the pound-mass as a unit of mass.

Epicenter: The surface of the Earth directly above the center of an earthquake.

Equation of state: An equation that can be used to calculate the expected change in one property that will result from changes to other properties.

Equilibrium: A condition in which a reaction and its reverse reaction occur at the same rate, resulting in a constant concentration of the reactants.

Equilibrium constant, K_w: The equilibrium constant of water is the product of the hydrogen and hydroxide concentrations at 25°C and equals 1.00×10^{-14}.

Equivalent resistance: The resistance equal to the resistance of a series or parallel combination of individual resistors.

Esterification: A chemical reaction between a carboxylic acid and an alcohol, that results in the formation of an ester.

Esters: An organic family characterized by the general formula R–CO–R′ , and responsible for the odors of many fruits.

Estimate: A number based on input in which there is a significant uncertainty.

Ether linkage: Two alkyl or aryl groups linked by, –O–.

Ether peroxides: Highly unstable compounds formed by the oxidation of ethers, which contain the ether peroxide linkage R–O–O–R.

Ether synthesis: The chemical reaction between two similar or two different alcohol molecules, that results in the formation of an ether.

Ethers: An organic family characterized by the general formula R–O–R.

Evaporation: The gradual transition of a liquid below its boiling point to the gaseous state.

Excess charge: Also called net charge; the amount of positive charge on an object minus the amount of negative charge on that object. Net charge can be a negative quantity.

Excited state: A higher than normal energy state. An energized atom can return to its normal state by releasing a high-energy photon in the form of a gamma ray.

Exothermic: A chemical reaction that produces energy, such as the combustion of a fuel.

Exponents: The power to which a number or unit is raised; that is, the number of times a number or unit is multiplied by itself.

Extensive properties: Those properties, such as mass, volume, and total energy, that depend upon the size of a system.

Extrapolation: The process of estimating data outside the range of measured values.

Fahrenheit: The temperature scale of the English system in which the freezing point of water is $32°$ and the boiling point of water is $212°$.

Family: A group of elements in a vertical column on the periodic table of elements that have isoelectric outer energy levels, and therefore, similar chemical properties.

Faraday's law of induction: Explains that when a permanent magnet is held close to a loop of wire and moved either toward or away from the loop, an induced current is produced.

Fatty acids: Monocarboxylic acids that contain ten or more carbon atoms in their longest carbon chain. They may be either saturated or polyunsaturated carbon chains.

Feedstock: The gaseous or liquid petroleum starting materials in an industrial manufacturing process.

Fehling's test: A reagent used to test for the presence of aldehydes and reducing sugars. It consists of two solutions, one of copper sulfate and the other of an alkaline tartrate. It is mixed just before using.

Fibrillation: The very rapid, irregular contractions of the heart muscle fibers that results in a lack of synchronism between heartbeat and pulse.

Fictitious force: A force, such as centrifugal force, that is the result of acceleration.

Field: An area of space that is influenced by an external force.

First law of thermodynamics: States that the total energy in the universe is constant and that energy can neither be created nor destroyed; also known as the law of conservation of energy.

Fission: The nuclear reaction caused by the absorption of a neutron by a heavy-mass nucleus and resulting in the splitting of the nucleus into two or more light-mass nuclei.

Fluid: A substance (liquid or gas) that will flow, or conform to the outline of its container.

Fluorescent radiation: The radiation emitted by an electron in a metastable state drops back to its ground state.

Flux: The number of radioactive particles per unit time per unit area.

FM: See frequency modulation.

Focal length: The distance from the center axis of a lens to the point at which the emerging rays meet.

Focal point: The point at which parallel rays passing through a convex lens meet.

Focus: The point at which light rays converge.

Force: A vector quantity that describes a push or a pull by indicating both magnitude and direction.

Forced convection: Convective heat transfer in which fluid motion is induced by some mechanical means.

Formula mass: The sum of the atomic masses of all atoms or ions present in one formula of a pure substance. One formula mass of any pure substance equals one mole of that substance.

Freezing point: The temperature at which a pure substance changes from a liquid to a solid.

Frequency: A property of a wave that describes how many wave patterns or cycles pass by in a period of time. Frequency is often measured in Hertz (Hz), where a wave with a frequency of 1 Hz will pass by at 1 cycle per second.

Frequency modulation: In FM radio, the frequency of the carrier wave changes to match changes in the audio wave from the station.

Friction: The static or kinetic force between two surfaces in contact, that resists relative motion.

Fulcrum: The point about which everything else rotates.

Function: A specific mathematical operation that, given an input, returns one and only one output.

Functional group: The unique arrangement of bonds or atoms that characterizes an organic family.

Functional isomers: Those molecules that have the same molecular formulas, but differ in the functional group present in the isomers. For example, dimethyl ether and ethyl alcohol are functional isomers.

Fundamental charge: The smallest amount of charge that has been isolated to date. It is the charge that is on one electron and equals 1.6×10^{-19} C.

Fundamental quantities: Seven quantities that can, collectively, be used to measure every known substance in the world. See SI system.

Fusion: The nuclear reaction that results in the joining of two light-mass atomic nuclei into one, heavier nucleus.

Galvanometer: An instrument designed for measuring small electric currents.

Gamma radiation: A type of nuclear radiation characterized by energetic photons.

Gas: The state of matter characterized by its compressibility and lack of both a definite shape and volume. Gases tend to fill their containers.

Gasohol: An oxygenated fuel blend based on 80–90% hydrocarbons and 10–20% methyl or ethyl alcohol.

Gauge pressure: A measurement of the difference between absolute and local atmospheric pressure.

Gauss: A secondary unit of magnetic strength, equal to 10^{-4} T.

Gaussmeters: Any instrument used for measuring the strength of magnetic fields.

General formula: A chemical formula that indicates the functional group, but replaces the alkyl or aryl portion of the molecule with R–, e.g. R–OH or R–O–R.

Glycerin: See glycerol.

Glycerol: Also known as 1,2,3-propanetriol, glycerol is an odorless, colorless, sweet tasting hygroscopic liquid.

Glycols: The common name for alcohols that contain two alcohol groups: diols.

GPR: See ground penetration radar.

Gram: The basic unit of mass in the metric system. Now defined as 1/1,000 of the SI kilogram.

Gravitational acceleration: The Earth's gravity force, which acts on other masses. Its accepted value is 9.81 m/s^2 or 32.2 ft/s^2.

Gravitational constant: A constant whose value is dependent on the units used. In the SI System, the value of G is 6.67×10^{-11} ($N \cdot m^2/kg^2$)

Gravitational potential energy: The stored energy resulting from lifting a mass to a higher position. It is the product of the object's mass, the acceleration of gravity, and height ($Ep = m \cdot g \cdot h$).

Gravity: Identified by Newton as a fundamental physical force (acceleration) between mutually attracting bodies that have mass.

Ground: See ground potential.

Ground penetration radar: A method for determining the various underlying layers of earth based on the reflection of microwaves.

Ground potential: Also known simply as ground, an arbitrary voltage defined to be zero.

Ground state: The condition in which all the electrons in an atom are occupying their lowest possible energy level.

Ground wave propagation: Radio waves that reach beyond the curvature of the earth because they bounce off the ionosphere and the earth alternately until they lose all their energy.

Group: See family.

Half-cell: One-half of the reaction occurring in an oxidation-reduction reaction. The oxidation half-cell generates a flow of electrons and the reduction half-cell accepts the electrons.

Half-life: The length of time required for one-half of a radioisotope's nuclei, in a given sample, to disintegrate.

Half-thickness: The thickness of a material that is required to stop, on average, one-half of the incident radiation flux.

Halocarbon: A hydrocarbon that has one or more hydrogen atoms replaced by halogen atoms.

Halogenated hydrocarbon: A hydrocarbon that has one or more of its hydrogen atoms replaced by a member of the halogen family (fluorine, chlorine, bromine, or iodine). See halocarbons.

Halogenation reaction: A chemical reaction in which one or more halogen atoms are added or substituted on a carbon chain.

Halogens: A group of elements on the right-hand side of the periodic table, next to the noble or inert gases. As a group, they are the most reactive nonmetals.

Halohydrocarbons: See halocarbons.

Hard magnetic material: Material that resists changes in its magnetization, and is typically used to make permanent magnets.

Hard sphere collision: The type of collision that occurs between neutrally charged spheres.

Heat capacity: The quantity of heat needed to raise the temperature of a system by one degree.

Heat engine: A device that converts thermal energy into other useful forms such as mechanical or electrical energy.

Heat flux: The rate at which energy flows or heat is transferred.

Heat sink: A substance or device used to absorb or dissipate unwanted heat away from a device such as a processor or other electronic device.

Heat transfer coefficient: A numerical value that quantifies the ability of a substance to transfer heat.

Heavy hydrogen: See deuterium.

Heavy metal oxide: Oxides of metals found in the lower portion of the periodic table, such as PbO or HgO.

Hemiacetal: Organic molecules characterized by having one alcohol and one ether linkage attached to the same carbon, R–CHOH–O-R. Glucose uses a hemiacetal bond to close its ring.

Hertz: The derived SI unit of frequency, defined as a frequency of one cycle per second.

Heterogeneous: Of a different composition throughout; dissimilar.

HF–VHF: High frequency and very high frequency radio waves, with wavelengths ranging from 3 to 300 m.

Homogeneous: Of the same composition throughout; uniform.

Horsepower: The unit of power equal to approximately 745.7 watts.

Humectant: A substance having an attraction for water and used to retain moisture in a product.

Hydrated: A substance that has combined with, or taken up water.

Hydrates: The products of hydration; pure substances such as salts, that contain a definite mole ratio of water. For example, $CuSO_4 \cdot 5H_2O$.

Hydration reactions: Type of chemical reaction in which water is added to a molecule.

Hydrocarbons: A broad category of organic compounds (alkanes, alkenes, alkynes, and aromatics) that contain only the elements hydrogen and carbon.

Hydrogen bonding: A weak electrical attraction between polar covalent molecules.

Hydrogenation: A type of chemical reaction in which hydrogen is added to an unsaturated hydrocarbon.

Hydrohalogenation: A type of chemical reaction in which hydrogen and a halogen (H–X) are added to an unsaturated hydrocarbon. More specifically, the names could be hydrobromonation or hydrochloronation.

Hydrolysis: A type of chemical reaction in which water splits a chemical bond in the presence of an acid or base.

Hydrolysis of salts: A reaction in which an ionic compound is broken into its component ions through the action of polar water molecules.

Hydronium ion: An ion (H_3O^+) formed by the attachment of a proton to a water molecule, a wet proton.

Hypothesis: A reasonable guess stated in such a manner that an experiment can be devised to test its validity.

Iceland spar: A form of calcite crystal that has the ability to doubly refract light.

Ideal gas equation: Also known as the ideal gas law, it states that at low pressures, the volume of a gas is proportional to the number of moles present and its absolute temperature, $PV = NRT$.

Ideal gas law: See ideal gas equation.

Image distance: The distance from the axis of a lens to the point where the image is formed.

Impulse: The product of the force and the time during which it acts: the change in momentum produced by a force.

Incident angle: The angle formed between a light ray and a line perpendicular (normal) to the reflective surface.

Index of refraction: The ratio of the speed of an electromagnetic wave in vacuum to its speed in a medium.

Induced current: The flow of electric charge in a conductor resulting from a change in the flux of a magnetic field.

Induced voltage: The voltage that exists across a load (light bulb or motor, etc.) as a result of an induced current.

Inductance: A quantity that measures how much energy an inductor can store in a magnetic field for a given amount of current flowing thorough it.

Inductor: An electrical device whose function is to store energy in a magnetic field when current is passed through it. The amount of energy an inductor can store for a given amount of current is measured by its inductance.

Inert gas: Any one of a group that includes helium, neon, argon, krypton, xenon, and radon that exhibits great stability and extremely low reaction rates.

Inertia: As described by Newton's first law, it is a property of matter by which it remains at rest, if at rest, or in motion, if in motion, in the same straight line, unless acted upon by an external force.

Infrared radiation: A form of electromagnetic radiation, with a wavelength of about 1 mm to 78 μm, that is invisible to the human eye, but perceived as heat.

Inorganic esters: Organic molecules containing an ester linkage, that are formed by reacting an alcohol with an inorganic acid.

Instantaneous velocity: The change in displacement or distance divided by a very tiny, almost zero, change in time.

Insulator: Materials in which charge carriers do not move readily. Most plastics are good insulators.

Intensive properties: Those properties that are independent of the size of the system such as temperature, pressure, and specific volume.

Interference fringes: The bright and dark bands resulting from constructive and destructive interference of overlapping waves of the same frequency.

Interference pattern: The pattern produced by two overlapping waves of the same frequency that combine constructively and destructively, depending on their relative phase.

Intermediate magnetic material: Material falling between soft and hard magnetic material, and used for recording media; e.g., audio tapes and computer storage.

Internal energy: All the energy belonging to a system while it is stationary including nuclear, chemical, and thermal.

International Union of Pure and Applied Chemistry: A voluntary nonprofit association of national organizations (based in Basel, Switzerland) representing chemists from across the world. Its main objective is to provide international agreement and uniform practice for all aspects of chemistry, including naming.

Interpolation: The process of estimating data between two sets of measured values.

Ionic bond: A major type of chemical bonding characterized by the electrical attraction between oppositely charged ions.

Ionic compound: Compound composed of oppositely charged ions and held in a crystalline form by ionic bonds.

Ionization: The process of breaking a polar molecule into ions aided by the action of polar water molecules.

Ionizing radiation: Any form of radiation capable of ionizing neutral matter; e.g., alpha particles.

Ion: An atom that has lost or gained one or more electrons, and carries a charge as a result.

Irreversible process: A process that cannot be reversed without leaving an effect on its surroundings, e.g. discharging and recharging a battery.

Isoelectronic: Two particles, or energy levels, that have the same electron configuration.

Isolated system: A system of fixed mass and constant internal energy.

Isomer: Two or more organic compounds having the same molecular formula but different structural formulas.

Isotopes: Atoms of the same element that vary in the number of neutrons they contain, and therefore their atomic masses.

Isotropic: Condition of uniformity in all directions.

IUPAC: See International Union of Pure and Applied Chemistry.

Joule: A derived unit of energy in the SI system equal to a kilogram meter squared per second squared.

k: A unit vector that signifies only direction.

Kelvin: The temperature scale of the SI system in which absolute zero is 0 and the freezing point of water is approximately 273 K. Also, the unit of measure on that scale.

Ketohexose: A six carbon sugar containing a ketone functional group, e.g. fructose.

Ketones: A family of organic compounds in which the carbonyl group ($>C=O$) is attached to two alkyl groups, e.g. dimethyl ketone or acetone.

Kilogram: The unit of mass in the SI system.

Kinetic friction: A frictional force that occurs between the surfaces of two objects that are moving with respect to one another.

Kinetic-molecular theory: It assumes that gases are composed of small particles that are in continuous and random motion and that the collisions they experience with the walls of the container and other particles are completely elastic.

Kirchoff's laws: A general reference to both Kirchoff's current law and Kirchoff's voltage law.

Kirchoff's current law: It states that the sum of all of the currents into a node must equal zero.

Kirchoff's voltage law: It states that the sum of the voltages around any loop in a circuit is zero.

Laser: An acronym for light amplification by stimulated emission of radiation.

Law: A natural law is a broad statement of fact that is invariable under the given set of conditions.

Law of conservation of energy: Energy cannot be created or destroyed, it can be transformed, however, from one form to another. See first law of thermodynamics.

Law of conservation of mass and energy: A combination of the two conservation laws that allows for the possibility of a conversion of mass to energy ($E = mc^2$).

Law of conservation of matter: Also known as the law of conservation of mass, it states that during ordinary chemical reactions, matter can be neither created nor destroyed.

Law of conservation of momentum: A statement that the momentum of a body, or a system of bodies, does not change except when an external force is applied.

Law of inertia: See Newton's first law of inertia.

Law of multiple proportions: Dalton's law of multiple proportions states that, "When two elements form a series of compounds, the ratios of the masses of the second element combine with a fixed mass of the first element can always be expressed as the ratio of small whole numbers."

Law of reflection: A law that states, in part, that the angle of incidence and reflection are always equal.

Law of universal gravitation: As determined by Newton, the law of universal gravitation states that the force of gravity is directly proportional to masses and inversely proportional to the square of the distance between them.

Leading zeros: Digits of 0 that appear before any significant figures in a number. For example, the number 0.00034 has four leading zeros.

Length: A measure of space in one dimension; the unit of length in the SI system is the meter.

Lever arm: As used in the calculation of torque, it is the distance of the force applied from the center of rotation.

Lewis acid: An ion or molecule that is electron deficient and can combine with another ion or molecule forming a covalent bond.

Lewis base: An ion or molecule that is electron rich and can combine with another ion or molecule, forming a covalent bond.

Light: Electromagnetic radiation that is visible to the human eye, and varies in wavelength from 780 nm (red) to 390 nm (violet).

Light absorption: The conversion of light energy into chemical energy and/or heat when it is absorbed by an opaque substance.

Light ray: A line drawn in space corresponding to the direction radiant energy is traveling.

Light-year: A unit of length that equals the distance that light traveling at 300,000 kilometers per second (671 million miles per hour) would travel in one year. One light-year therefore equals 9.46053×10^{12} km, or 5.880×10^{12} miles.

Limiting reagent: The reagent present in less than the stoichiometric amount to consume all of another reagent(s). It limits the amount of product(s) that can be formed.

Linear magnification: The relationship between the height of an object and the height its image.

Linear momentum: The product of an object's mass and straight-line velocity.

Linear scale: A scale in which the output is linearly proportional to the quantity that is being measured.

Linearly polarized wave: A light wave that is composed of waves vibrating in only one plane.

Line-of-sight: One antenna must be visible to another, without an intervening obstruction.

Liquid: The state of matter characterized by an incompressible, definite volume, that takes the shape of its container.

Logarithm: The exponent of base 10 that is equal to a number; e.g., since $100 = 10^2$, the $\log_{10} 100 = 2.00$.

Longitudinal waves: Waves, such as sound waves, in which the medium and the wave travel in the same direction. It results in the medium being compressed and decompressed.

Luminous intensity: A measure of the amount of energy per time that a light source emits into space. The unit of luminous intensity in the SI system is the candela.

Machine: A device for applying energy to do work.

Magnet wire: A special wire coated with very thin plastic or enamel insulation to prevent it from shorting out.

Magnetic bottles: Containers composed of strong magnetic fields that confine charged particles (plasma) by causing them to curve and thereby not contact the physical walls.

Magnetic field: The space near a magnetic or current-carrying conductor in which the magnetic forces can be detected.

Magnetic field lines: Lines that are drawn such that each line, as it passes through any point in space, is tangent to the magnetic field vector at that point.

Magnetic flux: The product of the surface area defined by a closed loop, and the average component of the magnetic field that is perpendicular to that surface.

Magnetic force: A force due to the presence of a magnetic field produced by charges in motion.

Magnetism: A physical phenomenon characterized by fields of force, associated with moving electricity, and exhibited by both magnets and electric currents.

Magnification: The apparent enlargement of an object by means of optical instrumentation.

Malleable: The quality of being capable of being shaped by beating with a hammer or pressure from rollers, and one of the characteristic properties of metals.

Manometer: A plastic or glass U-tube containing a fluid in which the pressure one side is compared to atmospheric pressure. The pressure is determined by the difference in the two liquid levels.

Markovnikov's rule: A general rule stating that whenever the reagent adding to an alkene release a hydrogen ion (H–Y) the entering hydrogen always attaches to the carbon that already has more hydrogen attached. Simply stated: Them that have – gets!

MASER: An acronym for microwave amplification by stimulated emission of radiation.

Mass: A measure of the amount of matter that is present, the unit of mass in the SI system is the kilogram.

Matter: Anything that occupies space and has mass.

Measuring system: An agreed upon method for expressing measured quantities, for example the English and metric systems.

Mechanical advantage: The ratio of the force lifted to the force used to do the lifting.

Mechanical energy: Sometimes defined as the type of energy that can be used to raise a weight, or the energy of motion. In its transitional form it is called work, and in its stored form it is called potential energy.

Melting point: This is the temperature at which a pure substance makes the transition from solid to liquid.

MEMS: Short for micro-electromechanical systems, a technique for building microscopic mechanical devices using the same semiconductor technology that has traditionally been used to construct electrical devices such as transistors.

Mercaptans: An organic family characterized by the general formula, R–SH. Many of the members of this family have disagreeable odors, such as that of skunk.

Metallic luster: The ability to reflect light: sheen. Most metals have a silvery-white metallic luster.

Metalloids: Also known as semi-conductors, this group of elements occupies positions on either side of the zigzag line on the periodic table of elements. They are generally characterized by having some metallic and some nonmetallic properties.

Metastable: Electrons that have been "excited" and are in a higher energy level than the ground state.

Meter: The unit of length in the SI system.

Metric system: A decimal system of weights and measures based on the meter and kilogram.

Micron: A unit of length equal to one millionth of a meter – also called a micrometer.

Microwaves: Electromagnetic radiation used for communications, radar, and cooking that range from about 30 cm to 1 mm in wavelength.

Mixtures: Combinations of pure substances that are not present in definite proportions: variable compositions. Each component in a mixture retains its own physical and chemical characteristics, which are sometimes used for their later separation.

MKS system: A self-consistent set of units in which the meter, kilogram, and second are the base units for length, mass, and time, respectively. Now referred to as SI units.

Models: A description or analogy used to help visualize something that cannot be directly observed; e.g., planetary model of the atom.

Moderator: A substance such as graphite or a hydrogen-rich substance such as water, heavy water, or the paraffins, that are used to slow down neutrons in a nuclear reactor.

Molality: A solution concentration that is equal to the number of moles of solute per kilogram of solvent.

Molarity: A solution concentration that is equal to the number of moles of solute per liter of solution.

Mole: The amount of a pure substance that contains the same number of atoms, molecules, ions, etc. as there are atoms in exactly 0.012 kg of carbon-12. A measure in the SI system that is equivalent to Avogadro's number (6.02×10^{23}).

Molecular equation: A chemical equation that expresses the formula for even ionized substances as molecules.

Molecular formulas: Formulas that show only the elements present and the ratio in which they occur, such as $C_{40}H_{82}$.

Molecule: The smallest unit of a pure substance that can exist independently and still exhibit all the properties of the substance; e.g., H_2O.

Moment of inertia: The tendency for a rotating object to continue rotating at the same speed and direction until acted upon by a force. The moment of inertia is dependent on both the mass and shape of an object. For very small masses, it is the product of its mass times its radius of rotation squared ($I = mr^2$).

Momentum: A property a moving body has by virtue of its mass and motion. It is equal to the product of the body's mass and change in velocity.

Monocarboxylic acids: Those carboxylic acids that contain only one carboxyl group, –COOH.

Monomers: Small molecules that can undergo a polymerization reaction resulting in the formation of a polymer; for example, ethylene is the monomer that polymerizes into polyethylene.

Monoprotic: This characteristic of some acids results in the production of only one hydrogen ion (H^+) when they ionize.

MTBE: Methyl tertiary butyl ether is an oxygenate used in gasoline to improve combustion.

Natural convection: The convective heat transfer in which fluid motion occurs naturally; e.g., due to density variation.

Natural philosophy: Any physical science, such as chemistry and physics, that deals primarily with nonliving materials and their interrelations.

Near point: The closest distance of an object upon which the tensed eye can focus; typically between 25 and 50 cm.

Negative charge: One of the types of electrical charges found in nature. The affect of negative charges is opposite the affect of positive charges.

Net charge: The amount of positive charge on an object minus the amount of negative charge on that object; also called excess charge. Net charge can be a negative quantity.

Net force: The sum of all the individual forces acting on a body.

Net ionic equation: A type of chemical equation in which the spectator ions have been eliminated, leaving only the participating ions.

Neutral wire: In home electrical wiring, it is one of the leads in a two-pronged electrical plug. The voltage on this wire is not the same as the ground of the circuit, but it is usually close.

Neutralization: A type of chemical reaction in which an acid and base are reacted, producing a salt and water.

Neutrino: A particle with no mass or charge, the neutrino is one of the products of beta decay.

Neutron: One of the three building parts of atoms; characterized as having no charge and a mass of 1 u.

Neutron activation: The capture of a neutron by the nucleus of a stable atom, which, in turn, causes it to become an unstable nucleus.

Newton: A derived unit of force in the SI system equal to a kilogram meter per second.

Newton's first law: Also known as the law of inertia, it states that a body continues in its state of constant velocity (which may be zero) unless it is acted upon by an external force.

Newton's second law: If an unbalanced force is acting on a body, the acceleration produced is proportional to the force; the constant of proportionality is the inertial mass of the body.

Newton's third law: It states that for every acting force there is a reacting force that is equal in magnitude, but opposite in direction.

Newton's universal law of gravitation: Two bodies attract each other with equal and opposite forces. The magnitude of this force is proportional to the product of the two masses and inverse to the square of the distance between the two body's center of mass.

Nitration: A chemical reaction in which a nitro group, $-NO_2$, is added to an organic molecule.

Nobel gases: See Inert gases

Node: A location in which two or more wires join in an electrical circuit.

Nonpolar covalent: A covalent bond is nonpolar when the nuclei share the electrons equally. The center of the positive charges coincides with the center of the negative charges.

Normal: A name given a carbon chain that contains no branching, and which the may be included as a part of a common name. For example, n-pentane.

Normal force: The force perpendicular to a surface at a point of tangency.

North seeking pole: The end of the needle in a compass that points northward (north pole).

n-Type: A type of semiconductor material containing impurities (or dopants) that give it excess unbound electrons.

Nucleons: The elementary particles (protons and neutrons) comprising the nucleus of an atom.

Nucleus: The small ($\approx 1 \times 10^{-13}$ cm) area in the center of an atom that is occupied by protons and neutrons.

Nuclides: Radioisotopes that are characterized by the mass, charge, and energy of their nucleus; radioactive.

Object distance: In optics, the distance between an object and the lens.

Objective lens: In a microscope, the lens that is closest to the object.

Octet rule: A rule that states (with the exception of the first energy shell) that atoms are the most stable when they have eight electrons in their outer-most energy shell.

Ocular lens: Also known as the eyepiece, it is the lens closest to the eye in a microscope.

Offset: A common difference between the true and measured values. An offset error is an example of an error that can be corrected for in a systematic way.

Ohm's law: Ohm's law states that the voltage, current, and resistance are related by the equation, $V = IR$.

Olefins: Another name for the alkene family.

One over r-squared law: A whimsical law that refers to a series of equations in which the r^2 factor appears in the denominator. This includes exposure to nuclear radiation, the intensity of a light bulb, and the sound energy from a jackhammer. In these equations, the amount of energy varies as the square of the distance from the source.

Opaque: A substance that has the ability to block the passage of radiant energy.

Open circuit: An electrical circuit around which current cannot flow because the path is interrupted.

Open system: A system in which mass may flow across the boundary.

Optical density: A measure of the speed with which light travels through a substance.

Optics: The study of the behavior of radiant energy.

Organic bases: An electron-rich family called as *amines*; Lewis bases.

Organic chemistry: The study of naturally occurring and man-made carbon-containing compounds.

Organic families: Subdivisions of organic compounds based on similar structural features (functional groups) and properties.

Outlier: A data point that is statistically distant from the average. Outliers can cause a significant error in a measurement if they are not taken out of a data set.

Oxidation: The process of losing electrons that occurs during oxidation-reduction reactions.

Oxidation half-cell: The oxidation half-cell generates the flow of electrons in an oxidation-reduction (redox) reaction. It may be expressed by the oxidation half-cell equation; for example $Zn \longrightarrow Zn^{2+} + 2\ e-$.

Oxidation number: The number of electrons lost, gained, or unequally shared in a chemical bond.

Oxidation-reduction: A chemical reaction in which one substance loses electrons (oxidation) that are gained (reduction) by another substance.

Oxidizing agent: The substance that gains electron(s) and is reduced in an oxidation-reduction (redox) reaction. Oxygen and fluorine are examples of strong oxidizing agents.

Oxyacid: An acid that contains oxygen as one of the elements; for example H_2SO_4.

Oxygenated fuel: A blend of hydrocarbons designed to promote more complete combustion and containing one or more oxygen-containing compounds, such as an alcohol or ether.

Ozone hole: Areas principally near the south pole of the Earth, where the ozone layer becomes seasonally depleted of ozone.

Paraffin wax: A white, tasteless, odorless solid consisting of a mixture of alkanes having 20 or more carbons in their longest chains.

Paraffins: Another name for alkanes.

Parallel combination: Two or more resistors connected "side by side" so that current flows through all of them at the same time.

Pascal's law: It states that the pressure applied to an enclosed fluid is transmitted undiminished to every portion of the fluid and the walls of the containing vessel.

Peptide bond: Also known as an amide bond.

Percent concentrations: There are three different percent concentration terms in use: $\%_{(wt/vol)}$, $\%_{(wt/wt)}$, and $\%_{(vol/vol)}$. Of these, only $\%_{(wt/wt)}$ is a true percent, calculated by dividing the grams of solute by the mass of the solution, times 100.

Percent error: The absolute difference between the theoretical and actual yield, divided by the theoretical yield, times 100.

Percent yield: The actual yield divided by the theoretical yield, times 100.

Period: A horizontal row of elements on the periodic table of the elements characterized by increasing atomic numbers and, generally, increasing atomic masses.

Period cyclic: The time interval for a cyclic motion or phenomenon to complete a cycle and begin to repeat itself. It is the reciprocal of the frequency.

Periodic law: When elements are arranged in the order of their increasing atomic numbers, they show a periodic repetition in their properties.

Periodic table of the elements: An arrangement of elements designed to align elements in different periods so that those with similar properties form vertical groups or families.

pH: A numerical scale for expressing the acidity or alkalinity of aqueous solutions, found by applying the formula $pH = -\log[H^+]$.

Photoelectric effect: The emission of electrons from the surface of a metal when struck by electromagnetic radiation of a sufficiently high frequency.

Photon detectors: Also known as quantum detectors, they contain a semiconductor that raises an electron to its conduction band where it is free to move along a circuit when struck by infrared radiation.

Photon: The small packets of energy with no rest mass, that move at the speed of light; the unit of electromagnetic radiation.

Physical changes: Those changes that do not produce a new substance; e.g., grinding, melting.

Physical properties: Those characteristics that describe the physical nature of a substance, such as color, odor, melting and boiling points, hardness, electrical conductivity, and density.

Physics: The branch of physical science that deals with matter and energy and their interactions.

Pickling: An industrial process that uses an acid to clean the surface of metals prior to other manufacturing operations.

Planck's constant: A constant, h, that relates the frequency, f, and energy, E, of electromagnetic waves, $E = h f$, where, h, equals 6.626×10^{-34} J/Hz or 4.136×10^{-15} eV/Hz.

Planetary model: Also known as the Bohr model, it depicts atoms as though they are miniature solar systems. The nucleus is in the center, surrounded by electrons in concentric energy levels.

Plasma: An ionized gas composed of a neutrally charged collection of positive ions and electrons.

pOH: A numerical scale for expressing the acidity or alkalinity of aqueous solutions, found by applying the formula, $pOH = -\log [OH-]$.

Point mass: An assumed point, with no spatial dimension, at which all of an object's mass is located.

Point source: A source that is considered as a single point.

Polar covalent: A covalent bond in which the shared pair of electrons is shifted closer to one of the nuclei than the other. This results in a bond that has partial positive charge at one end and a partial negative charge at the other.

Polarizer: A material that has the ability to produce linearly polarized light that is composed of waves vibrating in only one plane.

Polyatomic ions: A covalently bonded group of atoms, with a charge, that act as a unit during chemical reactions.

Polyenes: Molecules that contain many double bonds.

Polymerization: The chemical reaction characterized by taking many smaller molecules, linking them together, and forming a polymer.

Polymer: A huge molecule created by bonding many smaller molecules (monomers). Polymers may contain only one type of repeating monomer (–A–A–A–) or co-polymers containing two types of repeating monomers (–A–B–A–B–).

Polyunsaturated: Characterized by molecules that contain many multiple carbon-to-carbon bonds.

Population inversion: A state in which there are more electrons in the metastable state than in the ground state.

Positional isomers: Molecules in which the molecular formula and functional group are the same, but the position of the functional group along their carbon chain differs.

Positive charge: One of the types of electrical charges found in nature. Positive charges have the opposite effect of negative charges.

Positron: A positively charged particle with the same mass and properties as an electron; a type of antimatter.

Pound-force: A unit of force in the English system that is the product of mass times gravitational force.

Pound-mass: A unit of mass in the English system.

Precipitate: An insoluble salt or particle that settles from a liquid or gaseous solution by the force of gravity.

Precision: The amount of random uncertainty in a measurement.

Primary colors: The primary light colors are red, green, and blue. By mixing these three primary colors, all other colors can be produced.

Principal axis: In optics, the horizontal line that passes through the center of a lens.

Problem: As used in scientific inquiry, a problem is anything that catches your attention and causes you to wonder, "Why?"

Products: The general term used for the one or more substances that are the result of a chemical reaction.

Property: A characteristic of a system, such as mass, volume, temperature, internal energy, and electric resistivity.

Protons: The positively charged subatomic particles in the nucleus of an atom. These particles determine the element's name.

Pseudoscience: A set of ideas or hypothesis put forth as scientifically valid when, in fact, they are not.

p-Type: A type of semiconductor material containing impurities (or dopants) that give the material a lack of bound electrons.

Puffs: Technical slang for the electrical property of capacitance that is measured in picofarads.

Pure substance: A subdivision of matter characterized by constant composition. Elements and compounds are both pure substances.

P-waves: Longitudinal waves associated with earthquakes that travel faster through the Earth's crust than do compression waves.

Quantum: Also called a photon, it is the smallest packet of energy.

Quantum detectors: See Photon detectors.

Quantum dots: Proposed building blocks for a future computer industry, quantum dots are nanometer-scale "boxes" for selectively holding or releasing electrons.

Rad: The unit of radiation dose; equivalent to 0.01 joules per kilogram of absorbing tissue.

Radians: The supplementary SI unit of angular measure, defined as the central angle of a circle whose subtended arc is equal to the radius of the circle (2π radians = $360°$).

Radiation: Heat transfer that occurs in the form of long wavelength electromagnetic radiation.

Radical: An organic molecule missing one of its hydrogen atoms and attached to a carbon chain. For example methane, CH_4, becomes a methyl group, CH_3-. Also known as a side chain.

Radicals: See Polyatomic Ions.

Radio frequency: Very long wavelength electromagnetic radiation, ranging from power line radiation to AM, FM, HF, and VHF radio and television broadcast signals.

Radioactive decay: The random breakdown of the unstable nucleus of a radioisotope, during which particles and energy are released.

Radioisotopes: Isotopes of an element that have unstable nuclei and are, therefore, radioactive.

Radon: A radioactive inert gas that is particularly dangerous to human health because it can release alpha particles deep in the lung.

Random error: An unrepeatable error in a measurement; single random error cannot be corrected, but the effects of random errors can be reduced by multiple measurements.

Rankine: An absolute temperature scale on which the unit of measurement equals a Fahrenheit degree and on which the freezing point of water is $491.67°$ and the boiling point $671.67°$.

Ray diagram: A line drawing that illustrates the path of the light rays.

RBE: See Relative biological effectiveness.

Reactants: The general term used for the one or more substances that enter a chemical reaction.

Real image: The image that can be formed on a screen and is produced by converging light rays.

Redox: Another name for oxidation-reduction reactions.

Reducing agent: The substance that loses electrons and is oxidized during an oxidation-reduction reaction. Hydrogen, for example, $H_2 \longrightarrow 2H^+ + 2e-$ is a typical reducing agent.

Reduction: The process of gaining electrons during an oxidation-reduction reaction.

Reduction half-cell: The reduction half-cell is the one accepting the flow of electrons in an oxidation-reduction (redox) reaction. It may be expressed by the reduction half-cell equation; for example $2 H^+ + 2 e- \longrightarrow H_2$.

Reflected angle: The angle formed between a reflected light ray and a line perpendicular (normal) to the reflective surface.

Refraction: The bending of light waves as they pass from one medium into another with differing refractive indices.

Relative biological effectiveness: The relative effect that a given type of radiation has on the human body. Gamma radiation was assigned a relative value of one.

Rem: A unit of dose that takes into consideration the amount of damage to living tissue caused by that type. This is accomplished by the use of a factor called the relative biological effectiveness. The conversion between rad and rem, is 1 rem $= 1$ rad $\times$ RBE.

Reproducibility: The ability to repeatedly perform an experiment and obtain the same results.

Resistance: A quantity that defines how effectively a device called a resistor impedes the flow of current through it.

Resistivity: The property of a material that determines, along with its shape, how much resistance a resistor made of that material will have.

Resistor: An electrical device whose main purpose is to impede the flow of electrical current through it.

Resonance: The movement of electrons between two or more possible contributing structures. The actual molecular structure lies somewhere between the contributing resonance structures, as in the benzene ring.

Resonance hybrid: The compromise structure that results from the sum of the contributing resonance structures.

Resonance structures: One of two or more contributing structures to the resonance hybrid. In the case of the benzene ring, it would be one of the rings with the double bonds at the 1, 5, and 9 o'clock positions.

Reversible process: A process that can be reversed without leaving any trace on the surroundings.

RF: See Radio frequency.

Right-hand rule: A method used to determine the direction of a resulting vector.

RMS voltage: See Root Mean Squared Voltage.

Root mean squared voltage: A type of average voltage calculated by taking the square root of the average of the square of a voltage that is changing with time.

Salt bridge: A solution containing ions (electrolyte) that is placed between the oxidation and reduction half-cell solutions in an electrolytic cell. Its function is to maintain the electrical balance of both solutions.

Salts: Chemical substances that dissolve in water producing ions, but the positive ion cannot be H^+ and the negative ion cannot be OH^-.

Sampling: The process of measuring a portion of a system in order to determine a quantity pertaining to the whole system.

Saponification reaction: A reaction with a strong base, that results in the formation of a salt that has a long hydrocarbon-like tail and is, therefore, a soap. Carboxylic acids, esters, and amides can undergo saponification reactions.

Saturated hydrocarbons: Another name for the alkane family, the molecules cannot add any more hydrogen; therefore, they are considered saturated.

Saturated solution: A solution that has dissolved all of a substance that it can at a given temperature.

Scatter plot: Another name for an X-Y graph.

Scattering: The ability of atoms and molecules to absorb and randomly emit photons of equal frequencies.

Scientific inquiry: The systematic pursuit of knowledge involving the recognition of a problem, the collection of data through observation and experimentation, and the formulation and testing of hypotheses and theories.

Scientific notation: The mathematical form for expressing a number as the multiple of one number times a power of 10. For example, in scientific notation $233.6 = 2.336 \times 10^2$.

Scintillator: A device that detects radiation by counting the flickers of light produced when radiation passes through it.

Sea of electrons: The loosely held valence electrons present in metals that can easily move from place to place between the atoms.

Second: The unit of time in the SI system.

Second law of thermodynamics: It states that heat always flows spontaneously from an object at higher temperature to a cooler object, and that the amount of disorder in the universe always increases.

Self-consistent set of units: A set of true units making up a system of measurements; e.g., SI System. When only these base units are used in calculations, the results will have the appropriate unit.

Semiconductor: A material in which charge carriers move more readily than they do in an insulator, but not as readily as in a conductor.

Series: See Period.

Series combination: Two or more resistors connected "end to end" so that current flows through them one at a time.

SI fundamental units: See SI system.

SI system: The system of units based on the meter as the unit of length, the kilogram as the unit of mass, the second as the unit of time, and four other base units.

Side chain: Another name for an alkyl group. See alkyl group.

Significant figures: The number of digits in a number that carry information.

Silver mirror test: See Tollen's test

Single replacement reactions: A type of chemical reaction in which a more reactive element attempts to replace a less reactive element in a compound. Elements can successfully replace those that are below them in the electromotive series.

Sky wave propagation: Radio waves that radiate outward and upward toward the ionosphere, where they are reflected back to the Earth some distance away.

Slug: In the English system, the mass of a body whose acceleration is 1 ft/s^2, when the resultant force on the body is 1 pound.

Snell's law: The relationship between the angles of incidence and refraction; it is dependent upon the ratio of the two indices of refraction of the materials the light is passing through.

Snubber diodes: Inline reverse-biased diodes that stop the line spike resulting from suddenly stopping a device, such as an electric motor.

Soft magnetic material: Material that can be magnetized and demagnetized easily.

Solids: One of the three states of matter characterized as non-compressible, with a definite shape and volume.

Solute: The substance dissolved or present in the lesser amount in a solution.

Solution: A homogeneous mixture of solute and solvent in any combination of solids, liquids, and gases.

Solvated: The clustering of solvent molecules, usually water, around charged particles such as ions. Solvated ions are indicated by $Na^+_{(aq)}$, for example.

Solvent: The substance doing the dissolving or present in the greater amount in a solution.

South seeking pole: The end of the needle in compass that points southward (south pole).

Specific heat: The amount of energy needed to increase the temperature of a unit mass of a substance by one degree.

Spectator ions: The ions that are not changed during a chemical process: onlookers.

Spectroscope: An instrument designed for dispersing light and displaying the spectra of an element.

Specular reflection: Radiant energy reflected from a smooth surface forming a single well-defined reflected light beam.

Spherical aberration: The failure of light rays passing through a lens nearer its edge and those passing through the principal axis to meet at a single point.

Spike (voltage): The induced voltage that results when a switch quickly interrupts the current to a device such as an electric motor.

Spontaneous emission: The random emission of a photon by a metastable electron.

Standard: A reliably reproducible physical quantity to which other measurements can be compared.

Standard solution: A solution of known concentration used as a standard in volumetric analysis.

Standing wave: A single-frequency mode of vibration in which the amplitude varies from place to place, is constantly zero at fixed points, and has maxima at other points.

States of matter: There are three generally recognized states of matter: solids, liquids and gases. Plasmas are sometimes considered a fourth state of matter.

Static electricity: It is not a scientific term, but refers to the buildup of electrical charges on an object or person.

Static friction: A frictional force that occurs between surfaces to two objects that are not moving with respect to one another. It is the force that must be overcome to start the object moving with respect to the other surface.

Steady state: A process during which the properties of a system do not change.

Stefan's law: An equation, $\dot{Q}\,rad = \varepsilon\sigma AT4$, which represents the power radiated by a body.

Steradian: A solid angle at the center of a sphere subtending a section on the surface equal in area to the square of the radius of the sphere.

Stimulated emission: The emission of a photon by a metastable electron triggered by the presence of a photon of the proper frequency.

Stoichiometry: The area of science dealing with quantities of substances entering and produced by a chemical reaction.

Stored energy: Energy that resides in some form but can be made available to do work.

Strong acids: Those acids that ionize nearly 100% in solution; hydrochloric acid, HCl, and sulfuric acid, H_2SO_4, are two examples.

Strong nuclear force: The stronger of the two elementary forces that act at short range within the nucleus.

Structural formulas: Formulas written in either full or condensed structural form that indicate the position of key atoms or groups.

Structural isomers: Two or more compounds that have the same molecular formula but differ in the branching of the carbon chain.

Sublimation: The direct passage of a substance from the solid state to the gaseous state, without transitioning through the liquid state.

Substitution reaction: Those reactions that are characterized by the removal of one atom or group before a second atom or group is added to the molecule.

Sulfhydryl group: Another name for the mercaptan group, –SH.

Sulfonation: A chemical reaction that occurs between an organic molecule and sulfuric acid, H_2SO_4.

Superconductors: Materials that greatly reduce electrical resistance, usually only at very low temperatures.

Supersaturated solution: A solution that holds more solute than it should at a given temperature.

Surface tension: The attractive force exerted by molecules below the surface on those at the surface. This results in a concentration of molecules at the surface. In the case of water, for example, it causes it to bead on a non-wetting surface.

Surfactant: A detergent or other compound that has an effect on surface tension.

Surroundings: The region outside a system and separated from it by a real or imaginary surface called a boundary.

Suspensions: An unstable heterogeneous solution in which the solute particles can be removed by filtration.

S-waves: The transverse or shear waves that are generated in the Earth's crust as the result of an earthquake.

Symmetrical: A molecule that has two equal halves is said to be symmetrical; e.g., H-H or Br-Br.

Symmetrical additions: The addition of a symmetrical reagent to a compound containing a double carbon to carbon bond.

Synthetic compounds: Man-made or artificially produced compounds, as opposed to those produced by nature.

System: A region in space separated from its surroundings by a boundary.

System of units: A set of measurement units that are related to each other, and comprised of base (fundamental) units and derived units that are defined by multiplying and dividing the base units.

Systematic error: A repeatable error in a measurement that can be corrected after-the-fact.

Technician: A person specializing in the technical details of a subject or occupation. See Technology professional.

Technologist: A person who makes practical application of knowledge in a particular area. See Technology professional.

Technology: The practical application of science to provide the conveniences used in everyday life to facilitate work and play.

Technology professional: An individual with a one-year certificate, two-year associate degree, or a four-year bachelors degree in a technology and that makes practical application of that knowledge.

Temperature: A measure of the amount of energy in matter devoted to random motion of its constituent particles. The unit of temperature in the SI system is the Kelvin.

Temperature gradient: The derivative, dT/dx, that represents the change in temperature across some distance.

Temporal: A series of measurements taken at different times.

Tensile strength: The rupture strength per unit area of a material subjected to a specified dynamic load.

Terminal velocity: The speed that a free falling object cannot exceed because the frictional force of the air equals the gravitational force.

Tesla: The SI unit of magnetic field strength; equal to 1 newton-second per coulomb-meter.

Theoretical yield: The maximum amount of a product that can be produced if all conditions are ideal.

Theory: A scientifically acceptable general principle that explains everything you know and that appears to be true about a problem.

Thermal conductivity: A measure of how easily a material allows heat to flow through it.

Thermal contact: The condition that permits two objects to exchange heat.

Thermal detector: A device that converts radiated power into a measurable parameter such as temperature.

Thermal efficiency: The ratio of net work done by an engine to the heat added or thermal energy absorbed by the engine.

Thermal energy: Energy that is associated with the motion of the atoms and molecules of a system, and is related to the mass of the system.

Thermal equilibrium: The condition in which two objects in thermal contact are at the same temperature.

Thermal imager: A device that uses an array of IR detectors to produce a thermogram or heat picture.

Thermal radiation: See Radiation

Thermodynamic system: A quantity of matter or arbitrary region of space separated from its surroundings and chosen for study.

Thermodynamics: The study of processes involving energy transformations and heat.

Thermometric property: An observable or measurable property that changes in a predictable manner with respect to temperature.

Thermopile: A thermal detector made of a series of thermocouples that generate an increasing voltage as their temperature increases.

Thin lens: A lens with a thickness that is small compared to the distances associated with its optical properties.

Thioalcohols: Another name for mercaptans, R–SH.

Thyristor: Any of several semiconductor devices that act as switches, rectifiers, or voltage regulators.

Time: A measure of the ordering of events in the universe. The unit of time in the SI system is the second.

Time-varying magnetic flux: A change in the magnetic field over a given time period; its derivative is symbolized as $\Delta\Phi_B/\Delta t$.

Tolerance: A statement of the precision of a measurement, usually indicated by the use of a plus or minus, $\pm$, sign.

Tollen's test: A mild oxidizing solution containing ammoniacal silver nitrate used as a test for aldehydes.

Torque: A force that tends to produce a rotation. It is determined by the product of the force and the perpendicular distance from the line of action (lever arm) of the force to the axis of rotation.

Total ionic equation: An equation that shows everything in its correct form: molecules as molecules and ionic substances as ions.

Trailing zeros: Digits of 0 that appear after any significant figures in a number. For example, the number 34500 contains two trailing zeros.

Transistor: A semiconductor device in which the flow of current is between two inputs and is controlled by a third input, made by sandwiching one layer of either n-type or p-type material between two layers of the opposite type.

Transition elements: The elements located in the center ten columns in the periodic table.

Transitional energy: Energy in motion that can move across a system boundary and transform as work is done.

Translucent: A substance that permits the passage of radiant energy.

Transmutation: The conversion of one element into another, such as the conversion of lead to gold.

Transverse waves: A wave in which the movement of the medium is perpendicular to the movement of the wave.

Trendline: A line or curve passing through a two-dimensional set of data points that indicates the relationship of one variable to another, ignoring uncertainties in the data.

Triads: Groups of three elements that have similar properties; e.g., fluorine, bromine, and iodine or iron, cobalt, and nickel.

Tritium: The radioactive isotope of hydrogen that is composed of 1 proton, 2 neutrons, and 1 electron.

Trivial names: Names for organic compounds that fall outside any organized naming system, such as aspirin.

True solution: A homogeneous solution that is transparent, may be colored, stable, and cannot be separated by filtration. The solute particles are typically ions, atoms, or small molecules.

Ultraviolet radiation: A form of electromagnetic radiation, with a wavelength of about 390–100 nm, that is invisible to the human eye. It is sometimes further subdivided into UV-A, UV-B and UV-C.

Unbalanced chemical equation: A chemical equation that indicates the reactant(s) and product(s), but is in violation of the conservation laws.

Uncertainty principle: It acknowledges that the act of taking a measurement or observation sometimes alters the system being measured.

Unit: A fixed amount of a given property, used as a basis of comparison for measurements.

Universal gas constant: A constant of proportionality used in the ideal gas equation, whose value is unit-dependent. Some common R values are 8.314 J/(mol·K); 0.08206 liters atm/mole K; and 1.987 cal/mole K.

Unsaturated fatty acids: Monocarboxylic acids that generally contain more than ten carbon atoms and multiple double bonds along the carbon chain.

Unsaturated hydrocarbons: Any hydrocarbon that can add more hydrogen, including alkenes, alkynes, and aromatics.

Unsaturated solution: A solution capable of dissolving additional solute at a given temperature.

Unsymmetrical: Molecules that lack equal halves, such as H–OH or H–Cl.

Unsymmetrical additions: The addition of an unsymmetrical reagent to a compound containing a double carbon to carbon bond. The "favored" product of the addition is predicted by Markovnikov's Rule.

UV-A: The region of the UV spectra that is closest to violet light; also known as near-UV or black light, with wavelengths ranging from 320 to 390 nm.

UV-B: The region of the UV spectra referred to as erythemal UV with wavelengths ranging from 290 to 320 nm.

UV-C: The most powerful region of the UV spectra, called germicidal UV, with wavelengths ranging from 160 to 290 nm.

Vacuum pressure: The name assigned absolute pressures that are lower in value than local atmospheric pressure.

Valence electrons: The electrons in the outer, incompletely filled shell of electrons around the nucleus.

Valence shell: The outer incompletely filled shell of electrons around the nucleus.

Vapor pressure: The pressure of a gas in equilibrium with its liquid or solid form at a given temperature.

Vector quantity: A quantity that has magnitude and direction and is commonly represented by a directed line segment whose length represents the magnitude and whose orientation in space represents the direction.

Velocity: A vector quantity indicating both speed and the direction of movement.

Virtual image: An image that appears to be equal distance behind the reflective surface but is not actually there.

Viscosity: A measure of a fluid's internal resistance to flow.

Vital force theory: A theory based on many failed attempts to create organic compounds in the laboratory. It stated that a vital force that only exists in organisms must be present to make an organic compound. The theory was abandon in 1828, with the production of urea starting with a nonliving source.

Voltage: Also known as electrostatic potential, a property of space in which there is an accumulation of excess charge. Positive charges are attracted toward regions of lower voltage and away from regions of higher voltage. The opposite is true of negative charges.

Voltage-regulated power supply: An electronic component that provides a nearly constant voltage. Such a power supply requires a source of power, such as electrical power from a wall socket.

Watt: The watt is used to measure power or the rate of doing work. One watt is a power of 1 joule per second.

Wave: A form of periodic motion or vibrational energy of the particles in an elastic body; it travels through a medium in alternately opposite directions from the medium.

Waveband: A range of wavelengths of interest.

Wavelength: The distance from one phase of a wave to the next corresponding phase.

Wave-particle duality: A basic quantum mechanical principle, which asserts that a photon must have the properties of particles along with its known wave nature.

Weak acids: Acids that ionize less than one percent in water; acetic acid, $HC_2H_3O_2$ and carbonic acid, H_2CO_3, are examples.

Weak nuclear force: The weaker of the two elementary forces that act at short range within the nucleus. Recently shown to be another manifestation of the electromagnetic force.

Weber: The SI unit of magnetic flux and defined as the unit of magnetic field strength, T, times the area in m^2 ($1 \text{ Wb} = 1 \text{ Tm}^2$).

Weighted average: An average that is based on the percent occurrence.

Word equation: An equation in which the reactant(s) and product(s) are expressed in words; for example, acid + base $\rightarrow$ salt + water.

Work: The change in the energy of a system resulting from the application of a force acting over a distance.

Index